$1 atm = 14.7 \ psia$

P.G $\quad$ if $\quad 2 < \dfrac{I}{T_c} < 5 \quad$ & $\quad \dfrac{P}{P_c} < 1$

Air $\quad : T_c = 132.41°K$

$\qquad P_c = 547.575 \ psi$

$W = \dfrac{P_2 V_2 - P_1 V_1}{n-1} \qquad$ on the gas

$n = \dfrac{\ln\left(\dfrac{P_1}{P_2}\right)}{\ln\left(\dfrac{V_2}{V_1}\right)}$

· Assumption

1) Frictionless

2). PG

3) $\Delta KE, \Delta PE = 0$

1) Simple Pure substance

2) Linear Interpolation valid

3) $\Delta KE, \Delta PE = 0$

THERMODYNAMICS

Third Edition

J. P. Holman

Professor of Mechanical Engineering
Southern Methodist University

McGraw-Hill Book Company

New York St. Louis San Francisco Auckland Bogotá Hamburg
Johannesburg London Madrid Mexico Montreal New Delhi
Panama Paris São Paulo Singapore Sydney Tokyo Toronto

Library of Congress Cataloging in Publication Data

Holman, Jack Phillip.
     Thermodynamics.

     Bibliography: p.
     Includes index.
     1. Thermodynamics.   I   Title.
QC311.H76   1980   536'.7   79-20868
ISBN 0-07-029625-1

THERMODYNAMICS

34567890   DODO   8987654321

This book was set in Electra by Progressive Typographers.
The editors were Frank J. Cerra and James W. Bradley;
the designer was Charles A. Carson;
the production supervisor was John Mancia.
New drawings were done by J & R Services, Inc.
The cover was designed by John Hite.
R. R. Donnelley & Sons Company was printer and binder.

contents

preface

This book is intended for use in the first course in thermodynamics taken by students in engineering. Naturally, not all the material presented can be covered in one semester, and different courses may demand different emphases. For those programs which require it, the text contains sufficient material for a two-semester sequence.

Some difference of opinion exists as to the subjects to be covered in a first course in engineering thermodynamics. Some instructors prefer a strictly classical or macroscopic approach, with varying emphasis on properties, chemical thermodynamics, and power cycles. Others advocate a strong input of microscopic concepts. The student interested in solid-state electronics will find the sections on applications of statistical thermodynamics useful in later work but, it is hoped, will also recognize the importance of classical thermodynamics in developing a proper foundation for the microscopic models of substances. The student interested in power cycles and energy conversion will profit most from the macroscopic thermodynamics.

In this third edition the addition of new materials has been concentrated in the chapters dealing with macroscopic thermodynamics because of the continuing interest in such applications and increasing severity of problems associated with economical and pollution-free production of power.

Some discussion of the chapter sequence is in order at this point. Chapters 1 to 4 present a classical development of the first law of thermodynamics, properties of pure substances, and energy analysis of open systems. Chapter 5 introduces the subject of statistical thermodynamics and can be omitted in a course primarily concerned with macroscopic applications or deferred until the end of the course. Chapter 6 presents the second law of thermodynamics and the concept of available energy from a macroscopic viewpoint. Chapter 7 gives a classical presentation of equations of state and generalized compressibility relations. Chapter 8 illustrates the applicability of statistical thermodynamics to the calculation of properties of gaseous and solid materials, while Chapter 9 develops the basic molecular model for kinetic theory of gases and transport phenomena. Both Chapters 8 and 9 can be omitted in a course devoted strictly to macroscopic thermodynamics.

Chapter 10 considers gaseous mixtures from the macroscopic viewpoint, with substantial emphasis placed on air conditioning applications. The calculation of properties of real-gas mixtures is also discussed but can be omitted if sufficient time is not available to cover this material. Chapter 11 gives a classical development of chemical thermodynamics and equilibrium, including energy analyses of such systems. Chapter 12 discusses a broad variety of power cycles, with strong emphasis given to the limitations imposed on efficiency by the second law. Chapter 13 gives an abbreviated treatment of the thermodynamics of compressible flow, while Chapters 14 and 15 consider some specialized applications of irreversible thermodynamics and direct energy conversion.

While the text allows for considerable flexibility in topic coverage, one must insist upon the inclusion of the materials in Chapters 1 to 4 and 6 in order to provide a proper basis for further study. The course, with these materials as a base, can then move in the direction of applications most appropriate to those students enrolled.

Mention should be made of the material which has been added in this edition. Additional discussion of unit systems and increased emphasis for the SI system has been added to Chapter 1, and throughout the text. In all chapters examples and problems have been presented in both English and SI units. Some ideal-gas relations previously contained in Chapters 3 and 4 have been moved back to Chapter 2 to provide a more coherent presentation. The section on availability and irreversibility in steady flow has been expanded in Chapter 6, and applications of this material have been added in later chapters. Chapter 10 has a new section on cooling towers, and calculation charts and tables have been expanded for psychrometric mixtures. Chapter 11 includes new examples of irreversibility analysis in combustion reactions. Chapter 12 includes new sections on the Stirling cycle, hybrid cycles, and heat pumps as well as examples of availability and irreversibility analysis to power cycles.

It is appropriate to offer remarks regarding the progress of conversion to SI units. We may expect that the most rapid progress, albeit slow, will be in those industries with more exotic thermodynamic applications like aerospace, direct energy conversion, etc. The old established industries in the power generation and the heating, ventilating, and air conditioning fields move much slower. In fact, it is difficult to find practicing engineers in these fields who use SI units at all. So, the approach to units in this text has a very pragmatic objective: to prepare students for whatever practice they may encounter. Energy balance and power cycle analyses are presented in both SI and English unit formats. Compressible flow, irreversibility analysis (a more exotic application), and direct energy conversion have more SI emphasis. It is an injustice to educate the student only in SI units, because industry simply does not operate this way. But it is equally unfair to ignore the obvious calculation advantages of SI units and the input which new graduates will have in promulgating their use. Some further comments in this regard are presented in Chapter 1.

The author is grateful for the comments and suggestions received from many individuals who have used the first and second editions. Of especial importance have been those penetrating questions of students which strike at the heart of a subject and demand clarification and explanation. As new materials have been added, and clarifying remarks expanded, it is hoped that the new materials will answer many of the student questions and, most importantly, serve to stimulate further inquiry.

J. P. Holman

chapter one
Introduction

1-1 THE NATURE OF THERMODYNAMICS

Energy propels society. The matchless economic and technological advances of the civilized world are traceable directly to an increasing amount of energy available to perform the tasks previously performed through human muscular effort. The availability of goods and services, and industrial productivity in general, are directly related to per capita energy consumption.

Thermodynamics is the study of energy and its transformation. This statement may seem rather aspiring, because it could be interpreted to mean that thermodynamics is the one science that is most strongly related to man's societal needs because of his increasing consumption of energy to produce goods and services. There are many different types of energy: the frictional work of a block sliding on a plane, electric energy, magnetic energy, nuclear energy, the energy stored in a quantum of light, the chemical energy of a petroleum fuel, and others. All of these types of energy can fall in the province of thermodynamic analysis and we shall examine a variety of applications as the subject develops. As we shall see later, the laws of thermodynamics limit the amount of energy which is available to us for performing useful work.

Various sources of energy are becoming increasingly expensive to produce and one must be concerned with environmental factors involved in all conversion schemes. For these reasons, more and more attention must be focused on the efficiency of energy utilization. Thermodynamics furnishes the scientific basis for analysis of energy-conversion schemes and thus is central to an understanding of future energy-consumption trends and their social and economic impact.

Generally speaking, most studies of thermodynamics are primarily concerned with two forms of energy: heat and work. The principal objectives of the study are to develop basic principles describing these types of energy and to become conversant with the language surrounding these basic principles. As in all such studies the first step is to build a vocabulary of definitions and terms which may be used to conserve thought as the development becomes more complex. To define rigorously the various concepts of thermodynamics requires considerable space and effort, and will, of course, form a large part of the discussion in this book. In this introductory chapter we seek to give a brief qualitative picture of the broad subject of thermodynamics to achieve a perspective for detailed studies in subsequent chapters. In this respect it is well to note that many of the qualitative discussions are offered on the basis of physical reasonableness and should be accepted with the view that more rigorous definitions and developments will be presented later. The objective of this chapter is to achieve an overall picture of the scope of thermodynamics.

1-2 RELATION BETWEEN CLASSICAL MECHANICS AND THERMODYNAMICS

The study of classical mechanics involves concepts of force, mass, distance, and time. A force has a physical meaning of a "push or pull" which may be represented mathematically as a vector with a point of application. Mechanics is developed through the application of Newton's laws of motion and, particularly, the second law which states that the summation of forces acting on a particle is proportional to the time rate of change of momentum,

$$\sum F = \frac{d}{d\tau} (mv)$$

For purposes of analyzing mechanical systems a *free body* is used whereby a definite portion of a mechanism is broken away and all forces acting on this mechanism are specified for use with Newton's second law. It is important to realize that the mechanical *system* is specified in terms of its coordinates of space and velocity. The behavior of the mechanical system is further described in terms of its interaction with its surroundings through the application of various forces. We say that the *state* of the system may be specified with its space and velocity coordinates and its *behavior*; i.e., its change from one state to another is described in terms of its interactions with adjoining mechanisms or surroundings. It may be observed that the mechanical system will not change its state, i.e., its position in space and/or its velocity, unless it is acted upon by some net external force. The important point of this brief reference to classical mechanics is that the concept of a system (free body) and specification of the *state* of a system through the use of space or velocity coordinates are already familiar to those readers with experience in classical mechanics.

While we are concerned with dynamical quantities in mechanics, the analysis of thermodynamic systems is concerned with energy quantities. A system is described in thermodynamics by breaking away a certain quantity of matter similar to the free-body technique in mechanics. The matter outside this system is termed the *surroundings*, and the separation between the system and surroundings is called the *boundary* of the system. As an example of a thermodynamic system consider a mass of air contained under pressure in a steel tank. The boundary of the system would be the inside surface of the tank and the surroundings would consist of the tank and the medium outside the tank. It is well to mention that the boundary of a system may be either a real or imaginary surface. The air-tank system is shown in Fig. 1-1.

In mechanics the state of a system is specified by its space and velocity coordinates. The state of a thermodynamic system is described by specifying its thermodynamic coordinates. We cannot describe all thermodynamic

FIG. 1-1   *Compressed air in a tank as a thermodynamic system.*

coordinates at this point but may note that temperature, pressure, chemical energy content, etc., are typical examples. These coordinates are usually denoted as *properties* of the system. In mechanics we noted that a system will not change its state unless there is some interaction with its surroundings to change its spatial position and/or velocity. This interaction usually takes the form of an energy transfer into or out of the system.

When a thermodynamic system changes from one state to another, it is said to execute a *process*.

In the study of thermodynamics we are interested in the changes which a system may undergo as it executes various processes. Clearly, we must be able to define the state of the system or its thermodynamic coordinates if we are to meet with success in describing processes which the system may undergo, for, if we are to describe the process, we must say what is happening each step of the way. The state of the system must be described at each point in the process. To do this, a very fundamental concept must be introduced—the concept of equilibrium.

A system is said to be in equilibrium when its pressure, temperature, and density are uniform (we shall employ a more rigorous definition of *equilibrium* later on). The system consisting of a mass of air in a tank would be in equilibrium if its pressure, temperature, and density were uniform throughout; however, if heat is applied to the tank such that the temperature at one end is higher than at the other end, the system will not be in equilibrium. When a system is in equilibrium, its thermodynamic coordinates are related in a definite way. Notice the importance of the concept of equilibrium. The system must be in equilibrium if we are to define its thermodynamic coordinates and their interrelationships.

Obviously we are interested in the process that is a succession of equilibrium states, because in this type of process we can define the state of the system each step along the way. We call this type of process *reversible*. The meaning of this term will be demonstrated later; nevertheless this type of process is physically described as a succession of equilibrium states and is sometimes called a *quasistatic* process because it is composed of a series of nearly static equilibrium states.

The preceding paragraphs have linked the analysis methods of mechanics with those of thermodynamics; in particular, it has been implied that the free body and thermodynamic system represent similar analysis techniques. They are similar in that in both cases certain quantities of matter are broken away and a study is made of the interactions with the surroundings. There is one important difference, however. In the free body of mechanics we are always careful to study the forces which act on the free body. In thermodynamic system analysis we study the effect of the forces as they interact with the system and its surroundings and focus our attention on energy quantities. The two kinds of interaction energies, i.e., energies which leave or enter the system, are work and heat. We shall see that there are forms of energy "contained" within the system which must also be considered.

1-3 TEMPERATURE, HEAT, AND THE ZEROTH LAW OF THERMODYNAMICS

The terms *temperature* and *heat* are normally used with the implication that the reader understands the meaning of these concepts. The contrary is more likely, however, because a precise understanding of heat and temperature is one of the objectives of the study of thermodynamics. Intuitively, the physical meaning of temperature is that it describes whether a body is "hot or cold." For example, we touch a block of metal at 120°F and conclude that it is hotter than a block of ice. The reason for this conclusion is that the hot block of metal gives up heat energy to the hand whereas the cold block of ice extracts energy. Notice that this intuitive concept of temperature is based upon an energy-transfer process which we might simply describe as *heat exchange*. It might therefore be possible to conclude that if two bodies at the same temperature are brought into contact no heat will be exchanged between the two. This serves to define *equality of temperature* but cannot establish an absolute scale of temperature which may only be accomplished with the second law of thermodynamics to be discussed in subsequent chapters. The concept of equality of temperature may be stated in the form:

> *Two bodies, each in thermal equilibrium with a third body, are in thermal equilibrium with each other.*

This statement is sometimes referred to as the *zeroth law of thermodynamics*.

It may be noted that for two bodies to be in complete equilibrium it is necessary to specify that more than the temperature of the two bodies must be equal; i.e., the pressure must also be equal and the two bodies must be so constituted that chemical change does not take place when they are brought into contact.

Insofar as everyday observations are concerned, we usually take as an indication of the temperature of a system the effect on some easily observable property of a measuring device. Thus a mercury-in-glass thermometer identifies temperature level with the relative expansion of the mercury and glass, an electrical resistance thermometer identifies temperature level with the electrical resistance of some particular metal (usually platinum), a thermocouple identifies temperature with the electromotive force generated at the junction of two dissimilar metals, and so on. In all these cases temperature, a thermodynamic property (or thermodynamic coordinate), is expressed in terms of other properties of either a liquid or metallic substance.

In accordance with this brief discussion we may tentatively conclude that heat is a form of energy which flows from one body to another as a result of a temperature difference. Note the almost circular definitions: equality of temperature is related to a condition of zero heat flow—heat is related to temperature difference. We shall clarify this matter later on.

1-4 TEMPERATURE SCALES

The two temperature scales normally employed for measurement purposes are the Fahrenheit and Celsius scales. These scales are based on a specification of the number of increments between the freezing point and boiling point of water at standard atmospheric pressure. The Celsius scale has 100 units between these points, whereas the Fahrenheit scale has 180 units. The zero points on the scales are arbitrary.

It will be shown that the second law of thermodynamics serves to define an absolute *thermodynamic* temperature scale having only positive values. The absolute Celsius scale is called the *Kelvin scale*, and the absolute Fahrenheit scale is termed the *Rankine scale*. The zero points on both *absolute* scales represent the same physical state, and the ratio of two values is the same regardless of the scale used, i.e.,

$$\left(\frac{T_2}{T_1}\right)_{\text{Rankine}} = \left(\frac{T_2}{T_1}\right)_{\text{Kelvin}} \tag{1-1}$$

The boiling point of water is arbitrarily taken as 100° on the Celsius scale and 212° on the Fahrenheit scale. The relationship between the scales is indicated in Fig. 1-2, and it is evident that the following relations apply:

$$°F = 32.0 + \tfrac{9}{5}°C \tag{1-2}$$
$$°R = \tfrac{9}{5} K$$

K	°C	°F	°R
2273.15	2000	3632	4091.67
1773.15	1500	2732	3191.67
1273.15	1000	1832	2291.67
773.15	500	932	1391.67
673.15	400	752	1211.67
573.15	300	572	1031.67
473.15	200	392	851.67
373.15	100	212.0	671.67
273.15	0	32.0	491.67
233.15	−40	−40	419.67
173.15	−100	−148	311.67

FIG. 1-2 *Relationship between Fahrenheit and Celsius temperature scales.*

$$°R = °F + 459.67 \tag{1-3}$$
$$K = °C + 273.15$$

To perform a measurement of temperature it is necessary to set up standards which may be employed for calibration of various thermometer devices. The boiling and freezing points of water are two such "standards," but they certainly do not encompass the whole range of temperatures of interest in experimental measurements. The International Practical Temperature Scale of 1968 [5]† serves to set up standards check points over a wide range of temperatures. The primary and secondary fixed points for this scale are given in Tables 1-1 and 1-2. In addition to these fixed points the international temperature scale also establishes precise procedures for interpolating between these values.

The reader should realize that we still have not given a precise *definition* of temperature. For a satisfactory analytical definition we must enter into a full discussion of the second law of thermodynamics, whereas experimental measurements (or definitions) of temperature require a set of standards like the ones described previously. Detailed techniques for measurement of temperature are described in [2]. Precise experimental measurements of temperature require considerable laboratory experience, and new techniques are introduced periodically.

† Numbers in brackets pertain to references at end of each chapter.

TABLE 1-1

Primary points for the International Practical Temperature Scale of 1968 [5]

Point Normal pressure = 14.6959 psia = 1.0132 × 10⁵ N/m²	Temperature	
	°C	°F
Triple point of equilibrium hydrogen	−259.34	−434.81
Boiling point of equilibrium hydrogen at $\frac{25}{76}$ normal pressure	−256.108	−428.99
Normal boiling point (1 atm) of equilibrium hydrogen	−252.87	−423.17
Normal boiling point of neon	−246.048	−410.89
Triple point of oxygen	−218.789	−361.82
Normal boiling point of oxygen	−182.962	−297.33
Triple point of water	0.01	32.018
Normal boiling point of water	100	212.00
Normal freezing point of zinc	419.58	787.24
Normal freezing point of silver	961.93	1763.47
Normal freezing point of gold	1064.43	1947.97

TABLE 1-2

Secondary fixed points for the International
Practical Temperature Scale of 1968 [5]

Point	Temperature, °C
Triple point, normal H_2	−259.194
Boiling point, normal H_2	−252.753
Triple point, Ne	−248.595
Triple point, N_2	−210.002
Boiling point, N_2	−195.802
Sublimation point, CO_2 (normal)	−78.476
Freezing point, Hg	−38.862
Ice point	0
Triple point, phenoxibenzene	26.87
Triple point, benzoic acid	122.37
Freezing point, In	156.634
Freezing point, Bi	271.442
Freezing point, Cd	321.108
Freezing point, Pb	327.502
Boiling point, Hg	356.66
Boiling point, S	444.674
Freezing point, Cu-Al eutectic	548.23
Freezing point, Sb	630.74
Freezing point, Al	660.74
Freezing point, Cu	1084.5
Freezing point, Ni	1455
Freezing point, Co	1494
Freezing point, Pd	1554
Freezing point, Pt	1772
Freezing point, Rh	1963
Freezing point, Ir	2447
Freezing point, W	3387

1-5 THE STATE PRINCIPLE

From the brief preceding discussion we may easily see that certain properties of matter are functionally related: thermal expansion is related to temperature, the pressure of a gas is related to temperature and volume, and so on. Mention has been made of the fact that our main interest is with equilibrium states because it is in these states that we are able to define system properties. The primary problem is to be able to define the state of the system, and the question immediately arises as to how many properties (or thermodynamic coordinates) will be necessary to accomplish this. Once again we may draw upon experience with mechanics to illustrate the nature of the problem. In planar motion we need to specify only two coordinates, whereas in spatial motion three coordinates are needed to specify the position of a particle (the particle is the system in this case). Once the particle position is specified in cartesian coordinates, no additional information is required to know the position. Furthermore, any additional coordinates which may be supplied do not furnish us with any additional information about the particle position. For example, we may specify the particle position in cartesian coordinates and also give its cylindrical and spherical coordinates. These additional coordinates convey no new information; the *state* or position of the particle is completely specified with the cartesian coordinates, and we might consider the other coordinates or properties as functionally dependent on the cartesian coordinates, as we know very well they are. It may be noted, of course, that a complete specification of the *dynamical state* of the particle will require a specification of the velocity coordinates as well as the position coordinates.

In thermodynamic systems we would expect to have certain primary properties which are necessary to define the state of a system, whereas other properties would be functionally dependent on these primary state variables. Although the simple particle system previously considered is quite easy to analyze insofar as a specification of position coordinates is concerned, even the simplest thermodynamic system involves quite a bit more thought. Our problem is to discover the thermodynamic properties *necessary* to define the state of the system; they are known as the *primary properties*. The other thermodynamic properties will then be functionally related to the primary properties. As in mechanics, the primary properties are treated as independent variables while the remaining properties are the dependent variables for use in mathematical analysis. In general, the number of primary variables for each thermodynamic system may be determined only from experiment and experience. As an example, an ideal gas requires only two properties to define the state of the system. Any two of pressure, volume, or temperature may be used.

The state principle unites the experimental axiom that the properties of matter are functionally related and the intuitive concepts of primary and dependent properties. Note that the principle is one we accept without proof; i.e., it is based on a large number of experimental observations and has no other justification. We shall have more to say about the specification of the state of a system later; however, we can clearly see that variables such as chemical composition and the number of components present in the system can have a strong influence on the problem. For now, the reader should consider the state principle as a definition of the meaning and concept of primary and dependent properties of a system. We shall give a more precise statement of the principle in Chap. 3.

1-6 WORK

Work is defined as the energy expended by a force acting through a displacement and is expressed mathematically as

$$\text{Work} = \int_c \mathbf{F} \cdot ds \tag{1-4}$$

where the line integral indicates that we consider only the product of force and displacement *in the direction of the force*, as illustrated in Fig. 1-3. To calculate work we only need to be able to specify the force and its movement; if this information is available we need to know nothing about thermal properties of the system. Why then should work be considered in the study of thermodynamics? The answer to this question is that work is an energy quantity and as such falls in the province of thermodynamic analysis. We shall see later how work and heat are related.

1-7 MICROSCOPIC AND MACROSCOPIC THERMODYNAMICS

The discussion so far has centered around consideration of thermodynamic systems composed of quantities of matter of a finite size, such as the compressed air in the steel tank or a hot block of metal. Both of these systems fall

FIG. 1-3 *Work expressed as a line integral or path function.*

into the subject of *macroscopic* thermodynamics. In the analysis of such systems only the bulk nature and properties of matter are considered, and the detailed molecular and atomic structures of the substance are ignored. This type of analysis is sometimes called *classical* thermodynamics.

When we consider the detailed molecular and atomic nature of matter, the analysis is termed *microscopic* thermodynamics. Microscopic thermodynamics is considerably more complex than classical thermodynamics, but it has the advantage that it affords the engineer additional tools for extending the methods of analysis used in classical thermodynamics.

Microscopic thermodynamic analysis could proceed by first analyzing the behavior of gas molecules on the basis of classical mechanics. This development is called *kinetic theory*. In this analysis the laws of mechanics are accepted as axiomatic, and certain special macroscopic thermodynamic phenomena are shown to follow from the detailed microscopic analysis.

It is well known that the laws of classical mechanics must be supplemented by quantum theory when operating on an atomic scale of matter. The number of particles which are considered in a microscopic analysis is very large (of the order of 10^{20} cm^{-3}); hence the application of statistical techniques to describe the most probable distribution of the particles among certain energy and momentum states would be expected to meet with good success. This is precisely the case. When statistical techniques are coupled with the limitations imposed by the quantum theory, the general method of analysis is called *statistical mechanics* or *statistical thermodynamics*.

Microscopic thermodynamics has two objectives:

1 To show how energy is distributed at the molecular and atomic level and to indicate calculations of macroscopic energy properties from this distribution.
2 To show how a knowledge of energy distribution at the microscopic level may be employed for a calculation of transport rates. These transport rates may involve energy, momentum, or mass quantities and are important in a variety of engineering applications.

Our analysis in subsequent chapters will fulfill these objectives in a partial way for certain types of thermal systems.

To illustrate further the nature of the difference between classical macroscopic thermodynamic analyses and those which deal with the microscopic nature of matter, consider the case of a volume of gas, say 1 liter, at room temperature and pressure. We know that this volume contains something on the order of 10^{23} molecules which are moving randomly in all directions, hitting each other and the walls of the vessel in which they are contained. Let us suppose that these molecules behave like perfectly elastic spheres in their collision processes. Presumably, we could apply newtonian dynamics to these particles in order to determine their motions in space. For this analysis we would need to write Newton's second law of motion

$$\sum F = \frac{d}{d\tau}(mv)$$

for *each* particle. All of the equations thus obtained would have to be solved simultaneously in order to determine the various particle velocities. What an enormous task that would be! The solution of such a large set of equations is far beyond the reach of even the largest computers. If we complicate the problem by considering intermolecular force fields and the quantization of particle energies, the solution appears even more hopeless. Why, then, have we said that some thermodynamic analyses are performed on a microscopic basis, i.e., by considering the detailed molecular behavior? The answer is simple: we do not solve this formidable set of equations, but instead rely on other methods to infer the molecular behavior. These other methods are statistical in nature.

As a matter of utter necessity, we shall be forced to describe the microscopic behavior of matter on the basis of what mathematical statistics tells us *should* happen with a large number of particles. The particular statistical techniques must be adjusted to match what we believe to be a correct physical model for the element of matter under consideration (gas, liquid, or solid). Eventually the validity of the statistical approach must be established by experiment, as must all physical theories.

Statistical thermodynamics, then, is based on what we believe to be most probable at the microscopic level. It does not supplant the basic conservation of energy principle, the notions of mechanics (as modified by quantum considerations), or other basic concepts.

With these brief introductory remarks regarding the overall subject of thermodynamics, we now begin a more precise definition of terms and concepts to be used in future developments.

1-8   THERMODYNAMIC PROPERTIES

A thermodynamic property may be described as any observable characteristic of a system. The general classification of thermodynamic properties may be divided into the specific classes of *intensive* and *extensive* properties. Intensive properties are those which are independent of the quantity of matter enclosed by the boundary of the system. Extensive properties are those whose values are directly proportional to the mass of the system. Thus the total volume of a system would be an extensive property, whereas the specific volume (volume per unit mass) would be an intensive property. Temperature and pressure are other examples of intensive properties. In general, an extensive property when divided by the mass of the system becomes an intensive property.

A thermodynamic system may contain more than one *phase*, such as a mixture of water and water vapor in which the liquid and vapor phases are present. A mixture of air, water droplets, and ice particles would be a three-phase mixture. A thermodynamic system may also consist of more than one *component*. The mixture of water and water vapor contains only one component, whereas the air-water-ice mixture contains two components: water and air. Note the distinction between phase and component. A mixture of nitrogen and carbon dioxide at atmospheric pressure and temperature would contain two components but only one phase—the gaseous phase. If the components and phases are uniformly distributed throughout the volume of the system, the system is said to be *homogeneous*. If the components and phases are not distributed uniformly, the system is *heterogeneous*. A water and water-vapor mixture may be homogeneous if the water is in the form of tiny liquid droplets uniformly distributed throughout the system volume or heterogeneous if the liquid is contained as a pool in the bottom of a pan with the vapor above it. The mixture of nitrogen and carbon dioxide would be another example of a homogeneous system.

The thermodynamic properties of a system obviously depend on its number of phases and components and whether the system is homogeneous or not.

1-9 FUNDAMENTAL THERMODYNAMIC LAWS

Without exploring any details, we may at least state the fundamental principles of thermodynamics at this point so that the reader can grasp in a qualitative way an overall perspective of the subject. These principles are accepted as axioms: as fundamental physical laws because they explain such a large body of experimental observations.

The first law of thermodynamics states:

The energy of an isolated system remains constant.

It is seen that the first law is nothing more than the law of conservation of energy which is familiar even to the neophyte scientist or engineer. The term *isolated system* describes one which does not exchange energy with its surroundings. Notice that this law is accepted as being axiomatic because it explains such a wide variety of experimental observations in physical science.

The second law of thermodynamics accepts as axiomatic several equivalent experimental observations, a few of which are stated in the following:

1   Heat flows from a high temperature to a low temperature in the absence of other effects. This means that a hot body will cool down when brought into contact with a body at a lower temperature, and not the opposite.
2   Two gases, when placed in an isolated chamber, will mix uniformly

throughout the chamber but will not separate spontaneously once mixed.

3 A battery will discharge through a resistor releasing a certain amount of energy, but it is not possible to make the reverse of this happen, i.e., to add energy to the resistor by heating and thus cause the battery to charge itself.

4 It is not possible to construct a machine or device which will operate continuously while receiving heat from a single reservoir and producing an equivalent amount of work.

We see that the second law of thermodynamics relates to the *direction* in which an energy transfer or conversion may take place; some transformations are allowed whereas others are not. All of the phenomena described in the foregoing are consequences of the second law, although the most widely used statements are those of observations 1 and 4. Note that observation 4 states that we cannot build an engine (or cycle) which is 100 percent efficient in converting heat from a single reservoir to mechanical energy.[†] Later on we shall see that the second law has a very definite meaning in terms of the distribution of energy at the molecular level. In essence, the possible directions for energy changes on the macroscopic level are related to the manner in which the energy is distributed at the microscopic level.

The third law of thermodynamics pertains to the properties of substances at absolute zero of temperature and must be developed after an analytical exposition of the first and second laws.

1-10 APPLICATIONS OF THERMODYNAMICS

Because thermodynamics is a study of energy and its transformation, it is not surprising that the most widely used applications of the science are those involving beneficial energy-conversion schemes. The large steam power plant is a typical example which we have illustrated in Fig. 1-4. High-pressure water is supplied to a boiler where heat is added to produce high-pressure, high-temperature steam. The heat may be furnished by converting the chemical energy of a fossil fuel (natural gas, oil, coal) in a combustion process, or as the heat release from a nuclear reaction. The steam is then forced through a turbine which drives an electric generator. During this flow process through the turbine, energy is removed from the steam and, as a result, it leaves the turbine at a much lower temperature and pressure. The low-pressure steam is then fed to a large vessel which contains hundreds of tubes carrying cooling water. The steam condenses on the outside of the tubes and collects at the bottom of the vessel where it is picked up by the pump and returned to the boiler at high pressure.

† A pithy expression with appeal to most students is: The first law says you cannot get something for nothing; the second law says you cannot even break even.

FIG. 1-4 *Schematic of steam power plant.*

In the steam power plant we see the characteristic elements of a *power cycle* or *heat engine*:

1   Heat addition at high temperature (boiler)
2   Heat rejection at low temperature (condenser)
3   Net work output (work out of turbine less work into pump)

The *thermal efficiency* of a heat engine is defined as

$$\text{Thermal efficiency} = \frac{\text{net work output}}{\text{heat input at high temperature}} \qquad (1\text{-}5)$$

This efficiency has definite economic significance because the heat input at the high temperature represents the energy which must be purchased (fuel, uranium, etc.) and the net work output represents what we get for the purchase. Large steam power plants can achieve efficiencies on the order of 40 percent.

In Fig. 1-5 we have shown a schematic for a typical air conditioner for a home. In this application a fluid, typically Freon, at low pressure and temperature is allowed to enter the finned tubes of the evaporator. Air from the room is blown across the tubes thereby adding energy to the Freon. As a result of this heat addition the Freon vaporizes (or boils), and leaves the evaporator in a vapor or gaseous state. It is then compressed to high pressure and temperature by adding work in the compressor. This high-temperature vapor is now at a higher temperature than the outside air so it is next passed through another finned tube device where the outside air removes sufficient heat to condense the vapor into a liquid at high pressure. The temperature at which a liquid will boil or condense depends on its pressure; the lower the

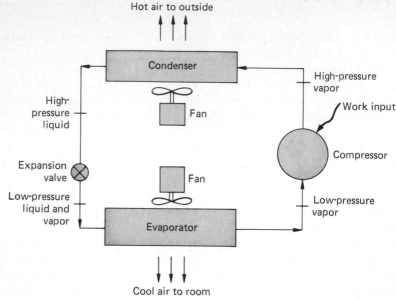

FIG. 1-5 *Schematic of air conditioning or refrigeration system.*

pressure, the lower the evaporation temperature. Thus, the next step in this cycle is to allow the high-pressure liquid to flow through an expansion valve which lowers the pressure, thereby producing the low temperature entering the evaporator.

The air conditioner or refrigeration cycle has an entirely different energy-conversion objective from the steam power plant. In the power cycle the objective is to obtain a work output from a heat input. For the refrigeration cycle our objective is the cooling effect for the room, while the energy we must purchase is the work input to the compressor. Thus, we measure our objective in this case by a *coefficient of performance*:

$$\text{Coefficient of performance (COP)} = \frac{\text{cooling effect}}{\text{work input}} \qquad (1\text{-}6)$$

Refrigeration devices can have a COP on the order of 3.0. Sometimes the refrigeration cycle is called a *heat pump* because it takes the heat at the low room temperature and "pumps" it up to the high temperature outside.

There are many other energy-conversion devices (fuel cells, thermoelectric generators, internal combustion engines) which make use of thermodynamic principles, and we shall discuss these applications after the basic principles have been developed. For now, our purpose is to alert the reader to the fact that these applications exist and furnish motivation for careful study of the analysis techniques which will be employed in later chapters.

1-11   UNIT SYSTEMS

Despite strong efforts in the professional scientific and engineering community to standardize units with a single international system, a multitude of systems are still employed in practice. Because of its simplicity and adaptability, we shall stress the international system as the most desirable one to use; however, full recognition must be given to the fact that the reader of this text will encounter other unit systems. It would be a disservice, then, not to offer explanations of the other systems in wide use and display proper conversion factors which must be employed to obtain correct answers in technical calculations. The main difficulties arise in mechanical and thermal units because electrical units have been standardized for some time. The SI (Système Internationale d'Unités) set of units will eventually prevail, and we shall point toward use of this system wherever appropriate.

The purpose of a unit system is to affix specific numerical values to observable physical phenomena so that these phenomena may be described analytically. A *dimension* is a physical variable used to specify the behavior or nature of a particular system, and a *unit* is the term used to measure the dimension. A length dimension, for example, may be measured in units of inches, feet, centimeters, etc.

The choice and specification of a particular unit system are usually made so that they fit nicely with some basic physical principle. The basic physical dimensions we shall use to describe all physical phenomena are:

L = length τ = time
M = mass T = temperature
F = force q = electric charge

The symbol q will also be used to designate heat, a form of energy, in subsequent chapters. For now we use it to designate charge. Mechanical systems of units are based on two physical principles: Newton's second law of motion,

Force ~ time rate of change of momentum

and Newton's gravitational attraction principle,

Force ~ $\dfrac{m_1 m_2}{r^2}$

Rewriting these principles in equation form,

$$F = k_N \frac{d(mv)}{d\tau} \tag{1-7}$$

$$F \longrightarrow \bigcirc \quad F = k_N \frac{d(mv)}{d\tau}$$

FIG. 1-6   *Newton's second law of motion.*

$$F = k_G \frac{m_1 m_2}{r^2} \tag{1-8}$$

where the k_N and k_G are proportionality constants. The Newtonian laws are illustrated in Figs. 1-6 and 1-7. Consider first Eq. (1-7) as applied to a particle of constant mass so that

$$F = k_N ma \tag{1-9}$$

where now a is the acceleration $dv/d\tau$. We may choose any system of units we like for this equation just so the value of k_N is properly chosen. It is worthwhile to mention, however, that something must be said to define the size of each unit. The most commonly used systems are specified with the following statements:

1 1 pound force (lbf) will accelerate 1 pound mass (lbm) 32.174 feet per second squared.
2 1 pound force will accelerate 1 slug mass 1 foot per second squared.
3 1 dyne force will accelerate 1 gram mass 1 centimeter per second squared.
4 1 newton (N) force will accelerate 1 kilogram mass 1 meter per second squared.
5 1 kilogram force will accelerate 1 kilogram mass 9.806 65 meters per second squared.

These statements serve to relate the quantities $FLM\tau$ through the dynamical Eq. (1-9). We might ask the question: How big is a newton? The answer is that it is big enough to accelerate 1 kg 1 m/s². The next questions are: How big is a kilogram, a meter, a second? The answer is that these quantities are maintained as standards at various national laboratories throughout the world—in the United States at the National Bureau of Standards. Other standard units are similarly maintained. The standard unit of time, the second, is defined as the duration of 9 192 631 770 periods of the radiation cor-

FIG. 1-7   *Newton's gravitational law.*

responding to the transition between the two hyperfine levels of the fundamental state of the atom of cesium 133. Eventually all unit systems must rely on these standards, and others, but the subject of standards is beyond the scope of our discussion.

For convenience the constant k_N is sometimes written as

$$k_N = \frac{1}{g_c} \tag{1-10}$$

where g_c is now a new constant having the dimensions of $MLF^{-1}\tau^{-2}$.

The weight of a body is defined as the force exerted on the body as a result of the acceleration of gravity. Thus

$$W = \frac{g}{g_c} m \tag{1-11}$$

where g is the acceleration of gravity. Note that the weight of a body has the dimensions of a force.

Unfortunately the concept of g_c is confusing to most students. In the first system of units discussed in the foregoing it has the value

1 $g_c = 32.174 \text{ lbm·ft/lbf·s}^2$

In this system g_c has the same numerical value as the acceleration of gravity at sea level, but *it is not the acceleration of gravity*. Rather, it is a dimensional constant employed to facilitate the use of Newton's second law of motion with this particular system of units. In the other unit systems noted previously, g_c has the following values:

2 $g_c = 1 \text{ slug·ft/lbf·s}^2$
3 $g_c = 1 \text{ kg mass·m/N·s}^2$
4 $g_c = 1 \text{ g mass·cm/dyn·s}^2$
5 $g_c = 9.8066 \text{ kg mass·m/kg force·s}^2$

The *kilopond force* (kp) is an alternative nomenclature for kilogram force and is employed widely in Europe.

In various equations used throughout this book the reader should exercise due caution to ensure that the units of various terms are consistent. If the term involves a dynamical quantity based on Newton's second law, then the g_c constant is understood to be included in the proper value for the unit system under consideration. We shall adopt the practice of omitting it from many equations in order to be consistent with the nomenclature followed in mechanics. Thus the equation

$$F = ma$$

is understood to include the $1/g_c$ term even though it is not written. In another example, kinetic energy will be written as

$$K.E. = \tfrac{1}{2}mV^2$$

but it is understood to mean that

$$K.E. = \frac{1}{2g_c}mV^2$$

The reader should note that in three of the five unit systems previously mentioned the numerical value of g_c is unity. This fact accounts for their wide usage for problems in mechanics. The pound mass (lbm) system is used extensively in many thermodynamic problems and, for this reason, due caution must be exercised when working with g_c.

ENERGY UNITS

Work and energy, of course, have the dimensions of a force times a distance and would be expressed as pound force-foot, dyne-centimeter, or newton-meter with the definitions:

1 dyne-centimeter (dyn·cm) = 1 erg
1 newton-meter (N·m) = 1 joule (J)

Power is the time rate of doing work and some appropriate units are

1 watt (W) = 1 J/s
1 horsepower (hp) = 550 ft·lbf/s = 745.7 watts (W)

Some units of energy peculiar to the study of thermodynamics are the so-called *thermal units*—the British thermal unit (Btu) and the calorie (cal):

1 Btu = energy required to raise 1 lbm of water 1°F at 68°F
1 cal = energy required to raise 1 g of water 1°C at 20°C
1 kcal = energy required to raise 1 kg of water 1°C at 20°C

The electronvolt (eV) is a unit of energy defined as the work required to move an electron through a potential difference of 1 volt (V). This unit may be converted to joules by observing that the charge on the electron, q_e, is

$$q_e = 1.602\ 189 \times 10^{-19} \text{ coulomb (C)}$$

But the volt is defined as the work to displace a unit charge so we obtain immediately that

$$1 \text{ eV} = 1.602\ 189 \times 10^{-19} \text{ J}$$

We can see that unit systems 3, 4, and 5 applied to Newton's second law of motion are "metric" but they employ different units for force, mass, and distance. The SI system is essentially that of number 4 and does not admit use of the thermal units of energy. In other words, the joule is the *only* energy unit which is allowed. Note that the calorie and kilocalorie are "metric" but they are not SI. In addition, the concept of g_c is not normally used in the SI system and the newton is *defined* as

$$1 \text{ newton (N)} \equiv 1 \text{ kilogram-meter per second squared (kg·m/s}^2)$$
$$(1\text{-}12)$$

Basic and derived SI units are given in Tables A-3 and A-4 (Appendix). Standardized multiplier prefixes are also specified as shown in Table 1-3. Thus, the centimeter is cm, the kilojoule is kJ, and a meganewton would be MN. Wherever possible, numerical results should use the standard prefixes. For example, we would write 2.6×10^5 J as either 260 kJ or 0.26 MJ.

Conversion factors for various units are given in the Appendix in Table A-2.

TABLE 1-3
Standard prefixes and multiples in SI units

Multiples and submultiples	Prefixes	Symbols	Pronunciations
10^{12}	tera	T	tĕr′à
10^{9}	giga	G	jĭ′gà
10^{6}	mega	M	mĕg′à
10^{3}	kilo	k	kĭl′ô
10^{2}	hecto	h	hek′tô
10	deka	da	dĕk′à
10^{-1}	deci	d	dĕs′ĭ
10^{-2}	centi	c	sĕn′tĭ
10^{-3}	milli	m	mĭl ĭ
10^{-6}	micro	μ	mī′ krô
10^{-9}	nano	n	năn′ô
10^{-12}	pico	p	pē′ cô
10^{-15}	femto	f	fĕm′ tô
10^{-18}	atto	a	ăt′ tô

EXAMPLE 1-1

A 10-kilopond (kp) force is applied to a mass of 3 lbm. Calculate the acceleration which results.

SOLUTION   We can perform the calculation in whatever unit system we choose, as long as we are consistent. Using the N-m-kg-s system, we observe

$$10 \text{ kp} = 98.066 \text{ N}$$
$$3 \text{ lbm} = (3)(0.454) \text{ kg} = 1.362 \text{ kg}$$

Then, from Eq. (1-9)

$$a = \frac{Fg_c}{m} = \frac{(98.066 \text{ N})(1.0 \text{ kg·m/N·s}^2)}{1.362 \text{ kg}}$$
$$= 72.00 \text{ m/s}^2 = 236.3 \text{ ft/s}^2$$

If we choose the lbm-lbf system

$$10 \text{ kp} = (10)(2.2046) = 22.046 \text{ lbf}$$
$$a = \frac{Fg_c}{m} = \frac{(22.046 \text{ lbf})(32.174 \text{ lbm·ft/lbf·s}^2)}{3 \text{ lbm}}$$
$$= 236.3 \text{ ft/s}^2 = 72.00 \text{ m/s}^2$$

The gravitation law [Eq. (1-8)] expresses the gravitational attraction force between two masses in terms of their separation distance r and k_G, *the universal gravitational constant*. In the SI system the value of k_G is

$$k_G = 6.672 \times 10^{-11} \text{ N·m}^2/\text{kg}^2 \tag{1-13}$$

This value is determined experimentally and is subject to slight modification as better experimental results become available.

Electrostatic units are based on Coulomb's law, which states

$$F = k_c \frac{qq'}{r^2} \tag{1-14}$$

where F is the force exerted between the two charges q and q' and r is the separation distance as shown in Fig. 1-8. The value of the proportionality constant k_c is 9×10^9 N·m^2/C^2, when the charge is expressed in coulombs and the force in newtons. As a matter of convention it is standard practice to let

$$k_c = \frac{1}{4\pi\epsilon_0} \tag{1-15}$$

FIG. 1-8   *Coulomb's electrostatic force law.*

where now $\epsilon_0 = 8.854\ 187\ 818 \times 10^{-12}$ is a new constant called the *permittivity of free space.*

Ampere's law states that the force between two parallel conductors, as shown in Fig. 1-9, is proportional to the currents in the conductors according to

$$F = k_A II' \frac{LL'}{r^2} \tag{1-16}$$

where I and I' are the currents in the conductors, L and L' are the lengths of the conductors, and r is the separation distance. The force is attractive if the currents are in the same direction. The proportionality constant k_A has the value

$$k_A = 10^{-7}\ \text{N/A}^2 \tag{1-17}$$

when the current is in amperes and the force in newtons. The ampere is defined by

$$1 \text{ ampere (A)} = 1 \text{ C/s} \tag{1-18}$$

For convenience the following substitution is sometimes made:

$$k_A = \frac{\mu_0}{4\pi} \tag{1-19}$$

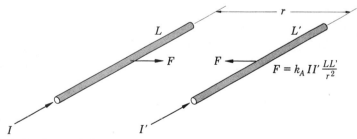

FIG. 1-9   *Ampere's law.*

where now μ_0 is a new constant called the *magnetic permittivity of free space*.

Ampere's law clearly relates a force to a movement of charge (current). If a charge of q' moving with a velocity v experiences a force in the absence of any electrostatic field, the force is said to be due to a magnetic field B. The force is related by

$$\mathbf{F} = q'\mathbf{v} \times \mathbf{B} \tag{1-20}$$

The vector equation (1-20) serves to define the *magnetic induction* or magnetic flux density $\mathbf{B}$. We see that a consistent set of units for $\mathbf{B}$ would be newton-second per coulomb-meter. It is common practice to introduce the new unit defined by

$$1 \text{ weber (Wb)} = 1 \text{ N·s·m/C} \tag{1-21}$$

Thus the units for $\mathbf{B}$ could be expressed as webers per square meter.

The *electric intensity* $\mathbf{E}$ is defined as the force exerted per unit charge resulting from an electric field. Thus

$$\mathbf{E} = \frac{\mathbf{F}}{q'} \tag{1-22}$$

The *electric potential* V is defined as the work necessary to displace a charge per unit charge. A typical set of units for the electric potential would be joules per coulomb (newton-meter per coulomb). This is defined as

$$1 \text{ volt (V)} = 1 \text{ J/C} \tag{1-23}$$

1-12 MOVEMENT TOWARD SI UNIT SYSTEM

There is a strong movement to adopt the SI unit system as the standard for the United States as it has become widely accepted throughout the world. Considerable benefit will be derived from standardization of units, but we may expect workers in industry, and some research workers as well, to continue to employ a multitude of units for an extended period of time. In other words, units like the Btu, calorie, and horsepower will still be used in technical discussions for the indefinite future. For this reason, we shall use them freely in this text. To give the reader a feel for the magnitudes of the various units we shall express the answers to example problems in several alternate ways when it appears beneficial to do so.

Undoubtedly, considerable confusion can arise when working with different unit systems, and different people prefer to use different systems. Furthermore, there can be vexing arguments over what represents a "funda-

TABLE 1-4
Typical energy quantities

	English	SI
Heating value of gallon of fuel oil	149 000 Btu	157 MJ
Heating value of bituminous coal	12 900 Btu/lbm	30 MJ/kg
Solar energy on earth, clear day	330 Btu/hr·ft²	1040 W/m²
Heating value, 1000 ft³ natural gas	10^6 Btu	1.05 GJ
Nuclear energy, uranium	3.2×10^{10} Btu/lbm	74×10^6 MJ/kg
Air conditioning cooling load for a typical dwelling, hot summer day	50 000 Btu/h	14.6 kW
Power output, typical large steam power plant	3.41×10^9 Btu/h	1000 MW
Typical body heat dissipation	450 Btu/h	132 W
Energy to vaporize water	1000 Btu/lbm	2326 kJ/kg

mental" physical quantity. For example, our senses respond to forces and we sense mass through the inertial force exerted by the mass. On the other hand, mass represents a "quantity of matter," whereas force has no such quantitative interpretation, except through an indirect measurement process based on a physical principle relating force to "quantitative" properties of matter such as Eqs. (1-7) and (1-8). The purpose of this brief discussion of units and dimensions has been to show that the selection of unit systems is made on the basis of convenience so that they will be easy to use when applied to fundamental physical phenomena.

We have stated that thermodynamics is the study of energy and its transformation. Since we have now defined the various units of energy, it is worthwhile to give some typical values which may be of interest to the reader. Table 1-4 gives a few such quantities in both English and SI units.

EXAMPLE 1-2

Two large conductors 3 ft long supplying current to an electric arc facility each carry a current of 1500 A and are separated by a distance of 3 in. Calculate the force acting on the conductors.

SOLUTION   Ampere's law is employed for this calculation.

$$F = k_A II' \frac{LL'}{r^2}$$

with

$$L = L' = 3 \text{ ft} = 0.9144 \text{ m}$$
$$I = I' = 1500 \text{ A}$$
$$r = 3 \text{ in} = 0.0762 \text{ m}$$

Thus

$$F = \frac{(10^{-7} \text{ N/A}^2)(1500 \text{ A})^2(0.9144)^2}{(0.0762 \text{ m})^2}$$
$$= 32.4 \text{ N} = 7.28 \text{ lbf}$$

EXAMPLE 1-3

A man circling the earth in a spaceship is weighed with a device which indicates his weight as 275 newtons. From navigational measurements it is known that the gravitational acceleration at this particular location is 11.0 ft/s². Calculate the mass of the man and his weight at sea level on the earth.

SOLUTION   The value of g is 11.0 ft/s² = 3.3528 m/s² and we use Eq. (1-11):

$$W = m \frac{g}{g_c}$$

$$275 \text{ N} = \frac{m(3.3528 \text{ m/s}^2)}{1.0 \text{ kg·m/N·s}^2}$$

which yields

$$m = 82.021 \text{ kg} = 180.83 \text{ lbm}$$

At sea level we have $g = 32.174 \text{ ft/s}^2 = 9.8066 \text{ m/s}^2$ so that the weight would be

$$W = \frac{mg}{g_c} = \frac{(82.021 \text{ kg})(9.8066 \text{ m/s}^2)}{1.0 \text{ N·m/kg·s}^2}$$
$$= 804.34 \text{ N} = 180.83 \text{ lbf}$$

We notice, of course, that 1 lbm weighs 1 lbf at sea level.

1-13   PRESSURE AND THE CONTINUUM

Most readers are familiar with the atomic and molecular nature of matter and know that a finite volume of liquid or gas, large enough to be visible to the eye, contains a very large number of particles. The density of such a finite volume ρ is defined as the mass per unit volume or

$$\rho = \frac{m}{V} \tag{1-24}$$

This definition assumes that the volume is sufficiently large that the total number of particles in the volume does not experience significant fluctuations with time. In this case we say that the system can be treated as a *continuum*.

Pressure is defined as the normal component of force per unit area exerted by a fluid on a boundary. The pressure is defined only for an area element sufficiently large so that the fluid may be treated as a continuum. Thus

$$p = \lim_{\Delta A \to \Delta A'} \frac{\Delta F_n}{\Delta A} \tag{1-25}$$

where $\Delta A'$ is the minimum area for which the fluid behaves as a continuum. In general, continuum behavior is experienced so long as the average distance that a fluid molecule travels between collisions is small compared with a boundary dimension—in this case, to a side of the area element $\Delta A'$.

The total pressure exerted on a boundary wall is called the *absolute pressure*. The pressure exerted on a wall by the atmosphere is called the *atmospheric pressure* and varies with location and elevation on the earth's surface. The atmospheric pressure is the result of the weight of air at the particular location. *Gage pressure* designates the difference between absolute and atmospheric pressure in a particular system and is normally measured by an instrument which has atmospheric pressure as a reference. *Vacuum* represents the amount by which the atmospheric pressure exceeds the absolute pressure of a system. The term *vacuum* is synonymous with the term *negative gage pressure*. From these definitions we see that the absolute pressure may not be negative and the vacuum may not be greater than the local atmospheric pressure. The various pressure terms are illustrated in Fig. 1-10.

In the U.S. engineering system of units, pressure is usually expressed in pounds per square inch absolute (psia). Gage pressure carrying the same units is expressed in pounds per square inch gage (psig). Pressure is fre-

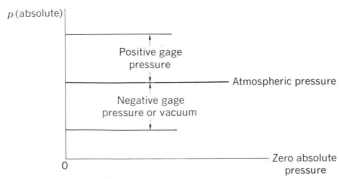

FIG. 1-10 *Relationship between pressure terms.*

quently expressed in terms of the height of a column of mercury (Hg) which it will support at a temperature of 20°C (68°F). At standard atmospheric pressure of 14.696 psia this height is 760 mm with the density of mercury taken as 13.5951 g/cm³. The SI unit for pressure is the newton per square meter or pascal (Pa). Conversion factors for pressure are given in the Appendix. Some other units commonly used are:

1 microbar (μbar) = 1 dyn/cm² = 10^{-6} bar = 0.1 N/m²
1 millimeter = 1 mm = 1333.22 microbars
1 micronmeter = 1 μm = 10^{-6} mHg = 10^{-3} mmHg
1 torr = 1 mmHg

Some of the common units for standard atmospheric pressure are:

$$1 \text{ standard atmosphere} = \begin{cases} 14.696 \text{ lbf/in}^2 \text{ (psi)} \\ 29.92 \text{ inHg at } 32°F \\ 1.01325 \times 10^5 \text{ N/m}^2 \\ 1.0332 \text{ kgf/cm}^2 \end{cases}$$

We should note that 1 bar (10^6 dyn/cm²) carries the connotation that it is "one *bar*ometric pressure," but it is in fact about one percent less than one standard atmosphere, as noted above.

1-14   THE IDEAL GAS

A *mole* (mol) is a quantity of a substance having a mass numerically equal to its molecular weight. One kg mol of oxygen, for example, would be 32.000 kg and 1 lbm mol would be 32.000 lbm. We shall designate the molecular weight with M and the number of moles with η. The mass of the substance m is, therefore,

$$m = \eta M \tag{1-26}$$

The mass per molecule is designated by M^* so that the mass per mole is

$$\frac{m}{\eta} = M^* N_0 = M \tag{1-27}$$

where N_0 is the number of molecules per mole. Thus

$$N_0 = \frac{M}{M^*} \tag{1-28}$$

This ratio is a constant called Avogadro's number and has the value

$$N_0 = 6.022\ 045 \times 10^{23} \text{ molecules/g mol} \qquad (1\text{-}29)$$

The total volume is designated by the symbol V and the *specific* volume (volume per unit mass) by v. The molal specific volume (volume per mole) is designated by $\bar{v}$.

Now suppose a series of experiments are conducted with different gases. Pressure, volume, and temperature are measured for one mole of each gas at various pressures and temperatures. The results of the experiments might be plotted as shown in Fig. 1-11 for a single gas. Regardless of the gas, it is found experimentally that lines of constant temperature on this chart all converge to a single point for the limit of zero pressure. This value is defined as the *universal gas constant*.

$$\mathscr{R} = \lim_{p \to 0} \frac{p\bar{v}}{T} \qquad (1\text{-}30)$$

and has the numerical value of 1545.35 ft·lbf/lbm·mol·°R or 8314.41 J/kg·mol·K.

To a good approximation many gases behave according to the simple equation

$$p\bar{v} = \mathscr{R}T \qquad (1\text{-}31)$$

over a rather wide range of temperatures and pressures. Equation (1-31) is called the *equation of state* of an ideal gas. The term *equation of state* means that it establishes the relationships between the thermodynamic properties necessary to define the state of the system.

It is to be noted that the temperature in Eq. (1-31) must be expressed in absolute units, i.e., in either Rankine or Kelvin units.

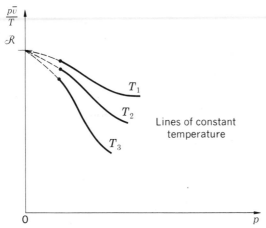

FIG. 1-11   *Determination of the universal gas constant.*

We have used the indefinite phrase "rather wide range" to describe the applicability of the ideal-gas equation. Hopefully, this will not cause confusion. It turns out that gases like oxygen, nitrogen, air, helium, carbon dioxide, etc., at room temperature do obey Eq. (1-31) up to pressures of tens of atmospheres. In Chap. 7 we shall examine some explicit techniques for handling nonideal gases. In the meantime, when working the problems, the reader should assume that the foregoing gases do exhibit ideal behavior.

We can write Eq. (1-31) in several different ways. Since $\bar{v} = V/\eta$, we have

$$pV = \eta \mathscr{R} T \tag{1-32}$$

Also, since $V = mv$, we have

$$pv = \frac{\eta}{m} \mathscr{R} T \tag{1-33}$$

But $m/\eta = M$, so that

$$pv = \frac{\mathscr{R}}{M} T = RT$$

The ratio $\mathscr{R}/M = R$ is called the *gas constant* for the particular gas. We may also write

$$p V = \frac{m}{M} \mathscr{R} T$$

$$pV = mRT \tag{1-34}$$

In Eq. (1.24) we defined the density of a finite volume as the mass per unit volume, or

$$\rho = \frac{m}{V} = \frac{1}{v} \tag{1-35}$$

Thus, the ideal-gas equation could also be written in the alternative form:

$$p v = RT$$

$$p = \rho RT \tag{1-36}$$

It is purely a matter of convenience as to which of the foregoing relations shall be employed for calculation purposes.

EXAMPLE 1-4

A certain automobile tire may be approximated as a torus 20 cm in diameter with an inside radius of 20 cm. Calculate the mass of air contained in the tire at 20°C and 2.0 atm *gage* pressure. The equivalent molecular weight of air is 28.97.

SOLUTION   The mass of air is calculated by inserting appropriate values in Eq. (1-34). The volume is calculated as

$$V = \frac{\pi(20)^2}{4} (2\pi)(20 + 10) = 59\ 218 \text{ cm}^3 = 0.0592 \text{ m}^3$$

The gas constant for air is calculated as

$$R = \frac{\mathscr{R}}{M} = \frac{8314 \text{ J/kg·mol·K}}{28.97 \text{ kg/kg mol}} = 287 \text{ J/kg·K}$$

The term *gage pressure* refers to the differential pressure above atmospheric. Thus, the absolute pressure is

$$p = 2 \text{ atm} + 1 \text{ atm} = 3 \text{ atm} = 3.04 \times 10^5 \text{ Pa}$$

From Eq. (1-34) the mass is

$$m = \frac{pV}{RT} = \frac{(3.04 \times 10^5)(0.0592)}{(287)(20 + 273)} = 0.214 \text{ kg}$$

EXAMPLE 1-5

Calculate the number of oxygen (O_2) molecules contained in a 170-cm³ container at 2 atm and 250°C.

SOLUTION   For this problem we first calculate the number of killogram-moles and then make use of Avogardro's number in Eq. (1-29).

$$V = 170 \text{ cm}^3 = 1.7 \times 10^{-4} \text{ m}^3$$
$$p = 2 \text{ atm} = 2.0264 \times 10^5 \text{ Pa (N/m}^2\text{)}$$
$$T = 250°C = 523 \text{ K}$$

The number of kilogram-moles may now be calculated with Eq. (1-32) using $\mathscr{R} = 8314$ J/kg·mol·K.

$$\eta = \frac{pV}{\mathscr{R}T} = \frac{(2.0264 \times 10^5)(1.7 \times 10^{-4})}{(8314)(523)} = 7.92 \times 10^{-6} \text{ kg·mol}$$

The number of molecules is thus

$$N = \eta N_0 = (7.92 \times 10^{-6})(6.022 \times 10^{26} \text{ molecules/kg·mol})$$
$$= 4.77 \times 10^{21} \text{ molecules}$$

Note that the type of gas (oxygen) does not influence the answer. If we wish to calculate the mass of oxygen we may use

$$m = \eta M = (7.92 \times 10^{-6} \text{ kg·mol})(32.00 \text{ kg/kg·mol}) = 2.53 \times 10^{-4} \text{ kg}$$

1-15 THE IDEAL-GAS THERMOMETER

The foregoing discussion furnishes the basic principles for a temperature-measurement device which may serve as a secondary experimental standard. A fixed volume is filled with gas and exposed to the temperature to be measured, as shown in Fig. 1-12. At the temperature T the gas system pressure is measured. Next the volume is exposed to a standard reference temperature (as discussed in Sec. 1-4) and the pressure is measured under these conditions. According to Eq. (1-31) at constant volume,

$$T = T_{\text{ref}} \left(\frac{p}{p_{\text{ref}}}\right)_{\text{const. vol.}} \tag{1-37}$$

Now suppose that some of the gas is removed from the volume and the pressure measurements are repeated. In general, there will be a slight difference in the pressure ratio in Eq. (1-37) as the quantity of gas is varied. However, regardless of the gas used, the series of measurements may be repeated and the results plotted, as shown in Fig. 1-13. When the curve is extrapolated to zero pressure, the true temperature as defined by the ideal-gas equation of state will be obtained. A gas thermometer may be used to measure temperatures as low as 1 K by extrapolation.

It is most important to realize that a reference temperature must be established in order to utilize the gas-thermometer principle (or any other

Unknown
temperature
T

Volume
V

Pressure
gage

FIG. 1-12 *Schematic of ideal-gas thermometer.*

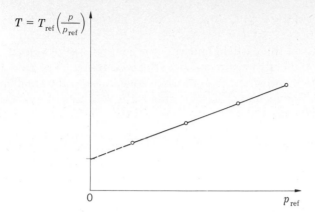

$$T = T_{ref}\left(\frac{p}{p_{ref}}\right)$$

0 p_{ref}

FIG. 1-13 *Results of measurements with ideal-gas thermometer.*

thermometer for that matter). This reference temperature may be established with the international temperature scale, as discussed in Sec. 1-4.

1-16 SIMPLE KINETIC THEORY OF AN IDEAL GAS

To give a preliminary analytical indication of the relation between macroscopic and microscopic thermodynamics, let us consider a very simple kinetic analysis of gas molecules. We assume that all molecules are spherical in shape and perfectly resilient in their collision processes.

The pressure of the gas is identified with the force exerted on a containing wall resulting from molecular collisions, as shown in Fig. 1-14. The impulse imparted to the wall by each particle is given by

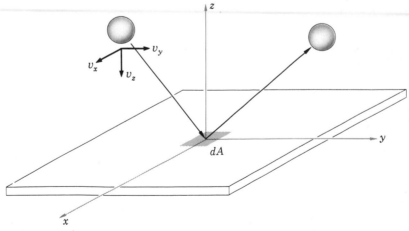

FIG. 1-14 *Kinetic theory interpretation of pressure of an ideal gas.*

$$F \, d\tau = M^* v_z - M^*(-v_z) = 2M^* v_z \tag{1-38}$$

where M^* is the mass of the particle and v_z is the component of velocity in the direction. For a perfectly elastic collision this component will have the same magnitude before and after the collision. Now consider the small area dA. The rate at which molecules strike this area is the *flow rate* of molecules in the z direction or

$$\dot{n} = n \, dA v_z \qquad \text{molecules/s} \tag{1-39}$$

where n is the molecular density (molecules/unit volume). If we assume random molecular motion, only *half* of the molecules will be headed downward so that the molecular flow rate of interest is

$$\dot{n}_{-z} = \tfrac{1}{2} n \, dA v_z \tag{1-40}$$

The total impulse imparted to the surface in time $d\tau$ is therefore given as a product of Eqs. (1-38) and (1-40).

$$F \, d\tau = (2M^* v_z)(\tfrac{1}{2} n v_2 \, dA \, d\tau)$$

The force per unit area, or pressure, is therefore

$$\frac{F}{dA} = p = M^* n v_z^2 \tag{1-41}$$

In reality the molecules do not all move with the same speed, and the velocity factor in Eq. (1-41) must be an average value. It is reasonable to assume that the molecules move randomly in all directions so that one component is just as likely as another. We have

$$v^2 = v_x^2 + v_y^2 + v_z^2$$

The "randomness" assumption means that

$$\overline{v_x^2} = \overline{v_y^2} = \overline{v_z^2} = \tfrac{1}{3} \overline{v^2} \tag{1-42}$$

where the bar sign denotes an average value. Rewriting Eq. (1-41) in accordance with Eq. (1-42) gives

$$p = \tfrac{1}{3} n M^* \overline{v^2} \tag{1-43}$$

The product nM^* represents the mass per unit volume, which, of course, is the reciprocal of the specific volume (volume per unit mass). Accordingly Eq. (1-43) may be written as

$$pv = \tfrac{1}{3}\overline{v^2}$$

or

$$pV = m\tfrac{1}{3}\overline{v^2} \tag{1-44}$$

Equation (1-44) is the same as Eq. (1-34) if we identify

$$\tfrac{1}{3}\overline{v^2} = RT \tag{1-45}$$

Thus, in this kinetic interpretation of an ideal gas, temperature is found to be a measure of the average kinetic energy of the gas molecules.

EXAMPLE 1-6

Calculate the rms (root-mean-square) velocity of oxygen molecules at 14.7 psia and 70°F. The rms velocity is defined as

$$v_{\text{rms}} = (\overline{v^2})^{1/2}$$

SOLUTION   We use Eq. (1-45) to calculate the mean square velocity and, thence, the value of v_{rms}.

$$\overline{v^2} = 3RT \tag{a}$$

The gas constant for oxygen is calculated as

$$R_{O_2} = \frac{\mathscr{R}}{M_{O_2}} = \frac{1545}{32} = 48.3 \text{ ft·lbf/lbm·°R} = 259.9 \text{ J/kg·K} \tag{b}$$

At this point we remind the reader that g_c must be properly used when dynamical quantities are involved. Since we are working in the lbm-lbf system of units, $g_c = 32.174$ lbm·ft/lbf·s² and the kinetic energy is understood to be written as

$$\text{K.E.} = \frac{1}{2g_c} mv^2$$

This means that relation (a) is understood to contain g_c since $\overline{v^2}$ is representative of kinetic energy. We have

$$\overline{v^2} = 3RT = (3)(48.3)(460 + 70) = 7.67 \times 10^4 \text{ ft·lbf/lbm}$$

Note that a simple multiplication by g_c gives the proper units for $\overline{v^2}$:

$$\overline{v^2} = (7.67 \times 10^4 \text{ ft·lbf/lbm})(32.174 \text{ lbm·ft/lbf·s}^2)$$
$$\overline{v^2} = 2.47 \times 10^6 \text{ ft}^2/\text{s}^2$$

The rms velocity is, thus,

$$v_{rms} = (\overline{v^2})^{1/2} = (2.47 \times 10^6)^{1/2} = 1570 \text{ ft/s} = 478.5 \text{ m/s}$$

There are several restrictions that must be imposed on the foregoing analysis which we shall discuss in later chapters. For now, the reader should glean the fact that microscopic analysis may be employed to yield valuable physical interpretations of macroscopic phenomena. In even this simple case it has (a) identified temperature with molecular kinetic energy, (b) modeled an "ideal gas" as one having perfectly elastic collisions, and (c) modeled an ideal gas as one having perfectly random molecular motion. The results in (b) and (c) were initial assumptions that were made to formulate the model. The fact that these assumptions lead to a prediction of the ideal-gas law lends credence to their validity.

These are not insignificant points, and it should be clear to the reader that more comprehensive microscopic analyses should be able to provide increasingly valuable information and physical insight.

REVIEW QUESTIONS

1   Discuss the following terms:
   (a) System   (b) State   (c) Property
   (d) Process   (e) Surroundings   (f) Quasistatic
   (g) Work and heat   (h) Boundary   (i) Phase
   (j) Component   (k) Dimensions and units

2   How are temperature scales defined for experimental work?

3   How is work defined?

4   What is the first law of thermodynamics?

5   Describe a physical phenomenon which is a result of the second law of thermodynamics?

6   What are the fundamental units in the SI system?

7   How does a pound mass differ from a pound force?

8   Which is larger: a kilopond or a newton?

9   Describe the so-called "thermal" units of energy.

10   Why isn't g_c the acceleration of gravity?

11   What is a continuum?

12   What is an ideal gas?

13 What is meant by the term *pressure?*

14 What is a mole?

15 What do the terms *gage pressure* and *vacuum* mean?

16 Why is the concept of equilibrium important in describing thermodynamic systems?

17 Discuss the similarity between a free body and the concept of a thermodynamic system.

18 What is a quasistatic process?

19 How are heat and temperature related?

20 Why are properties needed?

21 What determines the state of a thermodynamic system?

22 What is meant by the term *specific volume?*

23 How is thermal efficiency of a heat engine defined?

24 How does one characterize the performance of a refrigeration device?

25 What is Newton's second law of motion?

26 What is a Btu?

27 What is a kilopond force?

PROBLEMS   (Mixed units)

1-1 A force of 1 N is applied to a mass of 3 lbm. Calculate the acceleration of the mass in ft/s². Also determine a value of g_c which would be applicable for this particular set of units.

1-2 Two 1-lbm masses are separated by a distance of 4 in. Calculate the gravitational attraction force between the masses. What acceleration would this force produce in ft/s²?

1-3 Assuming that the diameter of the earth is 8000 mi, calculate an equivalent mass of the earth by using the acceleration of gravity at sea level (32.174 ft/s²) and Newton's law of gravitation.

1-4 What is the weight of a 1 slug mass at a location where the acceleration of gravity is 10 ft/s²?

1-5 Two 1-ft parallel conductors each have a current of 1000 A and are separated by a distance of 1.5 in. Calculate the attraction force between the conductors in units of lbf.

1-6 A mass having a weight of 1 lbf at a location where $g = 11$ ft/s² is acted on by a force of 1 N. Calculate the acceleration.

1-7 What force in newtons is necessary to accelerate 1 slug 1 ft/s²?

1-8   5 lbm of nitrogen are contained in a vessel at 10 psia and 500°F. Calculate the volume of the container and the number of molecules enclosed. Also calculate the rms velocity for the molecules.

1-9   Air ($M = 28.97$) is compressed from standard atmospheric temperature and pressure (14.7 psia, 70°F) to 300 psia and 300°F. The initial volume of the air is 1.0 ft³. Calculate the volume under the high-pressure conditions.

1-10   Nitrogen is contained in a 3.5-m³ tank at a pressure of 10 MPa and a temperature of 40°C. Calculate the mass in kilograms.

1-11   A small block of metal is weighed in a satellite where the acceleration of gravity is 3 m/s². The observed weight is 10 N. Calculate the mass in kilograms and also in some other units.

1-12   Calculate the rms velocity of nitrogen molecules at a temperature of 100°C.

1-13   A quantity of oxygen is compressed from 100 kPa and 50°C to 200 kPa and a temperature of 100°C. Calculate the percentage change in volume which results from this compression.

1-14   How many molecules are contained in 1 lbm·mol of air?

1-15   Convert the following temperatures to °C.
   (a) −40°F   (b) 0°F   (c) 1000°F   (d) 2500°R   (e) 1300 K

1-16   Calculate the density of the following gases at the specified conditions. Express in units of lbm/ft³ and kg/m³.
   (a) Air at 1 atm and 68°F
   (b) Oxygen at 500 psia and 300°F
   (c) Oxygen at 3 MPa and 200°C
   (d) Helium ($M = 4$) at 50 psia and 400°F
   (e) Helium at 100 kPa and 20°C
   (f) Carbon dioxide ($M = 44$) at 3 psia and 100°F
   (g) Carbon dioxide at 10 kPa and 50°C

1-17   A certain container is connected to a vacuum pump and evacuated to an absolute pressure of 10^{-9} atm. Calculate the number of air molecules per unit volume at this low pressure and 70°F.

1-18   The buoyancy force on a submerged object is equal to the weight of displaced fluid. Consider a 10-ft-diameter balloon filled with helium at 3-psig pressure and 70°F. Calculate the net lifting force of the balloon when immersed in standard atmospheric air at 14.7 psia and 70°F. Neglect the weight of the balloon.

1-19   Calculate the rms velocity of hydrogen molecules in a gas at 1000 K.

1-20   Calculate the number of air molecules in a 20 by 20 by 10 ft room.

1-21   Calculate the pressure exerted by a 1-ft column of mercury.

1-22   A special manometer fluid has a specific gravity of 2.95 and is used to measure a pressure of 17.5 psia at a location where the barometric pressure is 28.9 inHg. What height will the manometer fluid register?

1-23 A mass of 3 kg is accelerated to a velocity of 20 ft/s. Calculate the work required to accomplish this acceleration and express in units of (a) ft·lbf, (b) joules, and (c) Btu.

1-24 Assuming that two large masses of 1000 kg each obey Newton's gravitational attraction principle in free space, calculate the velocity the two masses would assume when separated by a distance of 50 ft if they are released from rest when separated by a distance of 300 ft.

1-25 Calculate the attractive force between two parallel 6-ft conductors carrying currents of 10 000 A and separated by a distance of 2 in. Express the answer in lbf.

1-26 Calculate the number of O_2 molecules striking the wall per unit area and per unit time in a container where the gas is maintained at a pressure of 80 psia and a temperature of 500°F.

1-27 Repeat Prob. 1-26 for a pressure of 10^{-5} atm and a temperature of $-100°F$.

1-28 The solar constant is the solar energy flux which arrives at the outer edge of the atmosphere and has a value of 444.2 Btu/hr·ft². Express this constant in units of W/cm², cal/in², and J/hr·in².

1-29 In the economic evaluation of energy systems the cost per million Btu of energy is a figure commonly used to measure alternatives. Express the cost of energy in this unit for each of the following sources:

Source	Unit cost	Energy content
Natural gas	$3.25/1000 ft³	1000 Btu/ft³
Gasoline	$1.50/gal	140 000 Btu/gal
Electricity	$0.08/kWh	

1-30 Suppose an automobile consumes gasoline at the rate of 12 mi/gal when moving at a speed of 60 mi/h. Using the data of Prob. 1-29 calculate the rate of expenditure of energy, and fuel cost per mile.

1-31 Nitrogen is maintained in a vessel at a pressure of 25 atm and a temperature of 75°C. Calculate the mass of gas in a 1000-cm³ vessel.

1-32 How much work is required to accelerate a 4800-lbm automobile to a speed of 60 mi/h.

1-33 Approximately 80 cal are required to melt 1 g of ice. A commonly used unit to describe refrigeration capacity is the "ton" defined as the energy required to freeze 1 ton of water during a 24-hr period. Express the value of the ton in Btu/h and also in kW.

1-34 In a certain large power plant the heat which must be dissipated in a nearby lake is equivalent to 2800 MW (1 MW = 10^6 watts). Water from the lake is used to accomplish the cooling, and the water is allowed to rise 12°F as it circulates through the power plant. If each gallon of water has a mass of 8.33 lbm and 1 Btu is required to raise each lbm 1°F, calculate the number of gallons of cooling water required each minute.

1-35 An alternate system to that of Prob. 1-34 is proposed whereby a large cooling tower is used to remove the heat by evaporating water. If 1040 Btu are required to vaporize each lbm of water, how many gallons would have to be evaporated each minute to dissipate the 2800 MW?

1-36 An average person consumes about 3000 kcal of food energy each day. Suppose solar energy falls on the earth's surface at the rate of 200 Btu/hr·ft² after attenuation and absorption in the atmosphere and that this figure is averaged over an 8-h period. How many ft² would be necessary to capture the solar energy equivalent of a man's daily consumption of food energy?

1-37 A steam power cycle is designed to produce a power output of 300 000 kW, and the energy supplied to the cycle is 13 500 Btu/kWh of power output. Consider two fuels: (a) coal with an energy content of 12 500 Btu/lbm and a cost of $35.00/ton and (b) natural gas with an energy content of 1000 Btu/ft³ at a cost of $2.50/1000 ft³. Calculate the operating cost per hour for each fuel.

REFERENCES

1 Stimson, H. F.: The International Temperature Scale of 1948, *J. Res. Natl. Bur. Std.* (paper, 1962), vol. 42, p. 211, March 1949.

2 Holman, J. P.: "Experimental Methods for Engineers," 3d ed., McGraw-Hill Book Company, New York, 1978.

3 Hatsopoulos, G. N., and J. H. Keenen: "Principles of General Thermodynamics," John Wiley & Sons, Inc., New York, 1965.

4 Sears, F. W.: "Thermodynamics," 2d ed., Addison-Wesley Publishing Company, Inc., Reading, Mass., 1953.

5 Barber, C. R.: The International Practical Temperature Scale of 1968, *Metrologia*, vol. 5, no. 2, p. 35, 1969.

6 Cohen, E. R., and B. N. Taylor: "The 1973 Least-Squares Adjustment of the Fundamental Constants," *J. Phys. Chem. Ref. Data.*, vol. 2, no. 4, 1973.

7 "The Metric System of Measurement: Interpretation and Modification of the International System of Units for the United States," *Federal Register*, vol. 41, no. 239, pp. 54018–54019, Dec. 10, 1976.

8 Mechtly, E. A.: "The International System of Units, Physical Constants, and Conversion Factors, Revised," NASA SP-7012, 1973.

chapter two
The first law of thermodynamics

2-1 INTRODUCTION

The first law of thermodynamics has been stated as the conservation of energy principle. This statement is a simple one, but a general application of the principle can be quite complex in practice because we must be careful to identify all types of energy involved in a process. Even with all the types of energy identified, the analysis of the problem is not necessarily simple.

In this chapter we seek to amplify the conservation of energy principle so that it may be expressed in explicit analytical form. An analysis of systems on the basis of the first law of thermodynamics takes the form of calculation of various energy quantities and a balance of these quantities in accordance with the conservation idea. At times we are tempted to say that the first law is given by this equation or that equation, and the tendency is to fall into an analytical vertigo while the simple physical principle is at least momentarily forgotten. At the risk of seeming repetitious, it is well to remind the reader that all of the analytical formalisms are only mathematical statements of a physical principle. The conservation idea is the basic underlying notion, and should uncertainties about an analysis procedure occur, the procedure should be examined to be sure that it is consistent with the physical principle.

Once the first law is expressed in analytical form, application to specific engineering systems of interest may begin immediately. As the application of the energy conservation principle progresses, it will be necessary to introduce information regarding thermodynamic properties and their interrelationship in order to study a variety of processes used in engineering applications.

2-2 WORK

Work is defined as the energy expended by a force acting through a distance. It is expressed mathematically as

$$W = \int_c \mathbf{F} \cdot d\mathbf{s} = \int F \cos \theta \, ds \qquad (2\text{-}1)$$

where the dot product of the force vector and elemental displacement vector indicates that the work is computed by considering only the component of force in the direction of the displacement. This calculation is indicated schematically in Fig. 2-1 where a force moves along a specified path. The total work done is given by the line integral along the path. We immediately sense the notion that work is dependent on the path, for it is a path integral which must be evaluated in order to calculate the work.

It is important to realize the sign convention for work which is implied by Eq. (2-1). A *positive* work is obtained when both the force and displace-

FIG. 2-1 *Definition of work.*

ment vectors are in the same directions. A *negative* work is obtained when they are in opposite directions. In other words, a force acting to the right, and displaced to the right, does positive work while a force acting to the right, and displaced to the *left*, does negative work.

Equation (2-1) is the fundamental analytical definition of work and may always be used as a basis for calculating work. To clarify this calculation, we now consider some applications of Eq. (2-1) to several simple processes.

2-3 EXPANSION OR COMPRESSION WORK IN A CYLINDER

First consider the expansion of a gas behind a piston in a cylinder, as shown in Fig. 2-2. The pressure in the cylinder is p and the volume of the gas is denoted by V. The force exerted on the piston is

$$F = pA$$

where A is the surface area of the piston exposed to the gas. The motion of the piston is in the direction of the applied force, and a differential displacement ds may be expressed in terms of the change in volume of gas dV as

FIG. 2-2 *Expansion work in a cylinder.*

$$ds = \frac{dV}{A}$$

Combining the force and displacement functions, the amount of work done by the force acting *on the piston* is

$$dW = pA \frac{dV}{A}$$

or

$$W = \int_{V_1}^{V_2} p \, dV \qquad \text{work done on face of piston} \qquad (2\text{-}2)$$

Note that the work done on the face of the piston is positive when the piston moves in the direction of the force, i.e., when dV is positive. This is the work done *on* the piston *by* the pressure force of the gas and may be illustrated graphically as the area under the curve in Fig. 2-3. Now let us think in terms of thermodynamic systems for a moment. If we take the piston as a thermodynamic system (or free body), the work delivered to the boundary of that system by the pressure forces of the gas is given by Eq. (2-2). On the other hand, if we took the *gas in the cylinder* as the thermodynamic system (or free body), the work delivered to the boundary of that system would be negative (assuming dV positive) because the force and displacement are in opposite directions. Considering the gas as a thermodynamic system we could write Eq. (2-2) as

$$W = - \int_{V_1}^{V_2} p \, dV \qquad \text{work done on gas by piston} \qquad (2\text{-}2a)$$

So, we see from this simple example that the sign one attaches to the work

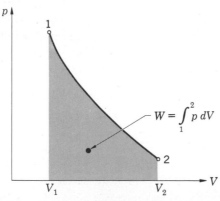

FIG. 2-3 *Graphical representation of expansion work.*

term depends on the system selected. Consistent with the sign conventions of mechanics we conclude that the work done on a system is positive because that is what we experience when the force acting on the system and the displacement are in the same direction. Conversely, the work done by the system is negative. As we have mentioned previously, we shall tend to focus our discussion in thermodynamics on *work delivered to the boundary of a system*, rather than forces and displacements. Nevertheless, one should keep in mind that it is the notion of a force acting through a displacement that represents the fundamental concept of work.

OPEN AND CLOSED SYSTEMS

In the piston-cylinder arrangement the fluid is taken as the thermodynamic system. We may note that this is a *closed system*, or one which does not admit the transfer of mass across the system boundary. Analysis techniques for *open systems* where mass may cross the boundary will be developed in Chap. 4.

NONEQUILIBRIUM EFFECTS

We have already described the quasistatic process in Chap. 1 as a series of equilibrium states. For the gas in the cylinder such a process would be experienced when the expansion occurs slowly and the pressure and temperature remain uniform during each step of the process. If, however, the expansion (or compression) occurs very rapidly the pressure may be nonuniform in the cylinder and the pressure force at the face of the piston may not be indicative of the pressure at other locations in the confined space. The *work at the boundary* is still calculated from an integration of the forces and displacements on the boundary, but for the nonequilibrium case there is no simple relation to the thermodynamic state of the gas in the cylinder. From this discussion we see once again that work is an effect which occurs at the boundary of a system and that it is strongly dependent on the particular way the thermodynamic process is executed.

Now consider the effect friction might have on the calculation of the expansion work in the cylinder. Friction can occur at the contact surface between the piston and cylinder and acts in a direction opposite to the motion of the piston. Thus, friction acts to reduce the work delivered to other things outside the piston-cylinder arrangement even if the process involving the gas is a nice quasistatic one. Sometimes one may speak of a thermodynamic system consisting of the gas *and* piston because it is the work output of this combination which is of frequent interest. For such a combined system a quasistatic process would only be experienced when the piston is slow-moving and frictionless. To evaluate the work, we need only to establish a free body and calculate the work done by the various forces acting on the free body. The calculation is made with Eq. (2-1).

Our discussion of the expansion work in the piston cylinder assumed a gas as the medium contained in the cylinder. We should note at this point that Eq. (2-2a) would apply equally well to a liquid or even to a solid; however, in these cases very large pressures may be required to effect any appreciable change in volume.

2-4 WORK IN A GRAVITATIONAL FIELD

Now consider a system consisting of a weight. We wish to compute the work expended in raising the weight through a height h in a gravitational field. Again, the force and displacement are specified in order to calculate the work. The force is equal to the weight, as shown in Fig. 2-4, and is given by

$$F = w = m\frac{g}{g_c}$$

The work done by the force is given by Eq. (2-1) as

$$W = \int \mathbf{F} \cdot d\mathbf{s} = \int_0^h w \cdot ds = wh \qquad (2\text{-}3)$$

Note the sign convention for the force and displacement, as shown in Fig. 2-4, and the fact that the work done on the system is positive.

FIG. 2-4 *Work to displace weight in gravitational field.*

2-5 ELECTRICAL AND MAGNETIC WORK

Consider a simple electric system as shown in Fig. 2-5. We take the battery as a thermodynamic system and wish to compute the work done as it discharges through the resistance R. The physical meaning of work in this case relates to the fact that an electric force is necessary to drive the charges through the resistor. We may recall that the electric potential V is the potential for doing work per unit charge. The electric current i is the time rate of change of charge or

$$q = \int i \, d\tau = i \, \Delta\tau$$

where $\Delta\tau$ is some specified time increment. Thus the electrical work expended during a time increment $\Delta\tau$ is

$$W = Vi \, \Delta\tau \qquad (2\text{-}4)$$

In Eq. (2-4) the work W is the work that the thermodynamic system (the battery) does on the resistor since the calculation is performed by analyzing the electric forces acting on the charges external to the battery. The sign convention adopted previously would require that this work carry a negative sign since it is done *by* the system (the battery). If we were considering the resistor as a thermodynamic system, the work would be positive. In simple terms, the battery does work on the resistor through the action of electric forces.

Finally, we consider a very simple system where magnetic work is involved. We wish to calculate the work done by moving a conductor through a magnetic field $\mathbf{B}$ with a velocity $\mathbf{v}$, as shown in Fig. 2-6. The force exerted on a moving charge placed in a magnetic field is

$$\mathbf{F} = q\mathbf{v} \times \mathbf{B} \qquad (2\text{-}5)$$

Of course, this force tends to displace the charges in the conductor and create a current flow in the direction shown. The movement of the charges

FIG. 2-5 *Electrical work.*

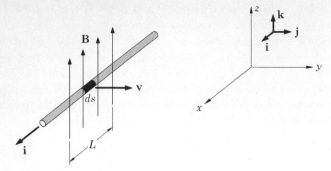

FIG. 2-6 *Work performed by conductor moving in magnetic field.*

in the direction of the current then produces the restraining force exerted on the conductor. For an element of the conductor of length ds we have

$$q\mathbf{v} = \int i \, d\tau \frac{ds}{d\tau} = \int i \, ds \tag{2-6}$$

Inserting Eq. (2-6) into Eq. (2-5) gives

$$\mathbf{F} = \int_0^L i \, ds \times \mathbf{B} = -iBL\mathbf{j} \tag{2-7}$$

The rate of doing work is

$$W = \mathbf{F} \cdot \mathbf{v} = \left(\int_0^L i \, ds \times \mathbf{B} \right) \cdot \mathbf{v}$$

For the coordinate orientation shown in Fig. 2-6 this becomes

$$W = -(iBL\mathbf{j}) \cdot (v\mathbf{j}) = -iLBv \tag{2-8}$$

In this example we are considering the conductor (set of charges) as the thermodynamic system and computing the work done on the conductor by the magnetic field. The force shown in Eq. (2-7) is the restraining force exerted on the conductor by the field. To move the conductor in the field, we would have to apply an external force equal in magnitude and opposite in direction. The work done on the conductor by this external force would then be

$$W = +iLBv \tag{2-8a}$$

This example is analogous to the problem of calculating the work necessary to raise a weight in a gravitational field. In this example the work of the external force is stored in the magnetic field. In the gravitational problem the work of the external force is stored in the gravitational field.

Another system where magnetic work is involved is shown in Fig. 2-7. The cross-sectional area of the toroidal ring is A and its circumferential length is L. The magnetic flux density is denoted by **B**, the number of turns is N, and the current is i. We wish to calculate the work done as the current is changed in the windings. From Faraday's law of electromagnetic induction the induced electromotive force (emf) E is given by

$$E = N \frac{d(BA)}{d\tau} \tag{2-9}$$

and the work done in the time $d\tau$ is

$$dW = Ei \, d\tau = NAi \, dB$$

The magnetic intensity H is calculated from

$$H = \frac{Ni}{L}$$

so that

$$dW = ALH \, dB \tag{2-10}$$

We note that AL is the total volume of the coil so that the work could be written as

$$dW = VH \, dB \tag{2-11}$$

In this case we observe that work producing force is that required to overcome the back emf of the coil. The calculation has been performed for a

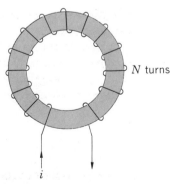

N turns

i

FIG. 2-7 *Magnetic work on toroidal ring.*

toroidal coil, but it would also be valid for long, slender coils of uniform cross section.

EXAMPLE 2-1 Expansion of gas in cylinder

A gas expands in a cylinder according to the relation $pV^{1.3} = \text{const}$. The initial volume of the gas is 1 ft³ and the initial pressure is 200 psia. The final pressure is 15 psia. Calculate the work done on the face of the piston by the pressure forces of the gas.

SOLUTION Equation (2-2) applies to this problem so that

$$W = \int_{V_1}^{V_2} p \, dV$$

From the given process relation, $pV^{1.3} = C$ or $p = CV^{-1.3}$. Thus

$$W = \int_{V_1}^{V_2} CV^{-1.3} \, dV = \frac{-C}{0.3}[V^{-0.3}]_{V_1}^{V_2} = -\frac{C}{0.3}\left(V_2^{-0.3} - V_1^{-0.3}\right) \quad (a)$$

But $C = p_1 V_1^{1.3} = p_2 V_2^{1.3}$, so that

$$W = \frac{p_2 V_2 - p_1 V_1}{-0.3} \quad\quad (b)$$

The final volume is

$$V_2 = V_1 \left(\frac{p_1}{p_2}\right)^{1/1.3} = 1 \left(\frac{200}{15}\right)^{1/1.3} = 7.32 \text{ ft}^3 \quad (0.2073 \text{ m}^3)$$

Inserting this value in Eq. (b) gives

$$W = -\frac{(144)[(15)(7.32) - (200)(1)]}{0.3}$$

$$= 4.33 \times 10^4 \text{ ft·lbf} \quad (5.87 \times 10^4 \text{ J})$$

EXAMPLE 2-2 Magnetic work

A 1-ft conductor carrying a current of 10 A is moved through a magnetic field having a magnetic flux density of 1 Wb/m². The field and conductor are orthogonal as shown in Fig. 2-6. Calculate the force necessary to move the conductor and the rate of doing work when the velocity is 15 ft/s.

SOLUTION The first step in the solution of this problem is to get all the physical quantities into a consistent set of units. We choose the metric system since it is simplest to use.

$L = 1 \text{ ft} = 0.305 \text{ m}$
$v = 15 \text{ ft/s} = 4.58 \text{ m/s}$

From Eq. (2-7) the force is

$\mathbf{F} = iBL\mathbf{j}$
$= (10)(1)(0.305) = 3.05 \text{ N}$

The work per unit time is then calculated from Eq. (2-8a) as

$W = iBLv = (10)(1)(0.305)(4.58)$
$= 13.98 \text{ N·m/s} = 13.98 \text{ J/s}$
$= 13.98 \text{ W}$

2-6 INTERNAL ENERGY

In the foregoing we have shown how the work done on a system may be evaluated by calculating the energy expended by a force acting through a distance. Work, then, is one form of energy which may cross the boundary of a system. It may be remarked again that there must be a force interaction between the system and its surroundings in order for work to be present. We thus sense once again that work is dependent on the process because it is only observed as the system changes from one state to another.

The concept of potential energy is a familiar one which arises in mechanics, electromagnetics, and other subjects. A review of this concept is appropriate because it bears on a more general type of energy which is treated in thermodynamics. It will be recalled that gravitational potential energy is defined as the work necessary to raise a weight to a height above a particular reference level. Kinetic energy is evaluated by determining the work required to accelerate a body from rest to a certain velocity. Thus,

$$F = \frac{1}{g_c} ma = \frac{1}{g_c} m \frac{dv}{d\tau} = \frac{m}{g_c} \frac{dv}{ds} \frac{ds}{d\tau} = \frac{m}{g_c} v \frac{dv}{ds}$$

and

$$W = \int \mathbf{F} \cdot d\mathbf{s} = \int \frac{m}{g_c} v \, dv = \frac{1}{2g_c} mv^2 = \text{K.E.} \qquad (2\text{-}12)$$

When the system changes from some initial velocity v_1 to a final velocity v_2 the corresponding *change* in kinetic energy is

$$\text{K.E.}_2 - \text{K.E.}_1 = \int_{v_1}^{v_2} \frac{m}{g_c} v \, dv = \frac{1}{2g_c} m(v_2^2 - v_1^2)$$

Thus, we imagine the acceleration work as being stored as kinetic energy. If the body were decelerated through the action of some restraining force this energy would be recovered as a work delivered to the restraining mechanism (assuming no losses due to friction).

In a similar fashion electric potential is defined as the work to displace a charge in an electric field. In all of these cases the concept of potential energy is identified with work. In all of the cases mentioned so far the potential energy is a function of the *state* of the system and does not depend on the particular process under consideration. Thus the kinetic energy of a body does not depend on the manner in which it is brought to rest or even on the direction of the velocity vector; the gravitational potential energy does not depend on the manner in which a weight is raised to a given elevation, but only on the value of the elevation, and so on. Note very carefully once again that the familiar concept of potential energy is defined in terms of work. Observe, furthermore, that the work done by the forces acting on the system may be expressed as

$$\text{P.E.}_{\text{final}} - \text{P.E.}_{\text{initial}} = \Delta \text{P.E.} = W \tag{2-13}$$

for the special class of systems discussed so far in which the only interaction of the system with its surroundings is in the form of work. We may remark that if the only types of work terms were those in which a potential-energy function could be defined, energy analysis of systems would be a simple matter indeed because the work would always be independent of the process. We note also that the technique of defining potential-energy terms automatically forces us to *leave out* of the work terms in Eq. (2-13) those values which may be expressed in terms of a potential function. It will be recalled from mechanics that if the only work done by a system is that which is independent of path, then the system is called *conservative*. The forces acting on such a system are called *conservative forces*. We may thus interpret Eq. (2-13) as:

Work done by conservative forces
$$= \text{work done by nonconservative forces}$$

Equation (2-13) is restrictive in that the only kinds of potential energy which it admits are those expressed through the conservative-force concept. The first law of thermodynamics permits a generalization of the concept of potential energy. To clarify this principle, imagine a system which changes from some state I to some state II, as shown in Fig. 2-8. Let us assume that the system is perfectly insulated so that no heat may be transferred to or from the system. We specify that the only interaction with the surroundings

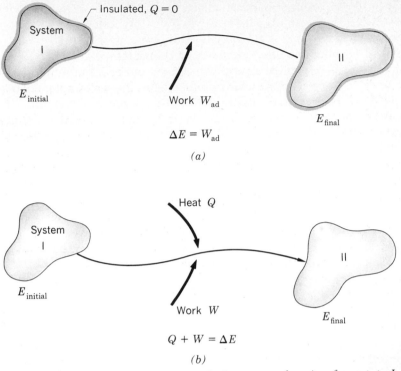

FIG. 2-8 *Conservation of energy principle for system changing from state I to II in (a) adiabatic process and (b) general process involving heat transfer.*

is in the form of work, and we exclude from this work all those forms that can be expressed as changes in potential functions. In this way we speak of the work that is, in fact, *delivered* to the system by the surroundings or some adjoining mechanism. As a result of this process, the principle of conservation of energy permits the definition of a *general internal energy function E,* such that

$$E_{final} - E_{initial} = \Delta E = W_{ad} \tag{2-14}$$

where W_{ad} is used to designate the work. The subscript ad indicates that the work is done in an *adiabatic process*, i.e., where no heat exchange occurs between the system and the surroundings.

Heat cannot be expressed as a force acting through a distance, and thus we are compelled to omit it from consideration for now. A verbal statement of Eq. (2-14) is:

Gain in internal energy = adiabatic work delivered to the system by
all forces acting on the system

It is found *by experiment* that the adiabatic work between two end states is always the same, regardless of the process. We thus conclude that the quantity E is a property, and is a function of only the state of the system. The quantity E is called the *internal energy* of the system.

In Chap. 1 we defined a property of a system as a function of state, a point function, or a quantity whose change between two end states does not depend on the process path used to effect the change. The quantity E fits this definition, so we call it a property of the system. Notice that the definition of E in terms of adiabatic work does not serve to establish an absolute scale for internal energy or a reference level for tabulation of properties of substances. Reference levels may be assigned arbitrarily, as in the case of reference points for the international temperature scale, but absolute values of the internal-energy function may only be obtained from detailed microscopic considerations. The important point is that a property E exists and that its existence is a matter of experimental evidence.

We shall have more to say about the constitution of the internal-energy function shortly, but now let us consider the effect of heat transfer in a process which the system undergoes. Suppose we let the system undergo a process between the same end states as before, but now the system is not insulated and it may exchange heat with the surroundings. The change in internal energy is the same as before since the end states are the same, but the work is altered according to the conservation of energy by the amount of heat energy added, Q. Thus,

$$W + Q = W_{ad} \tag{2-5}$$

where we define the heat energy as positive when it is added to the system and negative when it is given up by the system. Rewriting Eq. (2-15) according to Eq. (2-14),

$$Q + W = \Delta E \tag{2-16}$$

Equation (2-16) has a very simple physical interpretation. It states that the energy added at the boundary of the system as a result of forces acting on the boundary (work) plus all other energies added at the boundary (heat) must be experienced as an increase in the internal energy of the system. Or,

Energy added to system = accumulation of energy in the system

At this point we remind the reader that our discussion has focused on systems where mass does not cross the boundary, or *closed systems*. When mass is allowed to cross the boundary there can be a corresponding transport of energy carried by the mass. In Chap. 4 we shall see how such energy transport is analyzed.

2-7 THE FIRST LAW OF THERMODYNAMICS

Equation (2-16) is frequently given as an analytical statement of the first law of thermodynamics, but it must be remembered that it is simply a statement of the conservation of energy principle where we recognize heat, work, and internal energy as different forms of energy. We might think of Eq. (2-16) as a definition of heat in terms of work and internal energy. In a physical sense heat is thought of as an energy exchange with the surroundings which does not take the form of a force acting through a distance. Notice that the conservation idea is abundantly clear when we consider an *isolated* system, i.e., one where no energy crosses the boundary of the system. Thus,

$$\Delta E_{\text{isolated}} = 0 \qquad\qquad (2\text{-}17)$$

or the total internal energy of an isolated system does not change. The preceding reasoning is illustrated in Fig. 2-8.

Let us now consider the nature of the internal-energy function. Clearly it incorporates the potential-energy functions such as gravitational and kinetic energy. But it also includes other types of internal energy. To illustrate this, consider the process indicated in Fig. 2-9. We do work by stirring the water in a container which is insulated with respect to the surroundings. After the stirring, the water is allowed to come to rest again. Obviously the gravitational and kinetic energies do not change in this process, but the internal energy must increase as a consequence of the stirring action because we have added energy to the water. A similar example is the compression of a gas in a cylinder. Once again, if the cylinder is insulated, the internal energy of the gas is increased. In both of the aforementioned systems the increase in total internal energy of the system is the result of a change in the *internal thermal energy* of the system. In a physical sense, this internal thermal energy results from the kinetic energy of the molecules of the substance which compose the system and is usually evidenced by a lower or higher temperature of the system. The internal thermal energy is denoted

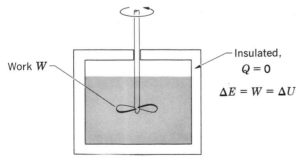

FIG. 2-9 *Process in which work goes to raise the internal thermal energy of the system.*

by the symbol U. We might also consider the inherent chemical energy of the substance which could produce external work, as in the case of a storage battery delivering electrical work to some external circuit. We could carry the consideration of total internal energy further by noting that the relativistic mass-energy relationship

$$E = mc^2$$

must be employed whenever nuclear reactions are involved.

Our discussion of the internal-energy function may now be summarized with the following expression:

$$E = U + \text{K.E.} + \text{P.E.} + \text{Ch.E.} + \cdots \tag{2-18}$$

where it is understood that electric, magnetic, and nuclear energies may also be included in this expression when appropriate. Capital letters (E and U) are employed to designate the *total* properties for an entire system. Lowercase characters (e and u) will be used to indicate the properties per unit mass of the system. In some circumstances we shall use $\bar{e}$ and $\bar{u}$ to designate properties per mole of substance. Typical units for e and u are Btu/lbm or kJ/kg.

We remark again that Eq. (2-16) represents the energy-transfer mechanisms in a process wherein the system changes from one state to another. Now suppose we allow the system to undergo a *cyclic* change, i.e., a continuous series of processes such that the system periodically returns to its initial state. If we consider a small change in the system then the first law may be written in differential form as

$$dQ + dW = dE$$

Clearly, since the internal energy E is a function only of state, its change must be zero over any complete cycle, and we have

$$\oint dQ + \oint dW = 0 \tag{2-19}$$

where the integral sign denotes a cyclic process.

It must be emphasized, in general, that both heat and work are path functions and, thus, to evaluate their magnitude we must take into consideration the process involved. On the other hand, internal energy is a state function and is characterized mathematically by

$$\oint dE = 0$$

Thus, in mathematical language, dE may be described as an exact differential whereas dQ and dW would be called *inexact differentials*. Hereafter we shall use the modified symbols $d'Q$ and $d'W$ to designate the fact that the work and heat differentials are, in general, functions of path.

2-8 HEAT AND SPECIFIC HEAT

We have seen that heat is an energy interaction between the thermodynamic system and its surroundings; however, this energy interaction may not be calculated by the application of a force through a distance as in the case of work. Heat and work, therefore, are fundamentally different types of energy. Intuitively, we associate heat with the temperature of a system because the temperature of the system usually rises when heat is added to it. A study of a historical development of the subject of thermodynamics will show that at one time scientists thought of heat as a substance called *caloric* which could be poured from one body into another, in the same way that hot water may be poured from one container into another. They therefore reasoned that heat was contained in a body. We now know that this idea is false and that *heat is not contained in a system* but is manifest only as an interaction of the system with its surroundings as the system changes from one state to another.

A holdover concept from the old caloric theory is that of heat capacity or, as it is now called, *specific heat*. If a quantity of heat $d'Q$ is added to a system resulting in a change in temperature dT, then the specific heat C is defined by

$$d'Q = C\ dT \tag{2-20}$$

Suppose we consider a thermodynamic system where the only changes in internal energy are present in the thermal energy function U. Furthermore, let us assume that the only work interactions with the surroundings are those which result from a change in volume of the system, as shown in Fig. 2-2. In this process we are assuming that other forms of work are not present and that the system volume increases. For this type of system we may write Eq. (2-16) as

$$d'Q + d'W = dU \tag{2-21}$$
$$d'Q - p\ dV = dU$$

where p is the system pressure. Recall the sign convention for work when developing Eq. (2-2a). We may then write for the heat added

$$d'Q = dU + p\ dV \tag{2-21a}$$

Now let this system undergo a process at constant volume so that $dV = 0$.

Thus we could write for the specific heat at constant volume, and per unit mass, recalling that uppercase letters denote total properties, while lowercase letters denote values per unit mass,

$$dQ_v = c_v \, dT_v$$

or

$$du_v = c_v \, dT_v$$

where the subscripts indicate a change at constant system volume. The specific heat per unit mass is thus written as

$$c_v = \left(\frac{\partial u}{\partial T}\right)_v \qquad (2\text{-}22)$$

Although the foregoing relation has been developed for a special system, we shall take Eq. (2-22) as the *defining* relation for specific heat at constant volume. At this point we note a departure from the old caloric theory. The modern concept is to treat specific heat as a *property* of the system which may be expressed in terms of other properties of the system as in Eq. (2-22) for the specific heat at constant volume. Although we now recognize the terms "specific heat" and "heat capacity" as misnomers, we cannot divorce these concepts from the historical treatment of thermodynamics where they are related to heat transfer in certain specific processes.

Now suppose we allow the simple system discussed previously to undergo a constant-pressure process. The heat transfer per unit mass is expressed as

$$d'Q_p = du_p + p \, dv_p \qquad (2\text{-}23)$$

Let us define a new property called the *enthalpy* by

$$h = u + pv \qquad (2\text{-}24)$$

The differential of the enthalpy is

$$dh = du + p \, dv + v \, dp$$

and, at constant pressure,

$$dh_p = du_p + p \, dv_p \qquad (2\text{-}25)$$

Thus

$$dQ_p = dh_p = c_p \, dT_p$$

and

$$c_p = \left(\frac{\partial h}{\partial T}\right)_p \qquad (2\text{-}26)$$

The specific heat at constant pressure is thus a property of the system and may be defined in terms of other properties as shown in Eq. (2-26).

Typical units for specific heat are Btu per pound mass-degree Fahrenheit, kilocalories per kilogram-degree Celsius, and joules per kilogram-degree Celsius. We shall use the symbols $\bar{c}_v$ and $\bar{c}_p$ to designate specific heats expressed on a molal basis.

Specific heats, like other thermodynamic properties, can vary strongly with temperature and pressure of the substance, and experimental data must be employed to obtain reliable values. For modest temperature ranges the values may frequently be assumed constant for calculation purposes. For liquids and solids c_p and c_v are almost equal while there is an appreciable difference for gases. Table 2-1 gives some typical values to illustrate the range which may be experienced.

TABLE 2-1
Typical values of specific heats

Substance	c_p		c_v	
	kJ/kg·°C	Btu/lbm·°F	kJ/kg·°C	Btu/lbm·°F
Solids (20°C)				
Aluminum	0.896	0.214		
Copper	0.383	0.091		
Iron	0.452	0.108		
Silver	0.234	0.056		
Tungsten	0.134	0.032		
Brick	0.84	0.201		
Glass wool	0.7	0.167		
Wood (pine)	2.8	0.669		
Liquids (20°C)				
Water	4.18	1.00		
Freon 12	0.966	0.231		
Ammonia	4.8	1.15		
Engine oil	1.9	0.45		
Mercury	0.14	0.033		
Ethylene glycol	2.38	0.57		
Gases (1 atm, 20°C)				
Air	1.005	0.24	0.718	0.1715
Hydrogen	14.32	3.42	10.17	2.43
Carbon dioxide	0.846	0.202	0.653	0.156

EXAMPLE 2-3   "Paddle-wheel" work

A 1-hp stirring motor is applied to a tank of water, as shown in Fig. 2-9. The tank contains 50 lbm of water, and the stirring action is applied for 1 hr. Assuming that the tank is perfectly insulated, calculate the change in internal energy of the water. Also calculate the rise in temperature of the water, assuming that the process occurs at constant volume and that c_v for water may be taken as 1.0 Btu/lbm·°F.

SOLUTION   Some appropriate conversion factors for use in this problem are:

1 hp = 550 ft·lbf/s = 2545 Btu/h
778.16 ft·lbf = 1 Btu = 1055 J

The increase in the internal energy of the water is clearly the work added during the 1-h time period, since $Q = 0$ for an insulated tank.

$$\Delta U = W = (550)(3600) = 1.98 \times 10^6 \text{ ft·lbf}$$
$$= 2545 \text{ Btu} \qquad (2.685 \times 10^6 \text{ J})$$

For this calculation we assume that the kinetic energies of the water in the tank are negligible. For the low velocities in the stirring process this is a reasonable expectation. From Eq. (2-22) we have

$$c_v = \left(\frac{\partial u}{\partial T}\right)_v$$

For a constant specific heat we may approximate the derivative by

$$c_v \approx \left(\frac{\Delta u}{\Delta T}\right)_v$$

or

$$\Delta u = c_v \, \Delta T$$

Multiplying by the mass, we obtain

$$m \, \Delta u = \Delta U = m c_v \, \Delta T$$

The increase in temperature is now calculated as

$$\Delta T = \frac{\Delta U}{m c_v} = \frac{2545}{(50)(1.0)} = 50.9°F \qquad (28.28°C)$$

2-9 SPECIFIC HEATS FOR IDEAL GASES

It is worthwhile to examine the specific heat behavior of gases at this time to provide a basis for calculations in the following chapters.

In Chap. 7 it will be shown that, if a gas obeys the equation of state given in Eq. (1-31), then the internal energy and enthalpy are functions of temperature alone, and we may integrate Eqs. (2-22) and (2-26) to give

$$\Delta u = u_2 - u_1 = \int_{T_1}^{T_2} c_v \, dT \qquad\qquad (2\text{-}27)$$

$$\Delta h = h_2 - h_1 = \int_{T_1}^{T_2} c_p \, dT \qquad\qquad (2\text{-}28)$$

Furthermore, if the specific heats are constant, we have the simple result

$$u_2 - u_1 = c_v(T_2 - T_1) \qquad\qquad (2\text{-}29)$$

$$h_2 - h_1 = c_p(T_2 - T_1) \qquad\qquad (2\text{-}30)$$

In a microscopic sense an ideal gas is one in which the molecules are sufficiently far apart that the internal energy of the gas may be calculated on the basis of the average kinetic energy of the molecules. It turns out, as illustrated in the simple analysis of Sec. 1-16, that this energy is a function of the macroscopic temperature of the gas. If a gas is compressed to a high pressure so that the molecules are forced into close proximity of one another and molecular-force field interactions become important, then the internal energy becomes a function of both pressure and temperature.

A casual reader may miss the important point regarding internal energy of an ideal gas, so we reiterate it here in boldface type to provide it with an increased emphasis:

Internal energy and enthalpy for an ideal gas are functions of temperature alone.

This important fact will be used over and over again in energy balance calculations.

Some differences in terminology for gases exist in the literature. A gas which obeys Eq. (1-31) is sometimes said to be *thermally perfect*. If it has constant specific heats, it is said to be *calorically perfect*. The terms *ideal gas* and *perfect gas* are used by various authors to indicate that the gas obeys the $pv = RT$ equation of state, and we shall employ this terminology in subsequent sections. For developments in this book the reader may consider the terms *ideal gas* and *perfect gas* as synonyms. The matter of specific heats will be taken into account with an appropriate statement such as "ideal gas with constant specific heats."

For gases at low pressures the specific heats are fairly constant and do not vary over modest temperature ranges. If calculations are to be performed on the basis of a constant specific heat assumption, then the appropriate values to be used are as given in Table 2-2.

We shall outline the microscopic analysis techniques for calculating specific heats of gases in Chap. 8.

One useful relation between c_v and c_p for an ideal gas may be derived as follows. We have

$$dh = c_p \, dT$$
$$du = c_v \, dT$$

Subtracting these expressions,

$$dh - du = (c_p - c_v) \, dT$$

But

$$dh = du + d(pv) = du + R \, dT$$

so that

$$R \, dT = (c_p - c_v) \, dT$$

and

$$R = c_p - c_v \qquad (2\text{-}31)$$

We thus have the result that the gas constant may be calculated as the difference between the constant-pressure and constant-volume specific heats. Reciprocally, if one of the specific heats is known along with the gas constant, then the other specific heat is easily calculated. Finally, if specific heats are expressed on a molal basis, we have

$$\mathscr{R} = \bar{c}_p - \bar{c}_v \qquad (2\text{-}32)$$

or the difference in molal specific heats for *all* ideal gases is a constant.

The reader may logically ask: When is a gas ideal? In a microscopic sense it is ideal when the molecules are so widely spaced that molecular force fields do not play a significant role in the pressure exerted by the gas on a containing vessel. This means that ideal-gas behavior is usually observed when the gas density is very low (a relatively small number of molecules per unit volume), but we have not yet developed a suitable means at our disposal to predict the pressure and temperature ranges for which the ideal-gas law is applicable. Section 7-13 will furnish us with this information. For now, the

TABLE 2-2
Properties of ideal gases at low pressures and normal room temperature (20°C)

Gas	Molecular weight	c_p		c_v		R		γ
		Btu/lbm·°F	kJ/kg·°C	Btu/lbm·°F	kJ/kg·°C	ft·lbf/lbm·°R	J/kg·K	
Air	28.97	0.240	1.005	0.1715	0.718	53.35	287.1	1.40
Hydrogen, H_2	2.016	3.42	14.32	2.43	10.17	767.0	4127	1.41
Helium, He	4.003	1.25	5.234	0.75	3.14	386.3	2078	1.66
Methane, CH_4	16.04	0.532	2.227	0.403	1.687	96.4	518.7	1.32
Water vapor, H_2O	18.02	0.446	1.867	0.336	1.407	85.6	460.6	1.33
Acetylene, C_2H_2	26.04	0.409	1.712	0.333	1.394	59.4	319.6	1.23
Carbon monoxide, CO	28.01	0.249	1.043	0.178	0.745	55.13	296.6	1.40
Nitrogen, N_2	28.02	0.248	1.038	0.177	0.741	55.12	296.6	1.40
Ethane, C_2H_6	30.07	0.422	1.767	0.357	1.495	51.3	276	1.18
Oxygen, O_2	32.00	0.219	0.917	0.156	0.653	48.24	259.6	1.40
Argon, A	39.94	0.123	0.515	0.074	0.310	38.65	208	1.67
Carbon dioxide, CO_2	44.01	0.202	0.846	0.156	0.653	35.1	188.9	1.30
Propane, C_3H_8	44.09	0.404	1.692	0.360	1.507	35.0	188.3	1.12
Isobutane, C_4H_{10}	58.12	0.420	1.758	0.387	1.62	26.6	143.1	1.09

63

TABLE 2-3
Molal constant-pressure specific heats for gases at very low pressures according to Ref. [1], English units

Gas or vapor	$\bar{c}_{po}$, Btu/lb·mol·°R T, °R	Range, °R	Max. error, %
O_2	$\bar{c}_{po} = 11.515 - \dfrac{172}{\sqrt{T}} + \dfrac{1530}{T}$	540–5000	1.1
	$= 11.515 - \dfrac{172}{\sqrt{T}} + \dfrac{1530}{T}$ $+ \dfrac{0.05}{1000}(T - 4000)$	5000–9000	0.3
N_2	$\bar{c}_{po} = 9.47 - \dfrac{3.47 \times 10^3}{T} + \dfrac{1.16 \times 10^6}{T^2}$	540–9000	1.7
CO	$\bar{c}_{po} = 9.46 - \dfrac{3.29 \times 10^3}{T} + \dfrac{1.07 \times 10^6}{T^2}$	540–9000	1.1
H_2	$\bar{c}_{po} = 5.76 + \dfrac{0.578}{1000}T + \dfrac{20}{\sqrt{T}}$	540–4000	0.8
	$= 5.76 + \dfrac{0.578}{1000}T + \dfrac{20}{\sqrt{T}}$ $- \dfrac{0.33}{1000}(T - 4000)$	4000–9000	1.4
H_2O	$\bar{c}_{po} = 19.86 - \dfrac{597}{\sqrt{T}} + \dfrac{7500}{T}$	540–5400	1.8
CO_2	$\bar{c}_{po} = 16.2 - \dfrac{6.53 \times 10^3}{T} + \dfrac{1.41 \times 10^6}{T^2}$	540–6300	0.8
CH_4	$\bar{c}_{po} = 4.52 + 0.007\,37T$	540–1500	1.2
C_2H_4	$\bar{c}_{po} = 4.23 + 0.011\,77T$	350–1100	1.5
C_2H_6	$\bar{c}_{po} = 4.01 + 0.016\,36T$	400–1100	1.5
C_8H_{18}	$\bar{c}_{po} = 7.92 + 0.0601T$	400–1100	est. 4
$C_{12}H_{26}$	$\bar{c}_{po} = 8.68 + 0.0889T$	400–1100	est. 4

reader can be assured that such gases as air, nitrogen, oxygen, helium, etc., do obey the ideal-gas equation of state at normal atmospheric temperature and pressure (20°C and 1 atm).

When the pressure of a gas is sufficiently low that the ideal-gas equation of state applies, but one wishes to take account of specific heat variation with temperature, the empirical relations of Table 2-3 are applicable. Due caution should be exercised to see that these expressions are used only when the gas is at low pressures. Water vapor, for example, is *not* an ideal gas at normal atmospheric pressure, and the relations of Tables 2-3 and 2-3M could not be used to calculate the specific heat at this pressure.

EXAMPLE 2-4 Constant volume heat addition

A 1-ft³ container is filled with air at 20 psia and 100°F. Calculate the final pressure in the container if 10 Btu of heat are added. Assume ideal-gas behavior, with constant specific heats.

TABLE 2-3M
Molal constant-pressure specific heats for gases at low pressures, adapted to SI units

Gas or vapor	$\bar{c}_{po}$, kJ/kg·mol·K T, K	Range, K
O_2	$\bar{c}_{po} = 48.212 - \dfrac{536.8}{\sqrt{T}} + \dfrac{3559}{T}$	300–2800
	$= 48.212 - \dfrac{536.8}{\sqrt{T}} + \dfrac{3559}{T}$	
	$+ 3.768 \times 10^{-4}(T - 2222)$	2800–5000
N_2	$\bar{c}_{po} = 39.65 - \dfrac{8071}{T} + \dfrac{1.5 \times 10^6}{T^2}$	300–5000
CO	$\bar{c}_{po} = 39.61 - \dfrac{7652}{T} + \dfrac{1.38 \times 10^6}{T^2}$	300–5000
H_2	$\bar{c}_{po} = 24.12 + 4.356 \times 10^{-3}T + \dfrac{62.41}{\sqrt{T}}$	300–2200
	$= 24.12 + 4.356 \times 10^{-3}T + \dfrac{62.41}{\sqrt{T}}$	
	$- 5.94 \times 10^{-4}(T - 2222)$	2200–5000
H_2O	$\bar{c}_{po} = 83.15 - \dfrac{1863}{\sqrt{T}} + \dfrac{17445}{T}$	300–3000
CO_2	$\bar{c}_{po} = 67.83 - \dfrac{15\,189}{T} + \dfrac{1.82 \times 10^6}{T^2}$	300–3500
CH_4	$\bar{c}_{po} = 18.92 + 0.055T$	300–830
C_2H_4	$\bar{c}_{po} = 17.71 + 0.0887T$	200–650
C_2H_6	$\bar{c}_{po} = 16.79 + 0.123T$	220–600
C_8H_{18}	$\bar{c}_{po} = 33.16 + 0.453T$	220–600
$C_{12}H_{26}$	$\bar{c}_{po} = 36.34 + 0.670T$	220–600

SOLUTION Since the container is rigid, there is no work because the pressure force cannot act through a displacement, and the first-law statement becomes

$$Q = \Delta U = m(u_2 - u_1)$$

For an ideal gas this becomes

$$Q = mc_v(T_2 - T_1) \qquad (a)$$

The mass of air is calculated with

$$m = \frac{pV}{RT} = \frac{(20)(144)(1)}{(53.35)(560)} = 0.0963 \text{ lbm} \qquad (0.043\,68 \text{ kg})$$

For air, $c_v = 0.1715$ Btu/lbm·°F, so that Eq. (a) becomes

$$10 = (0.0963)(0.1715)(T_2 - 100)$$

and

$$T_2 = 706°F = 1166°R \quad (374°C, 647 \text{ K})$$

The final pressure may now be calculated with

$$p = \frac{mRT}{V} \tag{b}$$

$$= \frac{(0.0963)(53.35)(1166)}{1}$$

$$= 5980 \text{ lbf/ft}^2 = 41 \text{ psia} \quad (2.827 \times 10^5 \text{ N/m}^3)$$

EXAMPLE 2-5

Nitrogen is to be heated at constant pressure from 100 to 3000°F. Calculate the percent error which would result from assuming constant specific heat behavior in a computation of the heat transfer.

SOLUTION   The "true" heat transfer per mole is given by

$$Q_p = \int_{T_1}^{T_2} \bar{c}_p \, dT \tag{a}$$

where the relation for $\bar{c}_p$ is taken from Table 2-3. This result is then to be compared with

$$Q_p = \bar{c}_p(T_2 - T_1) \tag{b}$$

where now $\bar{c}_p$ is a value calculated from the information in Table 2-2. From Table 2-3 we have, for nitrogen,

$$\bar{c}_p = 9.47 - \frac{3.47 \times 10^3}{T} + \frac{1.16 \times 10^6}{T^2} \text{ Btu/mol·°R}$$

where T is in degrees Rankine. Thus the "true" heat transfer is

$$Q_p = \int_{T_1}^{T_2} \left(9.47 - \frac{3.47 \times 10^3}{T} + \frac{1.16 \times 10^6}{T^2}\right) dT$$

$$= \left(9.47T - 3.47 \times 10^3 \ln T - \frac{1.16 \times 10^6}{T}\right]_{T_1}^{T_2} \tag{c}$$

where $T_1 = 560°R$ and $T_2 = 3460°R$. Evaluating Eq. (c) gives

$$Q_p = 22\ 920\ \text{Btu/mol}$$

Using the data of Table 2-2, we obtain

$$Q_p = (28.02)(0.248)(3460 - 560)$$
$$= 20\ 100\ \text{Btu/mol}$$

The error in assuming a constant specific heat is, therefore,

$$\text{Error} = \frac{22\ 920 - 20\ 100}{22\ 920} = 0.123 = 12.3\ \text{percent}$$

This example illustrates the necessity of taking specific heat variations into account when wide temperature ranges are encountered.

EXAMPLE 2-6

Air is contained in a piston-cylinder arrangement as shown with a cross-sectional area of 4 cm² and an initial volume of 20 cm³. The air is initially at 1 atm and 20°C. Connected to the piston is a spring having a deformation constant of $k_s = 100\ \text{N/cm}$ and the spring is initially undeformed. How much heat must be added to the air to increase the pressure to 3 atm? The apparatus is surrounded by atmospheric air.

SOLUTION The system is shown in the accompanying sketch. The force required to compress the spring is

$$F_s = k_s x \qquad\qquad (a)$$

where x is the displacement from the equilibrium position. The *work* to compress the spring is therefore

$$W_s = \int F_s\ dx = \int_0^x k_s x\ dx = \tfrac{1}{2} k_s x^2 = \Delta\text{P.E.}_s \qquad\qquad (b)$$

System

$$F_s = k_s\,x$$

Air

Q

FIG. EXAMPLE 2-6

Now consider the air as a thermodynamic system. The work consists of two parts: (1) that to compress the spring W_s and (2) that to displace the surrounding air at 1 atm; or $W_{surr} = -p \, \Delta V$ for constant pressure surroundings.

$$Q + W_s + W_{surr} = \Delta U_{air} \qquad (c)$$

For air $c_v \approx$ const. and

$$\Delta U_{air} = m_a c_v \, \Delta T_a \qquad (d)$$

We have $T_1 = 20°C = 293$ K

$$c_v = 0.1715 \text{ Btu/lbm·°F} = 718 \text{ J/kg·°C}$$
$$m_a = \frac{p_1 V_1 M_a}{\mathscr{R} T_1} = \frac{(1.0132 \times 10^5)(20 \times 10^{-6})(28.97)}{(8314)(293)}$$
$$= 2.41 \times 10^{-5} \text{ kg}$$

If we consider the piston as a free body we see that the initial pressure of the air in the cylinder is just balanced by the surrounding atmospheric pressure leaving a net force of zero exerted by the spring. At the final condition of 3 atm in the cylinder we still have a balancing pressure of 1 atm from the surroundings leaving the spring to provide 2 atm pressure over the 4 cm² face. Therefore,

$$F_{s2} = (2)(1.0132 \times 10^5 \text{ N/m}^2)(4.0 \times 10^{-4} \text{ m}^2)$$
$$= 81 \text{ N}$$

The final displacement of the spring is obtained from Eq. (a)

$$x = F_s k_s = (81 \text{ N})/100 \text{ N/cm} = 0.81 \text{ cm}$$

The additional volume swept out by the piston is $A \, \Delta x = (4 \text{ cm}^2)(0.81 \text{ cm}) = 3.24 \text{ cm}^3$ so the final volume of the air is

$$V_2 = V_1 + 3.24 = 20 + 3.24 = 23.24 \text{ cm}^3$$

The work to displace the surroundings is thus

$$W_{surr} = -p_{atm} \, \Delta V = -(1.0132 \times 10^5)(23.24 \times 10^{-6}) = -2.3546 \text{ J}$$

The final temperature is obtained from the ideal-gas law expressed as

$$\frac{p_2 V_2}{T_2} = \frac{p_1 V_1}{T_1}$$

so that

$$T_2 = \frac{(293)(3/1)(23.24)}{20}$$
$$= 1021 \text{ K} = 748°\text{C}$$

From Eq. (b)

$$W_s = -\tfrac{1}{2} k_s x^2 = \tfrac{1}{2} (100 \text{ N/cm})(0.81 \text{ cm})^2$$
$$= -32.81 \text{ N·cm} = -0.3281 \text{ J}$$

Also, from Eq. (d)

$$\Delta U_a = (2.41 \times 10^{-5} \text{ kg})(718 \text{ J/kg·°C})(748-20)$$
$$= 12.60 \text{ J}$$

The total heat added is now obtained from the energy relation of Eq. (c):

$$Q = 0.3281 + 2.3546 + 12.60 = 15.28 \text{ J} \qquad (0.0145 \text{ Btu})$$

In this example we see that most of the heat added goes to increase the temperature of the air enough to raise the pressure.

EXAMPLE 2-7   Constant-pressure heat addition
   Air at 1 atm and 20°C occupies an initial volume of 1000 cm³ in a cylinder. The air is confined by a piston which has a constant restraining force so the gas pressure always remains constant. Heat is added to the air until its temperature reaches 260°C. Calculate the heat added, the work the gas does on the piston, and the change in internal energy of the gas.

SOLUTION   The heat transfer is calculated with

$$Q_p = m c_p \, \Delta T_p \qquad\qquad\qquad (a)$$

Recalling that the molecular weight of air is 28.97 the gas constant for air is

$$R = \frac{\mathscr{R}}{M} = \frac{8314 \text{ J/kg·mol·K}}{28.97 \text{ kg/kg mol}} = 287 \text{ J/kg·K}$$

and the mass is calculated as

$$m = \frac{pV}{RT} = \frac{(1.0132 \times 10^5 \text{ N/m}^2)(1000 \times 10^{-6} \text{ m}^3)}{(287 \text{ J/kg·K})(293 \text{ K})} = 1.205 \times 10^{-3} \text{ kg}$$

The heat transfer is then

$$Q_p = (1.205 \times 10^{-3} \text{ kg})(1.005 \times 10^3 \text{ J/kg·°C})(260 - 20)$$
$$= 290.6 \text{ J} \quad (0.275 \text{ Btu})$$

The work done by the piston is given by Eq. (2-2):

$$W = \int p \, dV$$

For a constant-pressure process this becomes

$$W = p(V_2 - V_1) \tag{b}$$

The final volume is calculated from

$$\frac{p_1 V_1}{T_1} = mR = \frac{p_2 V_2}{T_1}$$

or

$$V_2 = V_1 \frac{p_1}{p_2} \frac{T_2}{T_1} = (1000)(1) \left(\frac{533}{293}\right) = 1819 \text{ cm}^3$$

Now, the work is calculated from Eq. (b) as

$$W = (1.0132 \times 10^5 \text{ N/m}^2)(1819 - 1000)(10^{-6})$$
$$= 82.98 \text{ J} \quad (0.079 \text{ Btu})$$

The work done *on the gas* is the negative of this quantity or

$$W \text{ (on gas)} = -82.98 \text{ J}$$

The change in internal energy of the air may now be calculated by applying Eq. (2-16) (neglecting all quantities in the E function except U):

$$Q + W = \Delta U$$
$$290.6 - 82.98 = \Delta U = 207.6 \text{ J} \quad (0.197 \text{ Btu})$$

To summarize, the 290.6 J of heat which is added to the air goes to raise the internal energy by 207.6 J and to deliver 82.98 J of work to the surroundings (the piston).

EXAMPLE 2-8   Constant-temperature compression

An ideal gas which obeys the equation $pV = mRT$ is compressed in a piston-cylinder arrangement such that the temperature remains constant (isothermal process). Derive an expression for the work done on the gas and calculate the quantity of work when 2 kg of helium is compressed from 1 atm, 20°C to 1 MPa, holding the temperature constant.

SOLUTION   We apply Eq. (2-2a), observing that

$$p = \frac{mRT}{V} \qquad\qquad (a)$$

so that

$$W = - \int_{V_1}^{V_2} \frac{mRT}{V} \, dV \qquad\qquad (b)$$

Because T = const., this reduces to

$$W = - mRT \ln \frac{V_2}{V_1} \qquad\qquad (c)$$

Again, because T = const.,

$$\frac{V_2}{V_1} = \frac{p_1}{p_2}$$

and we have the alternative form

$$W = - mRT \ln \frac{p_1}{p_2} \qquad\qquad (d)$$

We can now calculate the work for the helium. The gas constant is

$$R = \frac{\mathscr{R}}{M} = \frac{8314 \text{ J/kg·mol·K}}{4 \text{ kg/kg·mol}} = 2078.5 \, \frac{\text{J}}{\text{kg·K}}$$

and

$$W = -(2 \text{ kg}) \left(2078.5 \, \frac{\text{J}}{\text{kg·K}}\right) (293 \text{ K}) \ln \frac{1.0132 \times 10^5}{1.0 \times 10^6}$$

$$= 2.789 \text{ MJ}$$

EXAMPLE 2-9   Heat addition to room

Twenty people attend a cocktail party in a small room which measures 30 by 25 ft and has an 8-ft ceiling. Each person gives up about 450 Btu of heat per hour. Assuming that the room is completely sealed off and insulated, calculate the air temperature rise occurring within 15 min. Assume that each person occupies a volume of 2.5 ft³.

SOLUTION In this problem we take the air in the room as a thermodynamic system and assume that the people are adding heat to the air in a constant-volume process. The room volume is

$$V_{room} = (30)(25)(8) = 6000 \text{ ft}^3 \qquad (169.9 \text{ m}^3)$$

The volume of air is obtained by subtracting the volume occupied by the people:

$$V_{air} = 6000 - (20)(2.5) = 5950 \text{ ft}^3 \qquad (168.5 \text{ m}^3)$$

Assuming that the constant-volume heat-addition process starts at standard atmospheric conditions of 14.7 psia and 70°F, the mass of air is

$$m = \frac{pV}{RT} = \frac{(14.7)(144)(5950)}{(53.35)(530)} = 445.4 \text{ lbm} \qquad (202 \text{ kg})$$

For the constant-volume process,

$$c_v \approx \left(\frac{\Delta u}{\Delta T}\right)_v$$

and

$$\Delta U_v = m c_v \Delta T_v \qquad\qquad (a)$$

The change in internal energy of the air is equal to the heat added by the people, since

$$Q + W = \Delta U \qquad\qquad (b)$$

and there is presumably no work done on the air because the pressure force at the boundary does not act through a displacement. Thus

$$\Delta U = (20)(450) = 9000 \text{ Btu/h} \qquad (2637 \text{ W})$$

or, for a 15-min period,

$$\Delta U = \tfrac{15}{60}(9000) = 2250 \text{ Btu} \qquad (2133 \text{ kJ})$$

The temperature increase is now calculated from Eq. (*a*) as

$$\Delta T = \frac{\Delta U}{mc_v} = \frac{2250}{(445.4)(0.1715)} = 29.46 \; °F \qquad (16.37°C)$$

This problem illustrates the rather obvious need for adequate ventilation when a large number of people are confined in a small space.

2-10 PROCESSES FOR IDEAL GASES

In the foregoing examples we have already made considerable use of ideal-gas relations for a number of processes. Energy balances were made for the special cases of constant temperature (isothermal), constant volume, etc., to illustrate system behavior. We remind the reader at this point that we have only been examining *closed* systems where mass does not cross the boundary. We now wish to look at ideal-gas processes in a more general way. The analytical relationships which will be developed may be used for analysis of both closed systems, as we have done in this chapter, or open systems to be discussed in Chap. 4.

First let us summarize the terms which apply to special processes:

1 Constant temperature or *isothermal*
2 Constant pressure or *isobaric*
3 Constant volume is *isometric* or *isochoric*
4 Zero heat transfer or *adiabatic*

A *polytropic* process is one which may be represented by the relation

$$pv^n = \text{const.} = C \tag{2-33}$$

or, alternatively, in the logarithmic form,

$$\ln p = -n \ln v + \ln C \tag{2-34}$$

where n is appropriately called the *polytropic exponent*. For an ideal gas

$$pv = RT \tag{2-35}$$

and we may compare Eqs. (2-33) and (2-34) with (2-35) and show that our four special processes mentioned above can be fitted to the polytropic relation if we use the following values of n.

$$T = \text{const.} \qquad n = 1.0$$
$$p = \text{const.} \qquad n = 0$$
$$v = \text{const.} \qquad n = \pm \infty$$
$$Q = 0 \qquad n = \gamma = \frac{c_p}{c_v}$$

FIG. 2-10 *Polytropic process according to $pv^n = const.$*

These processes are plotted on a p-v diagram in Fig. 2-10, and the logarithmic relation of Eq. (2-34) is shown in Fig. 2-11. Note that $-n$ is the slope of the straight lines on this figure. We have not yet examined the adiabatic case to show that $n = \gamma$ but will do so in the following section.

We may write Eq. (2-33) between two end states 1 and 2 as

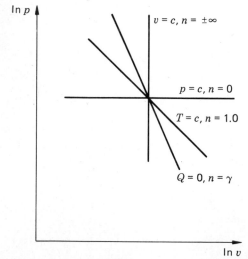

FIG. 2-11 *Logarithmic representation of polytropic process $pv^n = const.$*

$$p_1 v_1^n = p_2 v_2^n = C \qquad (2\text{-}36)$$

By manipulating this relation along with the ideal-gas law [Eq. (2-35)] we can obtain several alternate expressions for the end states in a polytropic process.

$$\frac{v_2}{v_1} = \left(\frac{p_1}{p_2}\right)^{1/n} \qquad (2\text{-}37)$$

$$\frac{T_2}{T_1} = \left(\frac{p_2}{p_1}\right)^{(n-1)/n} \qquad (2\text{-}38)$$

$$\frac{T_2}{T_1} = \left(\frac{v_1}{v_2}\right)^{n-1} \qquad (2\text{-}39)$$

Please note that Eqs. (2-38) and (2-39) apply *only for ideal gases*.

We may now consider the general case of ideal-gas compression work in a piston-cylinder arrangement which is given by

$$W = \int p \, dv \qquad (2\text{-}40)$$

Inserting the polytropic relation for p,

$$W = -\int_{v_1}^{v_2} Cv^{-n} \, dv = C \left[\frac{v^{1-n}}{n-1}\right]_{v_1}^{v_2} \qquad (2\text{-}41)$$

But $C = p_1 v_1^n = p_2 v_2^n$ so that

$$W = \frac{p_2 v_2 - p_1 v_1}{n-1} \qquad (2\text{-}42)$$

With Eq. (2-35) this may be written in an alternative form:

$$W = \frac{R(T_2 - T_1)}{n-1} \qquad (2\text{-}43)$$

For the special case of a constant-temperature process, $n = 1.0$ and Eqs. (2-42) and (2-43) are indeterminate; however, we may return to Eq. (2-41) with $n = 1$ and write

$$W = -\int_{v_1}^{v_2} Cv^{-1} \, dv = C \ln \frac{v_1}{v_2}$$

For $T = $ const.,

$$pv = C = RT$$

so that

$$W = p_1 v_1 \ln \frac{v_1}{v_2} = RT \ln \frac{v_1}{v_2} \qquad (2\text{-}44)$$

Again, because the temperature is constant $v_2/v_1 = p_1/p_2$ so that

$$W = RT \ln \frac{p_2}{p_1} \qquad (2\text{-}45)$$

2-11 QUASISTATIC-ADIABATIC PROCESS FOR IDEAL GAS

As promised in the preceding section we shall now examine the interesting case of a quasistatic-adiabatic process for an ideal gas. Recall that the term *quasistatic* means a series of equilibrium states, and, in a practical sense, generally refers to slow-moving frictionless processes. Later on, we shall also use the term *reversible* to describe such processes. For the adiabatic process we have

$$d'Q = 0 = du + p\, dv \qquad (2\text{-}46)$$

and for the ideal gas

$$du = c_v\, dT$$

so that

$$0 = c_v\, dT + p\, dv$$

and

$$c_v\, dT = -p\, dv \qquad (2\text{-}47)$$

Also, from the definition of enthalpy, and using Eq. (2-46),

$$dh = c_p\, dT = du + p\, dv + v\, dp$$

But $du + p\, dv = 0$, so that

$$c_p\, dT = v\, dp \qquad (2\text{-}48)$$

Dividing Eq. (2-48) by Eq. (2-47) gives

$$\frac{c_p}{c_v} = -\frac{v}{p}\frac{dp}{dv}$$

or

$$\frac{dp}{p} + \frac{c_p}{c_v}\frac{dv}{v} = 0 \tag{2-49}$$

The solution to Eq. (2-49) is

$$pv^\gamma = \text{const.} \tag{2-50}$$

where $\gamma = c_p/c_v$ is the ratio of specific heats and is called the *isentropic exponent*. The meaning of the word "isentropic" will become clear in later chapters. For now we may regard it simply as the value of the polytropic exponent when the process is quasistatic and adiabatic. The adiabatic process is illustrated in Figs. 2-10 and 2-11.

As a matter of interest, the gas constant may be expressed in terms of the adiabatic exponent in the following ways:

$$R = c_p - c_v = (\gamma - 1)c_v \tag{2-51}$$

$$R = \frac{\gamma - 1}{\gamma}c_p \tag{2-52}$$

The work to compress an ideal gas quasistatically and adiabatically in a cylinder may be calculated by applying Eq. (2-42) or (2-43) with $n = \gamma$.

The compression work in a cylinder, which is discussed in the foregoing paragraphs, is the work for a closed system. The mass of gas is fixed within the system, and the calculation is performed with the indicated equations. In an open, or flow, system one may still encounter a compression or expansion process. High-pressure air or steam may be used to drive a turbine and produce a useful work output. Conversely, a reciprocating piston compressor or turbine arrangement may be employed to compress air from a low to high pressure. We can have a quasistatic adiabatic process in flow systems as well as in the closed piston-cylinder arrangement. The derivation of Eq. (2-50) did not specify the type of system—open or closed; therefore, it may be used to describe the quasistatic adiabatic process for both types of systems.

Although the terms may be intuitively apparent at this time we should clarify the meaning of the words "compression" and "expansion." In a *compression* process the pressure of the system is raised, by whatever mechanism, while in an *expansion* process the pressure is lowered. Thus, we shall apply these terms to both flow and nonflow processes, and in various work and heat interaction situations.

EXAMPLE 2-10   Work in a cycle

A 1-g quantity of nitrogen undergoes the following sequence of quasi-static processes in a piston-cylinder arrangement:

(a) An adiabatic expansion in which the volume doubles
(b) A constant-pressure process in which the volume is reduced to its initial value
(c) A constant-volume compression back to the initial state

The nitrogen is initially at 150°C and 5 atm pressure. Calculate the net work done on the gas in this sequence of processes.

SOLUTION   The processes under consideration are illustrated in the accompanying figure. We have

$$p_1 = 5 \text{ atm} = 5.066 \times 10^5 \text{ N/m}^2$$
$$T_1 = 150°C = 423 \text{ K}$$
$$m = 1 \text{ g} = 10^{-3} \text{ kg}$$

For nitrogen the gas constant will be

$$R = \frac{\mathscr{R}}{M} = \frac{8314}{28} = 297 \text{ J/kg·K}$$

and the initial volume is calculated as

$$V_1 = \frac{mRT_1}{p_1} = \frac{(0.001)(297)(423)}{5.066 \times 10^5} = 2.48 \times 10^{-4} \text{ m}^3$$

We also have $V_2 = 2V_1 = 4.96 \times 10^{-4} \text{ m}^3$. From Eq. (2-37)

$$p_2 = p_1 \left(\frac{V_1}{V_2}\right)^\gamma = (5.066 \times 10^5)(\tfrac{1}{2})^{1.4} = 1.92 \times 10^5 \text{ N/m}^2$$

For process 1-2 the work done on the gas is calculated from Eq. (2-42):

$$W_{1-2} = \frac{p_2 V_2 - p_1 V_1}{\gamma - 1} = \frac{[(1.92)(4.96) - (5.066)(2.48)](10)}{1.4 - 1}$$
$$= -76.05 \text{ J}$$

In process 2-3 the pressure is constant and the work is obtained from

$$W_{2-3} = -\int_{v_2}^{v_3} p \, dV = -p_2(V_3 - V_2)$$
$$= -(1.92 \times 10^5)(2.48 - 4.96)(10^{-4})$$
$$= 47.62 \text{ J}$$

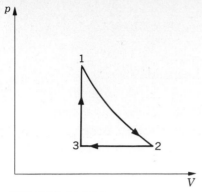

FIG. EXAMPLE 2-10

There is no work done in process 3-1 because $V_3 = V_1$. The total work in the sequence of processes is therefore

$$W = W_{1-2} + W_{2-3} + W_{3-1}$$
$$= -76.05 + 47.62 + 0 = -28.43 \text{ J}$$

The negative sign indicates that there is a net work *output* from the three-process cycle.

We can make some additional calculations which illustrate an important point. From Eq. (2-19) the sum of the heat flows in the cycle should equal the negative of the net work added in the cycle. Let us calculate these heat flows. The temperatures of interest are

$$T_2 = T_1 \left(\frac{V_1}{V_2}\right)^{\gamma-1} = (423)(\tfrac{1}{2})^{0.4} = 320.6 \text{ K}$$

Because $p_2 = p_3$

$$T_3 = T_2 \frac{V_3}{V_2} = (320.6)(\tfrac{1}{2}) = 160.3 \text{ K}$$

From Table 2-2 the specific heats for nitrogen are

$$c_v = 0.177 \text{ Btu/lbm·°F} = 741.1 \text{ J/kg·°C}$$
$$c_p = 0.248 \text{ Btu/lbm·°F} = 1038.3 \text{ J/kg·°C}$$

Process 1-2 is adiabatic so $Q_{1-2} = 0$. The heat added in the constant-pressure process 2-3 is

$$Q_{2-3} = mc_p(T_3 - T_2)$$
$$= (0.001 \text{ kg})(1038.3 \text{ J/kg·°C})(160.3 - 320.6)$$
$$= -166.4 \text{ J}$$

Process 3-1 is constant volume so the heat added is

$$Q_{3-1} = mc_v (T_1 - T_3)$$
$$= (0.001)(741.1)(423 - 160.3)$$
$$= 194.8 \text{ J}$$

The net heat added in the cycle is therefore

$$Q = Q_{1-2} + Q_{2-3} + Q_{3-1}$$
$$= 0 - 166.4 + 194.8 = 28.4 \text{ J}$$

and this, of course, is equal to the negative of the work added in the cycle.

EXAMPLE 2-11 Discharge from a tank
A 0.5-m³ tank contains air at $p = 7.0$ MPa and 250°C and is perfectly insulated from the surroundings. A valve on the tank is opened and the air discharged until the pressure drops to 400 kPa. Calculate the mass of air discharged from the tank.

SOLUTION The mass discharged is clearly the difference between the initial and final masses in the tank. Using $T_1 = 250 + 273 = 523$ K the initial mass is calculated as

$$m_1 = \frac{p_1 V}{RT_1} = \frac{(7 \times 10^6)(0.5)}{(287.1)(523)} = 23.31 \text{ kg} \tag{a}$$

To calculate the final mass we must determine the final temperature in the tank. This mass undergoes an adiabatic process (because the tank is insulated) and there is no evidence of frictional effects, and so we may assume that the process is quasistatic as well. Therefore, we may employ Eq. (2-37) with $n = \gamma$ to calculate the final temperature.

$$\frac{T_2}{T_1} = \left(\frac{p_2}{p_1}\right)^{(\gamma-1)/\gamma}$$

$\gamma = 1.4$ for air, and so

$$T_2 = (523) \left(\frac{0.4}{7.0}\right)^{(1.4-1)/1.4} = 230.7 \text{ K} \tag{b}$$

The final mass is now calculated as

$$m_2 = \frac{p_2 V}{RT_2} = \frac{(4 \times 10^5)(0.5)}{(287.1)(230.7)} = 3.02 \qquad (c)$$

and the mass lost is

$$\Delta m = m_1 - m_2 = 23.31 - 3.02 = 20.29 \text{ kg}$$

2-12 SUMMARY

This chapter has served to generalize the first law of thermodynamics as the conservation of energy principle. Work has been defined as energy expended by a force acting through a displacement. In thermodynamics this is a quantity which is delivered to the boundary of the system. The existence of a general internal-energy function has been demonstrated, which is dependent only on the state of the system; thus it is given a designation as a thermodynamic property. Heat has been defined as energy which may cross the boundary of a system but is not expressible as a force acting through a distance. Finally, all three types of energy are incorporated into the conservation idea, and an analytical statement of the first law is presented.

REVIEW QUESTIONS

1 How is work defined?

2 What is meant by expansion work?

3 What is meant by the term *potential energy?*

4 What is the first law of thermodynamics?

5 What is the general meaning attached to the term *internal energy?*

6 How does heat differ from work?

7 Why cannot heat be contained in a system?

8 Why is the term *specific heat* a misnomer?

9 What is a cyclic process?

10 How is enthalpy defined?

11 How would you explain the first law of thermodynamics to a layman?

12 How does work differ from power?

13 Upon what property is the internal energy and enthalpy of an ideal gas dependent?

14 Explain the sign convention for heat and work.

15   Why is work not a property of a system?

16   What is meant by a *closed* system?

17   What is an ideal gas?

18   Define the terms *isothermal* and *adiabatic*.

19   How are properties related in a quasistatic adiabatic process for an ideal gas?

20   How are specific heats and the gas constant related?

21   What is a polytropic process?

PROBLEMS   (English units)

2-1   A pressure of 342 psia acts on a piston having a diameter of 7.5 cm. Calculate the work done when this piston moves through a displacement of 3 in.

2-2   A large rocket can produce a thrust of 600 000 lbf at a velocity of 600 mph. Calculate the horsepower of the rocket and the energy dissipated in 1 s expressed in Btu units.

2-3   A small heat-power system receives a heat input of 100 000 Btu/h along with a work input of 1.5 hp. The system produces an electric energy output at the rate of 20 kW. Calculate the change in internal energy of the system during a 2-min time interval.

2-4   A certain battery is charged by applying a current of 40 A at 12 V for a 30-min time period. During the charging process the battery loses 200 Btu of heat to the surroundings. How much does the internal energy of the battery increase during the 30-min period?

2-5   Heat is added to a closed system which is so constructed that the pressure will remain constant. Calculate the change in enthalpy of the system when 800 Btu of heat are added and the system produces a work output of 200 Btu.

2-6   A spring is available which requires an applied force of 100 lbf to compress the spring 1 in. Calculate the change in internal energy of the spring when it is compressed 3 in. Neglect heat transfer.

2-7   Nitrogen expands in a cylinder from 100 psia and 500°F to 30 psia and 100°F. Assuming ideal-gas behavior ($pV = mRT$), calculate the work done on the face of the piston per pound of nitrogen. Assume pressure and volume are related by pV^n = const., where n is a suitable exponent.

2-8   A 1-ft³ rigid enclosure contains air at 20 psia and 70°F. Heat is added to the container until the pressure reaches 50 psia. Calculate the heat added, assuming that air behaves an an ideal gas with constant specific heat.

2-9   Calculate the temperature rise for the tank of water in Example 2-3 if the tank loses heat at the rate of 800 Btu/hr while the 1-hp stirring action is applied.

2-10 Suppose the tank of Example 2-3 is uninsulated and may lose heat to the surroundings according to the following relation: $q = 18 \, \Delta T$ Btu/h, where ΔT is the temperature difference between the tank and the surroundings and q is the heat transfer *rate*. Assuming that the stirring action starts when the tank and surroundings are at 70°F, calculate the equilibrium temperature of the water. Also calculate the temperature the water will attain after 15 min.

2-11 Air expands in a cylinder from 200 psia, 800°F, to 30 psia, 200°F. Calculate the work and heat transfer assuming that the process varies as $pv^n = $ const. and c_v for air is 0.1715 Btu/lbm·°F.

2-12 0.1-lbm of air is contained by a piston-cylinder arrangement at 25 psia and 100°F. Heat is added to the air so that it is allowed to expand at constant pressure until the volume doubles. Calculate the change in internal energy of the air and the heat added.

2-13 A small steel cylinder, $V = 0.1$ ft³, contains nitrogen at 2000 psia and 70°F. The cylinder is placed inside an evacuated and insulated chamber having a volume of 10 ft³. By some suitable means the small cylinder is allowed to discharge its contents into the large chamber. Calculate the final nitrogen pressure in the chamber when equilibrium conditions are established.

2-14 A certain 12-V battery carries a rating of 100 A·h. What is the energy stored in the battery expressed in Btu, J, and cal?

2-15 A window air conditioner is designed to remove 12,000 Btu/h from the room, and 1.2 hp of electrical work must be delivered to the unit to accomplish this cooling. How much heat is dissipated to the surroundings?

2-16 A certain automobile consumes fuel at the rate of 12 mi/gal at a speed of 60 mi/h, traveling on a level road. The internal energy of the fuel is 140 000 Btu/gal and the overall efficiency of the engine and drive mechanism is 13 percent; i.e., only 13 percent of the energy of the fuel is delivered to the road in the form of work, and the remainder is dissipated as heat. Calculate the overall drag force on the automobile using energy considerations. Suppose the same automobile, weighing 4400 lbm, is driven at the same speed up a grade inclined 5° with the horizontal. What will be its fuel consumption under these new conditions?

2-17 A 250-lb man is told that he can lose weight by increasing his energy consumption through exercise. He proposes to do this by climbing a 20-ft staircase several times a day. His current food consumption provides an energy input of 4000 kcal/day. What would you advise him regarding exercise versus a reduced food intake to effect a weight loss?

2-18 A thermodynamic system undergoes a cycle composed of a series of three processes for which $Q_1 = +10$ Btu, $Q_2 = +30$ Btu, $Q_3 = -5$ Btu. For the first process $\Delta E = +20$ Btu and for the third process $\Delta E = -20$ Btu. What is the work in the second process, and the net work *output* of the cycle?

2-19 4 lbm of oxygen are heated from 100 to 500°F in a constant-volume process. Calculate the change in enthalpy and heat transfer.

2-20   5 g of carbon dioxide are contained in a vessel at 10 psia and 100°F. Calculate the heat which must be added to raise the pressure to a value of 2 atm.

2-21   Calculate the heat necessary to raise 1 g mol of each of the following gases from 200 to 500°F at a constant pressure of 15 psia.
   (a) Hydrogen
   (b) Helium
   (c) Nitrogen
   (d) Oxygen
   (e) Argon
   (f) Air

2-22   A balloon having a volume of 25 ft³ is filled with air at 70°F and 14.696 psia. After exposure to solar radiation, the volume of the balloon increases to 30 ft³ and the pressure remains constant since the air is loosely contained. Calculate the heat added.

2-23   Consider a constant-pressure process in which hydrogen is heated from 300 to 4000°F in an electric-arc process. Calculate the heat transfer per pound mass. Suppose the specific heat of the hydrogen is evaluated at some mean temperature (2100°F) and this value is used to calculate the heat transfer using a constant specific heat analysis. What error would result if this latter technique were employed?

2-24   The *mean* specific heat for a gas over a given temperature range is defined as

$$c_{p,\,\text{mean}} = \frac{\int_{T_1}^{T_2} c_p \, dT}{T_2 - T_1}$$

Calculate such a mean specific heat for the hydrogen in Prob. 2-23. How does this value compare with that given in Table 2-2?

2-25   In a large majority of applicable problems air is assumed to behave as an ideal gas with constant specific heats for the temperature range 0 to 1000°F. Assuming that air behaves like nitrogen (air is 78 percent nitrogen), estimate the validity of this assumption.

2-26   Water vapor is heated at a constant pressure of 1 psia from 150 to 1000°F. Assume that the vapor obeys the ideal-gas equation of state in this region, and calculate the heat transfer by using the data of Table 2-3.

2-27   One exotic energy utilization scheme would employ a 5 by 5 mi solar collector in space to absorb the sun's energy in special solar cells. These cells convert the radiant energy directly into electric energy which can then be transmitted to the earth with a complicated microwave system. The efficiency of the solar cells is about 12 percent; i.e., only 12 percent of the incoming solar energy is converted to electricity, the remaining 88 percent is reradiated to space. Estimate the electric power output of the device using the data of Prob. 1-28.

2-28   A 1-ft³ rigid container is open to the atmosphere and heat is added to the bottom of the container until the air temperature inside reaches 850°F. The container is then quickly sealed, removed from the heating source, and allowed to cool to room temperature of 70°F. How much heat is lost during the cooling process?

2-29 An energy conservationist group claims that a substantial amount of electric power can be saved by using more efficient lighting systems. An incandescent light bulb is only about 5 percent efficient in converting electric energy into visible light; the remaining 95 percent is dissipated as heat in the room. Fluorescent lights are about 20 percent efficient. In modern buildings the heat dissipation must be removed by the air conditioning system which requires further power. About 1 kW of electric power is required to produce 10 000 Btu/h of cooling. What reduction in power consumption rate could be accomplished in a room currently using 2000 W of incandescent lights if they were all converted to fluorescents?

PROBLEMS   (Metric units)

2-1M   A pressure of 2.4 MPa acts on a piston having a diameter of 7.5 cm. Calculate the work done when this piston moves through a displacement of 8 cm.

2-2M   An electric potential of 110 V is impressed on a certain resistor such that a current of 12 A is drawn. Calculate the energy dissipated in a time of 3 min.

2-3M   A large rocket can produce a thrust of 2.7 MN at a velocity of 200 m/s. Calculate the power of the rocket and the energy dissipated in 1 s expressed in kJ.

2-4M   A small heat-power system receives a heat input of 30 kW along with a work input of 1.1 kW. The system produces an electric energy output at the rate of 20 kW. Calculate the change in internal energy of the system during a 2-min time interval.

2-5M   A certain battery is charged by applying a current of 40 A at 12 V for a 30-min time period. During the charging process the battery loses 200 kJ of heat to the surroundings. How much does the internal energy of the battery increase during the 30-min period?

2-6M   Heat is added to a closed system which is constructed so that the pressure will remain constant. Calculate the change in enthalpy of the system when 850 kJ of heat are added and the system produces a work output of 200 kJ.

2-7M   A spring is available which requires an applied force of 445 N to compress the spring 2.5 cm. Calculate the change in internal energy of the spring when it is compressed 7.5 cm. Neglect heat transfer.

2-8M   Nitrogen expands in a cylinder from 690 kPa and 260°C to 210 kPa and 40°C. Assuming ideal-gas behavior ($pV = mRT$), calculate the work done on the face of the piston per pound of nitrogen. Assume pressure and volume are related by $pV^n =$ const., where n is a suitable exponent.

2-9M   A 28-liter rigid enclosure contains air at 140 kPa and 20°C. Heat is added to the container until the pressure reaches 345 kPa. Calculate the heat added, assuming that air behaves as an ideal gas with constant specific heat, $c_v = 0.718$ kJ/kg·°C.

2-10M Calculate the temperature rise for the tank of water in Example 2-3 if the tank loses heat at the rate of 850 kJ/h while the 1-hp stirring action is applied.

2-11M Suppose the tank of Example 2-3 is uninsulated and may lose heat to the surroundings according to the following relation: $q = 18 \, \Delta T$ kJ/h, where ΔT is the temperature difference between the tank and the surroundings, and q is the heat transfer *rate*. Assume that the stirring action starts when the tank and the surroundings are at 20°C, and calculate the equilibrium temperature of the water. Also calculate the temperature the water will attain after 15 min.

2-12M Air expands in a cylinder from 1.4 MPa, 700 K, to 200 kPa, 370 K. Calculate the work and heat transfer assuming that the process varies as $pv^n = $ const., and c_v for air is 0.718 kJ/kg·°C.

2-13M 100 g of air is contained by a piston-cylinder arrangement at 170 kPa and 40°C. Heat is added to the air so that it is allowed to expand at constant pressure until the volume doubles. Calculate the change in internal energy of the air and the heat added.

2-14M A small steel cylinder, $V = 0.28$ liter, contains nitrogen at 14 MPa and 20°C. The cylinder is placed inside an evacuated and insulated chamber having a volume of 280 liters. By some suitable means the small cylinder is allowed to discharge its contents into the large chamber. Calculate the final nitrogen pressure in the chamber when equilibrium conditions are established.

2-15M A window air conditioner is designed to remove 13 kJ/h from the room, and 900 W of electrical work must be delivered to the unit to accomplish this cooling. How much heat is dissipated to the surroundings?

2-16M A certain automobile consumes fuel at the rate of 5 km/liter at a speed of 100 km/h, traveling on a level road. The internal energy of the fuel is 40 000 kJ/liter, and the overall efficiency of the engine and drive mechanism is 13 percent; i.e., only 13 percent of the energy of the fuel is delivered to the road in the form of work, and the remainder is dissipated as heat. Calculate the overall drag force on the automobile using energy considerations. Suppose the same automobile, weighing 2000 kg, is driven at the same speed up a grade inclined 5° with the horizontal. What will be its fuel consumption under these new conditions?

2-17M A 115-kg man is told that he can lose weight by increasing his energy consumption through exercise. He proposes to do this by climbing a 6-m staircase several times a day. His current food consumption provides an energy input of 17 kJ/day. What would you advise him regarding exercise versus a reduced food intake to effect a weight loss?

2-18M A thermodynamic system undergoes a cycle composed of a series of three processes for which $Q_1 = +10$ kJ, $Q_2 = +30$ kJ, $Q_3 = -5$ kJ. For the first process $\Delta E = +20$ kJ and for the third process $\Delta E = -20$ kJ. What is the work in the second process, and the net work *output* of the cycle?

2-19M One exotic energy utilization scheme would employ an 8 by 8 km solar collector in space to absorb the sun's energy in special solar cells. These cells convert the radiant energy directly into electric energy which can then be transmitted to the earth with a complicated microwave system. The efficiency of the solar cells is about 12 percent; i.e., only 12 percent of the incoming solar energy is converted to electricity, with the remaining 88 percent being reradiated to space. Estimate the electric power output of the device using the data of Prob. 1-28.

2-20M A 28-liter rigid container is open to the atmosphere, and heat is added to the bottom of the container until the air temperature inside reaches 450°C. The container is then quickly sealed, removed from the heating source, and allowed to cool to room temperature of 20°C. How much heat is lost during the cooling process?

2-21M An energy conservationist group claims that a substantial amount of electric power can be saved by using more efficient lighting systems. An incandescent light bulb is only about 5 percent efficient in converting electric energy into visible light; the remaining 95 percent is dissipated as heat in the room. Fluorescent lights are about 20 percent efficient. In modern buildings the heat dissipation must be removed by the air conditioning system which requires further power. About 1 kW of electric power is required to produce 10 MJ/h of cooling. What reduction in power consumption rate could be accomplished in a building currently using 2000 W of incandescent lights if they were all converted to fluorescents?

2-22M For the p-V diagram shown in the figure, a certain closed thermodynamic system may execute a cyclic process from point 1 through point 2 in a clockwise or counterclockwise manner. In the clockwise process the heat is $(Q_{1-2})_{CW} = 85$ kJ and the work is $(W_{1-2})_{CW} = -40$ kJ. When the system goes from state 2 to state 1, the heat is the same for either CW or CCW directions and is equal to $(Q_{2-1})_{CW} = (Q_{1-2})_{CCW} = -55$ kJ. Taking the counterclockwise path from 1 to 2 it is found that $(Q_{1-2})_{CCW} = 75$ kJ. Determine $(W_{1-2})_{CCW}$ and (W_{2-1}).

FIG. P 2-22M

2-23M   Oxygen is compressed quasistatically and adiabatically in a piston-cylinder device. Calculate the work required for compression from 20°C, 1 atm to $p = 500$ kPa. Also calculate the final temperature.

2-24M   A substance undergoes a certain quasistatic process behind a piston such that the pressure varies according to $p = a + bV$. Derive an expression for the work between two pressures p_1 and p_2.

2-25M   A balloon contains air at 120 kPa and 20°C and has a diameter of 1.0 m. Heat is added to the balloon until the diameter increases to 1.4 m. During this process the pressure is directly proportional to the diameter. Calculate the heat added.

2-26M   In a certain spring the force is related to displacement by $F = k_s x^2$. A force of 1000 N is required to compress the spring 1 cm. Calculate the work to compress the spring 3.0 cm. Express the answer in joules, ft·lbf, and Btu units.

2-27M   An experiment is conducted with a certain gas. It is found that when the gas is maintained in a constant volume container 70 kJ/kg of energy is required to raise the temperature 83°C. It is known that $\gamma = 1.50$ for the gas. Calculate the values of c_p and R for the gas.

2-28M   Hydroelectric power production is based on the conversion of gravitational potential energy of water into mechanical energy by means of large turbines. Suppose a dam facility is available whereby the water may fall through an elevation of 100 m. What water flow would be necessary to produce a power of 10 000 kW?

2-29M   The nuclear fission of one uranium 235 atom releases about 200 million electronvolts (MeV) of energy. Assuming that 30 percent of this energy could be converted to electricity in a power plant, how much U^{235} would be consumed in a year to produce 1000 MW continuously? (See Appendix for conversion factors.)

2-30M   A weather balloon is constructed of a nonstretchable plastic material to produce an inflated spherical shape of 10 m in diameter. Initially the balloon is collapsed and is to be filled with helium at a location where the pressure is 1 atm. How much work is done on the surroundings during the inflation process?

2-31M   A hydraulic lift for a service station uses compressed air for power. The air is stored at 7 atm and released into a cylinder which drives a 35-cm-diameter piston to raise the automobiles. Normally, the automobiles are lifted through a height of 2.3 m. The air supply is provided by a small electric motor-driven compressor which may run continuously if needed. Assuming that an adequate air storage tank is available, what horsepower motor would be required to perform 50 lifts per day on a 4000-lb vehicle?

2-32M   A large spring is to be designed to store energy which can be recovered at the rate of 300 W for a period of 10 min. The deformation of the spring is to be 30 cm. What must the spring constant be to accomplish this energy storage?

2-33M   A certain thermodynamic cycle is represented as a 5 by 5 cm square on a p-v diagram in which 1 cm = 1 m³ and 1 cm = 1 atm are the coordinate scales. If there is to be a net work input to the system during the cycle, calculate the overall heat transfer during the cycle.

2-34M   An electric resistance heater is placed inside a 1-liter rigid container filled with nitrogen at 50°C and 10 atm pressure. What electric power rate must be supplied to the heater to make the pressure rise to 15 atm in a period of 10 min, if the container is perfectly insulated?

2-35M   In an exotic power device, hydrogen is burned with oxygen to produce a very high combustion temperature. Prior to the burning the oxygen is heated from 20 to 1200°C at constant pressure of 130 kPa. How much heat is required for an oxygen consumption of 1 kg?

2-36M   Oxygen is compressed in a piston-cylinder arrangement from 160 kPa, 40°C, to 600 kPa, 90°C. Calculate the work done on the gas and the heat transfer if the process is quasistatic.

2-37M   An ideal gas undergoes a quasistatic process at constant pressure of 3 atm in a piston-cylinder arrangement. During the process 25 kJ of heat are added and the volume is reduced from 0.15 to 0.06 m³. Calculate the change in internal energy for this process.

2-38M   5 kg of nitrogen are compressed in such a manner that the internal energy increases by 160 kJ while 20 kJ of heat are added. Calculate the work that is supplied to the gas. Assume that the compression occurs in a closed system.

2-39M   A 1-kW-resistance heater is placed in a 3000-liter container filled with air at 100 kPa and 20°C. The heater is allowed to operate for 10 min and the container is insulated. Calculate the temperature and pressure of the air at the end of this time period.

2-40M   An ideal gas undergoes a quasistatic adiabatic expansion from conditions of 800 kPa, 0.7 liter, and 100°C to a final pressure of 200 kPa. Calculate the work done if the ratio of specific heats is 1.39. Also calculate values of the specific heats based on total mass.

2-41M   Proposals have been made to employ flywheels as energy storage devices in the future. Consult whatever references you feel appropriate and comment on such proposals.

2-42M   A tank having a volume of 0.1 m³ contains air at 14 MPa and 50°C. It is connected through a valve to a larger tank having a volume of 15 m³ which is completely evacuated. The entire assembly is completely insulated. The valve is opened and the gas allowed to come to equilibrium in both tanks. Calculate the final pressure.

2-43M   Suppose the valve in Prob. 2-42M is only opened long enough for the pressure to drop to 3.0 MPa in the small tank and that the process occurs rapidly. Calculate the final temperature in each tank.

2-44M   A closed thermodynamic system is designed to raise a 4.0-kg mass in a standard gravitational field while exchanging heat with the surroundings and receiving an electrical work input. In one cycle operation, the system receives 350 kJ of heat during one part of the cycle and gives up 150 kJ during another part of the cycle. For the overall cycle 50 J of electric energy are delivered to the system. Calculate the displacement of the 4.0-kg mass.

2-45M 2 kg of oxygen are heated from 40 to 200°C in a constant-volume process. Calculate the change in enthalpy and heat transfer.

2-46M 5 g of carbon dioxide are contained in a vessel at 70 kPa and 40°C. Calculate the heat which must be added to raise the pressure to a value of ≈ 200 kPa.

2-47M Calculate the heat necessary to raise 1 kg mol of each of the following gases from 90 to 260°C at a constant pressure of 100 kPa.
      (a) Hydrogen   (b) Helium
      (c) Nitrogen   (d) Oxygen
      (e) Argon   (f) Air

2-48M A balloon having a volume of 700 liters is filled with air at 20°C and 100 kPa. After exposure to solar radiation, the volume of the balloon increases to 850 liters and the pressure remains constant since the air is loosely contained. Calculate the heat added.

2-49M Consider a constant-pressure process in which hydrogen is heated from 150 to 220°C in an electric-arc process. Calculate the heat transfer per kilogram mass. Suppose the specific heat of the hydrogen is evaluated at some mean temperature (1200°C) and this value is used to calculate the heat transfer using a constant specific heat analysis. What error would result if this latter technique were employed?

2-50M In a large majority of applicable problems air is assumed to behave as an ideal gas with constant specific heats for the temperature range 255 to 810 K. Assuming that air behaves like nitrogen (air is 78 percent nitrogen), estimate the validity of this assumption.

2-51M 3.0 kg of argon are heated at constant pressure from 20 to 120°C. Calculate the heat required and the change in internal energy.

2-52M A certain gas undergoes a constant-pressure process in a piston-cylinder arrangement at 120 kPa. During this process the internal energy increases by 190 kJ/kg for a temperature rise of 75°C. At the same time the work delivered to the face of the piston is 100 kJ/kg. Calculate the values of c_v, c_p, and R for the gas.

2-53M 10.5 kg of air are heated from 15 to 100°C in a constant-pressure process. Calculate the change in internal energy.

2-54M An experiment is conducted with a closed system containing an unknown substance. As the system is allowed to change from state 1 to state 2, 15 kJ of heat are removed while the work output of the system is observed to be 38 kJ. The system is then restored back to state 1 in a different process in which 28 kJ of heat are added along with a work quantity. Calculate the value of the work in this second process.

2-55M A closed system undergoes a polytropic process in which the heat added is 17 kJ and the internal energy increases by 48 kJ. The initial state is 130 kPa and 0.15 m³, and the final pressure is 800 kPa. Calculate the final volume in the process. If the system were air what mass would be required?

2-56M   11.6 kg of air are compressed isothermally with a piston cylinder from 110 kPa, 30°C, to 450 kPa. Calculate the work required and the heat added.

REFERENCES

1 Callen, H. B.: "Thermodynamics," John Wiley & Sons, Inc., New York, 1960.

2 Hatsopoulos, G. N., and J. H. Keenan: "Principles of General Thermodynamics," John Wiley & Sons, Inc., New York, 1965.

3 Reynolds, W. C.: "Thermodynamics," 2d ed., McGraw-Hill Book Company, New York, 1968.

4 Sears, F. W.: "Thermodynamics," 2d ed., Addison-Wesley Publishing Company, Inc., Reading, Mass., 1953.

5 Van Wylen, G. J., and R. E. Sonntag: "Fundamentals of Classical Thermodynamics," John Wiley & Sons, Inc., New York, 1965.

6 Sweigert, R. L., and M. W. Beardsley: *Bull. No. 2*, Georgia School of Technology, 1938.

chapter three
Macroscopic properties of pure substances

3-1   INTRODUCTION

The properties of a substance depend on its constitution. The number of phases, the homogeneous or heterogeneous nature of the substance, and the types of pertinent energy quantities all have a bearing on the number and types of properties which may be required for analysis purposes. If, for example, we want to calculate the energy release in the fission of a U^{235} atom, we need to know the proton-neutron makeup and the binding energy of the nucleus. To determine the energy necessary to boil water, we need to know the energy of vaporization for water. To analyze a substance where electrical or magnetic work is involved, we need information concerning the electrical or magnetic properties of the substance.

The number of properties required to specify the state of a substance obviously depends on the nature of the energy interactions to be studied. There can be, of course, only heat and work interactions with a substance. But the work interactions can take several different forms. Consider an ionized gas, for example. One way of doing work on the gas is to compress it. Another way is to subject it to an electric potential so that electrical work is done to move the charged particles. Still another possible way to perform work is to move the gas through a magnetic field or to alter the orientation of magnetic dipoles in the gas. To calculate the compression work, we would need information about pressure and volume at various temperatures; to calculate the electrical work, we would need information about the electric conductivity as a function of pressure and temperature (and possibly as a function of electric field strength); and to calculate the magnetic work, we would need information on the conductivity and magnetic dipole moment of the gas. The total number of independent properties needed to specify the state of the gas under these circumstances may be obtained with just a little reasoning.

We are basically interested in determining the internal thermal energy of the substance u. For each possible work mode (volume change, electrical, magnetic, etc.) there is associated a characteristic *intensive* property (pressure, electric potential, magnetic potential) which describes the ability of a unit mass or unit volume of the system to accomplish the particular work mode. For each work mode there is also some characteristic *extensive* property (volume, conductivity, magnetic dipole moment) which describes how much total work will be performed in each mode. The intensive property represents the *force* and the extensive property represents the *displacement* in each work mode. The internal energy is a function of the possible work interactions and the heat interaction. The intensive property that describes the heat interaction is the temperature.[†]

In accordance with this brief discussion, we might conclude that in

---

[†] We have omitted kinetic and gravitational potential energies in this consideration. They are important, of course, but are normally treated as separate terms in the general internal-energy function E.

order to specify the state of a substance we need an intensive property for each work mode plus an intensive property to describe heat interactions. This principle is restricted to systems in equilibrium, which, of course, are our main concern. It is required that these be *independent* properties.

This brief discussion specifies the state principle of Chap. 1 in explicit form, which may be expressed as follows:

> *The number of independent properties required to specify the thermody-namic state of a system is equal to the number of possible work modes plus one.*

There are other considerations involved in the state principle which are discussed by Kline and Koenig [1] and Hatsopoulos and Keenan [2].

A *simple substance* involves only one possible work mode. A nonion-ized gas with no magnetic dipole moment is such a substance. In this case we have only one possible work mode, a volume change, so that only two independent properties are required to specify the state of the gas. Any two of pressure, temperature, or specific volume will do.

The term *pure substance* designates a substance which is homogeneous and has the same chemical composition in all phases. Water is a pure sub-stance since its chemical composition is the same in all phases, even for a mixture of liquid and vapor. A mixture of gases, like air, behaves as a pure substance, but if the mixture is cooled such that some of the components condense out into a liquid phase, the mixture may no longer be considered as a pure substance because it does not have the same chemical composition in all phases. The pure substance is of significant interest for many practical applications and, thus, deserves much of our attention.

The purpose of this chapter is to describe the thermodynamic proper-ties of some characteristic simple substances. Specific numerical data are presented in the Appendix so that they may be applied to problem solutions in the subsequent chapters. At this point the properties are presented in a purely empirical sense; i.e., they are based on experimental measurements. The calculation of some types of thermodynamic properties from funda-mental microscopic considerations will be discussed in Chap. 8.

REVIEW THOUGHTS

What do the following terms mean: *work modes, intensive and extensive properties, state principle, simple substance, pure substance?*

3-2 PROPERTIES OF PURE SUBSTANCES

Let us now examine the behavior of a pure substance which has work in-teractions involving only the compression (volume change) work mode. For such a substance only two independent properties are required to define the state of the system. To illustrate the physical behavior of such substances, let us consider the behavior of one of the most common substances of all— ordinary water. The substance may exist in several *phases*:

1   A pure solid phase commonly known as ice
2   A pure liquid phase
3   A pure vapor phase commonly known as steam
4   An equilibrium mixture of liquid and vapor phases
5   An equilibrium mixture of liquid and solid phases
6   An equilibrium mixture of solid and vapor phases

Figure 3-1 is a pressure-temperature diagram illustrating these various phase regions for water.

When a solid changes to a liquid, we say that a *fusion* process has taken place; when a liquid changes to a vapor, there is *vaporization*; and, finally, when a solid changes directly to a vapor, a *sublimation* process is said to have been executed. In each of these processes energy must be added to the substance to effect the change in phase. The temperature at which these changes will occur is dependent on the pressure exerted on the substance. Carbon dioxide at normal atmospheric pressure will sublime; i.e., the solid will change directly to vapor with no intermediate change to liquid. The sublimation temperature for CO_2 at atmospheric pressure is approximately $-78°C$.

Water at atmospheric pressure may be made to exist in all three phases, as depicted by the constant-pressure line AB in Fig. 3-1. If ice is heated at

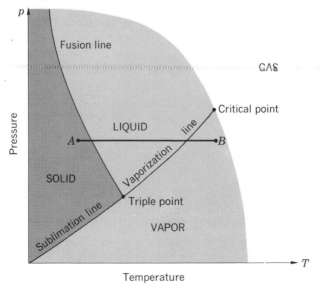

FIG. 3-1   *Pressure-temperature diagram for a substance that expands on freezing.*

constant pressure the temperature increases until the fusion line is encountered and melting starts. The melting process proceeds at constant temperature until all of the solid ice is transformed to liquid. Further heating then increases the liquid temperature until the vaporization line is reached. Further addition of heat progressively transforms the liquid into vapor and the temperature remains constant in the vaporization process. Once all the liquid is converted to vapor additional heating causes the temperature to rise once again. At atmospheric pressure the vaporization temperature for water is 100°C (212°F) and the fusion temperature is approximately 0°C (32°F). We also see from Fig. 3-1 that a change in pressure will alter the freezing or vaporization point of the liquid; for water, an increase in pressure raises the vaporization temperature and lowers the freezing temperature. It should be noted that Fig. 3-1 is not drawn to scale. In the actual case the fusion line would be almost vertical and very large increases in pressure would be required to lower the melting temperature substantially.

The fusion line represents the solid-liquid mixture, the vaporization line the liquid-vapor mixture, and the sublimation line the solid-vapor mixture.

The *triple point* is the state where it is possible to maintain an equilibrium mixture of all three phases.

The *critical point* is the state where the pure vapor phase has identical properties with a pure liquid phase at the same pressure and temperature. It is not possible to observe a distinction between liquid and vapor phases at supercritical pressures and temperatures. In other words, the surface meniscus separating liquid and vapor phases vanishes at the critical point.

3-3   SATURATION REGIONS

The three equilibrium lines in Fig. 3-1 (fusion, vaporization, and sublimation) are said to designate *saturation* regions. Thus we may say that the vaporization line represents the saturation region between liquid and vapor. The vapor which is present in such a mixture is called a *saturated vapor*, and the liquid present in the mixture is called a *saturated liquid*.

To illustrate the properties further, consider the constant-pressure process AB of Fig. 3-1 as plotted on the p-v diagram of Fig. 3-2 and the surface of Fig. 3-3. Here we see the nature of the saturation regions more clearly. Each region represents a circumstance where pressure and temperature remain constant while there is a significant change in specific volume. In the liquid-vapor saturation region the increase in specific volume (volume per unit mass) results from an increase in the fraction of the total mass which is present in the vapor form. The *quality* x represents the fraction of mass present in the vapor phase, and the *moisture* represents the fraction of mass present in the liquid phase. Clearly,

$$\text{Moisture} = 1 - x \qquad\qquad (3\text{-}1)$$

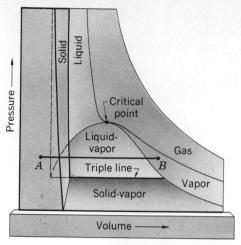

FIG. 3-2 *Pressure-volume diagram for a substance that expands on freezing.*

The term *wet mixture* is widely used to designate a liquid-vapor mixture with a quality of less than 100 percent. In designating the saturation properties of substances, the subscripts *i*, *f*, and *g* are used for solid, liquid, and vapor phases. Thus the specific volume of a saturated liquid would be v_f, and the specific volume of a saturated vapor would be v_g. The specific volume of an equilibrium mixture of liquid and vapor would be given by

$$v = \frac{V}{m} = \frac{1}{m}(m_f v_f + m_g v_g)$$

$$= \frac{m_f}{m} v_f + x v_g$$

$$= (1 - x)v_f + x v_g$$

Introducing the definition

$$v_{fg} = v_g - v_f \qquad\qquad (3\text{-}2)$$

we obtain

$$v = v_f + x v_{fg} \qquad\qquad (3\text{-}3)$$

Several other important terms are defined as follows:

A vapor is said to be *superheated* if it exists at a temperature greater than the saturation temperature corresponding to its pressure. A *subcooled liquid* or *compressed liquid* is one which exists at a temperature lower than the saturation temperature corresponding to its pressure *or* at a pressure greater than the saturation pressure corresponding to its temperature. The two terms are synonymous.

FIG. 3-3 *p-v-T surface for a substance that expands on freezing.*

A three-dimensional *p-v-T* surface for a substance like water (one that expands upon freezing) is shown in Fig. 3-3. A similar figure for a substance which contracts upon freezing is shown in Fig. 3-4.

We have already said that only two independent properties are necessary to define the state of the simple, pure substance previously described. What two properties shall we use? Temperature is one logical choice. It should be obvious that the other choice cannot be pressure when the substance is in one of the saturation regions, because pressure and temperature are not independent in these regions; i.e., a line of constant temperature in a saturation region is also a line of constant pressure; however, specific volume is independent of pressure and temperature in this region, so it may be selected for the other independent property. In a single-phase region, temperature and pressure may be selected as the independent properties.

We shall be interested in the energy properties of substances. The specific internal energy u and specific enthalpy h are properties commonly tabulated for pure substances. From the foregoing discussion we may anticipate that these energy properties may be expressed as functions of temperature and specific volume:

FIG. 3-4 *p-v-T surface for a substance that contracts on freezing.*

$$u = u(T,v)$$
$$h = h(T,v)$$

In the wet mixture region these properties are determined in a fashion analogous to that used for specific volume, i.e.,

$$u = u_f + xu_{fg}$$
$$h = h_f + xh_{fg}$$

where $u_{fg} = u_g - u_f$ and $h_{fg} = h_g - h_f$ are the internal energy and enthalpy of vaporization, respectively.

Another property called the *entropy s* is useful in solution of practical problems, as we shall see in Chap. 6, and is commonly tabulated along with internal energy and enthalpy. In the two-phase evaporation region it, too, takes the characteristic form

$$s = s_f + xs_{fg}$$

where $s_{fg} = s_g - s_f$ is the entropy of vaporization.

Properties for several substances are tabulated in the Appendix, the most extensive tabulation being that for water and its vapor. In the compressed-liquid region the properties for water are tabulated in Table A-11 as corrections to the saturated-liquid properties at the liquid temperature. In general, the corrections are not large as we can see by looking at an example of compressed-liquid water at 100°F and 1000 psia. At 100°F we have (see Table A-7 or A-11)

$$v_f = 0.016\ 132\ \text{ft}^3/\text{lbm}$$
$$h_f = 67.97\ \text{Btu}/\text{lbm}$$
$$p_{sat} = 0.9492\ \text{psia}$$

Entering Table A-11 we find the corrections to these values as

$$(v - v_f) \times 10^5 = -5.2$$
$$h - h_f = +2.64$$

so that the properties for the compressed liquid at 100°F and 1000 psia are

$$v = 0.016\ 132 - 5.2 \times 10^{-5} = 0.016\ 08\ \text{ft}^3/\text{lbm}$$
$$h = 67.97 + 2.64 = 70.61\ \text{Btu}/\text{lbm}$$

In most cases it will be inconvenient to use Table A-11 for the compressed liquid, and Table A-10 furnishes a direct tabulation of compressed-liquid properties. It should be noted that the lowest pressure entry in Table A-11 is 200 psia. Below this value the corrections are so small that they can usually be neglected and the properties of the compressed liquid can be taken as those of the saturated liquid at the liquid temperature.

At this point we note that the properties of liquids are *primarily* a function of the temperature, with pressure exerting a minor effect as was illustrated in the foregoing calculation.

EXAMPLE 3-1

Determine the specific volume v, enthalpy h, and entropy s of water under the following conditions:

(a) 7.5 MPa, 100°C
(b) 4 MPa, 360°C
(c) 150 kPa, 10% quality
(d) 300 kPa, 10% moisture
(e) Saturated liquid at 100 kPa
(f) Saturated vapor at 100 kPa

SOLUTION

(a) This is a compressed-liquid condition because 7.5 MPa is greater than the saturation pressure corresponding to 100°C (101.4 kPa). Therefore, Table A-10M is consulted to find

$v = 1.0397 \times 10^{-3} \text{ m}^3/\text{kg}$
$h = 424.62 \text{ kJ}/\text{kg}$
$s = 1.3011 \text{ kJ}/\text{kg·K}$

(b) 360°C is greater than the saturation temperature for 4 MPa (250.4°C) so this is superheated and Table A-9M is consulted to find

$v = 0.067\ 88 \text{ m}^3/\text{kg}$
$h = 3117.2 \text{ kJ}/\text{kg}$
$s = 6.6215 \text{ kJ}/\text{kg·K}$

(c) The 10 percent quality tells us that a saturation condition is encountered at 150 kPa so Table A-8M (saturation-*pressure* table) is consulted to obtain

$v_f = 1.0528 \times 10^{-3}$ $v_g = 1.159 \text{ m}^3/\text{kg}$
$h_f = 467.11$ $h_{fg} = 2226.5 \text{ kJ}/\text{kg}$
$s_f = 1.4336$ $s_g = 7.2233 \text{ kJ}/\text{kg·K}$

We are given the quality as $x = 0.10$ so that

$v = v_f + x v_{fg} = 1.0528 \times 10^{-3} + (0.10)(1.159 - 1.0528 \times 10^{-3})$
$\qquad\qquad = 0.116\ 85 \text{ m}^3/\text{kg}$
$h = h_f + x h_{fg} = 467.11 + (0.1)(2226.5) = 689.76 \text{ kJ}/\text{kg}$
$s = s_f + x s_{fg} = 1.4336 + (0.1)(7.2233 - 1.4336)$
$\qquad\qquad = 2.0126 \text{ kJ}/\text{kg·K}$

(d) This problem is like part (c) except we note that

Moisture = 1 − quality = 1 − x

so that

$x = 1 - 0.1 = 0.9$

We then enter Table A-8M at 300 kPa and can determine the values as

$v = 0.545\ 33 \text{ m}^3/\text{kg}$
$h = 2508.9 \text{ kJ}/\text{kg}$
$s = 6.4599 \text{ kJ}/\text{kg·K}$

(e and f). These properties are read directly from Table A-8M, at 100 kPa:

Saturated liquid	Saturated vapor
$v_f = 1.0432 \times 10^{-3}$ m³/kg	$v_g = 1.694$ m³/kg
$h_f = 417.46$ kJ/kg	$h_g = 2675.5$ kJ/kg
$s_f = 1.3026$ kJ/kg·K	$s_g = 7.3594$ kJ/kg·K

3-4 EQUATIONS OF STATE

An equation of state is an analytical expression which interrelates properties of substances. The ideal-gas expression derived in Sec. 1-14

$$pv = RT \tag{3-4}$$

is a very simple equation of state. Unfortunately, all equations of state are not so simple. In practice, the equation of state for a solid, liquid, or vapor is obtained by an empirical curve fit of experimental data. A different relationship is usually employed for each phase or saturation region, and each equation is quite long and complicated.

Tabular and graphical presentations of thermodynamic properties such as the ones given in the Appendix serve as substitutes for an analytical equation of state. We shall make use of this information for the present and defer further development of equations of state until Chap. 7.

EXAMPLE 3-2

A rigid 1-ft³ enclosure contains a mixture of liquid and vapor water at a pressure of 12 psia. The total mass of water in the container is 30 lbm. Calculate the heat added and quality of the mixture after sufficient heat is added to raise the pressure to 1000 psia.

SOLUTION   This problem requires an exercise of the first law of thermodynamics as well as information pertaining to properties of pure substances. Since the container is rigid, there can be no change in the system volume and, hence, no work done. The first-law statement then becomes

$$Q = \Delta U = m \, \Delta u \tag{a}$$

To calculate the internal energy, we need to establish the thermodynamic state explicitly. Since we have a mixture of liquid and vapor, pressure and temperature are not independent parameters. The specific volume of the mixture is

$$v = \frac{V}{m} = \frac{1}{30} = 0.0333 \text{ ft}^3/\text{lbm}$$

This volume is then expressed as

$$v = v_f + x v_{fg} \qquad\qquad (b)$$

in accordance with Eq. (3-3). For the initial state, designated with subscript 1,

$$v_{f_1} = 0.016\ 65 \qquad v_{fg_1} = 32.38\ \text{ft}^3/\text{lbm} \qquad (12\ \text{psia})$$

where the properties are obtained from the saturation-pressure table for steam given in Table A-8.
Inserting these values in Eq. (b) gives for the initial quality

$$x_1 = 0.000\ 515$$

The initial internal energy is thus

$$\begin{aligned} u_1 &= u_{f_1} + x_1\, u_{fg_1} \\ &= 169.92 + (0.000\ 515)(904.8) \\ &= 170.38\ \text{Btu}/\text{lbm} \end{aligned}$$

Since the container is rigid, the specific volume remains the same in the initial and final states. Designating the final state by subscript 2,

$$v_{f_2} = 0.0216 \qquad v_{fg_2} = 0.4240\ \text{ft}^3/\text{lbm} \qquad (1000\ \text{psia})$$

Again, the quality is calculated from Eq. (b) with

$$v_2 = v_1 = 0.0333\ \text{ft}^3/\text{lbm}$$

The result is

$$x_2 = 0.0292$$

The final internal energy is then calculated as

$$\begin{aligned} u_2 &= u_{f_2} + x_2\, u_{fg_2} \\ &= 538.4 + (0.0292)(571) \\ &= 555.0\ \text{Btu}/\text{lbm} \qquad (1.29 \times 10^6\ \text{J/kg}) \end{aligned}$$

The heat transfer is now calculated from Eq. (a) as

$$\begin{aligned} Q &= m(u_2 - u_1) \\ &= (30)(555.0 - 170.38) = 11\ 500\ \text{Btu} \qquad (1.213 \times 10^7\ \text{J}) \end{aligned}$$

EXAMPLE 3-3
Steam is contained at 100 psia and 50 percent quality in a 0.5-ft³ cylinder. The cylinder is fitted with a piston arrangement which maintains a

constant pressure. Calculate the amount of heat which must be added to the steam to increase the volume to 2.0 ft³.

SOLUTION We first write the first law for this system:

$$Q + W = \Delta U \qquad (a)$$

For the constant-pressure process the work done *on* the gas is

$$W = -\int_{V_1}^{V_2} p\, dV = -p(V_2 - V_1)$$
$$= -(100)(144)(2.0 - 0.5) = -21\ 600 \text{ ft·lbf} \qquad (-29\ 285 \text{ J}) \qquad (b)$$

To apply Eq. (*a*), we need to calculate the change in internal energy for the steam. This requires that the mass be determined from

$$m = \frac{V_1}{v_1}$$

$$v_1 = v_{f_1} + x_1 v_{fg_1}$$
$$= 0.01774 + (0.5)(4.434 - 0.017\ 74) = 2.226 \text{ ft}^3/\text{lbm}$$

$$m = \frac{0.5}{2.226} = 0.225 \text{ lbm} \qquad (0.1021 \text{ kg})$$

where the properties are obtained from the saturation-pressure table for steam at 100 psia. The final specific volume is

$$v_2 = \frac{V_2}{m} = \frac{2.0}{0.225} = 8.889 \text{ ft}^3/\text{lbm}$$

The final pressure is the same as the initial pressure, i.e., 100 psia. Consulting the saturation tables we find that v_g at 100 psia is only 4.434 ft³/lbm. Because v_2 is greater than v_g, the final state must be superheated. We therefore consult the superheat tables to find the final temperature at 100 psia and $v = 8.889$,

$$T_2 = 1038°F$$

and the final internal energy is

$$u_2 = 1387.7 \text{ Btu/lbm}$$

The initial internal energy is calculated from

$$u_1 = u_{f_1} + x_1 u_{fg_1}$$
$$= 298.3 + (0.5)(1105.8 - 298.3) = 702.05 \text{ Btu/lbm}$$

The internal energies and work are now inserted into Eq. (a) to determine the heat added:

$$Q - \frac{21\,600}{778} = m(u_2 - u_1) = (0.225)(1387.7 - 702.05)$$

$$Q = 182 \text{ Btu} \qquad (192 \text{ kJ})$$

EXAMPLE 3-4

A rigid enclosure, 50 cm on each side, contains a wet mixture of water vapor at 90°C and 20 percent quality. Heat is added until the pressure is raised to 500 kPa. Determine the final state and the quantity of heat added.

SOLUTION For a rigid container there is no change in volume and therefore no work done. The first-law statement then becomes

$$Q = \Delta U = m\,\Delta u = m(u_2 - u_1) \qquad (a)$$

Since both the total volume V and the mass remain constant, the specific volume must also remain constant.

$$v = \frac{V}{m} = v_1 = v_2 \qquad\qquad (b)$$

At the initial state

$$v = v_f + x v_{fg} \qquad\qquad (c)$$

Consulting the saturation tables at 90°C, we have

$$v_{f_1} = 1.0360 \times 10^{-3} \text{ m}^3/\text{kg} \qquad v_{g_1} = 2.361 \text{ m}^3/\text{kg}$$
$$u_{f_1} = 376.85 \text{ kJ/kg} \qquad u_{g_1} = 2494.5 \text{ kJ/kg}$$

Then, from Eq. (c),

$$v_1 = 0.001\,036 + (0.20)(2.361 - 0.001\,036) = 0.473 \text{ m}^3/\text{kg} \qquad (d)$$

We may now calculate the mass from Eq. (b) and the given total volume

$$m = \frac{V}{v} = \frac{(0.50)^3}{0.473} = 0.2643 \text{ kg} \qquad\qquad (e)$$

The final state is determined by consulting the steam tables at $v_2 = 473$ m³/kg and $p = 500$ kPa $= 5.0$ bar. At 500 kPa we find $v_g = 374.9$ m³/kg. We have a specific volume greater than this value so the final state must be superheated. Consulting the superheat table at $p = 500$ kPa, and interpolating, yields

$$T_2 = 244.3°C \qquad u_2 = 2721.4 \text{ kJ/kg}$$

The initial internal energy is calculated from

$$u_1 = u_{f_1} + x_1 u_{fg_1}$$
$$= 376.85 + (0.2)(2494.5 - 376.85)$$
$$= 800.38 \text{ kJ/kg}$$

We now insert the numerical values in Eq. (a) to calculate the heat transfer.

$$Q = (0.2643)(2721.4 - 800.38)$$
$$= 507.7 \text{ kJ} \qquad (481.3 \text{ Btu})$$

EXAMPLE 3-5

A rigid container is filled with steam at 700 kPa and 200°C. At what temperature will the steam start to condense when the container is cooled? To what temperature must the container be cooled to condense 50 percent of the steam mass?

SOLUTION   Because the container is rigid the specific volume must remain constant during the cooling process. The initial state is superheated so, from Table A-9M,

$$v_1 = 0.2999 \text{ m}^3/\text{kg} \qquad (4.804 \text{ ft}^3/\text{lbm}) \tag{a}$$

When the steam just starts to condense, this will be the specific volume of saturated vapor

$$v_1 = v_2 = v_{g_2} = 0.2999 \tag{b}$$

Interpolating in Table A-7M we find

$$T_2 = 161.1°C$$
$$p_2 = 636.9 \text{ kPa} \qquad (92.4 \text{ psia})$$

Cooling still further to some point 3 where $x_3 = 0.50$ means

$$v_3 = v_1 = 0.2999 = v_{f_3} + (0.50)v_{fg_3} \tag{c}$$

Most of the volume comes from the vapor so, very nearly,

$$v_3 \sim 0.5 v_{g_3} \qquad\qquad (d)$$

or

$$v_{g_3} \sim 0.5998$$

Again, consulting Table A-7M we find by interpolation that

$$T_3 = 134.3°C$$
$$p_3 = 309.3 \text{ kPa} \qquad (44.2 \text{ psia})$$

EXAMPLE 3-6

A small capsule has a volume of 0.3 in³ and contains liquid water at 1000 psia and 100°F. The capsule is placed inside a large container having a volume of 1 ft³. The large container is subsequently evacuated, and the capsule is broken by some appropriate mechanism so that the water is allowed to evaporate and fill the large container. Calculate the final quality of the water-vapor mixture, assuming that the large container is in heat communication with its surroundings and reaches a final equilibrium temperature of 100°F. Also determine the heat exchange with the surroundings.

SOLUTION   For this problem we take the boundary of the system as the inside surface of the large container. This surface is stationary, and thus there is no work done on the boundary of the system. From the first law,

$$Q + W = \Delta U \qquad\qquad (a)$$

Since $W = 0$, we have

$$Q = \Delta U = m(u_2 - u_1) \qquad\qquad (b)$$

We have $V_1 = 0.3$ in³ $= 1.735 \times 10^{-4}$ ft³. Using the compressed-liquid tables for water, we have

$$v_1 = 0.016\ 082 \text{ ft}^3/\text{lbm}$$
$$u_1 = 67.70 \text{ Btu}/\text{lbm}$$

The mass of water is

$$m = \frac{V_1}{v_1} = \frac{1.735 \times 10^{-4}}{1.608 \times 10^{-2}} = 0.0108 \text{ lbm}$$

The final specific volume is, thus,

$$v_2 = \frac{V_2}{m} = \frac{1}{0.0108} = 92.5 \text{ ft}^3/\text{lbm} \qquad (5.774 \text{ m}^3/\text{kg})$$

The final state is determined by entering the steam tables with the known values of T_2 (100°F) and v_2. At 100°F, v_g = 350.0 ft³/lbm, so that a mixture of vapor and liquid is obtained since $v_2 < v_g$. Thus,

$$v_2 = v_{f_2} + x_2 v_{fg_2}$$

Using the properties from the saturation-temperature table (Table A-7),

$$92.5 = 0.016\ 13 + x_2(350.0 - 0.016\ 13)$$

and

$$x_2 = 0.264$$

The final internal energy is now obtained from this same table.

$$\begin{aligned}
u_2 &= u_{f_2} + x_2 u_{fg_2} \\
&= 68.04 + (0.264)(1043.5 - 68.04) \\
&= 326 \text{ Btu/lbm}
\end{aligned}$$

The heat transfer is now calculated with Eq. (b).

$$\begin{aligned}
Q &= (0.0108)(326 - 67.7) \\
&= 2.79 \text{ Btu} \quad (2943 \text{ J})
\end{aligned}$$

The positive sign on the heat transfer indicates that 2.79 Btu of heat must be supplied to the large container to maintain the vapor at 100°F during the expansion-evaporation process.

3-5 PROPERTIES FOR SOLID-VAPOR SATURATION REGION

Properties for the solid-vapor saturation region for water may also be expressed in tabular form but in this region the vapor is at such a low pressure that it behaves very nearly as an ideal gas, and the properties may be expressed in fairly simple analytical form. Accordingly, the appropriate equations have been assembled in Table 3-1 in a convenient format. These equations apply for the saturation region from −40 to 0°C (−40 to 32°F) with an accuracy of approximately 1 percent. In Table 3-1 we have also given equations for a property called the entropy s. This property will not be defined until Chap. 6, but we list it here for completeness and later reference.

Although these equations pertain specifically to the solid-vapor saturation region they are applicable to other circumstances. Because the solid is only modestly compressible the internal energies and enthalpies given by

TABLE 3-1
Solid-vapor saturation properties for water

Property	Equation	Temperature range	Equation number
Saturation pressure	$p_g = 35.18 \exp\left(18.42 - \dfrac{6144}{T}\right)$ kPa; T in K	-40 to $0°C$	3-5a
	$p_g = 5.103 \exp\left(18.42 - \dfrac{11\,059}{T}\right)$ psia; T in °R	-40 to $32°F$	3-5b
Enthalpy and internal energy of saturated solid	$h_i = u_i = -334.6 + 1.96T$ kJ/kg; T in °C	-40 to $0°C$	3-6a
	$h_i = u_i = -158.9 + 0.467T$ Btu/lbm; T in °F	-40 to $32°F$	3-6b
Internal energy of saturated vapor	$u_g = 2374.9 + 1.403T$ kJ/kg; T in °C	-40 to $0°C$	3-7a
	$u_g = 1010.3 + 0.335T$ Btu/lbm; T in °F	-40 to $32°F$	3-7b
Enthalpy of saturated vapor	$h_g = 2501 + 1.863T$ kJ/kg; T in °C	-40 to $0°C$	3-8a
	$h_g = 1061.0 + 0.445T$ Btu/lbm; T in °F	-40 to $32°F$	3-8b
Specific volume of saturated solid	$v_i = 1.091 \times 10^{-3} + 2.5 \times 10^{-6}T$ m³/kg; T in °C	-40 to $0°C$	3-9a
	$v_i = 0.017\,43 + 1.39 \times 10^{-6}T$ ft³/lbm; T in °F	-40 to $32°F$	3-9b
Entropy of saturated vapor	$s_g = 1.863 \ln T + \dfrac{2833.6}{T} - 11.66$ kJ/kg·K; T in K	-40 to $0°C$	3-10a
	$s_g = 0.445 \ln T + \dfrac{1218.2}{T} - 3.046$ Btu/lbm·°R; T in °R	-40 to $32°F$	3-10b
Entropy of saturated solid	$s_i = -1.223 + 7.747 \times 10^{-3}T$ kJ/kg·K; T in °C	-40 to $0°C$	3-11a
	$s_i = -0.325 + 0.001\,028T$ Btu/lbm·°R; T in °F	-40 to $32°F$	3-11b

Eq. (3-6) can be used for pressures up to several atmospheres. Likewise, the specific volume of the solid is not too pressure-sensitive and Eq. (3-9) may be also used for higher pressures.

EXAMPLE 3-7

Using the equations given for the solid-vapor saturation region and the steam tables, estimate the energy required to melt 1 kg of ice at 0°C in a constant-pressure process.

SOLUTION   At 0°C = 32°F water can exist in all three phases: solid, liquid, and vapor. From the steam tables at 32°F

$$h_f = 0.00 \text{ Btu/lbm}$$

and from Eq. (3-6)

$$h_i = -158.9 + (0.467)(32) = -143.96 \text{ Btu/lbm} \tag{a}$$

For a constant-pressure process we have

$$
\begin{aligned}
dQ_p &= dh_p \\
Q_p &= \Delta h_p = h_f - h_i = 0 - (-143.96) \\
&= 143.96 \text{ Btu/lbm} \\
&= 317.09 \text{ Btu/kg} \\
&= 334.5 \text{ kJ/kg}
\end{aligned}
\tag{b}
$$

3-6   SPECIFIC HEATS

In Chap. 2 the constant-volume and constant-pressure specific heats were defined as

$$c_v = \left(\frac{\partial u}{\partial T}\right)_v \tag{3-12}$$

$$c_p = \left(\frac{\partial h}{\partial T}\right)_p \tag{3-13}$$

Thus, once tabular information on enthalpy and internal energy is available, we can evaluate the specific heats rather easily by performing the indicated differentiations. The following example illustrates such a calculation.

EXAMPLE 3-8

Using data from the superheat region of the steam tables, calculate the specific heat at constant pressure for water vapor at 400 psia and 600°F.

SOLUTION We shall employ a numerical approximation to the partial derivative relation for c_p. From the superheat tables at 400 psia

$$h = 1245.2 \text{ Btu/lbm} \quad \text{at } 500°F$$
$$h = 1362.5 \text{ Btu/lbm} \quad \text{at } 700°F$$

Then we write, approximately,

$$c_p = \left(\frac{\partial h}{\partial T}\right)_p \approx \left(\frac{\Delta h}{\Delta T}\right)_p = \frac{h_{700} - h_{500}}{700 - 500}$$

$$= \frac{1362.5 - 1245.2}{700 - 500} = 0.587 \text{ Btu/lbm·°F}$$

In this calculation we assume the specific heat value to apply midway between 500 and 700°F, or at the desired value of 600°F.

3-7 COEFFICIENT OF EXPANSION

Two additional properties of simple compressible substances are important. The *volume coefficient of expansion* β is defined as the rate of change of volume with temperature at constant pressure and per unit volume.

$$\beta = \frac{1}{v}\left(\frac{\partial v}{\partial T}\right)_p \tag{3-14}$$

The *isothermal compressibility* κ is defined as the rate of change of volume with pressure at constant temperature and per unit volume.

$$\kappa = -\frac{1}{v}\left(\frac{\partial v}{\partial p}\right)_T \tag{3-15}$$

These two properties are important because they may be determined fairly easily by experiment. For an ideal gas, very simple relations can be derived as follows. Since

$$v = \frac{RT}{p}$$

we have

$$\beta = \frac{1}{v}\left(\frac{\partial v}{\partial T}\right)_p = \frac{1}{v}\frac{R}{p} = \frac{1}{T} \tag{3-16}$$

$$\kappa = -\frac{1}{v}\left(\frac{\partial v}{\partial p}\right)_T = -\frac{1}{v}\left(-\frac{RT}{p^2}\right) = \frac{1}{p} \tag{3-17}$$

The isothermal compressibility is quite useful when employed in the calculation of work involved in the compression of liquids and solids. The work per unit mass is expressed as

$$w = -\int p \, dv \tag{3-18}$$

For an isothermal process the volume change is expressed in terms of κ by

$$dv_T = -\kappa v \, dp_T$$

Substituting this relation in Eq. (3-18) gives

$$w_T = \int_{p_1}^{p_2} p\kappa v \, dp_T \tag{3-19}$$

For many liquids and solids κ is very nearly constant over a fairly wide range of pressures. If the specific volume is considered constant for purposes of evaluating the integral in Eq. (3-19), we have

$$w_T = \frac{\kappa v}{2} (p_2^2 - p_1^2) \tag{3-20}$$

Equation (3-20) may thus be used to obtain an estimate of the work involved in compressing liquids and solids to elevated pressures while maintaining a constant-temperature condition.

Values of some typical coefficients of expansion appear in Table 3-2.

EXAMPLE 3-9

75 kg of water are compressed in a closed container from 140 kPa to 7.5 MPa while the temperature is maintained constant at 40°C. Calculate the isothermal work in this process.

SOLUTION   Equation (3-20) applies for this problem. To estimate the value of κ we make use of the compressed-liquid table and the approximation

$$\kappa = -\frac{1}{v} \left(\frac{\partial v}{\partial p} \right)_T \approx -\frac{1}{v} \left(\frac{\Delta v}{\Delta p} \right)_T \tag{a}$$

From the saturation-temperature table (Table A-7M, 40°C)

$$v_f = 1.0078 \text{ cm}^3/\text{g} = 1.0078 \times 10^{-3} \text{ m}^3/\text{kg}$$

From the compressed-liquid table we can evaluate the specific volume at 40°C and 7.5 MPa to obtain

$$v = 1.0045 \text{ cm}^3/\text{g}$$

TABLE 3-2
Volume coefficient of expansion for various substances at 25°C

Substance	β	
	°F^{-1} × 10^6	°C × 10^6
Solids		
Aluminum	39.9	71.8
Copper	27.6	49.7
Gold	23.7	42.7
Iron	19.5	35.1
Magnesium	42.0	75.6
Nickel	21.6	38.9
Silver	33.0	59.4
Steel, AISI 304 sheet	29.7	53.5
Tungsten	7.5	13.5
Pyrex glass	5.4	9.7
Ice	84.9	152.8
Wood, oak	8.1	14.6
Concrete	24.0	43.2
Liquids		
Acetic acid	600	1080
Ethyl alcohol	610	1098
Benzene	770	1386
Glycerin	280	504
Hydrochloric acid	270	486
Petroleum crude, approximate	450	810
Water	115	207

Then applying the approximation of Eq. (a)

$$\kappa \approx -\frac{1}{1.0078}\left(\frac{1.0045 - 1.0078}{7500 \text{ kPa} - 140 \text{ kPa}}\right)_{40°C}\left(\frac{1}{1000}\right)$$
$$= 4.45 \times 10^{-10} \text{ m}^2/\text{N} \qquad (\text{Pa}^{-1})$$

We now calculate the isothermal work per kilogram from Eq. (3-20)

$$w_T = \frac{\kappa v}{2}(p_2^2 - p_1^2)$$
$$= \frac{(4.45 \times 10^{-10} \text{ m}^2/\text{N})(1.0078 \times 10^{-3} \text{ m}^3/\text{kg})}{2}$$
$$\times [(7.5 \times 10^6)^2 - (140 \times 10^3)^2] \text{ N}^2/\text{m}^4$$
$$= 12.61 \text{ J/kg}$$

For the total mass of 75 kg the work is

$$W_T = mw_T = (75)(12.61) = 945.7 \text{ J}$$

EXAMPLE 3-10

At 300 K copper has the properties $\bar{v} = 70.062$ cm³/g mol and $\kappa = 0.776 \times 10^{-12}$ cm²/dyn with $M = 63.54$. Calculate the work to compress a block 2.0 cm on a side from 1 to 1000 atm pressure while maintaining the temperature constant at 300 K.

SOLUTION Again, we employ Eq. (3-20). On a molal basis

$$W_T = \frac{\kappa \bar{v}}{2} (p_2^2 - p_1^2)$$

with

$$\bar{v} = 7.062 \text{ cm}^3/\text{g mol} = 7.062 \times 10^{-3} \text{ m}^3/\text{kg mol}$$
$$\kappa = 0.776 \times 10^{-2} \text{ cm}^2/\text{dyn} = 0.776 \times 10^{-11} \text{ m}^2/\text{N}$$
$$W_T = \frac{(0.776 \times 10^{-11} \text{ m}^2/\text{N})(7.062 \times 10^{-3} \text{ m}^3/\text{kg mol})}{2}$$
$$\times [(1000)^2 - (1)^2](1.0132 \times 10^5 \text{ N/m}^2)^2$$
$$= 281.3 \text{ J/kg mol}$$

The number of moles is

$$\eta = \frac{V}{v} = \frac{(2.0)^3 \times 10^{-6}}{7.062 \times 10^{-3}} = 1.133 \times 10^{-3} \text{ kg mol}$$

so the total work is

$$W_T = (1.133 \times 10^{-3} \text{ kg mole})(281.3 \text{ J/kg mol})$$
$$= 0.319 \text{ J}$$

REVIEW QUESTIONS

1 What is a *pure substance?*

2 Distinguish between *intensive* and *extensive* properties.

3 Distinguish between the terms *pure substance* and *simple substance*.

4 Describe the physical phenomenon of vaporization at constant pressure.

5 How does the phenomenon of fusion help an ice skater perform his maneuvers?

6 How do you account for the fact that liquid CO_2 is not observed when dry ice (solid CO_2) is placed on a table at room temperature and pressure?

7 Why do you suppose foods cook faster in a pressure cooker (a device that allows the pressure to build up when water is boiled inside)?

8   What is meant by the term *saturation region?*

9   Distinguish between the terms *moisture* and *quality.* How do these terms relate to the terms *saturated liquid* and *saturated vapor?*

10   What is the quality of a superheated vapor?

11   Why are the terms *subcooled liquid* and *compressed liquid* synonymous?

12   What is an ideal gas?

13   Does an ideal gas have constant specific heats?

14   Is a wet mixture of water vapor an ideal gas? Why?

15   What is meant by the term *superheated vapor?*

16   What happens at the critical point?

17   What is the *state principle?*

18   What is a simple compressible substance?

19   How many properties are necessary to describe the state of a system?

20   What would be the specific heat of a wet mixture of steam?

21   Upon what variable are the properties of liquids primarily dependent?

PROBLEMS   (English units)

3-1   A container having a volume of 3 ft³ is filled with steam at 30 psia and 90 percent quality. Calculate the quantity of heat which must be removed from the container in order to reduce the pressure to 20 psia.

3-2   10 lbm of water vapor are contained at 25 psia and 90 percent quality in a suitable enclosure. Calculate the heat which must be added in order to just produce a saturated vapor. What will the pressure be at the end of the heating process?

3-3   Calculate the percent error which would result from assuming that water at 100°F and 1500 psia has a density of 62.4 lbm/ft³.

3-4   Calculate the specific volume for Freon 12 at 10 psia and 50 percent quality.

3-5   Suppose heat is added to 0.5 lbm of the Freon in Prob. 3-4 during a constant-pressure process. The heating proceeds until a temperature of 140°F is attained. Calculate the heat required and the work produced, assuming that the Freon is contained by a suitable cylinder-piston arrangement.

3-6   Calculate the percent increase in volume when a block of copper is heated from 100 to 1000°F.

3-7   In Table 3-2, β for water is given as 115×10^{-6} °F⁻¹. Verify this value by direct calculation from data in the steam tables for water at approximately 100°F.

3-8 With the use of the superheat steam tables, calculate the isothermal compressibility for water vapor under the following conditions:
  (a) 1 psia, 300°F
  (b) 500 psia, 500°F
Compare these values with that obtained by assuming that the steam obeys the ideal-gas equation of state. Comment on the comparison.

3-9 Repeat Prob. 3-8 for the volume coefficient of expansion.

3-10 Calculate the work required to compress 1 g of water from 50 psia, 100°F, to 1500 psia in an isothermal process.

3-11 A man who normally consumes energy at the rate of 85 kcal/h increases his consumption rate to 150 kcal/h during more vigorous activity. In a hot surrounding environment the body will perspire to cool the skin surface by evaporation. Assuming that the perspiration is evaporated at constant pressure and a constant skin temperature of 80°F, calculate the amount which must be evaporated per hour to make up for the difference in energy consumption rates stated above. (Ans.: 112 g)

3-12 A frost layer is formed in the freezing compartment of a refrigerator over a period of time so that it has a thickness of 1 cm. Because of its porous nature its overall density is only 30 percent of that for solid ice. The layer covers an area of 0.5 m². How much hot water at 140°F would have to be added to the frost to just produce water at 32°F? Assume that the melting process occurs in an insulated container.

3-13 1 lbm of steam is heated at a constant pressure of 14.696 psia from a temperature of 220 to 350°F. Calculate the heat required.

3-14 Calculate the energy required to vaporize 1 lbm of water at normal atmospheric pressure (14.696 psia). Compare this quantity with the amount of energy to vaporize ammonia or Freon 12 at this same pressure.

3-15 Calculate the mass of H_2O contained in a 2-ft³ vessel under the following conditions:
  (a) 14.696 psia, 100°F
  (b) 100 psia, 100°F
  (c) 1 psia, 500°F
  (d) 20 psia, 30 percent quality
  (e) 1000 psia, 1000°F
  (f) At the critical state

3-16 How much energy is required to vaporize 1 lbm of Freon 12 at room temperature of 70°F in a constant-pressure process? How does this compare with that required to vaporize water or ammonia under the same conditions?

3-17 Saturated water vapor at 1 atm is contained in a piston-cylinder arrangement, and the piston is restrained by a spring such that the pressure in the cylinder is proportional to the cylinder volume. How much heat must be added to the steam to raise the pressure to 20 psia?

3-18 Liquid water is contained in a cylinder at 1000 psia and 100°F. The cylinder has a diameter of 3 cm and contains 15 g of water. A piston has a constant restraining force imposed on it to keep the pressure constant at 1000 psia. How much heat must be added to the cylinder to move the piston through a distance of 3.5 cm?

3-19   Verify that it takes approximately 80 cal to melt 1 g of ice at 32°F.

3-20   Water vapor at 14.7 psia and 300°F is compressed isothermally until half the vapor has condensed. How much work must be performed on the steam in this compression process per lbm?

3-21   Water vapor at 14.7 psia and 300°F is contained in a rigid vessel. At what temperature will the vapor start to condense when the container is cooled slowly? How much heat will have been removed per lbm in this cooling process?

3-22   On a hot summer day in the south the solar irradiation on the earth is about 300 Btu/hr·ft², and about 75 percent of this energy will be absorbed by green grass. Assuming that all the energy absorbed goes to evaporate water at constant pressure and an average temperature of 100°F, calculate the amount of watering required each week for a 100 by 160 ft plot of land to just replenish the water lost by evaporation. For this calculation assume that the sun's radiation occurs over a 9-h period. Express the answer in gallons per week.

PROBLEMS   (Metric units)

3-1M   A container having a volume of 85 liters is filled with steam at 200 kPa and 90 percent quality. Calculate the quantity of heat which must be removed from the container in order to reduce the pressure to 150 kPa.

3-2M   5 kg of water vapor are contained at 150 kPa and 90 percent quality in a suitable enclosure. Calculate the heat which must be added in order to just produce a saturated vapor. What will the pressure be at the end of the heating process?

3-3M   Calculate the percent error which would result from assuming that water at 40°C and 10 MPa has a density of 999.6 kg/m³.

3-4M   Suppose heat is added to 0.5 kg of the Freon in Prob. 3-4 during a constant-pressure process. The heating proceeds until a temperature of 60°C is attained. Calculate the heat required and the work produced, assuming that the Freon is contained by a suitable piston-cylinder arrangement.

3-5M   1 kg of steam is heated at a constant pressure of 100.0 kPa from a temperature of 100 to 200°C. Calculate the heat required.

3-6M   Calculate the energy required to vaporize 1 kg of water at 100.0 kPa. Compare this quantity with the amount of energy to vaporize ammonia or Freon 12 at this same pressure.

3-7M   Calculate the mass of H_2O contained in a 60-liter vessel under the following conditions:
(a) 100 kPa, 40°C   (b) 700 kPa, 40°C
(c) 6 kPa, 240°C   (d) 150 kPa, 30 percent quality
(e) 8000 kPa, 520°C   (f) At the critical state

3-8M Water vapor is heated at a constant pressure of 7 kPa from 65 to 550°C. Assume that the vapor obeys the ideal-gas equation of state in this region and calculate the heat transfer by using the data of Table 2-3. Check this calculation by using the tabulated properties in the steam tables. Comment on the agreement or disagreement of these two calculations.

3-9M Calculate the percent increase in volume when a block of copper is heated from 40 to 540°C.

3-10M With the use of the superheat steam tables, calculate the isothermal compressibility for water vapor under the following conditions:
(a) 7 kPa, 160°C   (b) 3 MPa, 240°C

3-11M Calculate the work required to compress 1 kg of water from 350 kPa, 40°C, to 10 MPa in an isothermal process.

3-12M A man who normally consumes energy at the rate of 350 kJ/h increases his consumption rate to 630 kJ/h during more vigorous activity. In a hot surrounding environment the body will perspire to cool the skin surface by evaporation. Assuming that the perspiration is evaporated at constant pressure and a constant skin temperature of 25°C, calculate the amount which must be evaporated per hour to make up for the difference in energy consumption rates stated above. (*Ans.*: 112 g)

3-13M Water is contained in a piston-cylinder arrangement at 200 kPa and 40°C. The cylinder has a diameter of 15 cm and a height of 10 cm. How much work would be required to compress the water to a pressure of 15 MPa in an isothermal process?

3-14M How much energy is required to vaporize 1 kg of Freon 12 at room temperature of 20°C in a constant-pressure process? How does this compare with that required to vaporize water or ammonia under the same conditions?

3-15M Saturated water vapor at 100 kPa is contained in a piston-cylinder arrangement, and the piston is restrained by a spring such that the pressure in the cylinder is proportional to the cylinder volume. How much heat must be added to the steam to raise the pressure to 150 kPa?

3-16M Liquid water is contained in a cylinder at 7.5 MPa and 40°C. The cylinder has a diameter of 3 cm and contains 15 g of water. A piston has a constant restraining force imposed on it to keep the pressure constant at 7.5 MPa. How much heat must be added to the cylinder to move the piston through a distance of 3.5 cm?

3-17M Verify that it takes approximately 335 J to melt 1 g of ice at 0°C.

3-18M Water vapor at 100 kPa and 150°C is compressed isothermally until half the vapor has condensed. How much work must be performed on the steam in this compression process per kilogram?

3-19M Water vapor at 100 kPa and 150°C is contained in a rigid vessel. At what temperature will the vapor start to condense when the container is cooled slowly? How much heat will have to be removed per kilogram in this cooling process?

3-20M   Using SI units plot a curve of saturation pressure versus v_g for water for temperatures between 0 and 250°C.

3-21M   Using the appropriate tables for water and water vapor, fill in the blanks in the following table.

Temperature, °C	280		480		
Pressure, Pa		700 k		8.0 M	350 k
Mass, kg	1.0	2.0	1.3	3.5	3
Specific volume, m³/kg		0.3852			
Volume, m³					
Quality, %					60
Moisture, %				20	
Enthalpy, kJ/kg	2779.6		3005.4		
Internal energy, kJ/kg					

3-22M   Calculate the value of c_p for steam at $p = 500$ kPa and 240°C using the steam tables.

3-23M   At a certain state, steam has a temperature of 240°F and an entropy of 1.419 Btu/lbm·°F. Determine the enthalpy and specific volume at this point.

3-24M   A rigid tank having a total volume of 0.1 m³ is perfectly insulated and divided into two equal chambers separated by an adiabatic partition. One chamber contains steam at 100 kPa and 200°C and the other chamber contains steam at 1.5 MPa and 320°C. The partition is removed and the system allowed to reach equilibrium. Determine the pressure and temperature at the new equilibrium state.

3-25M   On a hot summer day in the South, the solar irradiation on the Earth is about 1 kW/m², and about 75 percent of this energy will be absorbed by green grass. Assuming that all the energy absorbed goes to evaporate water at constant pressure and an average temperature of 40°C, calculate the amount of watering required each week for a 30 by 50 m plot of land to just replenish the water lost by evaporation. For this calculation assume that the sun's radiation occurs over a 9-h period. Express the answer in liters per week.

3-26M   Calculate the enthalpy of steam at 10.0 MPa and 600°C using tabulated values of u and v and compare with the tabulated value for h.

3-27M   Water is contained in a piston-cylinder arrangement at 2 atm and 100°F. The cylinder has a diameter of 15 cm and a height of 10 cm. How much work would be required to compress the water to a pressure of 150 atm in an isothermal process?

3-28M   A thermometer is constructed of Pyrex glass and ethyl alcohol. The "bulb" portion has a volume of 0.4 cm³, and the column length is 20 cm. Assuming that only the bulb is used as the temperature sensor, what must be the cross-sectional area of the column in order for the thermometer to have a range of 10 to 40°C over the 20-cm length?

REFERENCES

1   Kline, S. J., and F. O. Koenig: The State Principle, *J. Appl. Mech.*, vol. 24, p. 29, 1957.

2   Hatsopoulos, G. N., and J. H. Keenan: "Principles of General Thermodynamics," John Wiley & Sons, Inc., New York, 1965.

3   Keenan, J. H., and F. G. Keyes: "Thermodynamic Properties of Steam," John Wiley & Sons, Inc., New York, 1936.

4   Van Wylen, G. J.: "Thermodynamics," John Wiley & Sons, Inc., New York, 1959.

chapter four
Energy analysis in open systems

4-1   INTRODUCTION

The two previous chapters have expressed the first law of thermodynamics in analytical form and presented methods for calculating thermodynamic properties of simple thermodynamic substances. In this chapter we wish to apply this information to the analysis of a wider class of problems: in particular, open systems where mass may cross the boundaries. In this way, we will be able to establish a rather full exploitation of the methodology of energy analysis.

It is to our advantage to adopt a plan of attack for performing an energy analysis of thermodynamic systems. The general procedure is basically a method of answering the following questions:

1   What is the thermodynamic system to be analyzed—how is it defined?
2   What is the thermodynamic substance to be considered in the analysis—are the necessary properties of this substance readily available?
3   What *type* of system shall be used for analysis—open or closed?
4   How shall we describe the thermodynamic process under consideration?
5   Are we considering a steady-state situation or is the system changing with time?

Once a clear answer is obtained for each of these questions, the solution of the problem is usually obtained through mathematical analysis; however, some experience is required to be efficient in performing this analysis. The questions are not necessarily posed in the order presented, and the answer to one question may be dependent on the answer to some of the other questions.

As has been our experience in the previous chapters, we shall find that the plan of attack described in the above questions can be developed into a more specific algorithm.

1   Sketch the system, and label all inlet and exit mass and energy quantities.
2   Write out the information which is available for analysis.
3   Write the appropriate energy balance in accordance with the labeling procedure followed in the system schematic.
4   Write the process description relations which apply.
5   Combine the above information to obtain the desired result.

4-2   ENERGY BALANCES AND SIGN CONVENTIONS

In Chap. 2 we developed general energy relationships for closed systems, i.e., systems which do not admit the transfer of mass across their boundaries. We adopted the sign convention that energy added to the system (in the form of work or heat) is positive and energy given up by the system to other things is negative. The main consideration is always the fact that *energy is conserved* or

Energy which goes into system = energy which leaves the system
+ energy which accumulates
in the system

Energy analysis, then, is basically an accounting procedure, and the equations of Chap. 2 express the analytical relations necessary to balance the books.

We could discuss energy balances at great length, but the important points of analysis techniques are best illustrated by the examples given in the following sections.

4-3 THE CLOSED SYSTEM

The general discussion of the first principle of thermodynamics as presented in Chap. 2 pertained to closed thermodynamic systems. These systems do not admit the transfer of mass across their boundaries. The only essential point is that all types of work and internal energy be considered in the energy balance

$$d'Q + d'W = dE \tag{4-1}$$

where the energy quantities have been written as differential values in order to apply the equation to any type of process. The energy balance for the closed system is illustrated in Fig. 4-1. Closed and open systems are contrasted in Fig. 4-2.

4-4 THE OPEN SYSTEM

We have seen that the closed system does not admit the transfer of mass across its boundaries; only energy may flow in and out of the system as it

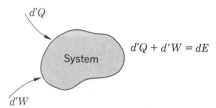

FIG. 4-1 *Energy balance for closed system.*

FIG. 4-2 (a) Closed container is a closed system with no mass transfer across the boundary. (b) Water flow through a heater is treated as an open system.

changes from one state to another. We now consider the open system where mass may enter and leave some volume in space. This may, at first, seem inconsistent with the meaning that has been attached to a thermodynamic system so far, since we were considering a quantity of matter broken away from its surroundings. This apparent anomaly will be resolved, however, as we relate the analysis of open systems directly to that of closed systems. In any event, the reader must realize that we are discussing analysis techniques which may be used to apply the conservation of energy principle to a variety of situations.

First, let us show how to calculate the mass flow through a channel as illustrated in Fig. 4-3. The rate at which the volume of fluid is swept through the channel is $A \, ds/d\tau$ where A is the flow cross-sectional area and ds is the displacement. The term $A \, ds$ is the volume displacement. The volume flow rate is designated by $\dot{V}$. The volume per unit mass is v, and so the rate at which the mass flows in the channel $\dot{m}$ is

$$\dot{m} = \frac{\dot{V}}{v} = \frac{1}{v}\frac{A \, ds}{d\tau}$$

But, $ds/d\tau = \bar{V}$ = channel flow velocity, so that

$$\dot{m} = \frac{A\bar{V}}{v} \tag{4-2}$$

$$F = pA$$
$$\Delta s = \frac{V}{A}$$
$$W = F \, \Delta s = pV$$

FIG. 4-3 Mass flow and flow work in a channel.

We recall that the fluid *density* is defined as the inverse of the specific volume, $\rho = 1/v$, so that

$$\dot{m} = \rho A \bar{V} \tag{4-3}$$

The $\bar{V}$ symbol is used here to distinguish the velocity from the symbol for volume.

4-5 THE CONTROL VOLUME

To study open systems, we introduce the concept of *a control volume*, as shown in Fig. 4-4. This volume is a region in space to be observed in respect to the matter and energy which cross its boundaries. Consider first the conservation of mass principle, which may be written as:

Mass flow into control volume = mass flow out of control volume
+ increase in mass inside
the control volume   (4-4)

or that which goes in either stays in or goes out. In accordance with Fig. 4-4*a*, we may write the mass conservation principle as

$$\dot{m}_i = \left(\frac{dm}{d\tau}\right)_\sigma + \dot{m}_e \tag{4-5}$$

where $\dot{m}_i$ is the mass flow per unit time into the control volume, $\dot{m}_e$ is the mass flow per unit time out of the control volume, and the subscript σ indicates the mass accumulation per unit time within the control volume.

   If multiple inlet and exit streams are involved, then we must perform a summation over all such flows to determine the mass balance:

$$\sum_i \dot{m}_i = \left(\frac{dm}{d\tau}\right)_\sigma + \sum_e \dot{m}_e \tag{4-6}$$

   As a preliminary to an energy analysis of a control volume, let us qualitatively consider what happens to a fixed quantity of mass as it moves through the control volume, i.e., the behavior of a *closed* thermodynamic system as it undergoes a process which causes it to move through the control volume. The closed thermodynamic system may undergo effects of pressure from the surroundings, the transfer of heat across its boundaries, and the action of various work-producing forces. The internal energy of the closed system could change as a result of its physical movement from one position to another and, perhaps, from a change in its velocity. Regardless of the effects observed, we could certainly analyze them with the conservation of en-

(a)

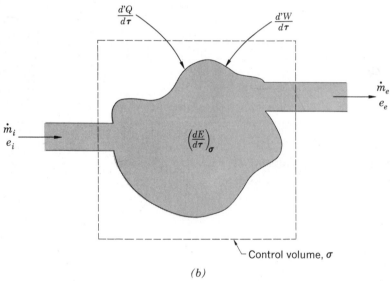

(b)

FIG. 4-4 *Open system balances. (a) Mass balance; (b) energy balance.*

ergy principle. Furthermore, the total mass flow into or out of the control volume may be thought of as a group of mass elements dm, in a sense, a group of small, closed thermodynamic systems. We may therefore consider that the mass flow in and out of the control volume transports internal energy across the boundaries of the control volume. Thus the energy conservation principle for this type system is:

Transport of internal energy into the control volume + heat added to control volume + work done on all elements as they pass through the control volume $=$ increase of internal energy in control volume + transport of internal energy out of the control volume (4-7)

In analytical form according to Fig. 4-4b,

$$\dot{E}_i + \frac{d'Q}{d\tau} + \frac{d'W}{d\tau} = \left(\frac{dE}{d\tau}\right)_\sigma + \dot{E}_e \qquad (4-8)$$

where $\dot{E}_i$ and $\dot{E}_e$ represent the energy transport per unit time at entrance and outlet, respectively, and the subscript σ again designates the change within the boundaries of the control volume. The internal-energy transports may be written as:

$$\dot{E}_i = \dot{m}_i e_i \qquad (4-9)$$
$$\dot{E}_e = \dot{m}_e e_e \qquad (4-10)$$

Equation (4-8) can be used to analyze open systems, but the work term is usually expressed in a more useful form for analysis purposes. As an intermediate step in developing such an expression, consider the very simple control volume, as shown in Fig. 4-3. Notice that, in order for the mass to move through the control volume, there must be some forcing function to push it through. This force is provided by the pressure of the system. We imagine a mass having an area of A and a length of Δs. To push this quantity of mass either in or out of the control volume, we must exert a force of pA through the distance of Δs. Regardless of the quantity of mass, Δs will be given by

$$\Delta s = \frac{V}{A}$$

so that the work to push the mass in or out is

$$W = \int F \, ds = F \, \Delta s = pA \frac{V}{A} = pV \qquad (4-11)$$

The net work done on the system as it moves from station 1 to 2 in Fig. 4-3, exclusive of external work, is

$$W_{net} = p_1 V_1 - p_2 V_2$$

where the p_1V_1 term is the work done on the fluid to force it into the control volume and the p_2V_2 is the work to force the fluid out of the control volume. The difference in the two terms is the net work added.

The work term designated by pV is called *flow work*, and it is customary to consider it separately from the work delivered to things outside the control volume. The energy equation may then be written in the form

$$\dot{m}_i(e_i + p_i v_i) + \frac{d'Q}{d\tau} + \frac{d'W_{\text{ext}}}{d\tau} = \left(\frac{dE}{d\tau}\right)_\sigma + \dot{m}_e(e_e + p_e v_e) \qquad (4\text{-}12)$$

where W_{ext} is the *work delivered to the control volume by external forces*. This quantity is sometimes called *shaft work* because in a practical sense it is frequently delivered by a turbine-shaft arrangement. Equation (4-12) represents the general energy balance for an open system. If the open system is operating in *steady-state* conditions, then there are no changes inside the control volume with time; thus, $(dE/d\tau)_\sigma = 0$ and $(dm/d\tau)_\sigma = 0$. A *steady-flow* situation is that circumstance in which the flow rates $\dot{m}_i$ and $\dot{m}_e$ do not change with time. For the special case of steady flow *and* steady state there is no accumulation of mass or energy inside the control volume and $\dot{m}_i = \dot{m}_e$ so that Eq. (4-12) reduces to

$$\frac{d'Q}{d\tau} + \frac{d'W_{\text{ext}}}{d\tau} = \dot{m}[(e_e + p_e v_e) - (e_i + p_i v_i)] \qquad (4\text{-}13)$$

We note once again that the internal energy e is composed of the internal thermal energy u and the gravitational potential energy, kinetic energy, etc. For convenience, it is customary to introduce the property enthalpy h, defined in Sec. 2-8 as

$$h = u + pv \qquad (4\text{-}14)$$

Enthalpy is a property because it is composed of other thermodynamic properties. Notice that the enthalpy has a physical meaning and significance when applied to open systems, but that this meaning is no longer valid when applied to a closed system because the pv term does not represent flow work in a closed system.

With the introduction of the enthalpy, our general energy equation for an open system may be written as

$$\dot{m}_i(h_i + \text{K.E.}_i + \text{Ch.E.}_i + \cdots) + \frac{d'Q}{d\tau} + \frac{d'W_{\text{ext}}}{d\tau}$$

$$= \left(\frac{dE}{d\tau}\right)_\sigma + \dot{m}_e(h_e + \text{K.E.}_e + \text{Ch.E.}_e + \cdots) \qquad (4\text{-}15)$$

If more than one entrance and exit streams are involved, a simple summation is performed over all streams supplying or removing energy from the control volume, and we would have

$$\sum_i \dot{m}_i(h_i + \text{K.E.}_i + \text{Ch.E.}_i + \cdots) + \frac{d'Q}{d\tau} + \frac{d'W_{\text{ext}}}{d\tau}$$

$$= \left(\frac{dE}{d\tau}\right)_\sigma + \sum_e \dot{m}_e(h_e + \text{K.E.}_e + \text{Ch.E.}_e + \cdots) \qquad (4\text{-}16)$$

The above equations for the general open system will appear rather cumbersome and foreboding to the reader becoming involved with thermodynamics for the first time. However, in most practical problems these equations will simplify considerably, and we will find that the energy analysis technique will be quite easy. For example, in our discussions of ideal gases and water-steam systems there was no need to be concerned with chemical energy, and so it will be possible to leave that term out in many cases.

At first glance, kinetic energy may seem to be a very important quantity in *flow* (open) systems. The fluid is moving so its kinetic energy must be important. For a large number of practical problems this is *not* the case because the kinetic energies are very small. To show that this is true let us calculate the velocity necessary to produce just 1.0 Btu/lbm or 1.0 kJ/kg of kinetic energy:

$$1.0 \text{ Btu/lbm} = 778 \text{ ft·lbf/lbm} = \frac{1}{2g_c}\bar{V}^2$$

With $g_c = 32.2$ lbm·ft/lbf·s^2 we obtain

$$\bar{V} = 224 \text{ ft/s}$$

In SI units, for K.E. $= 1.0$ kJ/kg,

$$1000 \text{ J/kg} = \tfrac{1}{2}\bar{V}^2$$

and

$$\bar{V} = 44.7 \text{ m/s}$$

These are much higher velocities than will be encountered in many practical problems. A notable exception is the flow in a jet engine or gas turbine; but, for flow of water, air, or steam in pipes, Freon in an air conditioner coil, and many other examples, the flow velocity is low enough that we may neglect kinetic energies in comparison to the respective enthalpies or internal energies of the flow streams.

4-6 SUMMARY OF OPEN-SYSTEM SPECIAL CASES

To aid the reader in applying open-system analysis we now give the general equations in reduced form for a number of special cases in Table 4-1. The purpose of this table is not to invoke a cookbook philosophy in the reader, but rather to show the simplicity which the relations can assume for a number of practical problems.

In summary, we may remark that the analysis of open systems deals with an energy balance, similar in many respects to the analysis for closed systems. The primary difference in the analysis for closed or open systems is that the work term is split into two parts in the open-system analysis; one part represents the work done external to the control volume, and the other part represents the work necessary to force the mass in and out of the control volume. The property of enthalpy is introduced as a matter of convenience for summing the internal thermal energy and flow work at a particular position. The following examples illustrate the use of the energy conservation principle for analysis of open systems.

EXAMPLE 4-1   Steady-flow steam turbine

One lbm/s of steam enters a steady-flow turbine at 500 psia and 700°F with a velocity of 800 ft/s. The steam leaves the turbine at 5 psia and 200°F with a velocity of 100 ft/s. The turbine is well insulated so that the process may be assumed to be adiabatic. Calculate the horsepower output of the turbine.

SOLUTION   This is a steady-flow–steady-state problem so that $\dot{m}_i = \dot{m}_e$ and $(dE/d\tau)_\sigma = 0$. Omitting the chemical energy terms from Eq. (4-15) and observing that $d'Q/d\tau = 0$, we have

$$\left(\frac{d'W}{d\tau}\right)_{ext} = \dot{m}(h_e - h_i) + \frac{\dot{m}}{2g_c}(V_e^2 - V_i^2)$$

From the superheat steam tables we have

$$h_i = 1357.0 \text{ Btu/lbm} \qquad 500 \text{ psia and } 700°F$$
$$h_e = 1148.8 \text{ Btu/lbm} \qquad 5 \text{ psia and } 200°F$$

Then

$$\left(\frac{d'W}{d\tau}\right)_{ext} = (1)(1148.8 - 1357.0) + \frac{(1)}{(2)(32.2)(778)}[(100)^2 - (800)^2]$$
$$= -221 \text{ Btu/s} = -7.96 \times 10^5 \text{ Btu/h}$$

Since 1 hp = 2545 Btu/h, the horsepower *output* is

TABLE 4-1
Special cases for open systems

Steady flow	Steady state	Multiple inlet streams	K.E.	Ch.E.	$\dfrac{d'Q}{d\tau}$	$\dfrac{d'W_{ext}}{d\tau}$	Energy equation
yes	yes	no	0	0	0	—	$\dfrac{d'W_{ext}}{d\tau} = \dot{m}(h_e - h_i)$
yes	yes	no	0	0	—	0	$\dfrac{d'Q}{d\tau} = \dot{m}(h_e - h_i)$
yes	no	no	0	0	—	—	$\dot{m}_i h_i + \dfrac{d'Q}{d\tau} + \dfrac{d'W_{ext}}{d\tau} = \dot{m}_e h_e + \left(\dfrac{dU}{d\tau}\right)_\sigma$
yes	yes	yes	0	0	0	0	$\Sigma \dot{m}_i h_i = \Sigma \dot{m}_e h_e$
yes	yes	yes	0	0	—	0	$\dfrac{d'Q}{d\tau} = \Sigma \dot{m}_e h_e - \Sigma \dot{m}_i h_i$
no	no	no	0	0	0	0	$\dot{m}_i h_i = \dot{m}_e h_e + \left(\dfrac{dU}{d\tau}\right)_\sigma$

$$hp = \frac{7.96 \times 10^5}{2545} = 313 \qquad (233.4 \text{ kW})$$

The negative sign on the work term above indicates that work is delivered to things external to the control volume.

EXAMPLE 4-2   Adiabatic pump work

Water at the rate of 80 lbm/min is compressed adiabatically from 100°F and 100 psia to 1000 psia in a steady-flow process. Calculate the horsepower required, assuming that the water is very nearly incompressible.

FIG. EXAMPLE 4-2

SOLUTION   A schematic for this process is indicated in the accompanying figure. We neglect the kinetic energy of the flow stream since the flow velocities are probably small, and we write Eq. (4-15) for steady flow and steady state,

$$\dot{m}(h_2 - h_1) = W \tag{a}$$

where W is the rate of doing work on the pump. From Chap. 2 we have

$$d'Q = du + p\,dv \tag{b}$$

for a simple compressible substance. Also,

$$dh = du + p\,dv + v\,dp \tag{c}$$

For an adiabatic process $d'Q = 0$, so that Eq. (c) becomes

$$dh_{\text{ad}} = v\,dp \tag{d}$$

Integrating Eq. (d), we have

$$(h_2 - h_1)_{\text{ad}} = \int_{p_1}^{p_2} v\,dp$$

If the fluid is incompressible, $v \approx$ const. and

$$(h_2 - h_1)_{ad} = v(p_2 - p_1) \tag{e}$$

The work is then calculated with Eq. (a) as

$$W = \dot{m}v(p_2 - p_1) \tag{f}$$

The specific volume is taken as the value for saturated liquid at 100°F:

$$v = 0.01613 \text{ ft}^3/\text{lbm}$$

and the work becomes

$$
\begin{aligned}
W &= (80)(0.016\ 13)(1000 - 100)(144) \\
&= 167\ 000 \text{ ft·lbf/min} \\
&= 5.06 \text{ hp} \quad (3775 \text{ W})
\end{aligned}
$$

The incompressibility assumption may be checked by consulting the compressed-liquid tables wherein it will be found that the change in specific volume due to the elevated pressure is only about 0.3 percent. Note the sign convention employed in writing Eq. (a). Because we seek the work delivered to the pump we draw the figure and take the work added as positive. The left side of the equation represents the difference between the energy leaving and entering the pump and thus must be equal to the work energy added.

EXAMPLE 4-3 Filling a tank

An evacuated tank is connected through an appropriate valve to an infinite source of steam at 100 psia and 500°F, as shown in the accompanying sketch. The valve is opened and the steam is allowed to flow into the tank until the pressure reaches 100 psia. Calculate the temperature

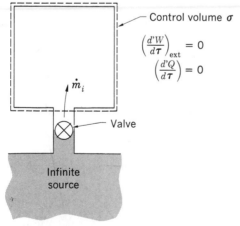

Control volume σ

$$\left(\frac{d'W}{d\tau}\right)_{ext} = 0$$

$$\left(\frac{d'Q}{d\tau}\right) = 0$$

$\dot{m}_i$

Valve

Infinite
source

FIG. EXAMPLE 4-3

of the steam in the tank when the pressure just reaches 100 psia. Assume that the tank is insulated and that there is no shaft work delivered
to the tank.

SOLUTION   This is a non-steady-flow and non-steady-state process, since
the flow rate into the tank and the mass of steam in the tank both vary with
time. We choose the control volume as shown by the dashed line in the figure. Neglecting kinetic energies, we may write Eq. (4-15) in integral form as

$$\int \dot{m}_i h_i \, d\tau = \int \left(\frac{dE}{d\tau}\right)_\sigma d\tau = m_2 u_2 - m_1 u_1 \tag{a}$$

where m_1 and m_2 are the initial and final masses in the tank. From the
problem statement, $m_1 = 0$, but the final mass in the tank is just

$$m_2 = \int \dot{m}_i \, d\tau$$

and so Eq. (a) becomes

$$h_i = u_2$$

From the superheat steam tables, we have

$$h_i = 1279.1 \text{ Btu/lbm} \qquad (2975 \text{ kJ/kg})$$

The final temperature corresponding to $u_2 = 1279.1$ Btu/lbm is found from
Table A-9. The result is

$$T_2 = 767°F \qquad (408.6°C)$$

EXAMPLE 4-4

The tank of Example 4-3 is reconstructed so that a piston is held in
place by a charge of air at 20 psia. The piston is constructed of a
non-heat-conducting material and the tank is insulated. A small cooling
coil is placed in the cylinder to maintain the air temperature constant at
100°F. The total volume of the tank is 3 ft³. The tank is again connected
to an infinite source of steam at 100 psia and 500°F, and the valve is
opened until the pressure in the tank reaches 100 psia. Calculate the
final temperature of the steam under these conditions.

SOLUTION   The control volume for this problem is shown as the dashed
line in the accompanying sketch. Note that the thermodynamic system is
chosen as the air, steam, and piston inside the cylinder. Even though the
tank is insulated, heat is being removed by the action of the cooling coil.
The amount of heat removed is calculated by determining the heat transfer

FIG. EXAMPLE 4-4

when the given quantity of air is compressed isothermally from 20 to 100 psia. The mass of air is

$$m_a = \frac{pV}{RT} = \frac{(20)(144)(3)}{(53.35)(560)} = 0.289 \text{ lbm} \qquad (131 \text{ g})$$

The work done on the gas is

$$W_a = \int -p \, dV = -\int_{V_1}^{V_2} mRT \frac{dV}{V} = -mRT \ln \frac{V_2}{V_1}$$

Since the temperature remains constant, $V_2/V_1 = p_1/p_2$, so that the work is

$$W_a = -mRT \ln \frac{p_1}{p_2}$$
$$= -(0.289)(53.35)(560) \ln \tfrac{20}{100}$$
$$= 13\,850 \text{ ft·lbf} = 17.8 \text{ Btu} \qquad (18.89 \text{ kJ})$$

For the air we may write

$$Q_a + W_a = \Delta U_a$$

But since $T_a = $ const., $\Delta U_a = 0$ and

$$Q_a = -W_a = -17.8 \text{ Btu} \qquad (-18.89 \text{ kJ})$$

The term *non-heat-conducting piston* means that the piston is a heat-insulating barrier between the hot steam and the cooler air, and the steam, by itself, undergoes an adiabatic process. Thus, the heat transfer from the air is the total heat loss from the control volume. We may now write Eq. (4-15) in integral form as

$$\int \dot{m}_i h_i \, d\tau + Q = \int \left(\frac{dE}{d\tau}\right)_\sigma d\tau = m_{2s} u_{2s} - m_{1s} u_{1s} \qquad (a)$$

since there is no change in internal energy of the air in the control volume. As in Example 4-3 we find that

$$\int \dot{m}_i \, d\tau = m_{2s}$$

so that, finally, since $m_{1s} = 0$,

$$m_{2s}(h_i - u_{2s}) = -Q \qquad (b)$$

The final volume of the steam is

$$V_{2s} = 3 - V_{a2} = 3.0 - \tfrac{3}{5} = 2.4 \text{ ft}^3 \qquad (67.96 \text{ liters})$$

and the mass of steam is

$$m_{2s} = \frac{V_{2s}}{v_{2s}}$$

Equation (b) thus becomes, with $h_i = 1279.1$ Btu/lbm and $Q = -17.8$ Btu,

$$\frac{2.4}{v_{2s}} (1279.1 - u_{2s}) = 17.8 \qquad (c)$$

Equation (c) must be solved by iteration with the use of the superheat steam tables. The result is

$$T_{2s} = 536°F \qquad (280°C)$$

The temperature of the steam is lower in this case than in Example 4-3 because part of the incoming energy is used to perform the compression work on the air. This compression work is subsequently lost from the control volume because of the cooling process.

EXAMPLE 4-5 Transient heating of water tank

Water at 80°F enters a tank at the rate of 10 lbm/min. The tank contains 300 lbm of water and is provided with an exit flow channel which discharges at the rate of 10 lbm/min. Water in the tank is well mixed. If the tank is insulated and an electric heater is suddenly placed in the tank generating heat at the constant rate of 20 000 Btu/h, calculate the time required for the exit temperature to reach 100°F. What will the exit temperature be when steady-state conditions are reached? Neglect the mass of the tank.

SOLUTION   This is an example of a problem which is steady *flow* but *not* steady *state*.

FIG. EXAMPLE 4-5

According to the problem statement and the schematic,

$$
\begin{aligned}
T_1 &= \text{const.} = 80°\text{F} \quad (26.67°\text{C}) \\
T_2 &= 80°\text{F} \quad \text{at } \tau = 0 \\
T_\sigma &= 80°\text{F} \quad \text{at } \tau = 0 \quad (\sigma \text{ refers to water inside tank}) \\
\dot{m}_1 &= \dot{m}_2 = 10 \text{ lbm/min} = 600 \text{ lbm/h} = \text{const.} \\
m_\sigma &= 300 \text{ lbm} = \text{const.} \\
Q &= 0 \quad (\text{insulated tank})
\end{aligned}
$$

$$
\frac{dW_{\text{ext}}}{d\tau} = +20\ 000 \text{ Btu/h starting at } \tau = 0
$$

(electric work crossing boundary of control volume)

The energy inside the tank is

$$
E_\sigma = m_\sigma e_\sigma = m_\sigma u_\sigma \quad \text{(negligible K.E.)} \tag{a}
$$

Because of the *well-mixed* condition in the tank, $T_\sigma = T_2$ at all times. For liquid water in this temperature range Δh and Δu are very closely approximated by

$$
\Delta h \cong (1.0)\Delta T \quad \text{Btu/lbm} \tag{b}
$$

$$
\Delta u \cong (1.0)\Delta T \quad \text{Btu/lbm} \tag{c}
$$

Checking Table A-7 at 80 and 100°F

h_f at 80°F = 48.09
h_f at 100°F = 68.05
$$\Delta h = 68.05 - 48.09 = 19.96 \cong (1.0)(100 - 80) \tag{d}$$

A similar calculation applies for Δu. Because this is an unsteady-state problem, we will need to calculate $(du/d\tau)_\sigma$, and we can now write

$$\left(\frac{dE}{d\tau}\right)_\sigma = \frac{d(m_\sigma u_\sigma)}{d\tau} = m_\sigma(1.0)\frac{dT_\sigma}{d\tau} \tag{e}$$

The general energy balance is now written as

$$\dot{m}_1(1.0)T_1 + 20\,000 = \dot{m}_2(1.0)T_2 + m_\sigma(1.0)\frac{dT_\sigma}{d\tau} \text{ Btu/h} \tag{f}$$

where τ is in hours because the energy input is per hour. Noting that $T_2 = T_\sigma$ and inserting the various numerical values gives the first-order differential equation:

$$300\frac{dT_2}{d\tau} + 600T_2 = 68\,000 \tag{g}$$

The solution to this equation is

$$T_2 - 113.33 = C_1 e^{-2\tau} \tag{h}$$

and the constant C_1 is evaluated from the initial condition

$$T_2 = 80°F \quad \text{at } \tau = 0 \tag{i}$$
$$C_1 = -33.33$$

so that

$$T_2 - 113.33 = -33.33 e^{-2\tau} \tag{j}$$

with τ in hours. Equation (j) may then be solved for the time when T_2 attains a value of 100°F as

$$\tau = 0.458 \text{ h} \quad \text{when } T_2 = 100°F$$

When steady-state conditions are reached, T_2 is no longer changing with time. Therefore, taking $dT_2/d\tau = 0$ in Eq. (g) yields

$$600T_2 = 68\,000 \quad \text{(steady state)}$$
$$T_2 \text{ at steady state} = 113.33°F \tag{k}$$

The same result can be obtained by evaluating Eq. (j) at a very long time, $\tau \rightarrow \infty$.

EXAMPLE 4-6   Closed feedwater heater

A closed feedwater heater for a steam power plant operates as shown in the accompanying figure.

FIG. EXAMPLE 4-6

High-pressure liquid water at 10 MPa and 30°C enters a series of heating tubes. Steam at 1.5 MPa and 95 percent quality is sprayed over the tubes and allowed to condense. The saturated liquid is subsequently allowed to leave the device at station 4. We assume that the high-pressure water is heated up to the saturation temperature of the condensing steam. We wish to calculate the mass of steam required per unit mass of incoming liquid. The heater is assumed to be well insulated (adiabatic).

SOLUTION   From the problem statement

$$m_1 = m_2 = 1.0$$
$$m_3 = m_4 = m_s$$
$$p_1 = p_2 = 10 \text{ MPa}$$
$$p_3 = p_4 = 1.5 \text{ MPa} \qquad x_3 = 0.95$$
$$T_3 = T_4 = T_2 = T_{\text{sat}} \qquad \text{at } 1.5 \text{ MPa} = 198.3°C \qquad (a)$$

Assuming negligible kinetic energies, the energy balance on the feedwater heater becomes

$$m_1 h_1 + m_3 h_3 = m_2 h_2 + m_4 h_4 \qquad (b)$$

or

$$h_2 - h_1 = m_s(h_3 - h_4) \qquad (c)$$

This equation simply states that the energy given up by the condensing steam is absorbed by the high-pressure water. We now determine the enthalpies from the steam table.

$$h_1 = 134.86 \text{ kJ/kg} \quad \text{(compressed liquid, 10 MPa, 30°C)}$$
$$h_2 = 848.54 \text{ kJ/kg} \quad \text{(compressed liquid, 10 MPa, 198.3°C)}$$
$$h_3 = h_{f_3} + x_3 h_{fg_3}$$
$$= 844.89 + (0.95)(1947.3)$$
$$= 2694.8 \text{ kJ/kg}$$
$$h_4 = h_f \text{ at } 1.5 \text{ MPa} = 844.89 \text{ kJ/kg}$$

Inserting these values in Eq. (c) gives

$$m_s = 0.3858 \text{ kg steam/kg water}$$

EXAMPLE 4-7   Adiabatic air compressor

An air compressor takes in air at 1 atm and 20°C and discharges into a line having an inside diameter of 1 cm. The average air velocity in the line at a point close to the discharge is 7 m/s, and the discharge pressure is 3.5 atm. Assuming that the compression occurs quasistatically and adiabatically, calculate the work input to the compressor. Assume that the inlet air velocity is very small.

FIG. EXAMPLE 4-7

SOLUTION   The schematic for the process is shown in the accompanying figure. We shall employ SI units for solution of the problem. The given properties are

$$p_1 = 1 \text{ atm} = 1.0132 \times 10^5 \text{ Pa}$$
$$p_2 = 3.5 \text{ atm} = 3.546 \times 10^5 \text{ Pa}$$
$$T_1 = 20°C = 293 \text{ K}$$
$$V_2 = 7 \text{ m/s}$$
$$V_1 \approx 0$$
$$\gamma = 1.4 \text{ for air}$$
$$d_2 = 1.0 \text{ cm} = 10^{-2} \text{ m}$$

Because of the quasistatic-adiabatic assumption

$$T_2 = T_1 \left(\frac{p_2}{p_1}\right)^{(\gamma-1)/\gamma}$$
$$= (293)(3.5)^{0.286} = 419 \text{ K}$$

At station 2 we may calculate the mass flow rate from Eq. (4-3), but first we need the air density at that point.

$$\rho_2 = \frac{1}{v_2} = \frac{p_2}{RT_2}$$

The gas constant for air is

$$R = \frac{\mathscr{R}}{M} = \frac{8315}{28.97} = 287 \text{ J/kg·K}$$

so that

$$\rho_2 = \frac{3.546 \times 10^5}{(287)(419)} = 2.949 \text{ kg/m}^3$$

The mass flow is then

$$\dot{m} = \rho_2 A_2 V_2 \qquad\qquad (a)$$
$$= \frac{(2.949) \, \pi \, (0.01)^2 \, (7)}{4}$$
$$= 1.621 \times 10^{-3} \text{ kg/s}$$

For steady flow and zero heat flow the energy equation is

$$\dot{m}\left(h_1 + \frac{1}{2g_c} V_1^2\right) + \frac{d'W_{ext}}{d\tau} = \dot{m}\left(h_2 + \frac{1}{2g_c} V_2^2\right) \qquad\qquad (b)$$

For air the value of c_p is

$$c_p = 0.24 \text{ Btu/lbm·°F} = 1004 \text{ J/kg·°C}$$

The K.E. at station 1 is zero by the problem statement; that at station 2 is

$$\text{K.E.}_2 = \frac{V_2^2}{2g_c} \frac{(7 \text{ m/s})^2}{(2)(1.0 \text{ kg·m/N·s}^2)} = 24.5 \text{ J/kg}$$

Taking air as an ideal gas the enthalpy difference is

$$h_2 - h_1 = c_p(T_2 - T_1)$$
$$= (1004 \text{ J/kg·°C})(419 - 293) \qquad\qquad (c)$$
$$= 1.265 \times 10^5 \text{ J/kg}$$

From these calculations we see that the kinetic energy at station 2 is negligible compared to the enthalpy difference. Now, inserting the numerical values into Eq. (b) we obtain

$$\frac{d'W_{\text{ext}}}{d\tau} = (1.621 \times 10^{-3} \text{ kg/s})(1.265 \times 10^5 + 24.50) \text{ J/kg}$$

$$= 205.1 \text{ J/s} = 205.1 \text{ W}$$

EXAMPLE 4-8   Helium turbine

Helium is expanded quasistatically and adiabatically in a turbine from 400 kPa and 260°C to 100 kPa. The turbine is insulated and the inlet velocity is small. The exit velocity is 200 m/s. Calculate the work output of the turbine per unit mass of helium flow.

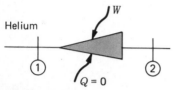

FIG. EXAMPLE 4-8

SOLUTION   The schematic diagram indicates the nomenclature for the problem. For a steady-flow adiabatic process the energy equation becomes, assuming unit mass flow,

$$h_1 + \frac{1}{2g_c} V_1^2 + W = h_2 + \frac{1}{2g_c} V_2^2 \tag{a}$$

Helium may be considered as an ideal gas, so that

$$h_2 - h_1 = c_p(T_2 - T_1) \tag{b}$$

The final temperature may be found from Eq. (2-38) with $n = \gamma$ for the quasistatic-adiabatic process:

$$\frac{T_2}{T_1} = \left(\frac{p_2}{p_1}\right)^{\gamma - 1/\gamma} \tag{c}$$

with

$$T_1 = 260 + 273 = 533 \text{ K}$$

From Table 2-2, for helium,

$$c_p = 5.234 \text{ kJ/kg·°C} \qquad \gamma = 1.66$$
$$T_2 = (260 + 273)(\tfrac{100}{400})^{(1.66-1)/1.66}$$
$$= 307 \text{ K}$$

Inserting the numerical values in Eq. (a), with $V_1 = 0$, gives

$$W = (5234)(307 - 533) + \frac{1}{(2)(1)} (200)^2$$

$$= -1.163 \text{ MJ/kg}$$

In writing Eq. (a) we have taken the work term as a positive quantity entering the control volume because it appears on the left side of the equation along with the enthalpy and kinetic energy entering the turbine. Thus, the negative sign in the answer indicates that there is a net work *output* from the turbine.

EXAMPLE 4-9 Adiabatic nozzle expansion

Air is expanded quasistatically and adiabatically in a nozzle from 200 psia and 300°F to a pressure of 100 psia. The inlet velocity to the nozzle is very small, and the process occurs under steady-flow conditions. Calculate the exit velocity from the nozzle.

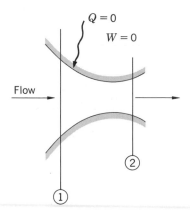

FIG. EXAMPLE 4-9

SOLUTION   We again make use of the steady-flow energy equation. For no heat transfer and zero work we have, for unit mass flow,

$$h_1 + \frac{1}{2g_c} V_1^2 = h_2 + \frac{1}{2g_c} V_2^2 \qquad (a)$$

where the subscripts 1 and 2 refer to entrance and exit conditions, as shown in the accompanying sketch. Since $V_1 \approx 0$ and enthalpy for an ideal gas is a function of temperature alone, Eq. (a) may be solved for the exit velocity as

$$V_2 = [2g_c(h_1 - h_2)]^{1/2} = [2g_c c_p(T_1 - T_2)]^{1/2} \qquad (b)$$

The final temperature is found by applying Eq. (2-38) with $n = \gamma$ to obtain

$$\frac{T_2}{T_1} = \left(\frac{p_2}{p_1}\right)^{(\gamma-1)/\gamma}$$
(c)

for the quasistatic adiabatic process. The final temperature is, therefore,

$$T_2 = (300 + 460)(\tfrac{100}{200})^{(1.4-1)/1.4}$$
$$= 625°R = 165°F \qquad (73.9°C)$$

Then the final velocity is calculated from Eq. (b) as

$$V_2 = [(2)(32.2)(778)(0.24)(300 - 165)]^{1/2}$$
$$= 1272 \text{ ft/s} \qquad (387.7 \text{ m/s})$$

EXAMPLE 4-10   Work in an air turbine

Air is expanded quasistatically and adiabatically in a turbine from 50 psia and 500°F to 15 psia. The turbine is insulated and the inlet velocity is small. The exit velocity is 500 ft/s. Calculate the work output of the turbine per unit mass of air flow.

FIG. EXAMPLE 4-10

SOLUTION   The schematic diagram indicates the nomenclature for the problem. For a steady-flow adiabatic process, the energy equation becomes, assuming unit mass flow,

$$h_1 + \frac{1}{2g_c} V_1^2 + W = h_2 + \frac{1}{2g_c} V_2^2$$
(a)

Air is considered an ideal gas, so that

$$h_2 - h_1 = c_p(T_2 - T_1)$$

The final temperature is calculated from Eq. (c) of Example 4-9.

$$T_2 = (500 + 460)(\tfrac{15}{50})^{(1.4-1)/1.4}$$
$$= 681°R = 221°F \quad (105°C)$$

Inserting the proper values in Eq. (*a*), with $V_1 = 0$, gives

$$W = (0.24)(221 - 500) + \frac{1}{(2)(32.2)(778)} (500)^2$$
$$= -61.9 \text{ Btu/lbm} \quad (-1.607 \times 10^5 \text{ J/kg})$$

In writing Eq. (*a*) we have taken the work term as a positive quantity en-tering the control volume because it appears on the left side of the equation along with the enthalpy and kinetic energy entering the turbine. Thus, the negative sign in the answer indicates that there is a net work *output* from the turbine.

EXAMPLE 4-11   Ammonia heater

Hot combustion gases at 400°F are used to produce superheated ammo-nia at 260 psia and 140°F from liquid ammonia at the same pressure and 80°F. The heat-exchange device is to handle 50 lbm/min of the ammo-nia, and the exit temperature of the combustion gases is to be 120°F. The velocity of the ammonia vapor leaving the device is not to exceed 20 ft/s, and the specific heat of the gases may be taken as 0.26 Btu/lbm·°F. Calculate the mass flow of combustion gases required and the diameter of the exit pipe for the ammonia vapor.

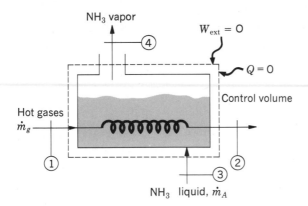

FIG. EXAMPLE 4-11

SOLUTION   The schematic for this heat exchange device is shown in the accompanying figure. If we assume that the overall heat-transfer device is in-sulated then $Q = 0$. No mention is made of work so $W_{ext} = 0$. Neglecting kinetic energies for the low velocities involved we then write the energy bal-ance as

Energy transported into heat exchanger $=$ energy transported out of heat exchanger

or

$$\dot{m}_g h_1 + \dot{m}_A h_3 = \dot{m}_g h_2 + \dot{m}_A h_4 \qquad (a)$$

The ammonia properties are obtained as

$$h_3 = 132.0 \text{ Btu/lbm} \qquad \text{(Table A-12, sat. liq. at 80°F)}$$
$$h_4 = 655.6 \text{ Btu/lbm} \qquad \text{(Table A-14, 260 psia, 140°F)}$$

If we assume ideal-gas behavior for the combustion gases

$$h_1 - h_2 = c_p(T_1 - T_2)$$
$$= (0.26 \text{ Btu/lbm·°F})(400 - 120)$$
$$= 72.8 \text{ Btu/lbm} \qquad (169.3 \text{ kJ/kg})$$

$\dot{m}_A$ is given as 50 lbm/min so we may insert the numerical values in Eq. (a) to obtain

$$\dot{m}_g(h_1 - h_2) = \dot{m}_A(h_4 - h_3) \qquad (b)$$
$$\dot{m}_g = \frac{(50)(655.6 - 132.0)}{72.8} = 359.6 \text{ lbm/min} \qquad (163.1 \text{ kg/min})$$

The specific volume of the superheated ammonia at exit from the exchanger is

$$v_4 = 1.257 \text{ ft}^3/\text{lbm} \qquad \text{(Table A-14, 260 psia, 140°F)}$$

We may now use Eq. (4-2) to calculate the flow area required for this vapor. We have $V_4 = 20$ ft/s so that

$$\dot{m}_A = \frac{A_4 V_4}{v_4} \qquad (c)$$

and

$$A_4 = \frac{(50 \text{ lbm/min})(1.257 \text{ ft}^3/\text{lbm})}{(20 \text{ ft/s})(60 \text{ s/min})}$$
$$= 0.052\,38 \text{ ft}^2 \qquad (48.66 \text{ cm}^2)$$

Assuming a circular pipe at exit,

$$A_4 = \frac{\pi d_4^2}{4}$$

and

$$d_4 = 0.258 \text{ ft} = 3.10 \text{ in} \qquad (7.87 \text{ cm})$$

EXAMPLE 4-12   Solar water heater

A simple solar water heater is designed to heat water from 30 to 90°C by collecting the sun's energy in the glass-covered device shown. The short wavelength radiation from the sun is transmitted through the glass and trapped inside. The bottom portion of the device is insulated but some energy may be radiated back out by the glass top, which for these temperatures would be at about 200 W/m². On a clear day the irradiation from the sun is about 1000 W/m². Calculate the surface area required to heat the water, and express as m²·s/liter. How many square feet would be required to produce a hot-water flow of 10 gal/min?

FIG. EXAMPLE 4-12

SOLUTION   The control volume for this problem is shown as the dashed line in the figure and the energy balance becomes

$$\dot{m}_w h_1 + Q_{sun} = \dot{m}_w h_2 + Q_{env} \qquad (a)$$

where $Q_{sun} = 1000 \text{ W/m}^2$ and $Q_{env} = 200 \text{ W/m}^2$. The bottom of the device is insulated so there are no other energy transfers. We could obtain the water enthalpies from the saturated-liquid tables at 30 and 90°C, but it is a bit more convenient to observe that in this range the value of c_p for the liquid is very nearly constant at 1.0 kcal/kg·°C so that, to a good approximation,

$$\begin{aligned} h_2 - h_1 &= c_p(T_2 - T_1) \qquad (b)\\ &= (1.0 \text{ kcal/kg·°C})(90 - 30)\\ &= 60 \text{ kcal/kg}\\ &= 2.51 \times 10^5 \text{ J/kg} \end{aligned}$$

At a mean temperature of 60°C = 140°F the specific volume of saturated liquid is obtained from Table A-7 as 0.016 29 ft³/lbm or

$$v = 0.016\,29 \text{ ft}^3/\text{lbm} = 0.4615 \text{ liter}/\text{lbm}$$ $$= 1.015 \text{ liters}/\text{kg}$$ (c)

We may now calculate the water flow rate from Eq. (a)

$$\dot{m}_w(h_2 - h_1) = Q_{sun} - Q_{env}$$ (d)
$$\dot{m}_w = \frac{(1000 - 200) \text{ W}/\text{m}^2}{2.51 \times 10^5 \text{ J}/\text{kg}}$$
$$= 3.187 \times 10^{-3} \text{ kg}/\text{s·m}^2$$

With the conversion to liters from (c) this becomes

$$\dot{m}_w = 3.235 \times 10^{-3} \text{ liter}/\text{s·m}^2$$

The inverse of this is the quantity we seek, or,

Area required $= 309.1 \text{ m}^2\text{·s}/\text{liter}$ (e)

To produce a hot-water flow of 10 gal/min (1 gal $= 231 \text{ in}^3$) we would have

$$\dot{m}_w = 2310 \text{ in}^3/\text{min} = 37.85 \text{ liters}/\text{min} = 0.6309 \text{ liter}/\text{s}$$

Using the relation in (e) the area is

$$A = (309.1 \text{ m}^2\text{·s}/\text{liter})(0.6309 \text{ liter}/\text{s})$$ $$= 195 \text{ m}^2$$ $$= 2099 \text{ ft}^2$$

4-7 THE JOULE-THOMSON COEFFICIENT

Consider the flow situation shown in Fig. 4-5. A *real* gas flows through the constant-area duct. Between sections 1 and 2 a porous restriction is placed in

FIG. 4-5 *Throttling process through a porous plug.*

the flow which causes an appropriate pressure drop. The process is called a *throttling process*, and if changes in kinetic and potential energies are negligible, the steady-flow energy equation will reduce to

$$h_1 = h_2$$

for an adiabatic process. Although the enthalpies at sections 1 and 2 are the same for the given conditions, the enthalpy of the flow may vary considerably as the fluid moves through the porous plug. This variation will depend on the size of the openings in the plug and the frictional resistance imparted to the fluid. The throttling process need not occur in a porous plug. It can also be observed as an adiabatic expansion of a fluid in a valve when the inlet and exit kinetic energies are negligible. If the kinetic energies are large enough to be considered in the analysis then the complete energy equation must be used and the term *throttling* is no longer applied.

If experimental data for real gases are plotted on a *T-p* diagram, a family of curves like those in Fig. 4-6 will result. The locus of the maximum points in the constant-enthalpy curves is called the *inversion curve*, and the maximum point in each curve is called an *inversion point*. The slope of an isenthalpic (constant enthalpy) curve is called the *Joule-Thomson coefficient* μ_J:

$$\mu_J = \left(\frac{\partial T}{\partial p}\right)_h \tag{4-17}$$

As we shall show in Chap. 7, and as has been assumed in Chap. 2, the enthalpy for an ideal gas is a function of temperature alone, so that, for an ideal gas, a line of constant enthalpy would also be a line of constant temperature. Thus

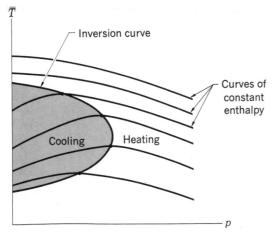

FIG. 4-6 *Inversion points for porous-plug experiment of Fig. 4-5.*

$$\mu_J = 0 \qquad \text{for an ideal gas} \tag{4-18}$$

The importance of the Joule-Thomson coefficient is that it may be mea-sured experimentally and subsequently used to calculate other thermody-namic properties of interest. We shall explore the matter of calculation of properties more thoroughly in Chap. 7, indicating the utility of a Joule-Thomson type of experiment at this point. If we assume that the enthalpy may be expressed as a function of the two independent variables p and T, then from the partial differential calculus we have

$$dh = \left(\frac{\partial h}{\partial p}\right)_T dp + \left(\frac{\partial h}{\partial T}\right)_p dT \tag{4-19}$$

Introducing the definition of c_p, we have

$$dh = \left(\frac{\partial h}{\partial p}\right)_T dp + c_p\, dT \tag{4-20}$$

For the adiabatic throttling process $dh = 0$, and Eq. (4-20) may be rear-ranged to give

$$0 = \left(\frac{\partial h}{\partial p}\right)_T \left(\frac{\partial p}{\partial T}\right)_h + c_p$$

or

$$c_p = -\frac{1}{\mu_J} \left(\frac{\partial h}{\partial p}\right)_T \tag{4-21}$$

The constant-temperature coefficient μ_T is defined as

$$\mu_T = \left(\frac{\partial h}{\partial p}\right)_T \tag{4-22}$$

It is readily measured by performing the porous-plug experiment at constant temperature for various pressures. The heat supplied or removed from the constant-area section (e.g., by an electric heater) represents the change in enthalpy of the gas since

$$d'Q = dh$$

for the process (no change in K.E., $W = 0$). Once μ_J and μ_T are determined, the constant-pressure specific heats may be evaluated with

$$c_p = -\frac{\mu_T}{\mu_J} \tag{4-23}$$

The knowledge of specific heats is a crucial matter in the calculation of thermodynamic properties, as we shall see in Chap. 7.

EXAMPLE 4-13   Throttling of Freon 12

Freon 12 is throttled in an expansion valve from saturated-liquid conditions at 150°F to a temperature of 40°F. Calculate the quality of the Freon vapor after the throttling process.

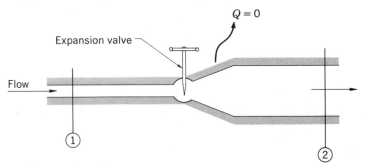

FIG. EXAMPLE 4-13

SOLUTION   The schematic for this process is shown in the accompanying figure. For an adiabatic throttling process, we have

$$h_1 = h_2$$

From the Freon table (Table A-15),

$$h_1 = 43.850 \text{ Btu/lbm} \qquad \text{saturated liquid at 150°F}$$

This value is then used to calculate the quality at the exit (40°F):

$$h_2 = h_1 = h_{f_2} + x_2 h_{fg_2}$$

or

$$43.850 = 17.273 + x_2(64.163)$$

and

$$x_2 = 0.414$$

This process is widely used in air conditioning and refrigeration systems, as we shall see in Chap. 12.

REVIEW QUESTIONS

1 Distinguish between open and closed systems.

2 Distinguish between the terms *steady flow* and *steady state*.

3 What is meant by the term *flow work?*

4 Why is the enthalpy function valuable in the analysis of open systems?

5 What is an ideal gas?

6 What equation may be used to describe a quasistatic adiabatic process for an ideal gas?

7 What is a throttling process? Under what conditions are the enthalpies at entrance and exit equal?

8 Why is it reasonable to assume that an expansion process in a nozzle occurs in an adiabatic manner?

9 In Example 4-3, the steam enters the tank from a source at 500°F. The tank is insulated so no heat is added. In a physical sense, how do you account for the rise in temperature to 767°F without heat addition or external work?

PROBLEMS   (English units)

4-1 10 lbm of nitrogen are compressed in such a manner that the internal energy increases by 150 Btu while 20 Btu of heat are added. Calculate the work that is supplied to the gas. Assume that the compression occurs in a closed system.

4-2 Rework Prob. 4-1, assuming that the compression occurs in a steady-flow open system.

4-3 A 1-kW-resistance heater is placed in a 100-ft³ container filled with air at 14.7 psia and 70°F. The heater is allowed to operate for 10 min and the container is insulated. Calculate the temperature and pressure of the air at the end of this time period.

4-4 Air flows through a turbine in a quasistatic-adiabatic process from 50 psia and 1000°F to 15 psia. The inlet velocity is 100 ft/s and the outlet velocity is 1000 ft/s. Calculate the work output of the turbine per pound of air.

4-5 Calculate the work required to pump 10 gal/min of water from 25 psia and 100°F to 1200 psia in an adiabatic process.

4-6 1 lbm of steam is heated in a constant-pressure process from saturated vapor at 100 psia to 500°F. Calculate the heat added in (*a*) a nonflow process in which the steam is contained behind a piston in a cylinder and (*b*) a steady-flow process for which there is no external work.

4-7 Rework Prob. 4-6 for a constant-volume process. Calculate the final pressure of the steam.

4-8 Saturated vapor (steam) is contained in a tank at 80 psia. Calculate the heat which must be removed to reduce the quality to 50 percent. Also calculate the final pressure for this process.

4-9 A 10-ft³ tank contains air at 14.7 psia and 70°F. A high-pressure line is connected to the tank until the pressure reaches 300 psia. Assuming that the tank and the connecting valve are adiabatic, calculate the final temperature in the tank if the conditions in the high-pressure line remain constant at 300 psia and 100°F. How much cooling has to be supplied to the tank to maintain the air temperature constant at 70°F?

4-10 In an open feedwater heater for a steam power plant, saturated vapor at 100 psia is mixed with subcooled liquid at 80°F and 100 psia. Just enough steam is supplied to ensure that the mixed stream leaving the heater will be saturated liquid at 100 psia. Assuming that the heater is a box which is perfectly insulated from its surroundings and that the process is steady flow, calculate the required flow rate of vapor per pound of entering subcooled liquid.

4-11 5 lbm/min of steam at 100 psia and 500°F are mixed with 10 lbm/min of steam at 100 psia and 50 percent quality in an adiabatic steady-flow device. Calculate the temperature or quality of the outlet mixture.

4-12 Steam enters a small insulated tube as saturated vapor at 60 psia. Because of the frictional resistance in the tube, the pressure of the steam will decrease to 45 psia over a sufficiently long length of the tube. Calculate the temperature of the steam when the 45-psia condition is reached.

4-13 Steam at 300 psia is throttled to 14.7 psia and 260°F. Calculate the quality of the steam at the 300-psia condition.

4-14 Freon 12 vapor enters a steady-flow compressor as saturated vapor at 40°F. The outlet conditions are 250 psia and 200°F, and the process is assumed to be adiabatic. Calculate the horsepower required if the Freon flow rate is 10 lbm/min. What diameter inlet tubing to the compressor is required if the inlet velocity shall not exceed 15 ft/s?

4-15 A heat exchanger is designed as shown in the accompanying sketch. Steam at 20 psia and 300°F is used to heat air from 40 to 85°F. Calculate the flow rate of steam required to heat 100 lbm/min of air with the exit conditions of the steam at 20 psia and 75 percent quality.

FIG. P 4-15

4-16 The cooling coil in a certain air conditioning system is designed to remove 60 000 Btu/h from 1500 ft³/min of air at 85°F and 14.7 psia. The cooling is accomplished in a heat exchanger like that indicated in Prob. 4-15, except that Freon 12 is evaporated at constant pressure from saturated-liquid conditions at 40°F to superheat at 43°F. Calculate the mass flow rate of Freon required and the exit air temperature.

4-17 Air at 3 psia and −30°F enters a diffuser-nozzle section at 900 ft/s. The air goes through a quasistatic-adiabatic process until its velocity is reduced to 100 ft/s. Calculate the final temperature and pressure.

4-18 Oxygen is compressed in a piston-cylinder arrangement from 23 psia, 100°F, to 85 psia, 200°F. Calculate the work done on the gas and the heat transfer if the process is quasistatic.

4-19 Saturated-liquid water at 500°F is throttled to atmospheric pressure. Calculate the final quality if saturated or the final temperature if superheated. Calculate the velocity of the steam in the final state for a flow rate of 1 lbm/min through a 1-in-diameter tube.

4-20 Calculate the inlet and exit areas required in Prob. 4-17 to accommodate a flow rate of 1 lbm/s.

4-21 What horsepower is required to pump 100 gal/min of liquid water at 70°F through a rise in elevation of 100 ft? Assume isothermal flow and negligible change in kinetic energy.

4-22 The horsepower of a jet engine may be roughly calculated in terms of the energy imparted to the air passing through it. Assuming that the exhaust gases have the same properties as air and ideal-gas behavior is experienced, calculate the flow rate of air necessary to produce 15 000 hp under the following conditions:

$$\text{Inlet air temperature} = -50°F$$
$$\text{Inlet air velocity} = 700 \text{ ft/s}$$
$$\text{Outlet air temperature} = 2000°F$$
$$\text{Outlet air velocity} = 1000 \text{ ft/s}$$

Also calculate the inlet flow area required for an inlet pressure of 3 psia.

4-23 A 0.5-hp fan is mounted inside a 100-ft³ box filled with air at 14.7 psia and 70°F. The box is perfectly insulated, and the fan is allowed to run for 10 min. Calculate the final temperature and pressure of the air.

4-24 Calculate the heat which must be removed from the box in Prob. 4-23 to have a final temperature of 110°F.

4-25 1 lbm of a certain gaseous fuel is burned with 50 lbm of air in a constant-pressure steady-flow process. The chemical energy of the fuel is 24 000 Btu/lbm, and both the fuel and air enter the burner at 70°F and negligible velocity. The products of combustion may be assumed to have the same properties as air. Assuming ideal-gas behavior and an insulated burner, calculate the temperature of the exit gases. For this calculation take the enthalpy of incoming air and fuel as zero at 70°F.

4-26 Show that the mass rate of flow for a nozzle like that in Example 4-9 is given by:

$$\dot{m}^2 = 2g_c A_2^2 \frac{\gamma}{\gamma - 1} \frac{p_1^2}{RT_1} \left[\left(\frac{p_2}{p_1} \right)^{2/\gamma} - \left(\frac{p_2}{p_1} \right)^{(\gamma+1)/\gamma} \right]$$

where A_2 is the exit area of the nozzle and $V_1 \approx 0$.

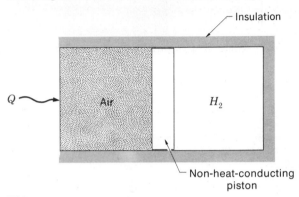

FIG. P 4-27

4-27 Consider the system shown in the accompanying figure. The two chambers initially have equal volumes of 1 ft³ and contain air and hydrogen, respectively. The chambers are separated by a frictionless piston which is non-heat-conducting. Both gases are initially at 20 psia and 100°F. Heat is added to the air side until the pressure of both gases reaches 40 psia. All outside walls of the chamber are insulated except for the surface where heat is added to the air. Calculate the final temperature of the air.

4-28 A small nozzle is constructed for use with liquid water at 100°F. The inlet pressure to the nozzle is 100 psia, and the discharge pressure is atmospheric. Assuming an adiabatic process, calculate the exit velocity. For an exit area of 0.05 in², calculate the flow rate in gal/min.

4-29 What horsepower pump would be required to produce the pressure differential and flow rate in Prob. 4-28?

4-30 Two 1-ft³ chambers are separated by a non-heat-conducting wall as shown (page 156.) The left chamber is filled with air at 100 psia and 150°F, and the right chamber is evacuated. A small valve is opened which allows the air to flow from the left to right chamber. The valve is closed when the pressures in the two chambers are equal. Calculate the final pressure and temperature for the air in each chamber. State assumptions necessary to solve the problem.

4-31 Repeat Prob. 4-30 for the case of a separating wall which is perfectly heat-conducting; i.e., the temperature is always the same on both sides of the wall. Again, state assumptions carefully.

4-32 Oxygen is contained at 30 atm and 50°C in a tank having a volume of 10 m³. A large valve is opened from the tank and the gas is suddenly discharged until the pressure drops to 15 atm. What is the quantity discharged in kilograms?

FIG. P 4-30

4-33 Steam enters a certain nozzle at 300 psia and 500°F with a velocity of 500 ft/s. During the expansion process the steam loses 15 Btu/lbm and the enthalpy drops to 1200 Btu/lbm. What is the exit velocity from the nozzle? What flow area at inlet would be required for a flow rate of 2 lbm/s?

4-34 A small tank having a volume of 500 cm³ contains Freon 12 in a saturation state at 70°F. Only 20 percent of the tank is filled with liquid. A small nozzle on the top of the can is used to spray the Freon in a room at atmospheric pressure. After an extended period of spraying the temperature in the can has dropped to 50°F, and the liquid occupies only 10 percent of the volume. Assuming that the nozzle is an adiabatic throttling device, how much energy is added to the can during the discharge process?

4-35 An industrial hot water heater uses a gaseous fuel with an internal chemical energy content of about 23 000 Btu/lbm. The heater is designed so that 18 lb of air at 70°F and 1 atm are supplied for each pound of fuel, and it is designed to produce 10 gal/min of hot water at 180°F from cold water at 60°F. The hot discharge gases are at 230°F. Calculate the quantity of fuel required.

4-36 A municipal incinerator is to burn garbage which has an average internal chemical energy of 3000 Btu/lbm. The hot gases from the burning process are

FIG. P 4-36

then used to preheat the air as it enters the incinerator. The internal construction of the incinerator also allows it to be used as a high-pressure hot water
heater for a nearby power plant as shown in the sketch. The respective temperatures are $T_1 = 100°F$, $T_3 = 1500°F$, $T_4 = 400°F$, $T_5 = 95°F$, $T_6 = 300°F$.
Assume that the hot gases behave as an ideal gas and that 15 lbm of air are required to burn each pound of garbage or each pound of an auxiliary fuel. The
auxiliary fuel has an internal chemical energy of 20 000 Btu/lbm and must be
used to get the burning temperature high enough for complete combustion.
How much total energy in the form of garbage and fuel must be supplied to
heat 3000 gal of water? If 1 ton of garbage is available for heating each 3000 gal
of water, how much auxiliary fuel would be required?

4-37   A small auxiliary power device is designed as shown in the schematic. Nitrogen is contained at 2000 psia and 70°F in the tank which is connected
through a valve to an adiabatic turbine which drives an electric generator. The
power output of the device is to be 75 W, and the pressure at point 1 is maintained constant at 100 psia by a pressure regulator which operates as a
throttling device. The nitrogen discharges from the turbine at atmospheric
pressure, and the turbine process may be assumed adiabatic. What size tank is
necessary to produce the power output for a period of 1 h? Assume the tank remains isothermal at 70°F and that power production ceases when the tank
pressure reaches 100 psia.

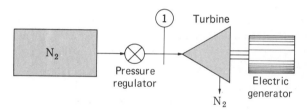

FIG. P 4-37

PROBLEMS   (Metric units)

4-1M   Air flows through a turbine in a quasistatic-adiabatic process from 350 kPa
and 550°C to 100 kPa. The inlet velocity is 30 m/s and the outlet velocity is
300 m/s. Calculate the work output of the turbine per kilogram of air.

4-2M   Calculate the work required to pump 40 liters/min of water from 170 kPa
and 40°C to 7.5 MPa in an adiabatic process.

4-3M   1 kg of steam is heated in a constant-pressure process from saturated vapor
at 700 kPa to 260°C. Calculate the heat added in (a) a nonflow process in
which the steam is contained behind a piston in a cylinder and (b) a
steady-flow process for which there is no external work.

4-4M   Saturated vapor (steam) is contained in a tank at 600 kPa. Calculate the heat
which must be removed to reduce the quality to 50 percent. Also calculate
the final pressure for this process.

4-5M   A high-speed flow of helium is to be heated in a duct. The gas enters the duct at 3 atm, 20°C, and a velocity of 150 m/s. While traversing the duct the pressure drops 0.15 atm and the velocity is reduced to 130 m/s. The exit temperature of the gas is 140°C. How much heat must be added for a mass flow of 100 g/s? What are the inlet and exit flow areas for the given conditions?

4-6M   High-pressure air at 5 atm and 30°C is to be used to drive a small turbine for powering a high-speed drill. The drill requires a power input of 746 W (1 hp), and the air may be assumed to discharge from the turbine at 1.8 atm. If the turbine process is quasistatic and adiabatic, calculate the mass flow rate of air required. If the inlet air line is 1 cm in diameter, calculate the entering air velocity.

4-7M   A 280-liter tank contains air at 100 kPa and 20°C. A high-pressure line is connected to the tank until the pressure reaches 2 MPa. Assuming that the tank and the connecting valve are adiabatic, calculate the final temperature in the tank if the conditions in the high-pressure line remain constant at 2 MPa and 40°C. How much cooling has to be supplied to the tank to maintain the air temperature constant at 20°C?

4-8M   In an open feedwater heater for a steam power plant, saturated water vapor at 700 kPa is mixed with subcooled liquid at 25°C and 700 kPa. Just enough steam is supplied to ensure that the mixed stream leaving the heater will be saturated liquid at 700 kPa. Assuming that the heater is a box which is perfectly insulated from its surroundings and that the process is steady flow, calculate the required flow rate of vapor per kilogram of entering subcooled liquid.

4-9M   2 kg/min of steam at 700 kPa and 260°C are mixed with 4 kg/min of steam at 700 kPa and 50 percent quality in an adiabatic steady-flow device. Calculate the temperature or quality of the outlet mixture.

4-10M   Steam enters a small insulated tube as saturated vapor at 400 kPa. Because of the frictional resistance in the tube, the pressure of the steam will decrease to 300 kPa over a sufficiently long length of the tube. Calculate the temperature of the steam when the 300-kPa condition is reached.

4-11M   Steam at 2 MPa is throttled to 100 kPa and 120°C. Calculate the quality of the steam at the 2-MPa condition.

4-12M   Freon 12 vapor enters a steady-flow compressor as saturated vapor at 40°F. The outlet conditions are 250 psia and 200°F, and the process is assumed to be adiabatic. Calculate the horsepower required if the Freon flow rate is 5 kg/min. What diameter inlet tubing to the compressor is required if the inlet velocity shall not exceed 5 m/s?

4-13M   A heat exchanger is designed as shown in the accompanying sketch. Steam at 150 kPa and 160°C is used to heat air from 5 to 30°C. Calculate the flow rate of steam required to heat 50 kg/min of air with the exit conditions of the steam at 150 kPa and 75 percent quality.

Steam

Air

FIG. P 4-13M

4-14M The cooling coil in a certain air conditioning system is designed to remove 18 kW from 40 000 liters/min of air at 30°C and 1 atm. The cooling is accomplished in a heat exchanger like that indicated in Prob. 4-13M, except that Freon 12 is evaporated at constant pressure from saturated-liquid conditions at 40°F to superheat at 43°F. Calculate the mass flow rate of Freon required and the exit air temperature.

4-15M Air at 20 kPa and -35°C enters a diffuser-nozzle section at 280 m/s. The air goes through a quasistatic-adiabatic process until its velocity is reduced to 30 m/s. Calculate the final temperature and pressure.

4-16M Calculate the inlet and exit areas required in Prob. 4-15M to accomodate a flow rate of 1 kg/min.

4-17M What power is required to pump 380 liters/min of liquid water at 20°C through a rise in elevation of 30 m? Assume isothermal flow and negligible change in kinetic energy.

4-18M The power of a jet engine may be roughly calculated in terms of the energy imparted to the air passing through it. Assuming that the exhaust gases have the same properties as air and ideal-gas behavior is experienced, calculate the flow rate of air necessary to produce 10 MW under the following conditions:

$$\text{Inlet air temperature} = -45°C$$
$$\text{Inlet air velocity} = 200 \text{ m/s}$$
$$\text{Outlet air temperature} = 1100°C$$
$$\text{Outlet air velocity} = 300 \text{ m/s}$$

Also calculate the inlet flow area required for an inlet pressure of 20 kPa.

4-19M A 350-W fan is mounted inside a 3000-liter box filled with air at 20 kPa and 20°C. The box is perfectly insulated, and the fan is allowed to run for 10 min. Calculate the final temperature and pressure of the air.

4-20M Calculate the heat which must be removed from the box in Prob. 4-19M to have a final temperature of 45°C.

4-21M   1 kg of a certain gaseous fuel is burned with 50 kg of air in a constant-pressure steady-flow process. The chemical energy of the fuel is 60 MJ/kg and both the fuel and air enter the burner at 22°C and negligible velocity. The products of combustion may be assumed to have the same properties as air. Assuming ideal-gas behavior and an insulated burner, calculate the temperature of the exit gases. For this calculation take the enthalpy of the incoming air and fuel as zero at 22°C.

4-22M   Consider the system shown in the accompanying figure. The two chambers initially have equal volumes of 28 liters and contain air and hydrogen, respectively. The chambers are separated by a frictionless piston which is non-heat-conducting. Both gases are initially at 140 kPa and 40°C. Heat is added to the air side until the pressure of both gases reaches 280 kPa. All outside walls of the chamber are insulated except for the surface where heat is added to the air. Calculate the final temperature of the air.

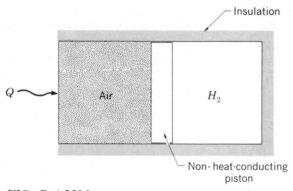

FIG. P 4-22M

4-23M   A small nozzle is constructed for use with liquid water at 40°C. The inlet pressure to the nozzle is 700 kPa, and the discharge pressure is 100 kPa. Assuming an adiabatic process, calculate the exit velocity. For an exit area of 0.3 cm², calculate the flow rate in liters/min.

4-24M   Two 28-liter chambers are separated by a non-heat-conducting wall as shown in the figure. The left chamber is filled with air at 700 kPa and 65°C, and the right chamber is evacuated. A small valve is opened which allows the air to flow from the left to the right chamber. The valve is closed when the pressures in the two chambers are equal. Calculate the final pressure and temperature for the air in each chamber. State assumptions necessary to solve the problem.

4-25M   Oxygen is contained at 3 MPa and 50°C in a tank having a volume of 10 m³. A large valve is opened from the tank and the gas is suddenly discharged until the pressure drops to 1.5 MPa. What is the quantity discharged in kilograms?

4-26M   Steam enters a certain nozzle at 2 MPa and 280°C with a velocity of 150 m/s. During the expansion process the steam loses 35 kJ/kg, and the

Insulation

Evacuated

Air

Non-heat-conducting wall

FIG. P 4-24M

enthalpy drops to 2.8 MJ/kg. What is the exit velocity from the nozzle? What flow area at inlet would be required for a flow rate of 1 kg/s?

4-27M   A high-speed flow of helium is to be heated in a duct. The gas enters the duct at 300 kPa, 20°C, and a velocity of 150 m/s. While traversing the duct the pressure drops 15 kPa and the velocity is reduced to 130 m/s. The exit temperature of the gas is 140°C. How much heat must be added for a mass flow of 100 g/s? What are the inlet- and exit-flow areas for the given conditions?

4-28M   High-pressure air at 500 kPa and 30°C is to be used to drive a small turbine for powering a high-speed drill. The drill requires a power input of 746 W (1 hp), and the air may be assumed to discharge from the turbine at 180 kPa. If the turbine process is quasistatic and adiabatic, calculate the mass flow rate of air required. If the inlet air line is 1 cm in diameter, calculate the entering air velocity.

4-29M   An industrial hot water heater uses a gaseous fuel with an internal chemical energy content of about 53 MJ/kg. The heater is designed so that 16 kg of air at 20°C and 1 atm are supplied for each kilogram of fuel, and it is designed to produce 38 liters/min of hot water at 80°C from cold water at 15°C. The hot discharge gases are at 110°C. Calculate the quantity of fuel required.

4-30M   A municipal incinerator is to burn garbage which has an average internal chemical energy of 7 MJ/kg. The hot gases from the burning process are then used to preheat the air as it enters the incinerator. The internal construction of the incinerator also allows it to be used as a high-pressure hot water heater for a nearby power plant, as shown in the sketch. The respective temperatures are $T_1 = 40°C$, $T_3 = 815°C$, $T_4 = 205°C$, $T_5 = 35°C$, $T_6 = 150°C$. Assume that the hot gases behave as an ideal gas and that 14 kg of air are required to burn each kilogram of garbage or each kilogram of an auxiliary fuel. The auxiliary fuel has an internal chemical energy of 46 MJ/kg and must be used to get the burning temperature high enough for complete combustion. How much total energy in the form of garbage and fuel must be supplied to heat 11 000 liters of water? If 900 kg of garbage are available for heating each 11 000 liters of water, how much auxiliary fuel would be required?

FIG. P 4-30M

4-31M A small auxiliary power device is designed as shown in the schematic. Nitrogen is contained at 14 MPa and 20°C in the tank which is connected
through a valve to an adiabatic turbine which drives an electric generator.
The power output of the device is to be 75 W, and the pressure at point 1 is
maintained constant at 700 kPa by a pressure regulator which operates as a
throttling device. The nitrogen discharges from the turbine at 100 kPa, and
the turbine process may be assumed adiabatic. What size tank is necessary to
produce the power output for a period of 1 h? Assume the tank remains
isothermal at 20°C and that power production ceases when the tank pressure
reaches 700 kPa.

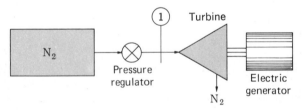

FIG. P 4-31M

4-32M A fluid undergoes a steady-flow process in which its enthalpy increases by 4.5
kJ/kg. The initial state of the fluid is 2 atm and 100°C, and the final state is
10 atm and 300°C. The initial and final velocities are 36 and 12 m/s, respectively, and the final state is at an elevation 25 m above the initial state.
25 MJ of heat are supplied for a flow rate of 0.75 kg/s. Calculate the work for
the process and express in units of W, Btu/h, and hp.

4-33M Steam undergoes an adiabatic and steady-flow process in a turbine from
6.0 MPa, 500°C, to a pressure of 10 kPa and a quality of 90 percent. Calculate
the work output for a flow of 1.0 kg/s. What flow rate would be required to
produce a work output of 400 kW?

4-34M A spherical tank is perfectly insulated and evacuated and exposed to the surrounding air at 1 atm and 35°C. A valve on the tank is opened on the side of

the tank and air is allowed to enter slowly. What is the temperature of the air in the tank when its pressure just reaches 1 atm?

4-35M A 1.0-m³ rigid vessel contains steam at 2.0 MPa and 320°C. A valve on the vessel is opened and steam is allowed to escape until the pressure drops to 100 kPa. During the process heat is added to maintain the temperature constant. Calculate the quantity of heat required.

4-36M Steam flows in a line at 1.5 MPa and 320°C. Connected to the line is an adiabatic cylinder with a frictionless piston as shown in the figure.

FIG. P 4-36M

One side of the cylinder contains 0.05 kg of air at 1 atm and 105°C, while the other side contains 0.03 kg of steam at 1 atm and 105°C. The piston is adiabatic. The valve is opened and steam is allowed to enter the cylinder until the pressure reaches 1.5 MPa, at which time the valve is closed. Calculate the mass of steam added, the final state of the steam in the cylinder, and the final air temperature.

4-37M A centrifugal blower receives air at 1 atm and 20°C in a volume flow rate of 0.7 m³/s. The air enters at 1.0 m/s and discharges at 10.0 m/s, and temperature is essentially constant. Calculate the input power requirements. State assumptions.

4-38M An alternative to an adiabatic throttling device for measuring steam quality is a setup in which the sample of steam is allowed to flow through a chamber where a measured quantity of electric heating (known) is applied. The flow rate of exit steam from the calorimeter is also measured. In one application, steam at 2.0 MPa is sampled at a flow rate of 3.8 g/s. 380 W of electric power are applied to the device, and the outlet conditions are measured as $p = 150$ kPa and $T = 180°C$. Calculate the quality of the steam at the 2.0-MPa condition.

4-39M A closed system undergoes a quasistatic process between two end states such that the enthalpy decreases by 110 kJ and the system gives up 10 kJ of heat to the surroundings. Calculate the work for this process. Then suppose the same system undergoes a steady-flow process between the same end states and also experiences an increase in kinetic energy of 5 kJ while experiencing the same heat loss. Determine the external work for this process.

4-40M Oxygen flows in a line at 70 m/s with conditions of 120 kPa and 15°C. Connected to the line is a 6.0-liter tank containing oxygen at 5°C and 1 atm. A valve is opened allowing oxygen to flow into the tank until it contains five times the original mass, at which time the valve is closed. The tank is adiabatic. Calculate the final pressure and temperature in the tank.

4-41M   Steam undergoes a process at constant enthalpy from 1.5 MPa, 360°C, to a pressure of 10 MPa. Find the final temperature in this process and the change in volume.

4-42M   Saturated-liquid water at 260°C is throttled to 100 kPa. Calculate the final quality if saturated or the final temperature if superheated. Calculate the velocity of the steam in the final state for a flow rate of 1 kg/min through a 5-cm-diameter tube.

4-43M   A tank having a volume of 0.1 m³ contains nitrogen at a pressure of 15 atm and 100°C and is perfectly insulated from the surroundings. A valve on the tank is opened and N_2 is discharged until the pressure drops to 2 atm. Calculate the mass of N_2 discharged from the tank.

REFERENCES

1   Hatsopoulos, G. N., and J. H. Keenan: "Principles of General Thermodynamics," John Wiley & Sons, Inc., New York, 1965.

2   Reynolds, W. C.: "Thermodynamics," 2d ed., McGraw-Hill Book Company, New York, 1968.

3   Sears, F. W.: "An Introduction to Thermodynamics, The Kinetic Theory of Gases, and Statistical Mechanics," Addison-Wesley Publishing Company, Inc., Reading, Mass., 1950.

4   Van Wylen, G. J., and R. E. Sonntag: "Fundamentals of Classical Thermodynamics," John Wiley & Sons, Inc., New York, 1965.

chapter five
Principles of statistical thermodynamics

5-1   INTRODUCTION

Up to this point we have been primarily concerned with macroscopic thermodynamics and the conservation of energy principle. Some preliminary remarks concerning microscopic thermodynamics were given in the introductory chapter, but we have not yet shown how these developments are related to the overall subject matter of thermodynamics. Some experimental consequences of the second law of thermodynamics were also stated in the first chapter, but their relationship to the microscopic behavior of matter has not been established.

The objective of this chapter is to establish some mathematical and physical modeling concepts for microscopic thermodynamics which may be invoked in later chapters concerned with specific applications. At this point the reader is cautioned against expecting final results from the exposition of this chapter. As we shall show, microscopic analysis can yield fruitful results only when combined with the natural macroscopic laws of thermodynamics. For this reason, we shall return to macroscopic analysis in Chaps. 6 and 7 in order to provide a proper link between the microscopic models of this chapter and the behavior of matter as observed in finite engineering systems. Finally, with the microscopic models connected to the finite world, we shall then begin to develop detailed calculation techniques in Chaps. 8 and 9 for the molecular behavior of gases and solids.

At some times the reader may begin to feel that the thread of continuity is wearing thin. In these circumstances he or she should recall that the final objective is to create a mathematical and physical model of the behavior of microscopic particles which is in accordance with phenomena which may be observed macroscopically. To accomplish this objective, we must bring all sources of analysis into play. This chapter furnishes some initial steps in the modeling procedures. To provide a proper focus of thought, the reader should continually ask the following questions as the development of Chaps. 5 to 9 progresses:

1   How does the establishment of a microscopic physical model enable us to calculate the macroscopic behavior of matter?
2   How do the laws of macroscopic thermodynamics help in the formulation of a microscopic physical model for a system?

5-2   PROBABILITY

For our purposes, probability will be related to the fraction of times a certain phenomenon is observed in a large number of occurrences. An unweighted coin will land on heads one-half of the time when flipped a large number of times, an unweighted die will land on each side one-sixth of the time in a large number of rolls, the probability of drawing a *particular* card from a deck of cards is $\frac{1}{52}$, and so on. Later on we shall be interested in counting the number of ways various objects and groups of objects may be arranged, subject to certain restrictions. This counting process is related to probability because the probability of occurrence of a certain event is the number of ways that event can occur divided by the total number of possible events. For example, consider the toss of two dice. We might want to know the probability that a seven will be obtained. There are six possible ways to obtain a seven and there are 36 possible arrangements. The probability of getting a seven is thus $\frac{6}{36}$ or $\frac{1}{6}$. If we know the probability that separate events will occur, the probability that either of the events will occur is the sum of the individual probabilities for the events. The probability for a double six to occur on the toss of two dice is $\frac{1}{36}$ (there is only one way this can occur), so that the probability of getting *either* a double six or seven would be

$$\frac{6}{36} + \frac{1}{36} = \frac{7}{36}$$

If several *independent* events occur at the same time such that each event has an independent probability p_i, the probability that *all* events will occur simultaneously is given as the product of the probabilities for the individual events. Thus

$$p = \prod_i p_i \tag{5-1}$$

where the $\prod$ sign denotes a product. A probability of unity corresponds to certainty.

It will be of interest to calculate the number of ways of arranging objects in sequence. Let us assume that we have N distinguishable objects and we want to calculate the number of different ways the objects may be arranged in sequence. For the first object in the sequence there are N choices, for the second object $N - 1$ choices, $N - 2$ choices for the third object, and so on. The total number of possible sequences is, therefore,

$$N(N - 1)(N - 2) \cdot \cdot \cdot N \cdot \cdot \cdot (N - 2)(1) = N! \tag{5-2}$$

It is important to note that all of these possible sequences would be identical if the objects were indistinguishable. Various sequences of a set of identical coins, for example, would all be the same. A sequence of playing cards, on the other hand, would have several possibilities.

Our later discussions will be concerned with the number of ways microscopic particles may be arranged among various energy states subject to specific restrictions such as conservation of energy and mass. Once we are able to predict the number of particles which occupy each energy state, then we may calculate the total energy of all particles by performing the summation

$$E = \Sigma n_i \epsilon_i$$

where ϵ_i is an energy level and n_i is the number of particles occupying this energy level. The allowable energy states which a particle may attain are determined from quantum mechanics.

5-3   PHASE SPACE

In the introductory chapter an analogy was drawn between mechanics and thermodynamics to the extent that the free body was likened to the system concept and thermodynamic properties were shown to be analogous to space and velocity (or momentum) coordinates in mechanics. It has been noted more than once that the observable thermodynamic properties, i.e., the macroscopic properties, are dependent on the microscopic state of the system. If we were able to say exactly what was happening at the microscopic level, then, presumably, we would be able to calculate the macroscopic thermodynamic properties.

Phase space is a six-dimensional space formed by the three Cartesian space coordinates x, y, and z, and the three momentum coordinates p_x, p_y, and p_z, where p denotes the momentum vector:

$$p_x = mv_x$$
$$p_y = mv_y$$
$$p_z = mv_z$$

There is really nothing new about phase space. It is a concept used in mechanics for the solution of even simple impulse-momentum problems. We are merely relating the fact that it is necessary to specify these six coordinates for a precise description of the dynamical state of a particle. We choose to call the region formed by these coordinates a phase space. The fact that a six-dimensional space is not readily accessible to intuitive visualization does not alter the basic concept. If we can establish the location of each microscopic particle in phase space, then we are able to specify the precise thermodynamic state of the system.

5-4   QUANTUM CONSIDERATIONS

It is well known that classical mechanics must be supplemented and modified by quantum theory when operating on a microscopic scale. The purpose of this brief discussion of quantum theory is to lay the foundation for a realistic physical model to describe the properties and behavior of matter on a microscopic scale and to show how the use of this physical model will result in calculation of macroscopic thermodynamic properties which are in agreement with experiment. Our objective is to use the principles of quantum theory to establish a physical model to be used for thermodynamic analysis. We shall, therefore, not go into a historical development of the subject but refer the reader to Ref. [1] for additional background information.

The basic premise of quantum theory is that energy occurs only in discrete quanta and that matter has a wave nature as well as a particle nature. The characteristic increment of energy for electromagnetic radiation is

$$E = h\nu \tag{5-3}$$

where ν is the frequency of the radiation and h is Planck's constant with the value of 6.625×10^{-34} J·s. Energy transfer at the microscopic level occurs as a result of the transition from one quantum state to another. In this transition process a photon having the energy given by Eq. (5-3) is frequently emitted or absorbed, depending on the direction of the energy transfer. As a matter of experimental evidence, the energy transitions occur in discrete increments and, thus, the quantum concepts are accepted as basic physical postulates.

Electromagnetic radiation travels at the speed of light c and its wavelength λ is defined as

$$c = \lambda\nu \tag{5-4}$$

According to the special theory of relativity, energy is related to mass with

$$E = mc^2 \tag{5-5}$$

The "mass" of a photon is therefore identified with

$$m = \frac{E}{c^2} = \frac{h\nu}{c^2} \tag{5-6}$$

and the momentum of a photon is calculated as

$$p = mc = \frac{h\nu}{c} = \frac{h}{\lambda} \tag{5-7}$$

All physical principles are based on experiment. This is true of quantum theory as well as of macroscopic energy conservation principles; however, an additional principle limits the precision with which measurements may be performed at the microscopic level. The *Heisenberg uncertainty principle* limits the specification of momentum and incremental displacement according to

$$\Delta p_x\, \Delta x \sim h \tag{5-8}$$

That is, the closer we measure the location of a particle, the less precise will be our measurement of its momentum. The net result of the uncertainty principle is that the best we can hope to do is to locate particles in small volumes of phase space having the dimensions of h^3. This is analogous to having several boxes all the same size with balls placed in the boxes. The lid is closed on each box so that we are unable to specify the exact location of each ball in the boxes. We are only allowed to specify the *number* of balls in each box.

At first glance, we might say that the uncertainty principle introduces no serious measurement problem because the allowable uncertainties are so small. It does not, for macroscopic systems; but with the very small dimensions and momenta involved in microscopic analysis, it represents a significant limitation.

Quantum theory postulates that particles as well as electromagnetic radiation have energies and momenta which are quantized and have wave properties. From Eq. (5-7), the characteristic wavelength† is expressed in terms of momentum by

$$\lambda = \frac{h}{p} \tag{5-9}$$

In general, we may expect that particles may only have speeds and momenta which result in integral numbers of characteristic wavelengths. This means that if the characteristic wavelength is very small compared to a characteristic dimension of the particle or some characteristic displacement encountered in the motion of the particle, then quantum effects are not too important, since a large number of wavelengths is available for the particle motion so that a continuous displacement spectrum is observed. We may see the nature of this argument by examining two simple cases.

Consider first the motion of a 1-kg mass at a velocity of 1 m/s. The characteristic wavelength for this motion is

$$\lambda = \frac{h}{p} = \frac{h}{mv} = \frac{6.625 \times 10^{-34}}{(1)(1)} = 6.625 \times 10^{-34} \text{ m}$$

Certainly there is no great restriction in saying that the motion of this mass may occur only in increments of 6×10^{-34} m. We may, therefore, conclude that quantum considerations are not important in this motion—as we knew from the start.

Now consider the motion of an electron having a mass of 9.106×10^{-31} kg at a velocity of 10^6 m/s. The characteristic wavelength in this case is (neglecting relativistic effects)

---

† This parameter is frequently termed the de Broglie wavelength.

$$\lambda = \frac{h}{mv} = \frac{6.625 \times 10^{-34}}{(9.106 \times 10^{-31})10^6} = 7.28 \times 10^{-10} \text{ m}$$

Although 7.28×10^{-10} m may seem a very small distance, it is quite large for the motion of microscopic particles. For comparison purposes, it may be noted that the radius of an oxygen molecule is about 2×10^{-10} m. The aforementioned wavelength is, therefore, rather large for electron motion, and we might expect that quantum effects could exert a significant influence on electron motion.

The purpose of this brief discussion has been solely to show that quantum effects can be significant when considering the motion, energy, and momentum distributions of microscopic particles. A development of precise calculation techniques for quantum-mechanical energy distributions is beyond the scope of our discussion.

5-5 DEGENERACY

We have already recognized that energy at the microscopic level is quantized. In addition to the notion of discrete energy levels we also need to recognize that a particular energy level may occur in more than one way and still satisfy the principles of quantum mechanics involved. For a given quantum energy level ϵ_i we shall designate the number of quantum *states* having the energy ϵ_i with the symbol g_i. If more than one quantum state with the energy ϵ_i is possible, then the system of particles is said to be *degenerate*. At the moment we have no specific indication as to the possible values of g_i for different energy levels, but we may obtain a physical understanding of the degeneracy idea by considering a single particle which may have three translational degrees of freedom. Let us choose some energy unit which is the minimum increment allowed according to quantum considerations and examine the possible number of ways by which four energy levels could be attained. The situations of interest are depicted in Fig. 5-1. This simple tabulation shows that the number of possible quantum states increases very rapidly with an increase in the energy level. We may remark that the minimum quantum increment of energy may not always be the same for all energy modes. For example, a diatomic gas molecule may involve translational, rotational, and vibrational energy modes, as illustrated in Fig. 5-2. The characteristic quantum energy increment for rotational energy may not be the same as for the translational modes, and we shall find it necessary to take this into account in our later calculations. Accordingly, the term *degeneracy* may be applied to several energy modes.

It is interesting to note at this point that for ideal gases at moderately high temperatures (room temperature is moderately high), the number of

1 energy unit (eu) = minimum quantum increment											
$\epsilon_i = 0$			$\epsilon_i = 1$ eu			$\epsilon_i = 2$ eu			$\epsilon_i = 3$ eu		
K.E.$_x$	K.E.$_y$	K.E.$_z$	K.E.$_x$	K.E.$_y$	K.E.$_z$	K.E.$_x$	K.E.$_y$	K.E.$_z$	K.E.$_x$	K.E.$_y$	K.E.$_z$
0	0	0	1	0	0	1	1	0	1	1	1
			0	1	0	0	1	1	1	2	0
			0	0	1	1	0	1	1	0	2
						2	0	0	2	1	0
						0	2	0	0	1	2
						0	0	2	2	0	1
									0	2	1
									3	0	0
									0	3	0
									0	0	3
$g_i = 1$			$g_i = 3$			$g_i = 6$			$g_i = 10$		

FIG. 5-1 *Illustration of degeneracy concept for simple translational kinetic energy model.*

available energy states is very large compared with the number of particles which are to occupy these states. Such a gas is highly degenerate because there are many states which have the same energy level.

The *nondegenerate* system is one for which $g_i = 1$, that is, each energy level may be attained in only one way.

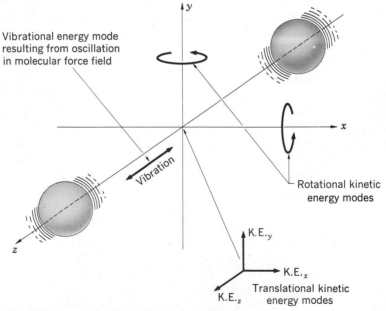

FIG. 5-2 *Energy modes for a diatomic molecule.*

5-6 MICROSTATES, MACROSTATES, AND THERMODYNAMIC PROBABILITY

For the moment let us consider a monatomic gas the molecules of which act as perfect billiard balls. We could presumably calculate the internal energy of the gas by taking the sum of the kinetic energies of all the molecules. The pressure of the gas could be expressed in terms of the momentum exchange of the gas molecules with a containing wall. To specify the *microstate* of such a system, we would have to specify the exact location and velocity of *each* particle in phase space, i.e., we would have to specify the location and velocity of every particle in the system. Such a detailed specification, however, is unnecessary for purposes of calculating the macroscopically observable thermodynamic properties; that is, to calculate the energy, we need to know only the *number* of particles which have a certain kinetic energy (or velocity). More specifically, we need to know the *number* of particles in each element of volume (a six-dimensional element) of phase space, and it is unnecessary to specify *which* particle is in each volume element. When the number of particles in each element of phase space is specified without regard to which particles are involved, we say that a *macrostate* has been determined. We shall have more to say about these concepts later, but the reader should realize that a determination of the macrostate of a system will be of central concern since it is related to observable thermodynamic properties.

Before we can analyze the physical models discussed above some mathematical preliminaries are required. These preliminaries are directed toward a calculation of the number of permutations and combinations which may be achieved by arranging balls in boxes. Later, the balls will represent particles and the boxes will represent energy levels and quantum states which the particles may occupy.

Our eventual objective is to determine the macrostate which is most probable and then to calculate the energy corresponding to this state. The most probable macrostate will be the one which has the largest number of possible microstates, i.e., the one that can occur in the largest number of ways. In the following calculations we shall determine a quantity Ω called the *thermodynamic probability*, which is the number of microstates corresponding to a macrostate. It is, in reality, a "number of arrangements" instead of a "probability." Again, using the throw of two dice as an example we might examine the macrostate which gives a total of seven on the throw. There are six ways to throw a seven so there are six microstates for the designated macrostate. In this case we would say that Ω is 6. The value of Ω for a throw of eleven would be two, etc. The important point to remember is that the number of arrangements is recognized to be proportional to probability, so that a study of the probability of observing certain arrangements may be made by examining the behavior of Ω.

The three mathematical cases corresponding to our later physical models are as follows:

Case 1 The number of ways a total of N *distinguishable* balls may be arranged in a set of large boxes with N_1 balls in the first box, N_2 balls in the second box, N_i balls in the ith box, and so on, with no restriction on the number of balls in each box (except that the number obviously cannot be greater than N, the total number of balls). Inside each ith large box there are g_i smaller boxes which are distinguishable and into which the N_i balls may be distributed. The number of small boxes g_i may be greater or less than the number of balls N_i which will be distributed among them, and the number of small boxes in each large box is not necessarily the same. This situation is depicted in Fig. 5-3.

Case 2 The same as case 1, except that the balls are *indistinguishable*.

Case 3 The same as case 2, except that a restriction is imposed on the system such that no more than *one* ball may be placed in each *small* box.

In these models the large boxes will correspond to energy levels, the small boxes to the number of quantum states in each energy level, or degeneracy, and the balls are the particles.

Case 1 may be calculated as follows. The *total* number of ways N distin-

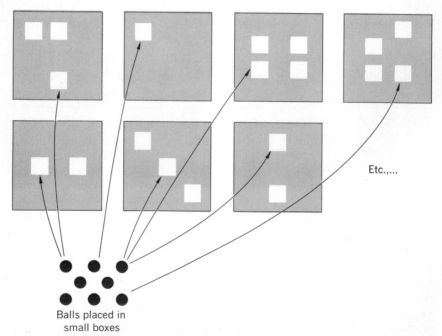

Balls placed in
small boxes

FIG. 5-3 *Balls and boxes for calculation of number of arrangements.*

Etc.,...

guishable balls may be arranged in sequence is N!. The number of ways of arranging the balls inside the large boxes is $N_1!$ in the first box, $N_2!$ in the second box, and so on. Thus, if we are to determine the number of ways of arranging the balls in the various large boxes, we must divide out the number of arrangements in each box by forming the quotient

$$\frac{N!}{N_1!N_2!N_3! \cdots N!}$$

or

$$\Omega = \frac{N!}{\prod_i N_i!} \tag{5-10}$$

where N_i designates the number of balls in the ith large box. We must also consider the number of possible ways the N_i balls may be arranged in the g_i small boxes contained in each large box. For each ball we have g_i choices, so that for all balls in the ith large box we have the following number of possibilities:

$$g_i \cdot g_i \cdot g_i \cdots = g_i^{N_i} \tag{5-11}$$

We thus have the result that there are $g_i^{N_i}$ possible ways to arrange the N_i balls in the g_i small boxes inside the ith large box. This means that Eq. (5-10) must be modified in the following way to account for all possible arrangements:

$$\Omega = \frac{N!}{\prod_i N_i!} g_1^{N_1} \cdot g_2^{N_2} \cdot g_3^{N_3} \cdots g_i^{N_i}$$

or

$$\Omega = N! \sum_i \frac{g_i^{N_i}}{N_i!} \tag{5-12}$$

Case 2 is somewhat more complicated. In this instance the balls are assumed to be indistinguishable so that the permutations of a number of balls inside a box do not alter the basic arrangement. There are g_i small boxes of the ith type and a total of N_i indistinguishable balls distributed among these boxes. Let us consider first the number of ways of arranging N_i indistinguishable balls in the g_i distinguishable boxes. Suppose that we start placing the N_i balls in the g_i small boxes one at a time, and *assume for the moment that the balls are distinguishable*. To show the nature of the counting process we shall designate the small boxes with numbers 1, 2, 3, etc., and the balls with the letters A, B, C, etc. For two boxes and two balls the possible arrangements are

```
1 A,   2 B        2 B,   1 A
1 AB, 2 0         2 0,   1 AB
1 BA, 2 0         2 0,   1 BA
1 B,   2 A        2 A,   1 B
1 0,   2 AB       2 AB, 1 0
1 0,   2 BA       2 BA, 1 0
```

where the zero indicates zero balls in the box. If we were to expand the listing with larger numbers of balls and boxes we would find that each possible arrangement could be designated by a sequence beginning with a number (a box) and including all the letters (balls) and other numbers. The total number of sequences of g_i and N_i distinguishable objects is

$$(g_i + N_i)! = (g_i + N_i)(g_i + N_i - 1)!$$

and the number always beginning with g_i (a box) is

$$g_i(g_i + N_i - 1)! \qquad (5\text{-}13)$$

In terms of our balls and boxes this is the number of ways of arranging N_i distinguishable balls in g_i distinguishable boxes. Because the balls are actually indistinguishable such pairs of sequences as

```
1 A,   2 B       1 B,   2 A
1 AB, 2 0        1 BA, 2 0
```

are actually the same. Now consider the two pairs:

```
1 AB, 2 0        2 0, 1 AB
1 B,   2 A       2 A, 1 B
```

In terms of a microstate of the system the arrangements in each pair are identical. It makes no difference how the small boxes (energy states) are arranged; the number of particles in each is what matters. We can take care of the indistinguishability of the balls and unimportance of ordering of small boxes by dividing the relation in Eq. (5-13) by $N_i!$ and $g_i!$, respectively, so that

$$\Omega_i = \frac{g_i(g_i + N_i - 1)!}{g_i! N_i!} = \frac{g_i(g_i + N_i - 1)!}{g_i(g_i - 1)! N_i!}$$

or

$$\Omega_i = \frac{(g_i + N_i - 1)!}{(g_i - 1)!N_i!} \tag{5-14}$$

The total thermodynamic probability is computed by taking the product of all such functions over all the "large" boxes.

$$\Omega = \prod_i \Omega_i = \prod_i \frac{(g_i + N_i - 1)!}{(g_i - 1)!N_i!} \tag{5-15}$$

In this equation we are making use of the product relation in Eq. (5-1) relating the total probability to the probability of independent events.

It is of interest to note that Eq. (5-15) simplifies to the following relation when $g_i \gg 1$:

$$\Omega = \prod_i \frac{(g_i + N_i)!}{g_i!N_i!} \tag{5-16}$$

We should note here that although individual quantum states are distinguishable (the g_i), they behave as if they were indistinguishable for determining the thermodynamic probability. Summarizing, we can see how this reasoning operates with the 1, 2, A, B sequence discussed above ($N_i = 2, g_i = 2$). For the four distinguishable objects the total number of sequences is 4! = 24, and the number of sequences starting with a number is $g(g + N - 1) = (2)(2 + 2 - 1)! = 12$. If the balls are indistinguishable we divide by 2! and obtain only six arrangements, and if the ordering of the boxes is not important we divide by another 2! and get only three different arrangements for computing the macrostate. These three arrangements are one ball in each box and both balls in either box.

Case 3 obviously involves the restriction that $N_i \leq g_i$; otherwise there would have to be more than one ball in each small box. Suppose we start placing the N_i balls in the g_i small boxes, one at a time. For the first ball there would be g_i open boxes, $g_i - 1$ for the second ball, $g_i - 2$ for the third ball, etc., so that the possible number of arrangements in the ith large box, if the balls were distinguishable, would be

$$g_i(g_i - 1)(g_i - 2) \cdots [g_i - (N_i - 1)] = \frac{g_i!}{(g_i - N_i)!}$$

Since the balls are indistinguishable, we must divide by $N_i!$ to obtain the desired number of arrangements. Thus

$$\Omega_i = \frac{g_i!}{N_i!(g_i - N_i)!} \tag{5-17}$$

The possible number of arrangements for all the ith boxes is, therefore,

$$\Omega = \prod_i \Omega_i = \prod_i \frac{g_i!}{N_i!(g_i - N_i)!} \tag{5-18}$$

The following examples illustrate the application of these formulas for calculation purposes. Subsequent sections will show how they are related to a microscopic physical model.

EXAMPLE 5-1

Calculate the number of ways of arranging six distinguishable balls in four boxes so that $N_1 = 1$, $N_2 = 1$, $N_3 = 3$, $N_4 = 1$.

SOLUTION   Equation (5-10) applies for this case, so that

$$\Omega = \frac{N!}{\prod_i N_i!} = \frac{6!}{1!1!3!1!} = 120$$

EXAMPLE 5-2

Calculate the number of ways of arranging six indistinguishable balls in four distinguishable boxes.

SOLUTION   For this problem we use Eq. (5-13) and take out the distinguishability of the balls by dividing by $N_i!$ so that

$$\Omega = \frac{g(g + N - 1)!}{N!} = \frac{4(4 + 6 - 1)!}{6!} = 2016$$

If the boxes were indistinguishable, i.e., their ordering were unimportant, Eq. (5-14) would apply and we would have

$$\Omega = \frac{(g + N - 1)!}{(g - 1)N!} = \frac{(4 + 6 - 1)!}{(4 - 1)!6!} = 84$$

EXAMPLE 5-3

Calculate the number of ways of arranging six indistinguishable balls in eight distinguishable boxes with no more than one ball per box.

SOLUTION   Equation (5-17) applies, so that

$$\Omega = \frac{g!}{N!(g - N)!} = \frac{8!}{6!(8 - 6)!} = 28$$

5-7   PHYSICAL MODELS

Let us now formulate several possible physical models for the microscopic behavior of matter. We shall, for the moment, consider only particulate matter (no electromagnetic radiation) and will limit our attention to isolated systems. Four fundamental assumptions are made for all the models.

1　The total energy of the system remains constant.
2　The total number of particles remains constant.
3　A sufficiently large number of particles is involved so that their behavior may be inferred from a statistical analysis. In other words, we assume the sample is large enough that statistical principles will apply.
4　All microstates are equally probable. In other words, a particle is just as likely to be in one volume element of phase space as in any other volume element.

The first two assumptions are necessary because the energy and mass of an isolated system remain constant while the last two assumptions govern the way we attack the problem in a statistical sense.

Three possible physical models are proposed:

1　*Maxwell-Boltzmann* (MB) *model.* The particles are distinguishable and distributed among various quantum energy levels designated by the subscript i. Thus there are N_i particles having an energy ϵ_i in the ith energy level. The values of the energies ϵ_i are quantized, and there are several ways that a particle may attain the energy ϵ_i. For example, a particle having only translational kinetic energy could have the energy distributed several different ways among each translational mode. Three possibilities for a particle having a total kinetic energy of six units could be:

$$\begin{Bmatrix} \text{K.E.}_x \\ \text{K.E.}_y \\ \text{K.E.}_z \end{Bmatrix} = \begin{Bmatrix} 1 \\ 2 \\ 3 \end{Bmatrix} \text{ or } \begin{Bmatrix} 3 \\ 0 \\ 3 \end{Bmatrix} \text{ or } \begin{Bmatrix} 2 \\ 1 \\ 3 \end{Bmatrix}$$

Thus, for generality, we must assume that each group of particles (in the ith energy level) could be composed of several quantum *states* g_i, all having the same energy ϵ_i. There are no restrictions on the number of particles which may occupy each energy level ϵ_i or on the number of quantum states within each energy level.
2　*Bose-Einstein* (BE) *model.* Here the physical situation is the same as the MB model except that the particles are *indistinguishable*. Still, there is no restriction on the number of particles which may occupy each energy level or on the various quantum states g_i.
3　*Fermi-Dirac* (FD) *model.* This model is the same as the BE model except that no more than one particle is allowed to occupy each quantum state. This restriction requires that $g_i \geqslant N_i$.

These three physical models serve to analyze a great body of microscopic phenomena. All three models assume that the particular quantum energy level may be attained in several different ways. The MB model as-

TABLE 5-1
Summary of characteristics of statistical models

Statistics	Type of particles	Energy quantized	No. of particles per quantum state
Maxwell-Boltzmann (MB)	Distinguishable	Yes	Any number
Bose-Einstein (BE)	Indistinguishable	Yes	Any number
Fermi-Dirac (FD)	Indistinguishable	Yes	One

sumes that each particle is different in some way, i.e., distinguishable, whereas the BE and FD models assume that they are indistinguishable. The characteristics of the three models are summarized in Table 5-1.

5-8 PURPOSE OF STATISTICAL ANALYSIS

At this time we shall not be concerned with the type of system (gas, liquid, or solid) for which each of the foregoing physical models applies. Our purpose is to obtain the equilibrium energy distribution for each of the physical models subject to constant total energy and number of particles restrictions. In other words, we want to determine the number of particles which will occupy each energy state (level), which, in turn, will give the number of particles in each energy level under the most probable conditions; i.e., we seek N_i as a function of ϵ_i. The most probable distribution is called the *equilibrium distribution*.

We are interested in the equilibrium distribution because this distribution governs the macroscopic properties of systems normally observed experimentally. Thus our analysis is being directed toward an eventual tie-in with macroscopic thermodynamics—as was promised in the opening remarks of this chapter.

We shall use the term *most probable* to designate the state which is attainable through the largest number of permutations. Since the objective of the analysis is to determine the number distribution of particles in various energy states, we are, in fact, seeking a specification of the most probable *macrostate* of the system. This macrostate may then be related to the macroscopically observable thermodynamic properties of the system. Once the energy distributions are obtained, we shall be able to apply the information to specific physical systems. Most of the applications of the statistical material will be presented in Chap. 8.

It should be fairly obvious that the three physical models discussed in the preceding section correspond *exactly* to the mathematical statistics models concerning balls and boxes discussed in Sec. 5-6. The thermodynamic probability or number of arrangements Ω is then the quantity we wish to maximize, subject to the constant energy and number of particles restrictions.

5-9 STIRLING'S APPROXIMATION

In subsequent sections we shall be dealing with factorials of large numbers. In particular we shall be interested in evaluating

$$\ln x! \qquad \text{for } x \gg 1$$

We may write

$$\ln x! = \ln 2 + \ln 3 + \cdots + \ln x = \sum_{i=1}^{n} \ln x_i \qquad (5\text{-}19)$$

This sum is approximated by the area under the curve shown in Fig. 5-4. The summation in Eq. (5-19) is the sum of the small rectangular areas shown in the figure. When x becomes very large, the sum of these rectangular areas is closely approximated by the area under the curve $y = \ln x$. Thus we may write

$$\ln x! \approx \int_{1}^{x} \ln x \, dx \qquad \text{for } x \gg 1$$

Evaluating the integral,

$$\ln x! \approx x \ln x - x - 1$$

Finally, for large x we may neglect the 1, so that

$$\ln x! \approx x \ln x - x \qquad \text{for } x \gg 1 \qquad (5\text{-}20)$$

Equation (5-20) is called Stirling's approximation for the factorial.

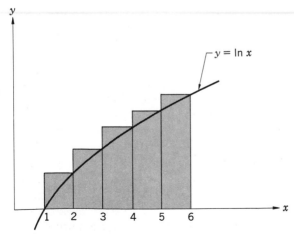

FIG. 5-4 *Stirling's approximation.*

5-10 BOSE-EINSTEIN STATISTICS

We shall now obtain the equilibrium distribution for the BE physical model. We have

$$\Omega = \prod_i \frac{(N_i + g_i - 1)!}{N_i!(g_i - 1)!} \tag{5-21}$$

and want to maximize Ω, subject to

$$N = \sum_i N_i \qquad \text{constant number of particles} \tag{5-22}$$

$$U = \sum_i \epsilon_i N_i \qquad \text{constant total internal energy} \tag{5-23}$$

Here the symbol ϵ_i designates the energy of each particle occupying the ith energy level.

We now want to maximize Ω with respect to the number of particles N_i having each energy ϵ_i. Thus we seek the distribution of particle energies which is most probable and, therefore, the one which we may call the equilibrium energy distribution. If Ω is maximized, so is $\ln \Omega$. It will be more convenient to work with the logarithm, so we write

$$\ln \Omega = \sum_i [\ln (N_i + g_i - 1)! - \ln N_i! - \ln (g_i - 1)!]$$

If both g_i and N_i are large compared to unity, ones may be neglected and Stirling's approximation may be applied to obtain

$$\ln \Omega = \sum_i [(N_i + g_i) \ln (N_i + g_i) - (N_i + g_i)$$
$$- N_i \ln N_i + N_i - g_i \ln g_i + g_i] \tag{5-24}$$

The maximization process is denoted by setting

$$\frac{\partial(\ln \Omega)}{\partial N_i} = 0 \tag{5-25}$$

This condition means that the variation in $\ln \Omega$ is zero for a small variation in N_i from the equilibrium distribution or

$$\delta(\ln \Omega) = 0 = \frac{\partial(\ln \Omega)}{\partial N_i} \delta N_i$$

Using Eq. (5-24), this condition becomes

$$\sum_i \ln \frac{N_i + g_i}{N_i} \, \delta N_i = 0 \qquad\qquad (5\text{-}26)$$

The constant-energy condition is

$$\delta U = 0 = \frac{\partial U}{\partial N_i} \, \delta N_i$$

or

$$\sum_i \epsilon_i \, \delta N_i = 0 \qquad\qquad (5\text{-}27)$$

Finally, the particle-conservation condition is

$$\delta N = 0 = \frac{\partial N}{\partial N_i} \, \delta N_i$$

or

$$\sum_i \delta N_i = 0 \qquad\qquad (5\text{-}28)$$

We thus have three conditions to satisfy in order to establish the maximum probability. If we were not bound by the constant energy and number of particles restrictions, then the number of particles in each energy state N_i could all be considered as independent variables. In this case we would only need to apply Eq. (5-26) and obtain

$$\ln \frac{N_i + g_i}{N_i} = 0$$

as the maximizing condition. Clearly, as soon as the total number of particles is specified, not all the N_i's are independent. Similarly, the energy-conservation condition imposes an additional restriction on the independence of the N_i's. The particular restrictions are governed by the particular total energy of the system and the particular total number of particles. Suppose Eq. (5-27) is multiplied by a constant $-\beta$, which is a function of the total system energy, and Eq. (5-28) is multiplied by another constant $-\alpha$, which is a function of the total number of particles. If the resulting equations are added to Eq. (5-26), we obtain

$$\sum_i \left[\ln \left(1 + \frac{g_i}{N_i} \right) - \beta \epsilon_i - \alpha \right] \delta N_i = 0 \qquad\qquad (5\text{-}29)$$

In effect, the constants β and α have incorporated the energy and number of particles restriction into Eq. (5-26) so that the N_i's may now be treated as independent. We thus obtain the condition

$$\ln \left(1 + \frac{g_i}{N_i}\right) - \beta \epsilon_i - \alpha = 0$$

or

$$\frac{N_i}{g_i} = \frac{1}{Ae^{\beta \epsilon_i} - 1} \tag{5-30}$$

where, now, A has been substituted for e^{α}. The constants α and β are called *undetermined multipliers* and the technique of maximizing Ω is called the *Lagrange multiplier method*. The constants α and β play a role similar to constants of integration in the solution of differential equations in that their values must be determined from appropriate physical boundary conditions. We shall determine the values of these constants in subsequent sections.

5-11 FERMI-DIRAC STATISTICS

The FD distribution is determined in an analogous way to the BE model. The thermodynamic probability distribution is given by Eq. (5-18), and we want to maximize $\ln \Omega$ subject to the restrictions of constant total energy and constant total number of particles. We have, after applying Stirling's approximation to $\ln \Omega$,

$$\delta(\ln \Omega) = \frac{\partial(\ln \Omega)}{\partial N_i} \delta N_i = \sum_i \ln \frac{g_i - N_i}{N_i} \delta N_i = 0 \tag{5-31}$$

$$\delta U = \frac{\partial U}{\partial N_i} \delta N_i = \sum_i \epsilon_i \, \delta N_i = 0 \tag{5-32}$$

$$\delta N = \sum_i \delta N_i = 0 \tag{5-33}$$

Using the undetermined-multiplier technique as before, we obtain

$$\sum_i \left[\ln \left(\frac{g_i}{N_i} - 1\right) - \alpha - \beta \epsilon_i\right] \delta N_i = 0 \tag{5-34}$$

There results

$$\frac{N_i}{g_i} = \frac{1}{Ae^{\beta \epsilon_i} + 1} \tag{5-35}$$

For the MB model we would obtain

$$\frac{N_i}{g_i} = \frac{1}{Ae^{\beta \epsilon_i}} \tag{5-36}$$

Equations (5-30), (5-35), and (5-36) are very similar in that they differ only in the way the factor of one appears in the denominator. Consider the physical circumstance where $N_i \ll g_i$; i.e., the number of particles is much smaller than the available number of quantum states for each energy level. In this case the factor of one in the denominator of Eqs. (5-30) and (5-35) is very small compared with the $Ae^{\beta \epsilon_i}$ term, and both the BE and FD distributions approach Eq. (5-36), the MB model. This limiting case is important because it will enable us to analyze indistinguishable particles with the simpler MB distribution for cases where $N_i \ll g_i$.

5-12   CLASSICAL MAXWELL-BOLTZMANN MODEL

The quantum nature of the statistical model can be removed in the following way. We examine the MB model for the case where $g_i = 1$ for all energy levels, and the energy is assumed to have a continuous distribution so that no quantum restrictions are placed on the energy levels. In this instance each energy level can be slightly different from others and there is no need to be concerned about different "kinds" of energy states, so we take $g_i = 1$ for each energy level. There is, in effect, an infinite number of possible energy states for such a continuous distribution. Such a model is called a *classical* statistical model because it ignores the possibility of different types of quantum states having the same energy. In this case the number of particles N_i having the energy ϵ_i is simply

$$N_i = \frac{1}{Ae^{\beta \epsilon_i}} \tag{5-37}$$

There is a problem with the MB model. Certain types of microscopic particles just are not distinguishable. It is not possible, for example, to tell one oxygen molecule from another or one electron from another. It will be seen later that this fact makes the MB model unsuitable for analyzing some substances, except when employed as a limiting case of the BE or FD statistics.

5-13   THE EQUILIBRIUM DISTRIBUTION

The foregoing analyses have served to establish a most probable distribution of particles among various energy states subject to restrictions of constant energy and number of particles. We have yet to determine the values of the constants A and β, but the functional form of the distribution has been established. This distribution represents, in fact, the most probable *macro-*

state. It defines the energy distribution which should be observed most often. It is of interest to know whether the distribution is just slightly more probable than some other distribution or whether it is overwhelmingly the most probable occurrence. We choose to call this most probable distribution the *equilibrium macrostate* of the system since it is the one we would expect to observe after the system is left alone for a while. If it is overwhelmingly the most probable occurrence, then it should be directly related to the *macroscopic* thermodynamic properties which are normally observed. For simplicity, let us consider the MB distribution. We have

$$\Omega = N! \prod_i \frac{g_i^N}{N_i!}$$

Forming the logarithm and using Stirling's approximation,

$$\ln \Omega = N \ln N + \sum_i N_i \ln g_i - \sum_i N_i \ln N_i \tag{5-38}$$

The most probable distribution is obtained by inserting Eq. (5-36) into Eq. (5-38). Thus

$$\ln \Omega_{max} = N \ln N + \sum_i N_i(\ln A + \beta\epsilon_i) \tag{5-39}$$

We now wish to examine the effect of a change in N_i on Ω. In particular, we want to compare $\Omega_{max} + \delta\Omega$ with Ω_{max}, where $\delta\Omega$ is the deviation from Ω_{max} resulting from some deviation δN_i from the most probable distribution. We can write

$$\ln (\Omega + \delta\Omega) = N \ln N + \sum_i (N_i + \delta N_i) \ln g_i$$
$$- \sum_i (N_i + \delta N_i) \ln (N_i + \delta N_i) \tag{5-40}$$

Subtracting Eq. (5-38) from Eq. (5-40),

$$\ln \frac{\Omega + \delta\Omega}{\Omega} = \sum_i \ln g_i \, \delta N_i - \sum_i N_i \ln \left(1 + \frac{\delta N_i}{N_i}\right)$$
$$- \sum_i \ln (N_i + \delta N_i) \, \delta N_i \tag{5-41}$$

One of the conditions for Ω_{max} is $\delta(\ln \Omega) = 0$ or

$$\sum_i (\ln g_i - \ln N_i) \, \delta N_i = 0 \tag{5-42}$$

Subtracting Eq. (5-42) from Eq. (5-41),

$$\ln \frac{\Omega_{max} + \delta\Omega}{\Omega_{max}} = - \sum_i N_i \ln \left(1 + \frac{\delta N_i}{N_i}\right) - \sum_i \ln \left(1 + \frac{\delta N_i}{N_i}\right) \delta N_i \quad (5\text{-}43)$$

With some algebraic manipulation Eq. (5-43) can be reduced to the following expression under the condition that δN_i be small compared with N_i.

$$\ln \frac{\Omega_{max} + \delta\Omega}{\Omega_{max}} \approx - \frac{1}{2} \sum_i \frac{(\delta N_i)^2}{N_i} \quad (5\text{-}44)$$

This relation provides a means for comparing the probability of a distribution deviating by an amount $\delta\Omega$ from the equilibrium distribution as a result of a change δN_i in the occupation of the ith energy level. As an example, let us consider just two energy cells both having the same energy and containing 10^{20} total particles between them. Since the energy of each cell is the same, the most probable distribution of particles is one of equal numbers of particles in each cell, which we take as 5×10^{19}. Let us investigate the likelihood of a deviation from this most probable distribution. Suppose that only 0.01 percent of the particles were to change cells. We have

$$N_1 = N_2 = 5 \times 10^{19}$$
$$\delta N_1 = -\delta N_2 = (0.0001)(5 \times 10^{19}) = 5 \times 10^{15}$$

Forming the sum in Eq. (5-44) gives

$$\ln \frac{\Omega_{max} + \delta\Omega}{\Omega_{max}} = - \frac{1}{2} \sum_{i=1}^{2} \frac{(\delta N_i)^2}{N_i}$$

$$= - \frac{1}{2} \left[\frac{(5 \times 10^{15})^2}{5 \times 10^{19}} + \frac{(-5 \times 10^{15})^2}{5 \times 10^{19}}\right]$$

$$= -5 \times 10^{11}$$

or

$$\frac{\Omega_{max} + \delta\Omega}{\Omega_{max}} \approx e^{-5\times10^{11}}$$

Thus we find that the probability of even this slight deviation from the most probable state is very small indeed.

We could carry this discussion further, but it should already be evident to the reader that the equilibrium distributions given in Eqs. (5-35) and (5-36) are not just the most probable ones—they are overwhelmingly so. Furthermore, they are so predominant that we should not expect to observe macroscopic properties which deviate from those calculated with these distributions, except in very special circumstances.

As mentioned previously, our analytical efforts are directed toward a calculation of macroscopic thermodynamic properties from microscopic considerations. We should mention, however, that there are important areas of fluctuation theory and transport phenomena where deviations from the most probable equilibrium distribution are very significant, but these subjects are beyond the scope of our discussion.

5-14 MICROSCOPIC INTERPRETATION OF HEAT AND WORK

We have already seen that the first law of thermodynamics can be written as

$$dU = d'Q + d'W \tag{5-45}$$

for a closed system. From Eq. (5-23) the change in internal energy can be written

$$dU = \sum_i \epsilon_i \, dN_i + \sum_i N_i \, d\epsilon_i \tag{5-46}$$

Now consider a simple system of noninteracting particles whose only work interaction occurs in the form of a compression (volume change) work mode. It can be shown from quantum mechanics that the allowable translational energy states of such a system are proportional to $V^{-2/3}$ where V is the system volume. When V is fixed the allowable energy states are fixed. We can therefore take $\epsilon_i = f(V)$ and rewrite Eq. (5-46) as

$$dU = \sum_i \epsilon_i \, dN_i + \sum_i N_i \left(\frac{d\epsilon_i}{dV}\right) dV \tag{5-47}$$

If we apply this equation to a quasistatic process for a simple compressible substance, we see that for zero volume change the work is zero and the internal energy must increase because of the heat interactions, so we identify

$$d'Q = \sum_i \epsilon_i \, dN_i \tag{5-48}$$

For zero heat transfer, but with work addition at the boundary, we interpret

$$d'W = \sum_i N_i \left(\frac{d\epsilon_i}{dV}\right) dV = \sum_i N_i \, d\epsilon_i \tag{5-49}$$

From this discussion we are able to make the following microscopic appraisal of the quasistatic process. A constant-volume heat addition is evidenced by a change in the population of the energy states (N_i) but the allowable energy levels are not changed; heat addition causes an increase in the

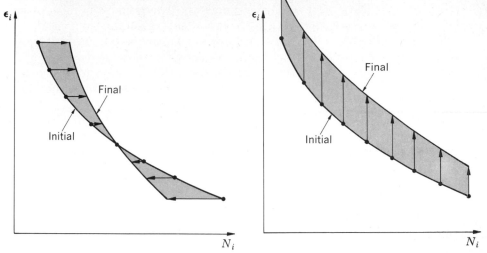

FIG. 5-5 *(Left) Constant-volume heat addition causes shift in populations; (right) adiabatic compression causes shift in energy levels.*

population of higher energy states and a corresponding decrease in population of lower energy states. In contrast, an adiabatic-compression process causes a shift upward in the energy levels ϵ_i while the population of each level remains the same. These two types of processes are illustrated in Fig. 5-5.

5-15 ENTROPY

Because the most probable distribution is so predominant, we shall specify Ω_{max} as a macroscopic property of the system. For convenience, we choose to work with $\ln \Omega_{max}$ and write

$$S = k \ln \Omega_{max} \tag{5-50}$$

where now k is some constant to be determined later and the property S is a *macroscopic equilibrium property* called the *entropy*.

For a large number of particles the states represented by the values of Ω other than Ω_{max} are much less likely than Ω_{max} but still represent *possible* physical states of the system. Accordingly, we could define the entropy for these possible states as

$$S = k \ln \Omega \tag{5-51}$$

A comparison of Eqs. (5-50) and (5-51) indicates that, as an *isolated* (constant energy and number of particles) system approaches its most likely state, its entropy approaches a maximum value. Furthermore, it is highly improbable that the entropy would ever decrease from its maximum because of the overwhelming nature of Ω_{max}. Note that such a decrease is *not* impossible, only highly improbable. The behavior of the property entropy is thus quite different from the internal-energy property for an isolated system. We accept the premise that it is *impossible* for the internal energy to change in such a system. This premise, of course, results from the first law of thermodynamics.

Another way of stating the foregoing remarks is to say that in any change involving an isolated system the entropy will most likely increase. Written in analytical form,

$$\Delta S_{isolated} \geq 0 \qquad \text{most probable} \tag{5-52}$$

5-16   THE SECOND LAW OF THERMODYNAMICS

In the introductory chapter we mentioned a few physical phenomena which were ascribed to be the result of a general physical notion called the second law of thermodynamics. We are now in a position to state that Eq. (5-52) is a suitable analytical statement of the second law. We hasten to add, however, that this fact will not be apparent at all to the reader at this point. In Chap. 6 we shall investigate this relation more thoroughly. For now the reader should interpret the microscopic relations as indicating that an isolated system will tend toward a condition of maximum entropy.

5-17   THE PARTITION FUNCTION

According to Eq. (5-36), we may write the following relation for the MB model:

$$N_i = \frac{1}{A} g_i e^{-\beta \epsilon_i} \tag{5-53}$$

The quantity g_i is the number of quantum states having an energy ϵ_i. If we assume a classical model such that a continuous distribution of energy is available, then the number of particles having an energy ϵ_i is given by Eq. (5-37):

$$N_i = \frac{1}{A} e^{-\beta \epsilon_i} \qquad g_i = 1 \text{ for classical model}$$

Now

$$N = \sum_i N_i = \frac{1}{A} \sum_i e^{-\beta \epsilon_i} \qquad (5\text{-}54)$$

The summation on the right side of this equation is defined as the *partition function* or *sum of state*† Z.

$$Z = \sum_i e^{-\beta \epsilon_i} \qquad (5\text{-}55)$$

It follows that we may rewrite Eq. (5-37) as

$$\frac{N_i}{N} = \frac{1}{Z} e^{-\beta \epsilon_i} \qquad (5\text{-}56)$$

The partition function is just what its name implies: a function related to the specific way particles are "partitioned" among the various energy states accessible to them.

The foregoing discussion is restricted in that we considered a simplified model, i.e., for which $g_i = 1$. Let us adopt a general definition of the partition function in accordance with Eq. (5-55), but require that the summation be taken over all energy *states* accessible to the particles. Then, using the subscript j to designate energy states and retaining the i subscript to denote energy levels, we obtain

$$Z = \sum_i e^{-\beta \epsilon_j} = \sum_i g_i e^{-\beta \epsilon_i} \qquad (5\text{-}57)$$

This expression results because there are g_i energy states having the value ϵ_i. We have already called g_i the degeneracy of the particular energy level. Other writers may prefer to call g_i the *statistical weight*, thus saying that certain energy levels may be more heavily "weighted" than others.

By adopting the definition of partition function given by Eq. (5-57), we may then examine the MB distribution without the restriction that $g_i = 1$. Thus

$$N = \sum_i N_i = \sum \frac{g_i e^{-\beta \epsilon_i}}{A} = \frac{Z}{A}$$

or

$$\frac{N_i}{N} = \frac{1}{Z} g_i e^{-\beta \epsilon_i} \qquad (5\text{-}58)$$

† Translation of German *Zustandssumme*.

EXAMPLE 5-4

For the four energy levels illustrated in Fig. 5-1 compute the value of the partition function, assuming that the particles obey MB statistics and $\beta = 1.0$ per energy unit. Subsequently calculate the fractional populations of the four energy levels.

SOLUTION To calculate the partition function, we make use of Eq. (5-57) and the information from Fig. 5-1.

$$Z = \sum_i g_i e^{-\beta \epsilon_i} = g_0 e^{-\beta \epsilon_0} + g_1 e^{-\beta \epsilon_1} + g_2 e^{-\beta \epsilon_2} + g_3 e^{-\beta \epsilon_3}$$
$$= (1)e^0 + (3)e^{-1} + (6)e^{-2} + (10)e^{-3}$$
$$= 1.0 + 1.104 + 0.813 + 0.498$$
$$Z = 3.415$$

The fractional populations of the four energy levels are calculated with Eq. (5-58).

$$\frac{N_i}{N} = \frac{1}{Z} g_i e^{-\beta \epsilon_i}$$

$$\frac{N_0}{N} = \frac{1.0}{3.415} = 0.293$$

$$\frac{N_1}{N} = \frac{1.104}{3.415} = 0.323$$

$$\frac{N_2}{N} = \frac{0.813}{3.415} = 0.238$$

$$\frac{N_3}{N} = \frac{0.498}{3.415} = 0.146$$

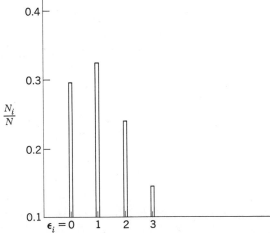

FIG. EXAMPLE 5-4

For this example it appears that the second energy level ($\epsilon_i = 1$ energy unit) is the most heavily populated. The populations are depicted in Fig. Example 5-4.

5-18 THE CONSTANTS A AND β

We have shown above how the constant A may be evaluated for the MB distribution [Eq. (5-58)]. The evaluation of A for the BE and FD distributions is not so simple and depends on the physical system (gas, liquid, or solid) to which these statistics are applied. We shall defer our evaluation for these cases until Chap. 8 where we apply the statistical distributions to some specific circumstances.

To evaluate the constant β, we need to enter into a fuller discussion of the second law of thermodynamics and some general thermodynamic property relations; however, we can at least state the result at this point. It will be shown that

$$\beta = \frac{1}{kT} \tag{5-59}$$

where k is the gas constant per molecule, or *Boltzmann's constant*,

$$k = 1.3807 \times 10^{-23} \text{ J/molecule·K} \tag{5-60}$$

and T is the absolute temperature. It will also be shown that Boltzmann's constant is the value used in Eq. (5-50) to define the entropy. The value of β is given by Eq. (5-59) for all three of the statistical distributions discussed previously. The Boltzmann constant is given in terms of the universal gas constant as

$$k = \frac{\mathcal{R}}{N_0} \tag{5-61}$$

where, as before, N_0 is Avogadro's number.

REVIEW QUESTIONS

1   What is phase space?

2   What is the de Broglie wavelength?

3   Why are quantum considerations unimportant for analysis of macroscopic systems?

4   What is meant by the term *degeneracy*?

5　Distinguish between quantum *state* and energy *level*.

6　Distinguish between the terms *microstate* and *macrostate*.

7　How is the concept of thermodynamic probability used to predict the macroscopic behavior of a system?

8　What are the assumptions pertaining to the MB, BE, and FD statistical models?

9　Describe heat and work addition to a simple compressible substance in terms of microscopic phenomena.

10　How is entropy defined?

11　Define the partition function.

12　What is meant by the term *equilibrium distribution?*

13　How is entropy related to thermodynamic probability?

14　What is a property?

15　How is the second law related to microscopic thermodynamic analysis?

16　How is the Boltzmann constant related to the universal gas constant?

PROBLEMS

5-1　Calculate the number of ways of arranging three distinguishable balls in two boxes so that $N_1 = 1$, $N_2 = 2$.

5-2　Calculate the number of ways of arranging eight indistinguishable balls in four indistinguishable boxes.

5-3　Calculate the number of ways of arranging 10 indistinguishable balls in 12 boxes with no more than one ball per box.

5-4　Calculate the thermodynamic probability for arranging five particles among the energy levels and states shown in Fig. 5-1 with $N_0 = 1$, $N_1 = 2$, $N_3 = 1$. Make the calculation for all three physical models.

5-5　Repeat Prob. 5-4 but with $N_0 = 1$, $N_1 = 1$, $N_2 = 2$, $N_3 = 1$.

5-6　Determine the degeneracy for a translational energy level of four energy units when the minimum quantum increment is two energy units. Repeat for an energy level of six energy units.

5-7　Determine the degeneracy for a translational energy level of four energy units when the minimum quantum increment is one energy unit.

5-8　Consider a system of particles for which the energy levels are quantized according to

$$\epsilon_i = (i + 1)i \quad \text{energy units (eu)} \quad i = 0, 1, 2, \ldots$$

and the degeneracy of each level is

$$g_i = 2i + 1$$

Assuming that the particles obey MB statistics and that $\beta = 1.0$ (eu)$^{-1}$, calculate the partition function for a group of particles occupying (a) the first four energy levels; (b) the first five energy levels. Subsequently calculate the fractional populations of the energy levels.

5-9 Consider a system of particles for which the energy levels are quantized according to the following for $\beta = 1.0$ (eu)$^{-1}$

$$\epsilon_i = (i + \tfrac{1}{2}) \text{ eu} \qquad i = 0, 1, 2, \ldots$$

and the energy levels are nondegenerate. Calculate and plot the fractional populations of the first 10 energy levels assuming that the particles obey MB statistics.

5-10 Four indistinguishable particles are to be placed in two energy levels, each of which has a degeneracy of three. Evaluate the thermodynamic probability for all possible arrangements. What is the most probable arrangement?

5-11 Calculate the thermodynamic probability for six indistinguishable balls to be placed in three boxes with:
(a) $N_1 = 1, N_2 = 4, N_3 = 1$
(b) $N_1 = N_2 = N_3 = 2$
(c) $N_1 = 1, N_2 = 3, N_3 = 2$

5-12 Calculate the maximum thermodynamic probability for 100 distinguishable particles distributed among the four energy levels illustrated in Fig. 5-1. Establish the number of particles in each level for this equilibrium distribution. Estimate the relative decrease in the thermodynamic probability resulting from a switch of one particle from the equilibrium distribution. $\beta = 1.0$ (eu)$^{-1}$.

5-13 Recalculate the maximum probability for Prob. 5-12 assuming that the particles are indistinguishable.

5-14 Repeat Prob. 5-12 for the first four energy levels as described in Prob. 5-8. Repeat the solution assuming that the particles are indistinguishable.

REFERENCES

1 Kaplan, I.: "Nuclear Physics," 2d ed., Addison-Wesley Publishing Company, Inc., Reading, Mass., 1963.

2 Allis, W. P., and M. A. Herlin: "Thermodynamics and Statistical Mechanics," McGraw-Hill Book Company, New York, 1952.

3 Callen, H. B.: "Thermodynamics," John Wiley & Sons, Inc., New York, 1960.

4 Crawford, F. H.: "Heat Thermodynamics, and Statistical Physics," Harcourt, Brace & World, New York, 1963.

5 Lee, J. F., F. W. Sears, and D. L. Turcotte: "Statistical Thermodynamics," Addison-Wesley Publishing Company, Inc., Reading, Mass., 1963.

6 Lewis, G. N., and M. Randall: "Thermodynamics," 2d ed. (revised by K. S. Pitzer and L. Brewer), McGraw-Hill Book Company, New York, 1961.

7 Mayer, J. E., and M. G. Mayer: "Statistical Mechanics," John Wiley & Sons, Inc., New York, 1940.

8 Reynolds, W. C.: "Thermodynamics," 2d ed., McGraw-Hill Book Company, New York, 1968.

9 Schrodinger, E.: "Statistical Thermodynamics," Cambridge University Press, Cambridge, 1952.

10 Sears, F. W.: "An Introduction to Thermodynamics, The Kinetic Theory of Gases, and Statistical Mechanics," Addison-Wesley Publishing Company, Inc., Reading, Mass., 1950.

11 Sommerfeld, A.: "Thermodynamics and Statistical Mechanics," Academic Press, Inc., New York, 1956.

12 Tolman, C.: "The Principles of Statistical Mechanics," Oxford University Press, London, 1938.

13 Hatsopoulos, G. N., and J. H. Keenan: "Principles of General Thermodynamics," John Wiley & Sons, Inc., New York, 1965.

chapter six
The second law of thermodynamics

6-1 INTRODUCTION

We have already noted in Chap. 1 that the second law of thermodynamics relates to the direction of energy exchange processes. Some energy transformations are allowed whereas others are not. In Chap. 2 a more precise explanation of the conservation of energy principle has been given and related to previous studies in mechanics. Insofar as the first law of thermodynamics is concerned, all forms of energy are equally useful; no preference is given for the different types of energy, and the only concern is that in an isolated system energy must be conserved. Even though no preference is given, the different types of energy are distinguished from one another: work is a force acting through a distance and represents an interaction between systems, internal energy is a function of state the change of which is defined in terms of work, and heat is subsequently defined in terms of work and internal energy. Nothing in the conservation of energy principle states that heat and work are not equivalent forms of energy; in fact, the cyclic equation

$$\oint d'Q + \oint d'W = 0 \tag{6-1}$$

carries a casual implication that they *are* equivalent. The second law of thermodynamics recognizes the fact that heat and work are not equivalent and eventually establishes a set of formal relationships which may be used to supplement the first law in the study of thermodynamic systems.

6-2 PHYSICAL DESCRIPTION OF THE SECOND LAW

A brief summary of a set of phenomena falling into the realm of the second principle has been given in Chap. 1. These phenomena are restated here for convenience.

1 Heat flows from a high temperature to a low temperature in the absence of other effects. This means that a hot body will cool down when brought into contact with a body at a lower temperature, and not the opposite.
2 Two gases, when placed in an isolated chamber, will mix uniformly throughout the chamber but will not separate spontaneously once mixed.
3 A battery will discharge through a resistor releasing a certain amount of energy, but it is not possible to make the reverse of this happen, i.e., to add energy to the resistor by heating and, thus, cause the battery to charge itself.
4 It is not possible to construct a machine or device which will operate

continuously while receiving heat from a single reservoir and producing an equivalent amount of work.

Let us examine statements 1 and 4 in detail. The circumstance in statement 1 is in accordance with our everyday experience, and at first thought it would seem to be a direct consequence of the conservation of energy. It is true that the energy given up by the hot block goes to raise the energy level of the cooler block such that the total energy of the set of blocks remains constant. But the total energy would also remain constant in the case where energy taken from the cooler block was used to raise the energy level of the hot block by an equivalent amount. This second process does not occur, even though it does not violate the conservation of energy principle. In another example, we do not expect a pan of water to boil when placed in a refrigerator as a result of heat transferred to it from the cold surroundings. This simply does not happen.

In statement 4 the considerations are a bit more abstract but still in the realm of normal experience. We would not expect to be able to build an engine operating continuously and in cyclic processes, which could extract energy from some readily available source such as the ocean or earth, and producing an equivalent amount of mechanical work. If it were possible to build such an engine, it would be a very fine thing indeed, for we would only have to connect it to the ocean and produce all the power needed for some time to come. Such an engine is called a *perpetual-motion machine of the second kind* because it succeeds in converting heat from a single reservoir into work on a continuous basis. A perpetual-motion machine of the first kind is one which *creates* energy in violation of the first law of thermodynamics.

In accordance with this brief discussion, we may tentatively conclude that the second law of thermodynamics has its primary emphasis in an acknowledgment of the unidirectional nature of heat transfer and certain types of energy conversion.

Our overall objective in this chapter is to develop satisfactory analytical descriptions of the second law of thermodynamics based strictly on macroscopic arguments which accept statement 4 as an empirical axiom. This macroscopic analysis ignores the microscopic structure of matter. For those readers who choose to study the microscopic thermodynamic presentations of Chaps. 5 and 8 it will be shown that the two types of analysis can be tied together neatly but that such material is unnecessary for an understanding of this chapter.

6-3 KELVIN–PLANCK AND CLAUSIUS STATEMENTS

Two commonly accepted statements of the second law of thermodynamics relate to the foregoing statements 1 and 4.

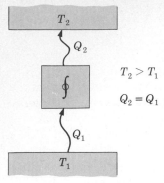

FIG. 6-1 *It is impossible to construct a device which operates in a cycle and whose sole effect is to transfer heat from a cooler body to a hotter body.*

> **Clausius statement** *It is impossible to construct a device which operates in a cycle and whose sole effect is to transfer heat from a cooler body to a hotter body.*
> **Kelvin-Planck statement** *It is impossible to construct a device which operates in a cycle and produces no other effect than the production of work and exchange of heat with a single reservoir.*

These statements are depicted in Figs. 6-1 and 6-2.

6-4   REVERSIBLE PROCESSES AND CYCLES

Let us now consider an idealized state path called the *reversible* process. A process is reversible if the initial state of the system can be restored with no observable effects in the system *and* its surroundings. Consider the expansion of a gas behind a piston in a cylinder. If the expansion occurs slowly and frictionlessly, the system (the gas) will remain essentially in equilibrium, and we might call the process *quasistatic* or a succession of equilibrium states, as discussed in Chap. 1. To find out if this process is reversible, we ask the question: Is it possible to run the process backward with no extra effects ob-

FIG. 6-2 *It is impossible to construct a device which operates in a cycle and produces no other effect than the production of work and exchange of heat with a single reservoir.*

FIG. 6-3 *Heat transfer across a finite temperature difference is irreversible.*

served in the system or surroundings? In other words, could we compress the gas back to its original state by applying *exactly* the same amount of work that was removed in the expansion process and adding *exactly* the same amount of heat that was removed in the expansion process? It turns out that the quasistatic process is, indeed, a reversible process. A general criterion for reversibility may be stated as follows:

> *If a hypothetical reversal of a thermodynamic process can be performed which does not violate the second law of thermodynamics, then the process is said to have been reversible.*

Two obvious applications of this criterion are shown in Figs. 6-3 and 6-4. If we try to reverse the heat-transfer process in Fig. 6-3, a violation of the Clausius statement of the second law results. If we try to reverse the situation in Fig. 6-4, the Kelvin-Planck statement is violated. We may conclude that transfer of heat across a finite temperature difference and the conversion of work into heat during a cyclic process are both irreversible processes. To make this conclusion, we define an *irreversible process as one which is not reversible*. This definition is, of course, rather obvious.

A *reversible cycle* is a succession of reversible processes such that the system periodically returns to its initial state. As previously mentioned, the cyclic engine, or device, depicted in Fig. 6-2 is called a perpetual-motion machine of the second kind. The term *second kind* means that it violates the second law of thermodynamics.

At this point we may digress for a moment to discuss the meaning of a *cyclic heat engine*. Work is a very useful form of energy in that it can be adapted to practical applications in a variety of ways. Electrical work can power numerous devices, and mechanical work can drive automobiles, machines, etc. Heat is not necessarily so useful, and the purpose of a heat

FIG. 6-4 *Conversion of work into heat is irreversible.*

engine is to convert heat into work. The notion of the cycle and cyclic process is introduced because energy-conversion processes of interest are those which can operate continuously. The internal combustion engine is a cyclic engine which sequentially inhales fuel and air, compresses and combusts the mixture, and produces work output while exhausting the products of combustion to the surroundings. A *power cycle* is one which produces a net work output (with some net heat input), and a *refrigeration cycle* involves a net work *input* and net heat output. A *reversible* power cycle can be changed to a reversible refrigeration cycle by just reversing all the heat flow and work quantities. To fix the proper directions and signs for the work and heat quantities we shall rely on proper schematic diagrams rather than a rigid sign convention.

The equivalence of the Clausius and Kelvin-Planck statements may be demonstrated rather quickly, as shown in Fig. 6-5. Let us suppose that it would be possible to transfer heat as in Fig. 6-5a in violation of the Clausius statement. To the system (a) we could add the reversible engine as in (b). From energy conservation, the work output of this engine is $W = Q_A - Q_B$. The second law, of course, does not prohibit this type of engine. However, the net result of adding the heat engine in (b) is that no net heat is exchanged with the reservoir at T_2 and the arrangement is equivalent to the one in (c). The engine in (c) is just the type of device excluded by the Kelvin-Planck statement. Thus we have shown that a violation of the Clausius statement results in a violation of the Kelvin-Planck statement as well.

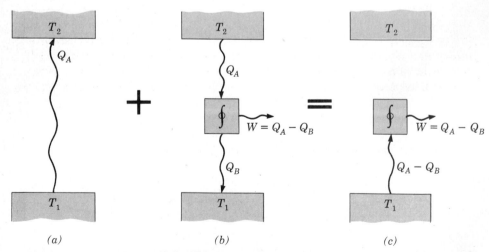

(a)   (b)   (c)

FIG. 6-5 *Equivalence of the Clausius and Kelvin-Planck statements of the second law.*

6-5 SIGN CONVENTION FOR HEAT AND WORK

In our discussion of the first law of thermodynamics and energy balances we employed a sign convention for heat and work such that energy added to the boundary of a system (heat or work) is positive. The equations that represent the energy balances were based on this convention. Now, as we start to examine thermal power cycles in the light of the second law, the reader will notice that schematic diagrams are drawn with heat and work "arrows" so that energies are both added and removed from the cycles. The purpose of such diagrams is to recognize, in advance, the directions of such energy transfers so that a better appreciation may be gleaned of the processes involved. From an energy-balance standpoint there is no change in fundamental concepts. We still say that

**Energy added to system = energy accumulated in system
+ energy removed from system**

By drawing the arrows in certain directions we are simply focusing our attention on energy quantities of practical interest. In a *power cycle* the object is to produce a work *output* as a result of a heat input at a high temperature; thus, we draw a schematic diagram such that the work is coming out of the cycle and heat is added. In a *refrigeration cycle* the object is to remove heat from some source at low temperature by adding work; so, we draw the schematic diagram with a heat addition at low temperature and a work *input*.

The important point of this chapter is that the second law places very definite restrictions on allowable conversions of heat into work, but the restrictions do not abrogate the basic notion of energy conservation in any way. As we have stated many times, energy-conversion analysis is basically an accounting procedure, and as long as it is performed correctly, the sign convention does not matter. As our exposition of the second law develops, the reader should concentrate on understanding the physical phenomena involved; with this concentration, sign convention will take care of itself.

6-6 THERMODYNAMIC TEMPERATURE

Let us suppose that we have at our disposal two reversible heat engines which operate in cyclic processes between temperature reservoirs, as shown in Fig. 6-6a. The thermal efficiency η_t of a heat engine is defined as

$$\eta_t = \frac{\text{useful energy effect}}{\text{energy that costs}} = \frac{\text{work output}}{\text{heat added}} \tag{6-2}$$

(a)

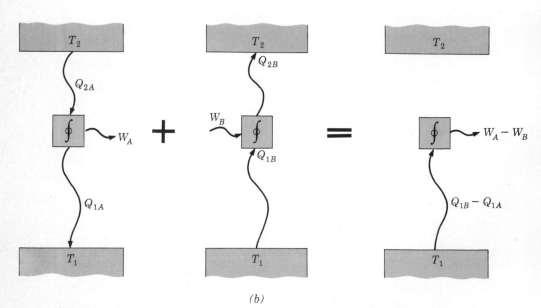

(b)

FIG. 6-6 *All reversible engines operating between the same temperature limits have the same efficiency.*

According to this definition the "energy that costs" is the heat added at the high temperature since this heat must be supplied by some fuel. Thus

$$\eta_t = \frac{W}{Q_2} \tag{6-3}$$

So far, we have said nothing about such an efficiency. The reversible engines A and B could be constructed differently and, at first glance, we might expect them to have different efficiencies. We are about to prove, however, that they have the *same* efficiency. To show this, let us tentatively assume that engine A has a greater efficiency than B and set

$$Q_{2A} = Q_{2B}$$

Then

$$W_A > W_B \quad \text{and} \quad Q_{1A} < Q_{1B}$$

Since the engines are reversible, engine B could be reversed and combined with engine A, as shown in Fig. 6-6b. But since $Q_{2A} = Q_{2B}$,

$$W_A - W_B = Q_{1B} - Q_{1A}$$

The arrangement in Fig. 6-6b is a violation of the Kelvin-Planck statement of the second law, so that the assumption that A has a greater efficiency than B must be false. We are, therefore, justified in concluding that *all reversible engines operating between the same temperature limits have the same efficiency.* This efficiency may be written as

$$\eta_t = \frac{W}{Q_2} = \frac{Q_2 - Q_1}{Q_2} = 1 - \frac{Q_1}{Q_2} \qquad (6\text{-}4)$$

Since the efficiency is the same for all reversible engines operating between the same temperature limits, it follows that the heat ratio Q_1/Q_2 must be a function of these two temperatures:

$$\frac{Q_1}{Q_2} = f(T_1, T_2) \qquad (6\text{-}5)$$

Suppose that some intermediate temperature T_3 is inserted between T_1 and T_2 with two new reversible heat engines as shown in Fig. 6-7. Because the two reversible engines A and B comprise a larger reversible engine operating between T_2 and T_1, the combination must have the same efficiency as engine C. From Eq. (6-5) we know that the heat ratios are functions of the temperature limits of the cycles:

$$\frac{Q_1}{Q_2} = f(T_1, T_2)$$

$$\frac{Q_3}{Q_2} = f(T_3, T_2)$$

$$\frac{Q_1}{Q_3} = f(T_1, T_3)$$

Combining these relations,

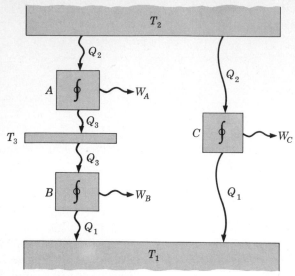

FIG. 6-7 *Thermodynamic temperature.*

$$\frac{Q_1}{Q_2} = f(T_1, T_2) = \frac{Q_1}{Q_3}\frac{Q_3}{Q_2} = f(T_1, T_3) \times f(T_3, T_2) \tag{6-6}$$

or

$$f(T_1, T_2) = f(T_1, T_3) \times f(T_3, T_2) \tag{6-7}$$

Because the left side of Eq. (6-7) does not involve T_3, we conclude that the f function must be of such form that the T_3 will cancel out of the product on the right side of Eq. (6-7). This is possible if

$$f(T_1, T_3) = \frac{\phi(T_1)}{\phi(T_3)}$$

$$f(T_3, T_2) = \frac{\phi(T_3)}{\phi(T_2)}$$

where ϕ is a new function. The net result is that we select for the heat ratios the functional form

$$\frac{Q_1}{Q_2} = \frac{\phi(T_1)}{\phi(T_2)} \tag{6-8}$$

It is possible, of course, to satisfy this equation with a number of temperature functions. The proposal was made by Kelvin that the temperature function be taken as

$$\frac{|Q_1|}{|Q_2|} = \frac{T_1}{T_2} \tag{6-9}$$

and that Eq. (6-9) serve as a *definition* of an *absolute thermodynamic temperature scale*. The efficiency of a reversible heat engine which operates between two reservoirs and exchanges heat only with these two reservoirs is, thus,

$$\eta_t = 1 - \frac{T_L}{T_H} \tag{6-10}$$

where T_L designates the low-temperature reservoir and T_H designates the high-temperature reservoir. This temperature scale also corresponds to the temperature measured by an ideal-gas thermometer as discussed in Chap. 1.

6-7 THE INEQUALITY OF CLAUSIUS

Consider the arrangement shown in Fig. 6-8. An amount of heat $d'Q_R$ from the constant-temperature reservoir T_R is transferred to the reversible engine and produces the amount of work $d'W_R$. The heat rejected from this engine is used to supply the other cyclic engine producing the work $d'W_E$. Taking *both* engines as a system, the total work *output* $d'W$ is

$$d'W = d'W_R + d'W_E$$

Also

FIG. 6-8 *Diagram for proof of inequality of Clausius.*

$$d'W_R = d'Q_R \left(1 - \frac{T}{T_R}\right) = d'Q \left(\frac{T_R}{T} - 1\right)$$

$$d'W_E = d'Q$$

so that

$$d'W = d'Q \left(\frac{T_R}{T} - 1 + 1\right) = \frac{T_R}{T} d'Q \tag{6-11}$$

Equation (6-11) may be written for the complete cycle as

$$\oint d'W = T_R \oint \frac{d'Q}{T} \tag{6-12}$$

The arrangement in Fig. 6-8 cannot produce a net work *output* because this would violate the second law as given by the Kelvin-Planck statement. Accordingly, the only way the arrangement can operate is with a cyclic work *input* and a cyclic heat flow into the reservoir. Mathematically, this means that

$$\oint d'W \leqslant 0 \tag{6-13}$$

where $d'W$ is the work *output*. Using Eq. (6-12), this condition becomes

$$\oint \frac{d'Q}{T} \leqslant 0 \tag{6-14}$$

The relation given in Eq. (6-14) is called the *inequality of Clausius*.

In the foregoing derivation no mention was made of the reversibility of engine E. Suppose this engine is reversible. Suppose further that

$$\oint d'W < 0 \tag{6-15}$$

for this circumstance. Since the engine is assumed reversible, we could reverse it and obtain

$$\oint d'W > 0$$

But this is impossible according to the second law because a perpetual-motion machine of the second kind would be created. The net result is that the inequality of Eq. (6-15) cannot hold for the case of a reversible engine E. It follows that the equality sign of Eq. (6-14) must hold in the reversible case, and we may write

$$\oint \left(\frac{d'Q}{T}\right)_{\text{rev}} = 0 \tag{6-16}$$

6-8 MACROSCOPIC DEFINITION OF ENTROPY

Mathematically, any quantity which may be represented as an exact differential is a property or point function; i.e., it is a function of the *state* of the system. The mathematical condition for an exact differential is that the cyclic integral of the differential be zero; i.e., if

$$\oint dx = 0$$

then x is a property for the region over which the cyclic integral is taken. In accordance with this definition, Eq. (6-16) yields

$$dS = \left(\frac{d'Q}{T}\right)_{rev} \tag{6-17}$$

where S is a property which we choose to call the *entropy*. *Equation (6-17) is the macroscopic definition of entropy*. Note that entropy is defined only for reversible processes and a change in entropy may be calculated with

$$\Delta S = S_2 - S_1 = \int_1^2 \left(\frac{d'Q}{T}\right)_{rev} \tag{6-18}$$

Only *changes* in entropy are defined by these relations. We have not yet established an absolute scale for entropy.

6-9 PRINCIPLE OF INCREASE OF ENTROPY

Consider the two state points shown in Fig. 6-9 and the two processes which comprise a cycle operating between the two points. R is a reversible process and I is an irreversible process, so that the overall cycle is irreversible. From the inequality of Clausius, we have

$$\oint \frac{d'Q}{T} = \int_1^2 \frac{d'Q_I}{T} + \int_2^1 \frac{d'Q_R}{T} < 0 \tag{6-19}$$

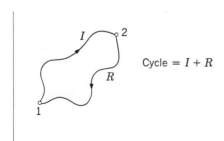

Cycle $= I + R$

FIG. 6-9 *Irreversible cycle.*

We have used the inequality sign since the cycle is irreversible. Observing that

$$\int_2^1 \frac{d'Q_R}{T} = S_1 - S_2$$

Eq. (6-19) may be written as

$$\int_1^2 \frac{d'Q_I}{T} + S_1 - S_2 < 0$$

or

$$S_2 - S_1 > \int_1^2 \frac{d'Q_I}{T}$$

In general then, we may write

$$S_2 - S_1 \geq \int_1^2 \frac{d'Q}{T} \tag{6-20}$$

where the equality holds for a reversible process and the inequality for an irreversible process. Equation (6-20) is called the *principle of increase of entropy*.

Suppose Eq. (6-20) is written for adiabatic processes, $d'Q = 0$:

$$S_2 - S_1 \geq 0 \qquad \text{adiabatic process} \tag{6-21}$$

If the process is a reversible adiabatic one, the entropy change will be zero, and we call it a constant-entropy or *isentropic* process. If the process is irreversible and adiabatic, then the entropy must increase. Suppose, furthermore, that we consider an isolated system. Then,

$$\Delta S_{\text{isolated system}} \geq 0 \tag{6-22}$$

or the entropy of an isolated system must increase if irreversible changes are taking place inside it. If only reversible changes are taking place within the isolated system, the entropy will remain constant.

No *real* process is reversible. All heat transfers occur across a finite-temperature difference. Frictional effects are always present to render processes involving work irreversible. Because of these facts we are forced to conclude that the entropy of the universe (presumably an isolated system) is always increasing. The postulate that the universe is "filling up with entropy" is a metaphysical matter of serious concern to philosophers and theologians. We shall not worry about the consequences of this, however, since it probably will not seriously affect engineering problems for some time.

Entropy is a property defined by Eq. (6-17). Its change may be calculated with Eq. (6-18), and that is all we need to know for most macroscopic considerations.

ENTROPY, A THERMODYNAMIC PROPERTY

The foregoing discussion has demonstrated that entropy is a property of a system. Thus, it may be expressed in terms of other thermodynamic properties and tabulated just like enthalpy or internal energy. It is indeed tabulated as a function of pressure and temperature for a wide variety of substances. The properties given in the Appendix furnish examples of such tabulations.

THE REVERSIBLE ADIABATIC AND ISOTHERMAL PROCESSES

It is clear from Eq. (6-17) that for a reversible adiabatic process $dQ_{rev} = 0$ and hence $dS = 0$, or the entropy remains constant. This type of process is thus given the name *isentropic*, designating constant entropy. Similarly, the *reversible isothermal* process can be represented by

$$Q_{rev\ isothermal} = T \Delta S \tag{6-23}$$

by an integration of Eq. (6-17).

THE *T-S* DIAGRAM AND IRREVERSIBLE PROCESSES

It is clear from Eq. (6-21) that the entropy for an irreversible adiabatic process must increase, but the amount of increase is indefinite. Physically, the reversible process is quasistatic; i.e., it is a succession of equilibrium states, and we are able to specify the state of the system each step along the way. The irreversible process is one in which the system departs substantially from equilibrium because of frictional effects, substantial temperature gradients within the system, or other effects. In the irreversible process the entropy change must be calculated by devising some reversible process between the given end states. The degree to which friction or temperature gradients affect the change of state of a system must be determined apart from conventional thermodynamic analysis, and is usually treated as a subject in transport phenomena. We shall give a brief introduction to the analysis of some transport processes in Chaps. 9 and 14. For the adiabatic process the reversible and irreversible processes will appear as shown in Fig. 6-10. The greater the increase in entropy for the irreversible process, the greater we say is its "irreversibility." Note that the irreversible process is drawn as a shaded line since it is a series of nonequilibrium states and, thus, is not strictly representable on a diagram of thermodynamic properties of equilibrium states.

FIG. 6-10 *Reversible and irreversible adiabatic processes on a temperature-entropy diagram. The greater the increase in entropy, the larger the irreversibility.*

6-10 PRACTICAL CAUSES OF IRREVERSIBILITIES

We have already shown two examples of irreversible processes in Figs. 6-3 and 6-4. Thus, if a hot block of metal is brought into thermal communication with a cooler block, and the two allowed to come to temperature equilibrium, we will experience an irreversible process. If the two blocks were taken as an overall adiabatic system, an increase in entropy would be observed for the combination. If one adds work in a paddle-wheel process as shown in Fig. 6-11, this too is an irreversible process because we cannot reverse the process by adding heat to the container. The block sliding against a frictional resistance shown in Fig. 6-12 also represents the same type of irreversible process because the work W done by the force F is eventually dissipated as heat Q into either the plane or the surrounding atmosphere. Again, we cannot reverse the situation and expect the block to move when heat is added to it. Fluid friction in a pipe is another similar example of an irreversible process. Work is required to move the fluid through the pipe against a wall frictional resistance, and we cannot accomplish the same effect by adding heat to the pipe.

FIG. 6-11 *Paddle-wheel work is irreversible.*

FIG. 6-12 *Sliding block is irreversible.*

Thus, whenever heat is transferred across a finite temperature difference or friction is encountered, the process will be rendered irreversible to some extent. All *real* processes are irreversible, some more than others. Later on, we shall find that certain reversible processes represent the best hoped-for situation in energy-conversion and power-production devices and thus merit our study and attention. The degree to which we are able to approach such ideal processes is a matter of the skill of engineering design in various items of equipment and the cost factors which can be tolerated in a particular economic climate.

6-11 ENTROPY OF A PURE SUBSTANCE

We have shown that entropy is a property of a system. It is an extensive property like the total internal energy or total enthalpy which may be calculated from *specific* entropies based on either a unit mass or unit molal quantity of the system so that

$$S = ms = n\bar{s}$$

For pure substances values of the specific entropy may be tabulated along with enthalpy, specific volume, and other thermodynamic properties of interest, as illustrated in the various tables of the Appendix. In the liquid-vapor saturation region the specific entropy is obtained from the saturation properties and quality in the same manner as other properties so that

$$s = s_f + x s_{fg} \qquad s_{fg} = s_g - s_f$$
$$s = (1 - x)s_f + x s_g$$

We have seen that the temperature-entropy diagram is a useful vehicle for studying thermodynamic processes, and the *T-s* diagram for a pure substance will be particularly helpful when we study practical power cycles. Figure 6-13 shows such a diagram for water and is illustrative of the general appearance of such diagrams for other substances. The figure shows the saturation line as a dome-shaped curve with the critical point at the peak, the

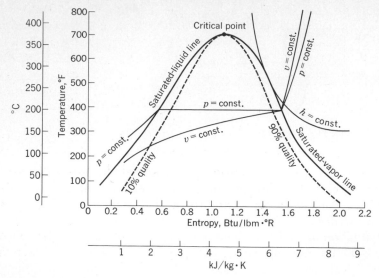

FIG. 6-13   *Temperature-entropy diagram for water.*

saturated-liquid line to the left, and the saturated-vapor line to the right. All of the area under the dome is *wet-mixture* region with appropriate lines of constant quality indicated. The region to the left of the saturated-liquid line is compressed liquid, while that to the right of the saturated-vapor line is the superheated vapor region. At temperatures above the critical point there is no distinction between liquid and vapor. Some typical lines of constant pres-

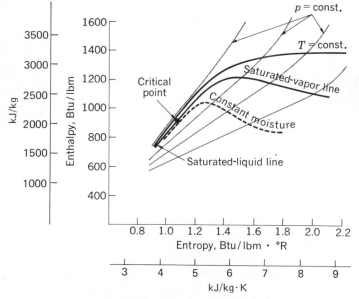

FIG. 6-14   *Enthalpy-entropy diagram for water.*

sure, constant volume, and constant enthalpy are also indicated to show the general shape of such curves. It should be noted that the line of constant pressure is also a line of constant temperature in the saturation region.

The *Mollier diagram* is a plot of enthalpy versus entropy, and a plot for steam is illustrated in Fig. 6-14. A large Mollier diagram for calculation purposes is presented as Fig. A-4 of the Appendix. The *h-s* diagram has an entirely different shape from the *T-s* diagram. The saturation line has a dome shape, but the critical point is located on the left side of the dome with the saturated-vapor line extending over the top and to the right of the dome. Lines of constant quality also have a dome shape, and lines of constant temperature in the superheat region behave as indicated. The general shape of constant pressure lines is also shown. The *h-s* diagram in the Appendix covers mainly the saturation and superheat regions since they are of most interest in practical calculations.

Formulas for calculation of the entropy for solid and vapor water below 32°F were given in Chap. 3 as Eqs. (3-10) and (3-11).

6-12   CHANGE IN ENTROPY FOR AN IDEAL GAS

Equation (6-18) furnishes a basis for calculating changes in entropy between any two end states provided a suitable reversible process can be devised between these states. An ideal gas offers a good example of the types of calculation which may be performed. The reader will recall that, for an ideal gas, enthalpy and internal energy are functions of temperature alone (we shall prove this in Chap. 7) and may be written as

$$du = c_v \, dT \tag{6-24}$$

$$dh = c_p \, dT \tag{6-25}$$

For a reversible process we may write the heat flow from Eq. (6-17) as

$$d'Q = T \, ds \tag{6-26}$$

An ideal gas is one class of simple compressible substance and, thus, we may write

$$d'Q = du + p \, dv = T \, ds$$

Thus,

$$ds = \frac{du}{T} + \frac{p}{T} \, dv \tag{6-27}$$

From the ideal-gas equation of state we have

$$\frac{p}{T} = \frac{R}{v} \tag{6-28}$$

so that, making use of Eqs. (6-24) and (6-28), Eq. (6-27) becomes

$$ds = c_v \frac{dT}{T} + R \frac{dv}{v} \tag{6-29}$$

For constant specific heats, Eq. (6-29) may be integrated between two end states to give

$$s_2 - s_1 = c_v \ln \frac{T_2}{T_1} + R \ln \frac{v_2}{v_1} \tag{6-30}$$

We may develop an alternative expression by noting that

$$d'Q = dh - v\, dp = T\, ds$$

Again using the ideal-gas equation of state along with Eq. (6-25),

$$ds = c_p \frac{dT}{T} - R \frac{dp}{p} \tag{6-31}$$

For constant specific heat this relation may be integrated to give

$$s_2 - s_1 = c_p \ln \frac{T_2}{T_1} - R \ln \frac{p_2}{p_1} \tag{6-32}$$

It should be evident that the change in entropy given by Eqs. (6-30) and (6-32) is calculated for a reversible process between the given end states (p_1, v_1, T_1) and (p_2, v_2, T_2). However, since entropy is a property, its change between any two end states must be independent of the process; thus, these relations apply to *any* ideal-gas process between these end states.

We have already shown in Chap. 2 that the *isentropic* process for an ideal gas with constant specific heats is given as

$$pv^\gamma = \text{const.}$$

where $\gamma = c_p/c_v$.

6-13 THE CARNOT CYCLE

We have already shown that the efficiency of all reversible cycles operating between the same temperature limits is the same and is given by the relation in Eq. (6-10). A particular cycle for which this relation applies is the *Carnot* cycle composed of the following four reversible processes indicated in Fig. 6-15:

1 A reversible isothermal heat addition, *ab*
2 A reversible adiabatic process in which work is done by the system, *bc*
3 A reversible isothermal heat rejection, *cd*
4 A reversible adiabatic process in which work is done on the system, *da*

The cycle may involve any working substance whatsoever, just as long as these four processes are involved in sequence such that the system always returns to its original state. On a temperature-entropy diagram the Carnot cycle has a rectangular shape as illustrated in Fig. 6-15a. In this diagram *da* and *bc* are the isentropic processes while *ab* and *cd* are the reversible heat-addition and -rejection processes, respectively. From our isothermal relation derived above, the heat flows may be written

$$Q_H = T_H \, \Delta S$$
$$Q_L = T_L \, \Delta S$$

The net work output of the cycle is the difference between the heat added Q_H and the heat rejected Q_L:

$$W_{net} = Q_H - Q_L = (T_H - T_L) \, \Delta S \tag{6-33}$$

and the thermal efficiency of the cycle may be written as

$$\eta_t = \frac{W_{net}}{Q_H} = \frac{(T_H - T_L) \, \Delta S}{T_H \, \Delta S} = 1 - \frac{T_L}{T_H}$$

This result agrees, of course, with Eq. (6-10), because our definition of entropy follows from the definition of the absolute temperature used to arrive at Eq. (6-10).

According to Eq. (6-33) the net work output of the Carnot cycle is given graphically as the enclosed area of the rectangle on a *T-S* diagram. It is easy to extend this statement to say that the net work output of *any reversible* cycle is the area enclosed on its corresponding *T-S* diagram.

If one were to execute the Carnot cycle with air or some other ideal gas as the working fluid, the pressure-volume diagram for the cycle would appear as in Fig. 6-15b. Heat addition would occur in the isothermal process from *a* to *b*, followed by the isentropic expansion from *b* to *c*. For a piston-cylinder arrangement there is work done on the face of the piston in both of these processes, and therefore a work *output*, because the force on the piston is moving in the same direction as the displacement. Carrying through the cycle, process *c-d* is the isothermal heat rejection and *d-a* is the

(a)

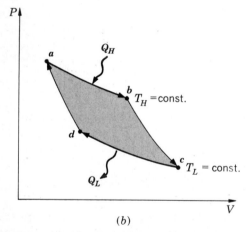

(b)

FIG. 6-15 *Graphical representation of a Carnot cycle. (a) T-S diagram; (b) p-V diagram.*

isentropic compression. In these latter two processes there is a work *input* to the gas. For a piston-cylinder arrangement the net work output of the cycle will be the enclosed area on the *p*-V diagram and must correspond to the enclosed area on the *T*-S diagram.

EXAMPLE 6-1 Entropy change in heating process

Five lbm of steam are heated from saturated liquid at 50 psia to a temperature of 500°F. The heating process occurs at constant pressure. Calculate the change in entropy.

SOLUTION   The solution to this problem involves the simple procedure of selecting values of entropy from the steam tables. The total change in entropy is then

$$\Delta S = m(s_2 - s_1)$$

At 50 psia, saturated liquid, $s_1 = 0.4113$ Btu/lbm·°R, and at 500°F the entropy is $s_2 = 1.7909$ Btu/lbm·°R. The change in entropy is

$$\Delta S = (5)(1.7909 - 0.4113) = 6.898 \text{ Btu/°R} \quad (13.10 \text{ kJ/K})$$

EXAMPLE 6-2   Reversible isothermal heating

Water is evaporated in a reversible isothermal process at 20 psia. Calculate the heat transfer per pound mass in this process.

SOLUTION   For the reversible isothermal process we have

$$Q = T \, \Delta S$$

and, for evaporation, this expression becomes

$$Q = T(s_g - s_f) = T s_{fg}$$

For saturation conditions at 20 psia we obtain from the steam tables

$$T = 227.96°F = (459.69 + 227.96)°R$$
$$= 687.65°R$$
$$s_{fg} = 1.3962 \text{ Btu/lbm·°R} \quad (5.8453 \text{ kJ/kg·K})$$

The heat transfer is, thus,

$$Q = (687.65)(1.3962) = 960.1 \text{ Btu/lbm} \quad (2233 \text{ kJ/kg})$$

EXAMPLE 6-3   Polytropic compression of air

Air is compressed in a reversible polytropic process from an initial state of 15 psia and 100°F to a final pressure of 50 psia. The polytropic exponent is $n = 1.3$. Calculate the change in entropy per pound mass of air.

SOLUTION   We may calculate the change in entropy from Eq. (6-32) once the final temperature is known. For the polytropic process we have,

$$pv^n = \text{const.}$$

Using the ideal-gas relation $v = RT/p$, this equation may be written as

$$\frac{T_2}{T_1} = \left(\frac{p_2}{p_1}\right)^{(n-1)/n}$$

For the conditions of this problem,

$$T_2 = (100 + 460)(\tfrac{50}{15})^{(1.3-1)/1.3}$$
$$= 739°R = 279°F \qquad (137.2°C)$$

Inserting this temperature in Eq. (6-32) gives

$$s_2 - s_1 = (0.24) \ln \frac{739}{560} - \frac{53.35}{778} \ln \frac{50}{15}$$
$$= -0.0159 \text{ Btu/lbm·°R} \qquad (-0.0666 \text{ kJ/kg·K})$$

EXAMPLE 6-4 Throttling process

Air is throttled through a porous plug like that shown in Fig. 4-5. The inlet conditions are 30 psia and 85°F, and the outlet pressure is 15 psia. The process is adiabatic and kinetic energies are negligible. Calculate the change in entropy per pound of air.

SOLUTION   This is an irreversible process, but the change in entropy may be calculated as if there were any reversible process between the given end states. For an ideal gas, enthalpy is a function of temperature alone, so that the relation

$$h_1 = h_2$$

reduces to

$$T_1 = T_2$$

We may then apply Eq. (6-32) as

$$s_2 - s_1 = c_p \ln 1.0 - R \ln \frac{p_2}{p_1}$$
$$= -R \ln \frac{p_2}{p_1}$$

Using the given data, this equation becomes

$$s_2 - s_1 = -\frac{53.35}{778} \ln \frac{15}{30}$$
$$= +0.0475 \text{ Btu/lbm·°R} \qquad (0.1989 \text{ kJ/kg·K})$$

Note that the entropy change is positive, as it must be for an irreversible adiabatic process.

EXAMPLE 6-5   Heat transfer across finite temperature difference
A heat reservoir at 1000°F is brought into thermal communication with a heat reservoir at 500°F. One thousand Btu of heat are transferred from the high-temperature reservoir to the low-temperature reservoir. Calculate the change in entropy of the universe resulting from this heat-exchange process.

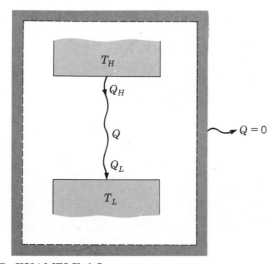

FIG. EXAMPLE 6-5

SOLUTION   We may treat each of the two reservoirs as a subsystem and calculate their changes in entropy separately. The change in entropy of the universe is then the algebraic sum of these quantities. We assume an isothermal process for both reservoirs. Thus

$$\Delta S_H = \frac{Q_H}{T_H}$$

$$\Delta S_L = \frac{Q_L}{T_L}$$

We have $Q_H = -1000$ Btu and $Q_L = +1000$ Btu. Thus

$$\Delta S_H = -\tfrac{1000}{1460} = -0.685 \text{ Btu/°R} \qquad (-1.301 \text{ kJ/K})$$
$$\Delta S_L = +\tfrac{1000}{960} = +1.041 \text{ Btu/°R} \qquad (1.977 \text{ kJ/K})$$

The change in entropy of the universe is

$$\Delta S_{universe} = \Delta S_H + \Delta S_L$$
$$= -0.685 + 1.041$$
$$= 0.356 \text{ Btu/°R} \quad (0.6761 \text{ kJ/K})$$

The schematic diagram for this process shows that the two reservoirs taken together comprise an adiabatic system. The heat-transfer process taking place inside this system is irreversible since it involves heat transfer across a finite temperature difference. We, therefore, have an irreversible adiabatic process insofar as the universe is concerned and experience an increase in entropy.

EXAMPLE 6-6   Isentropic steam turbine

Steam is expanded reversibly and adiabatically in a turbine from 400 psia and 800°F to 50 psia. Calculate the work done per pound of steam in a steady-flow process. Neglect changes in kinetic energy.

SOLUTION   For a reversible adiabatic process we have

$$s_1 = s_2$$

The steady-flow energy equation for this process becomes

$$W = h_2 - h_1 \tag{a}$$

The values of h_1 and s_1 may be obtained directly from the superheat steam tables.

$$h_1 = 1416.6 \text{ Btu/lbm} \quad (3295.0 \text{ kJ/kg})$$
$$s_1 = 1.6844 \text{ Btu/lbm·°R} \quad (7.0514 \text{ kJ/kg·K})$$

The final state of the flow process is now determined by the two properties

$$s_2 = s_1 = 1.6844$$
$$p_2 = 50 \text{ psia}$$

We thus consult the steam tables again (superheat tables) and find that

$$T_2 = 315.2°F \quad (157.3°C)$$
$$h_2 = 1192.1 \text{ Btu/lbm} \quad (2773 \text{ kJ/kg})$$

The work is now calculated from Eq. (*a*) as

$$W = h_2 - h_1$$
$$= 1192.1 - 1416.6$$
$$= -224.5 \text{ Btu/lbm} \quad (-522 \text{ kJ/kg}) \tag{b}$$

The negative sign indicates that the work is delivered to things external to
the turbine.

6-14 AVAILABLE AND UNAVAILABLE ENERGY

We have seen that there is some maximum efficiency for a heat engine
operating between given temperature limits. For a given heat added at the
upper temperature limit there is some maximum amount of work which we
can obtain as an output from the device. Of course, we can increase the
work output by either raising the upper temperature limit of the cycle or by
reducing the lower temperature limit. In a practical sense, the lower temperature limit is not a quantity which can be reduced indefinitely; to reduce it
below the surrounding atmospheric temperature we would have to provide a
refrigeration device with a work input which would penalize the performance of our overall system. For this reason it is customary to speak of a
lowest available temperature T_0 which is accessible to us for operation of a
heat engine. This temperature is usually the local atmospheric temperature
or that of some large heat sink like a lake or river.

Given some lowest available temperature T_0 at which heat can be rejected, a small reversible engine receiving heat at temperature T will produce, at the most, a work *output* of

$$d'W = \left(1 - \frac{T_0}{T}\right) d'Q \tag{6-34}$$

where $d'Q$ is the heat added at the upper temperature. The *available energy*
of a system is the portion of the heat added to the system which could be
converted to work in a series of reversible engines operating between the
temperature of the system and the lowest available temperature T_0. This
quantity is obtained by integration of Eq. (6-34) and is designated by W_{max}:

$$W_{max\ output} = \int \left(1 - \frac{T_0}{T}\right) d'Q \tag{6-35}$$

The *unavailable energy* is the difference between the total heat added and
the available energy.

Now let us suppose that we wish to determine the change in available
energy of a system as it changes from some state 1 to state 2. We would then
integrate Eq. (6-35) between these states to find the maximum increase in
work output. The maximum work output occurs when heat is added reversibly so we choose a reversible process between the states to evaluate the
integral

$$W_{\text{max output}} = \int_1^2 \left(1 - \frac{T_0}{T}\right) d'Q_{\text{rev}}$$

We may then recall the definition of entropy and write

$$W_{\text{max output}} = Q - T_0(S_2 - S_1) \tag{6-36}$$

where S_1 and S_2 are the initial and final entropies, respectively, and $W_{\text{max output}}$ represents the increase in available energy resulting from a reversible process between states 1 and 2.

Of course, the overriding notion in this discussion is the fact that a larger portion of heat energy is available for producing work, the higher the temperature of the system. From Eq. (6-36) we see that the *increase* in the *unavailable* energy is represented by the term $T_0(S_2 - S_1)$.

EXAMPLE 6-7   Available energy "loss"

How much available energy is "lost" in the process of Example 6-5, assuming $T_0 = 70°F = 530°R$?

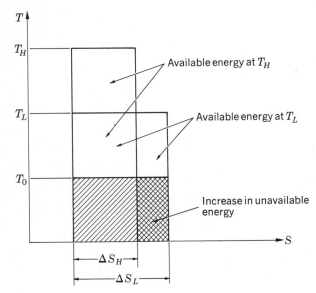

FIG. EXAMPLE 6-7

SOLUTION   If the 1000 Btu leaving the high-temperature reservoir (1000°F) were added to a reversible engine operating between that temperature and 70°F, the work produced would be

$$W = \left(1 - \frac{T_0}{T_H}\right) Q_H = \left(1 - \frac{530}{1460}\right)(1000) = 637 \text{ Btu} \qquad (672 \text{ kJ})$$

The 637 Btu is the "available" portion of the 1000 Btu at the high-temperature reservoir. However, if this same 1000 Btu were added to a reversible engine operating between 500 and 70°F, the work produced would be only

$$W = \left(1 - \frac{T_0}{T_L}\right) Q_L = \left(1 - \frac{530}{960}\right)(1000) = 448 \text{ Btu} \qquad (473 \text{ kJ})$$

Thus, as a result of the heat transfer of 1000 Btu of heat across the 500° temperature difference there has been a "loss" of available energy in the amount of 637–448 or 189 Btu. The available and unavailable energies at these two temperatures are displayed graphically in the accompanying figure. Of course, if the 1000 Btu were transferred directly to T_0, the available energy would be lost.

EXAMPLE 6-8 Heating increases available energy
How much is the available energy of 1 kg of air increased by heating it reversibly at a constant pressure of 1.5 atm from 40 to 250°C with a lowest available temperature of 20°C?

SOLUTION For this problem we find the increase in available energy by performing an integration of Eq. (6-35). For the constant-pressure process and unit mass,

$$d'Q = c_p \, dT$$

Then

$$W_{\text{max}} = \int_{T_1}^{T_2} \left(1 - \frac{T_0}{T}\right) c_p \, dT$$

$$= c_p(T_2 - T_1) - c_p T_0 \ln \frac{T_2}{T_1}$$

where we recognize that T_0 is a constant. We have $T_0 = 20°C = 293$ K; $T_1 = 40°C = 313$ K; $T_2 = 250°C = 523$ K; and c_p for air is 0.24 Btu/lbm·°F = 1005 J/kg·°C. Thus,

$$W_{\text{max}} = (1005)[(523 - 313) - (293) \ln \tfrac{523}{313}]$$
$$= 5.988 \times 10^4 \text{ J/kg}$$

The total heat added is $c_p(T_2 - T_1) = 2.11 \times 10^5$ J/kg so the increase in unavailable energy is $(21.1 - 5.988) \times 10^4 = 15.11 \times 10^4$ J/kg. Note that the heat-addition process causes an increase in both the available and unavailable energy of the air.

EXAMPLE 6-9 Work addition increases available energy

A 10-ft³ insulated rigid box contains air at 14.7 psia and 100°F. Paddle-wheel work is added to the box until the temperature rises to 400°F. Calculate the change in available energy of the air for $T_0 = 70°F$.

SOLUTION This example illustrates the important point that the available energy of a system may be changed by either heat *or* work additions. In other words, any change in the state of a system produces a change in the maximum amount of work which could be obtained when operating between the temperature of the system and T_0, the lowest available temperature. Let us first calculate some needed numerical quantities. The mass of the air is

$$m = \frac{p_1 V_1}{R T_1} = \frac{(14.7)(144)(10)}{(53.35)(560)} = 0.709 \text{ lbm} \qquad (0.322 \text{ kg})$$

From the first law

$$Q + W = \Delta U = m c_v (T_2 - T_1)$$

where W is the work *added*. The box is insulated so $Q = 0$ and

$$W = (0.709)(0.1715)(400 - 100) = 36.48 \text{ Btu} \qquad (38.49 \text{ kJ})$$

For constant-volume ideal-gas behavior

$$\frac{p_2}{p_1} = \frac{T_2}{T_1}$$

and

$$p_2 = (14.7)(\tfrac{860}{560}) = 22.58 \text{ psia} \qquad (155.7 \text{ kPa})$$

The entropy changes for the air can be calculated with Eq. (6-30),

$$s_2 - s_1 = c_v \ln \frac{T_2}{T_1} + R \ln \frac{v_2}{v_1}$$
$$= (0.1715) \ln \tfrac{860}{560} = 0.0736 \text{ Btu/lbm·°R}$$
$$S_2 - S_1 = m(s_2 - s_1) = (0.709)(0.0736) = 0.0522 \text{ Btu/°R}$$

Now, to evaluate the change in available energy we imagine a *reversible* process between the initial and final states, and then evaluate Eq. (6-36) accordingly. Clearly, the reversible process of interest is a constant-volume heating process in which

$$Q = mc_v(T_2 - T_1) = 36.48 \text{ Btu} \qquad (38.49 \text{ kJ})$$

Then, evaluating Eq. (6-36),

$$\begin{aligned}W_{\text{max output}} &= Q - T_0(s_2 - s_1)\\ &= 36.48 - (530)(0.0522)\\ &= 8.814 \text{ Btu} \qquad (9.299 \text{ kJ})\end{aligned}$$

The important point of this example is the fact that the available energy of a system may be changed by any action which changes the state of the system, whether the action occurs reversibly or not. In this case the change in state results from the irreversible paddle-wheel work addition, but the increase in available energy is the same as would have been experienced by a like amount of heat addition.

In summary, the 8.814 Btu represents the maximum amount of work which could be produced if the air were cooled reversibly and at constant volume from 400 to 100°F, in conjunction with a lowest available temperature of 70°F.

6-15 SECOND–LAW ANALYSIS FOR A CONTROL VOLUME

We have already discussed first-law energy analyses for a control volume (open system) in Chap. 4 and illustrated calculation techniques which may be applied to such systems. Let us now examine the control volume from the viewpoint of the second law as illustrated in Fig. 6-16. In this diagram the

FIG. 6-16 *Control volume for second-law analysis.*

fluid moves through the control volume from section i to section e while work is delivered external to the control volume. We assume that the boundary of the control volume is at the environment temperature T_0 and that all of the heat transfer Q occurs at this boundary. We have already noted that entropy is a property, so it may be transported just like enthalpy or internal energy. The entropy flow into the control volume resulting from mass transport is therefore $\dot{m}_i s_i$, and the entropy flow out of the control volume is $\dot{m}_e s_e$, assuming that the properties are uniform at sections i and e. Entropy may also be added to the control volume because of heat transfer at the boundary of the control volume. Realizing that heat may be added at a number of locations, and that the temperature may vary, we write

$$
\begin{array}{l}
\text{Entropy inflow} \\
\text{from heat transfer} \\
\text{at surface of} \\
\text{control volume}
\end{array}
= \int_{\text{surface}} \frac{1}{T_i} \frac{d'Q_1}{d\tau}
\tag{6-37}
$$

where the integral is taken over all the surface of the control volume where heat transfer is present and T_i is the temperature of the surface corresponding to each $d'Q_i$. In addition, there can be an increase in entropy in the control volume resulting from mass accumulation or internal irreversibilities, fluid friction, etc. We shall designate this storage or accumulation term as $(dS/d\tau)_\sigma$ using the σ subscript as before to designate a quantity internal to the control volume. Finally, we can recognize that there may be several streams entering or leaving the control volume, and we perform the following sum over all such streams to obtain the total *entropy production* for the control volume:

$$
\left(\frac{dS}{d\tau}\right)_{\text{prod}} = \underbrace{\sum_{\text{out}} \dot{m}_e s_e}_{} - \underbrace{\sum_{\text{in}} \dot{m}_i s_i}_{} - \underbrace{\int_{\text{surface}} \frac{1}{T_i} \frac{d'Q_i}{d\tau}}_{} + \underbrace{\left(\frac{dS}{d\tau}\right)_\sigma}_{}
\tag{6-38}
$$

$$
\text{entropy outflow } - \text{ entropy inflow } + \text{ entropy accumulation}
$$

If one imagines this flow process as happening over a small length of time $d\tau$, Eq. (6-38) is concerned with the mass elements involved at that instant, and the right side of Eq. (6-38) is equivalent to writing for a closed system consisting of all the mass elements

$$
(dS)_{\text{prod}} = \Delta S - \int \frac{d'Q}{T}
\tag{6-39}
$$

From the second law of thermodynamics this quantity must be greater than or equal to zero. Therefore, for Eq. (6-38) we write

$$
\left(\frac{dS}{d\tau}\right)_{\text{prod}} \geq 0
\tag{6-40}
$$

The equal sign applies to reversible processes while the inequality is applicable to irreversible processes. For steady flow and steady state there is no change with time inside the control volume, so $(dS/d\tau)_\sigma = 0$; $\dot{m}_i = \dot{m}_e$, and

$$\sum_{\text{out}} \dot{m}_e s_e - \sum_{\text{in}} \dot{m}_i s_i \geq \int_{\text{surface}} \frac{1}{T_i} \frac{d'Q_i}{d\tau} \qquad (6\text{-}41)$$

For an adiabatic, steady-flow, steady-state process, $d'Q_i/d\tau = 0$ and Eq. (6-41) becomes

$$s_e \geq s_i \qquad (6\text{-}42)$$

6-16 AVAILABILITY AND IRREVERSIBILITY IN STEADY FLOW

We have already seen the applicability of steady-flow energy analysis to practical power-production systems. The foregoing sections have shown how the principle of increase of entropy can place restrictions on the allowed directions of energy transfer in both flow and nonflow processes. Let us now use these principles to predict the maximum amount of work output we can obtain from a steady-flow device, as illustrated in Fig. 6-17. We have drawn the diagram with the work assumed positive leaving the control volume because we are interested in maximizing the power output. The steady-flow energy equation is therefore

$$\dot{m}(h_1 + \text{K.E.}_1 + \text{P.E.}_1 + \cdots) + \dot{Q}_0$$
$$= \dot{m}(h_2 + \text{K.E.}_2 + \text{P.E.}_2 + \cdots) + \dot{W} \qquad (6\text{-}43)$$

and the entropy production rate is obtained from Eq. (6-38) as

$$\frac{dS}{d\tau} = \dot{m}s_2 - \left(\dot{m}s_1 + \frac{\dot{Q}_0}{T_0}\right) \geq 0 \qquad (6\text{-}44)$$

FIG. 6-17 *Availability in steady flow.*

where we assume that all the heat Q_0 is obtained from the environment at T_0 and this is the temperature at the boundary of the control volume. Combining Eq. (6-43) and (6-44) gives

$$\dot{m}(s_2 - s_1)T_0 - \dot{m}[(h_2 - h_1) + \text{K.E.}_2 - \text{K.E.}_1 + \text{P.E.}_2 - \text{P.E.}_1 + \dot{W}] \geqslant 0$$

or

$$\dot{W} \leqslant \dot{m}[(h + \text{K.E.} + \text{P.E.} - T_0 s)_1 - (h + \text{K.E.} + \text{P.E.} - T_0 s)_2] \tag{6-45}$$

In many practical problems the kinetic and potential energy changes are negligible and we obtain

$$\dot{W} \leqslant \dot{m}[(h - T_0 s)_1 - (h - T_0 s)_2] \tag{6-46}$$

The quantity $h - T_0 s$ is called the *steady-flow availability function b,*

$$b = h - T_0 s \tag{6-47}$$

and the maximum work output per unit mass will be equal to the decrease in this function,

$$\dot{W}_{\text{max output}} = \dot{m}(b_1 - b_2) \tag{6-48}$$

The actual work output will be less than this value. The *irreversibility* is defined as the difference between the maximum possible work output and the actual work output, or

$$\dot{I} = \dot{W}_{\text{max}} - \dot{W}_{\text{act}} = \dot{m}(b_1 - b_2) - \dot{W}_{\text{act}} \tag{6-49}$$

If the process is adiabatic and kinetic energies are negligible, the actual work output must be equal to the decrease in enthalpy, or

$$\dot{W}_{\text{act}} = \dot{m}(h_1 - h_2) \tag{6-50}$$

Combining the above relations for an *adiabatic* process yields

$$\dot{I} = \dot{m}[(h - T_0 s)_1 - (h - T_0 s)_2] - \dot{m}(h_1 - h_2)$$
$$= \dot{m}T_0(s_2 - s_1) \tag{6-51}$$

When multiple streams enter and leave the control volume, as shown in Fig. 6-18, a summation must be performed over all the availability functions to yield

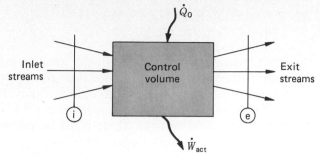

FIG. 6-18 *Irreversibilities for multiple inlet-exit streams.*

$$\dot{W}_{max} = \Sigma \, \dot{m}_i b_i - \Sigma \, \dot{m}_e b_e \qquad (6\text{-}52)$$

and

$$\dot{I} = \dot{W}_{max} - \dot{W}_{act} = \Sigma \, \dot{m}_i b_i - \Sigma \, \dot{m}_e b_e - \dot{W}_{act} \qquad (6\text{-}53)$$

For an adiabatic process this reduces to

$$\dot{I} = T_0[\Sigma \, \dot{m}_e s_e - \Sigma \, \dot{m}_i s_i] \qquad (6\text{-}54)$$

The availability function is useful for analyzing real processes to see how closely they approximate an ideal reversible process. We should note that this quantity is a function of both the fluid properties and the surroundings temperature, which comes into play because any heat transfer across a finite temperature difference is irreversible and generates an increase in entropy at the boundary of the control volume. In a reversible adiabatic process there is no heat transfer ($s_1 = s_2$), and the irreversibility is zero.

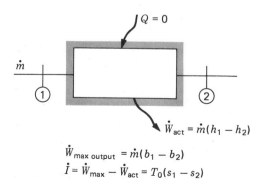

$$\dot{W}_{max\ output} = \dot{m}(b_1 - b_2)$$
$$\dot{I} = \dot{W}_{max} - \dot{W}_{act} = T_0(s_1 - s_2)$$

FIG. 6-19 *Availability decrease and irreversibility for adiabatic steady-flow process with* $\Delta K.E. \cong 0$.

6-17 PRACTICAL CONSIDERATIONS WITH AVAILABILITY

A few remarks concerning the practicality of irreversibility analysis are in order at this point. In choosing between alternatives for power-production or energy-conversion devices several factors must be brought into play. Naturally, one must be able to design and build the device so that it operates in a reliable fashion. Beyond this general statement, there must be considerations of initial cost of the equipment as well as the annual operating cost which is usually a strong function of energy costs. Capital costs depend on many factors such as material and fabrication expenses and, not the least, the care with which engineering design of equipment is effected. Costs of energy consumption are somewhat different. We can make a fairly general statement that energy-consumption rates in power-producing devices are generally lower, the lower the generation of irreversibility.

Thus, we may anticipate that a process or device which produces the least irreversibility will probably involve the least energy cost. This is not an absolute principle by any means, but it does provide us with a technique for evaluation of technical alternatives.

Obviously, there are many other factors to consider in the design of practical equipment including ease of maintainance, labor skills required for operation, and the *physical availability* of a particular energy source. For example, it would make no sense to consider natural gas as an energy source in some region where natural gas is unavailable.

The examples that follow illustrate how the concepts of availability and irreversibility may be used for analysis purposes.

EXAMPLE 6-10

The closed feedwater heater of Example 4-6 is to be analyzed in respect to the irreversibility which is generated in the heat-exchange process. We shall make the calculation for a lowest available temperature of 20°C.

FIG. EXAMPLE 6-10

SOLUTION   The schematic diagram of Example 4-6 is reproduced here for convenience. The actual work for this process is zero, and the maximum work output can be expressed in terms of the availability functions as

$$\dot{W}_{max} = \dot{m}_1(b_1 - b_2) + \dot{m}_s(b_3 - b_4) \qquad (a)$$

The entropies for these calculations are obtained from the steam tables as

$$s_1 = 0.4316 \text{ kJ/kg·K} \qquad \text{(compressed liquid, 10 MPa, 30°C)}$$
$$s_2 = 2.3016 \text{ kJ/kg·K} \qquad \text{(compressed liquid, 10 MPa, 198.3°C)}$$
$$s_3 = s_{f_3} + x_3 s_{fg_3}$$
$$\quad = 2.3150 + (0.95)(6.4448 - 2.3150)$$
$$\quad = 6.2383 \text{ kJ/kg·°C}$$
$$s_4 = s_f \text{ at } 1.5 \text{ MPa} = 2.3150 \text{ kJ/kg·°C}$$

Equation (a) may now be evaluated with $m_1 = 1.0$, $m_s = 0.3858$, and $T_0 = 20°C = 293$ K:

$$\dot{W}_{max} = [(h_1 - h_2) - T_0(s_1 - s_2)] + \dot{m}_s[(h_3 - h_4) - T_0(s_3 - s_4)]$$
$$\quad = [(134.86 - 848.54) - (293)(0.4316 - 2.3016)]$$
$$\quad + (0.3858)[(2694.8 - 844.89) - (293)(6.2383 - 2.3150)]$$
$$\quad = 104.44 \text{ kJ/kg water} \qquad (b)$$

Because $\dot{W}_{act} = 0$, the irreversibility is

$$\dot{I} = \dot{W}_{max} - \dot{W}_{act} = 104.44 \text{ kJ/kg water}$$

EXAMPLE 6-11   Alternatives for an air heater

We wish to evaluate the heating of 1.0 lbm/s of air in a constant-pressure process from 50°F to 120°F using four types of heaters:

(a) Condensing steam at 1 atm with a change from saturated vapor to saturated liquid
(b) Using hot water entering the heat exchanger at 200°F and leaving at 150°F
(c) Using hot water entering the heat exchanger at 150°F and leaving at 100°F
(d) Using an electric heater

For all the alternatives it is assumed that there are no heat losses from the heat exchanger. The schematic of the system is shown in the accompanying figure. We wish to evaluate the overall irreversibility for each alternative and comment on the advisability of using each in a practical situation. The lowest available temperature is taken as $T_0 = 70°F = 530°R$.

FIG. EXAMPLE 6-11

SOLUTION The heat which must be transferred to the air in each system is

$$q = \dot{m}_a(h_2 - h_1)_a = (1)(0.24)(120 - 50) = 16.8 \text{ Btu/s} \qquad (17.7 \text{ kW})$$

In this process the decrease of the availability of the air is

$$\dot{m}_a(-\Delta b_a) = \dot{m}_a[(h_1 - T_0 s) - (h_2 - T_0 s)]_a \qquad (a)$$

The entropy change for the air in a constant-pressure process is

$$s_2 - s_1 = c_p \ln \frac{T_2}{T_1} \qquad (b)$$

$$= 0.24 \ln \frac{580}{510} = 0.03087 \text{ Btu/lbm·°R}$$

$$= 0.1292 \text{ kJ/kg·K}$$

Then, the decrease in the availability is

$$-\Delta b_a = (1.0)[(0.24)(50 - 120) - (530)(-0.030\ 87)]$$
$$= -0.4389 \text{ Btu/s} \qquad (-0.463 \text{ kW}) \qquad (c)$$

This result must now be combined with the decrease in availability for the heater or heating fluid to evaluate the overall performance.

Case 1: Condensing steam The properties of steam at 1 atm of interest are

$$h_1 - h_2 = -h_{fg} = 970.4 \text{ Btu/lbm}$$
$$s_1 - s_2 = s_{fg} = 1.4446 \text{ Btu/lbm·°R}$$

The energy gained by the air is equal to the energy lost by the steam, and so

$$q = 16.8 = \dot{m}_s h_{fg} \qquad (d)$$

and

$$\dot{m}_s = 0.017\ 31\ \text{lbm/s} \qquad (0.00785\ \text{kg/s})$$

The decrease in availability of the steam is then

$$\dot{m}_s(-\Delta b_s) = (0.0173)[970.4 - (530)(1.4446)]$$
$$= +3.5424\ \text{Btu/s} \qquad (3.736\ \text{kW})$$

The overall decrease in availability is

$$\dot{m}_a(-\Delta b_a) + \dot{m}_s(-\Delta b_s) = -0.4389 + 3.5424$$
$$= +3.1035\ \text{Btu/s}$$

This is the maximum work *output*. The *actual* work output is zero, so the irreversibility is

$$\dot{I} = \dot{W}_{max} - \dot{W}_{act} = 3.1035 - 0 = +3.1035\ \text{Btu/s}$$
$$= 3.274\ \text{kW} \qquad\qquad (e)$$

Case 2: Hot water source of heating For this case the appropriate water properties are

$$h_1 = h_f \text{ at } 200°F = 168.07\ \text{Btu/lbm}$$
$$h_2 = h_f \text{ at } 150°F = 117.96\ \text{Btu/lbm}$$
$$s_1 = s_f \text{ at } 200°F = 0.2940\ \text{Btu/lbm·°R}$$
$$s_2 = s_f \text{ at } 150°F = 0.2150\ \text{Btu/lbm·°R}$$

The energy balance enables a calculation of the water mass flow from

$$q = 16.8\ \text{Btu/s} = \dot{m}_w(h_1 - h_2)_w$$
$$\dot{m}_w = 0.3353\ \text{lbm/s} \qquad (0.1521\ \text{kg/s})$$

The corresponding decrease in the availability of the water is

$$\dot{m}_w \neq (-\Delta b_w) = (0.3353)[(168.07 - 117.96) - (530)(0.2940 - 0.2150)]$$
$$= +2.7629\ \text{Btu/s} \qquad (2.91\ \text{kW})$$

Now the overall decrease in the availability is the sum of that for the water and that for the air, or

$$-0.4389 + 2.7629 = 2.3240\ \text{Btu/s} \qquad (2.451\ \text{kW})$$

As in case 1, this is the irreversibility:

$$\dot{I} = +2.3240\ \text{Btu/s} \qquad (2.451\ \text{kW}) \qquad\qquad (f)$$

Case 3: Hot water heating at lower temperature For this case we should remark that a larger heat exchanger would be required because of the lower temperature difference between the hot water and air. The water properties of interest are

$$h_1 = h_f \text{ at } 150°F = 117.96 \text{ Btu/lbm}$$
$$h_2 = h_f \text{ at } 100°F = 68.05 \text{ Btu/lbm}$$
$$s_1 = s_f \text{ at } 150°F = 0.2150 \text{ Btu/lbm·°R}$$
$$s_2 = s_f \text{ at } 100°F = 0.1296 \text{ Btu/lbm·°R}$$

As in case 2, we can use an energy balance and obtain the water mass flow as

$$\dot{m}_w = 0.3366 \text{ lbm/s}$$

The decrease in the availability for the water is then

$$
\begin{aligned}
\dot{m}_w(-\Delta b_w) &= \dot{m}_w[(h_1 - T_0 s_1) - (h_2 - T_0 s_2)]_w \\
&= (0.3366)[(117.96 - 68.05) - (530)(0.2150 - 0.1296)] \\
&= +1.5645 \text{ Btu/s} \quad (1.650 \text{ kW}) \tag{g}
\end{aligned}
$$

The overall decrease in availability is then the sum of that for the air and that for the water, and this is equal to the irreversibility:

$$-0.4389 + 1.5645 = +1.1256 \text{ Btu/s} = \dot{I} \tag{h}$$

Case 4: Electric heating For this case there is no heating fluid and the overall decrease in availability is that for the air, or

$$\dot{m}_a(-\Delta b_a) = -0.4389 \text{ Btu/s} \quad (-0.463 \text{ kW})$$

This represents the maximum work output for this heater system. Unlike the three previous cases, however, the actual work is no longer zero. Now, there is an electrical work *input* of 16.8 Btu/s (or an actual work *output* of *minus* 16.8 Btu/s). The irreversibility is thus

$$
\begin{aligned}
\dot{I} &= \dot{W}_{max} - \dot{W}_{act} \tag{i} \\
&= -0.4389 - (-16.8) \\
&= +16.3611 \text{ Btu/s} \quad (+17.258) \text{ kW}
\end{aligned}
$$

The four cases are summarized in the tabulation on the next page. The first three cases show a drop in irreversibility as the temperature of the heating fluid is dropped. This indicates that lower irreversibility is experienced when there is a smaller temperature difference across which the heat is to be transferred. The last case shows the high irreversibility which results from a conversion of work (in this case electrical work) to heat.

Example 6-11 illustrates one approach to a second-law analysis of en-

Heater	Decrease in availability, Btu/s	Actual work output, Btu/s	Irreversibility	
			Btu/s	kW
Steam, 1 atm, 212°F	3.1035	0	3.1035	3.274
Hot water, 200 to 150°F	2.3240	0	2.3240	2.451
Hot water, 150 to 100°F	1.1256	0	1.1256	1.187
Electric heater	−0.4389	−16.8	16.3611	17.258

ergy transfer schemes. We must caution the reader, however, that the analysis is incomplete. One should also evaluate the irreversibility which results from producing the steam or hot water and the losses and irreversibility creation which can take place at the power plant where electricity is generated. The purpose here has been to employ availability and irreversibility analysis as a tool to indicate the relative second-law effectiveness of heating schemes.

EXAMPLE 6-12   Irreversibility in a steam turbine

The adiabatic turbine of Example 6-6 is operated so that the steam enters at 800°F and 400 psia and exits at 50 psia and 400°F. Calculate the maximum work output which could be obtained between these end states and the irreversibility. Neglect kinetic energies. The lowest available temperature is 70°F.

SOLUTION   The properties of interest are

$$h_1 = 1416.6 \text{ Btu/lbm} \quad \text{at 800°F and 400 psia}$$
$$s_1 = 1.6844 \text{ Btu/lbm·°R}$$
$$h_2 = 1235.0 \text{ Btu/lbm} \quad \text{at 400°F and 50 psia}$$
$$s_2 = 1.7370 \text{ Btu/lbm·°R}$$

The actual work output is

$$\dot{W}_{act} = h_1 - h_2 = 181.6 \text{ Btu/lbm} \quad (422.4 \text{ kJ/kg})$$

while the maximum possible work output is obtained as the decrease in the availability function

$$\begin{aligned}
\dot{W}_{max\ output} &= (h - T_0 s)_1 - (h - T_0 s)_2 \\
&= h_1 - h_2 - T_0(s_1 - s_2) \\
&= 181.6 - (530)(1.6844 - 1.7370) \\
&= 209.48 \text{ Btu/lbm} \quad (487.25 \text{ kJ/kg})
\end{aligned}$$

The irreversibility is therefore

$$\dot{I} = \dot{W}_{max} - \dot{W}_{act}$$
$$= 209.48 - 181.6 = 27.88 \text{ Btu/lbm} \qquad (64.85 \text{ kJ/kg})$$

6-18 MICROSCOPIC INTERPRETATION OF ENTROPY

For those readers who choose to study the principles of microscopic thermo-dynamics in Chaps. 5 and 8 it is shown that there is a characteristic function of state which behaves like the macroscopic entropy we have described in this chapter. In an isolated system of particles (no mass or energy added to the system) it is found that the system tends to move toward a most probable distribution of the particles among the allowable energy states. It is also shown that deviations from the most probable state are very unlikely indeed. This is equivalent to saying that the entropy of an isolated system tends to increase to its maximum value, and it is highly improbable that it will de-crease from this maximum. The microscopic definition of entropy thus fur-nishes the same information as Eq. (6-22), except that the principle of in-crease of entropy becomes one which represents a *most probable* behavior. A further implication of the microscopic definition is that the second law of thermodynamics becomes a statement of what will most probably happen. But we are no longer permitted to use the word "impossible."

A further study of the distribution of energy at the microscopic level will show that our simplified kinetic theory model of an ideal gas presented in Chap. 1 turns out to be correct. For gases with wide spacing between mole-cules we are able to identify the temperature of the gas with the mean kinetic energy of the molecules, even when one takes into account the allowable quantum energy states for the gas molecules.

Therefore, we may be comfortable in stating that our development of the first and second laws of thermodynamics on a purely macroscopic basis will stand the test of a much more detailed analysis at the microscopic level.

6-19 MAXWELL'S DEMON AND INFORMATION THEORY

An interesting sidelight to the second law was given by J. C. Maxwell in 1871. Imagine the container shown in Fig. 6-20 with the two compartments A and B filled with identical gases at the same temperatures. Some of the molecules move fast and some move slowly, but the *average* kinetic energy of the molecules is the same in both chambers in accordance with our simple discussion in Sec. 1-14. Maxwell imagined what might happen if there were a small door in the separating partition and a small "demon" in-telligent and quick enough to open and close the door very rapidly so that

FIG. 6-20 *Maxwell's demon, "fast" molecules to one side, "slow" molecules to the other side.*

the faster molecules in section A would be allowed to pass through to section B while the slower molecules from B would be allowed to pass through into A. It is easy to see that the demon would be a remarkable person indeed if he were able to perform this task because the separation of fast and slow molecules would result in a temperature difference between the chambers. Such an occurrence is a violation of the second law of thermodynamics because it is, in effect, a transfer of heat "uphill on the temperature scale."

The concept of the Maxwell demon has been debated for years. The general tenor of the arguments is that the demon must be considered as part of the overall thermodynamic system under analysis. By some means or other, the demon must be able to recognize whether a molecule is moving "fast" or "slow." This requires that some measurement be performed and information about the measurement be communicated to the demon. The conclusion of a considerable body of opinion is that either (1) the demon will be unable to distinguish between fast and slow molecules because he is subjected to a random molecular bombardment or (2) the entropy of the demon must change to account for the "information" he receives in the measurement process. This latter argument pertaining to information has led to the use of the term *entropy* in the science of information theory, and has even prompted some authors to develop the entire subject of thermodynamics on the basis of concepts from information theory.† In this development, entropy is taken to be a measure of our "increasing lack of information."

In our discussion we shall prefer to think of the principles of thermodynamics as experimental axioms which are based on observations of the physical behavior of nature. We have no means of determining *why* nature behaves the way it does, but it would not seem reasonable to assume this behavior is caused by our lack of information. Such reasoning is analogous to an individual saying: "I wrote some checks which I forgot to list in my checkbook, so I do not have enough information to determine my bank balance." This individual's lack of information does not alter or affect in any way the accounting procedure at the bank, and if he tries to violate that procedure or "principle" by writing too many checks he will be in trouble,

† See Ref. 7.

regardless of his "lack of information" about his balance. If he wants to get the true story, he should experimentally observe the accounting procedure at the bank and duplicate it very carefully in his own checkbook. Such an observation-and-duplication process is exactly what we try to accomplish in thermodynamic analysis, although it is not so easy to perform as the checkbook accounting process. †

REVIEW QUESTIONS

1   What is the *first* law of thermodynamics?

2   Why is heat a different form of energy from work?

3   What is a perpetual motion machine of the second kind?

4   How would you describe the second law in simple terms to a layman?

5   What is a reversible process? What is the meaning of the term in a physical sense?

6   Why is the concept of a heat engine important?

7   Why is the second law necessary for a definition of an absolute temperature scale?

8   Is it possible to attain zero temperature on the absolute thermodynamic temperature scale? Why?

9   What is the *macroscopic* definition of entropy?

10   If entropy change is defined for a reversible process, how is it possible to make a calculation for irreversible processes?

11   Give an analytical statement of the second law in terms of entropy.

12   Why is a reversible adiabatic process one of constant entropy?

13   What is a Carnot cycle? What other kinds of cycles can be more efficient than the Carnot when operating between the same temperature limits?

14   What is meant by the term *available energy?* How does this differ from the concept of *availability* in steady-flow systems?

15   Is entropy conserved in an isolated system?

16   How would you explain entropy to a layman?

17   How is work defined?

18   What is meant by the thermal efficiency of a heat engine?

19   What is a coefficient of performance of a refrigeration process?

† Sometimes it can be a rather difficult process to balance a checkbook too.

20 What is meant by "a series of equilibrium states"?

21 What is a cycle?

22 What is the inequality of Clausius and how is it related to entropy?

23 What is a property?

24 How is the state of a system defined?

25 Is entropy a property?

26 How is reversible heat flow represented on a *T-S* diagram?

27 What is *irreversibility* in a steady-flow system?

28 What is thermodynamic probability?

PROBLEMS (English units)

6-1 In a system which undergoes a reversible process the work done is 5 Btu and the heat rejected is 7 Btu. The change in entropy is:
    (a) Positive
    (b) Negative
    (c) Zero
    (d) Can't tell

6-2 In a system which undergoes an irreversible process the work done is 5 Btu and the heat rejected is 7 Btu. The change in entropy is:
    (a) Positive
    (b) Negative
    (c) Zero
    (d) Can't tell

6-3 In a system which undergoes an irreversible process the work done is 5 Btu and the heat added is 7 Btu. The change in entropy is:
    (a) Positive
    (b) Negative
    (c) Zero
    (d) Can't tell

6-4 A Carnot cycle receives 600 Btu of heat at 1000°F and has an efficiency of 50 percent. Calculate work output and temperature at which heat is rejected.

6-5 A Carnot cycle receives 1000 Btu of heat at 800°C and rejects heat at 300°F. Calculate the work output.

6-6 A reversed Carnot cycle is to be used as a refrigerator operating between the temperature limits of 0 and 70°F. Calculate the horsepower input required to remove 12 000 Btu/h from the low-temperature reservoir.

6-7 A certain heat pump for a house may be approximated by a reversed Carnot cycle operating between the outside air temperature of 10°F and the inside air temperature of 80°F. Calculate the horsepower input required to the cycle to deliver 100 000 Btu/h to the high-temperature reservoir (the house).

6-8 Prospective inventors frequently devise machines which are in violation of the second law. Suppose an inventor asks you to analyze his "new" device which will receive heat from boiling water at atmospheric pressure (212°F) and reject heat at 70°F. He claims the device could achieve an efficiency of 75 percent. How do you rate this claim? What is the maximum efficiency you would think is realistic?

6-9 A Carnot cycle operates with air as the working fluid. The temperature limits of the cycle are 800 and 100°F, respectively, and the pressure at the start of heat addition is 50 psia. Calculate the mass of air required to produce a work output of 1 hp if the cycle is repeated 5000 times/min. What is the lowest pressure occurring in the cycle if the volume doubles in the heat addition?

6-10 Steam is expanded reversibly and adiabatically in a nozzle from 800 psia, 700°F, to 20 psia. Calculate the exit velocity of the steam if the inlet velocity is very small.

6-11 Steam is expanded in a reversible isothermal process from 100 psia, 400°F, to 50 psia in a piston-cylinder arrangement. Calculate the work and heat transfer for this process.

6-12 Calculate the change in entropy of the universe resulting from the mixing process of Prob. 4-10.

6-13 Steam enters a turbine at 500 psia, 500°F, and a velocity of 600 ft/s. The process is isentropic and the exit velocity is 200 ft/s. Calculate the work output of the turbine per pound mass of steam for discharge at 1 atm.

6-14 1 lbm of saturated liquid at 100 psia is mixed adiabatically with 1 lbm of saturated vapor at 100 psia. What is the change in entropy of the universe resulting from this process? Repeat but for liquid at 200°F, 100 psia, and saturated vapor at 100 psia. Assume steady flow.

6-15 2 lb of water at 150°F are mixed with 5 lb of water at 100°F in an adiabatic process. Calculate the change in entropy of the universe.

6-16 Two blocks of steel (c_p = 0.11 Btu/lbm·°F), one at 800°F and the other at 300°F, are brought into contact inside an insulated container and allowed to reach thermal equilibrium. What is the change in entropy of the universe?

6-17 Freon 12 is compressed isentropically from 25 psia, 40°F, to 150 psia in a piston-cylinder arrangement. Calculate the work per lbm of fluid. Repeat the calculation for compression in a steady-flow compressor device.

6-18 Calculate the change in entropy of the universe resulting from the heat-exchange process of Prob. 4-15. Assume that the air is heated at a constant pressure of 15 psia.

6-19 A 5-lbm block slides along a horizontal surface with a velocity of 10 ft/s. The coefficient of friction is 0.2 and the temperature of the surroundings is 70°F. Calculate the rate of entropy generation in this process. State assumptions.

6-20 Steam is throttled from 100 psia, 75 percent quality, to atmospheric pressure. Calculate the increase in entropy.

6-21 10 lbm of air at 50 psia and 300°F are mixed with 5 lbm of air at 30 psia and 500°F. The mixing takes place in an adiabatic chamber. Calculate the change in entropy if the final pressure is 25 psia.

6-22 Calculate the total change in entropy for the process in Prob. 4-27.

6-23 Calculate the total change in entropy for the process in Prob. 4-30.

6-24 Steam at 20 psia and 300°F enters an insulated capillary. The pressure drops to 16 psia as a result of fluid friction. Calculate the change in entropy per pound mass of steam.

6-25 Air is expanded adiabatically in a turbine from 1000°F and 200 psia to 400°F. If the process had been reversible, the final temperature would have been 350°F for discharge at the same pressure. Calculate the change in entropy and the work produced per lbm of air.

6-26 A 1-kW electric heater is placed in a 300-ft³ insulated room filled with air at 14.7 psia and 70°F. The heater is allowed to operate for 15 min. Calculate the change in entropy of the air.

6-27 An electric heater is used to heat 30 gal of water in an insulated container. The temperature of the heater is constant at 200°F and the water is heated from 60 to 120°F. Calculate the change in entropy of the universe resulting from this heating process.

6-28 A Carnot cycle receives heat from a constant-temperature reservoir at 800°F and rejects heat to a 10 lbm of water which is initially saturated liquid at 15 psia. As the engine operates, the water is heated at constant pressure until its temperature reaches 300°F. Calculate the work output of the Carnot engine.

6-29 Repeat Prob. 6-28, assuming that the reservoir from which heat is supplied is a 10 lbm of steam initially at 800°F and 800 psia, and that its pressure remains constant during the heat-removal process.

6-30 A 1-lbm piece of putty is dropped from a height of 10 ft onto a flat surface. It immediately sticks to the surface. Calculate the change in entropy of the universe. State assumptions.

6-31 Show that the irreversibility for an adiabatic steady-flow process is

$$I = \dot{m}T_0(s_2 - s_1)$$

6-32 A Carnot cycle is to be designed to operate with an efficiency of 30 percent with steam as the working fluid. Heat addition occurs at 500°F and is accomplished by changing the water from saturated liquid to saturated vapor. Calculate the temperature, entropy, and quality at each point in the cycle and sketch the cycle on a T-S diagram. What is the work output per pound mass of steam?

6-33 A reversed Carnot cycle is to be used as a refrigeration device with Freon 12 as the working fluid. The device is to operate between 50 and 120°F, and heat rejection (at the *upper* temperature) occurs as the fluid is changed from saturated vapor to saturated liquid. Calculate the entropy and quality at each point in the cycle, the refrigeration effect, and the work input per pound mass of Freon.

6-34 Steam is contained in a perfectly insulated 10-ft³ tank at 700°F and 50 psia. A valve on the side of the tank is opened and the steam allowed to discharge to the atmosphere. The valve is closed when the pressure in the tank just reaches 14.7 psia. Calculate the mass of steam which escapes in this process and the final temperature in the tank.

6-35 A small spray can having a volume of 20 in³ contains Freon 12 at 70°F, and 30 percent of the volume is occupied by liquid. A small nozzle is attached to the top of the can which operates as a throttling device for discharge to the surroundings at 14.7 psia and 70°F. For a slow discharge process the temperature in the can remains constant at 70°F. Calculate the heat added to the can and change in entropy inside the can for a slow discharge process in which half the initial mass is lost. Also calculate the change in entropy of the vapor which is lost.

6-36 A 2000-ft³ tank contains water in the saturated vapor state at 300 psia. Suppose the tank is to serve as a heat source, giving up energy as the steam is condensed to the liquid state. How much "available energy" does it have relative to an environment temperature of 70°F?

6-37 Water vapor is cooled at constant pressure of 200 psia from the saturated-vapor state to the saturated-liquid state. How much has the available energy decreased for this system, assuming a lowest available temperature of 70°F?

6-38 A tank of high-pressure steam having a volume of 200 ft³ is available to drive a small turbine which may be assumed to operate reversibly and adiabatically. Initially, the steam is at 1000 psia and 1000°F, and the tank is perfectly insulated. Discharge from the turbine is at 100 psia. A pressure control valve on the discharge from the turbine throttles the steam to atmospheric pressure. The turbine is allowed to run until the tank pressure reaches 300 psia at which time the discharge valve is shut off. Calculate the work output of the turbine for this process.

6-39 A heat exchanger is designed which will condense steam from the saturated-vapor to saturated-liquid state at 15 psia. Air serves as the cooling medium and enters and leaves the exchanger at 70°F and 150°F, respectively. The device is insulated so that all of the heat lost by the steam is gained by the air. Steady-flow operation may be assumed. For each pound mass of steam condensed calculate:
    (a) The mass flow of air required
    (b) The loss of availability of the steam
    (c) The gain of availability of the air (assume $T_0 = 70°F$)
    (d) The steady-flow irreversibility of the process
    (e) The change in entropy of the universe resulting from the process

6-40 Steam at 20 psia undergoes a constant-pressure steady-flow heating process from a quality of 50 percent to superheated vapor at 500°F. Heat for the process is obtained from a constant-temperature reservoir at 600°F, and the flow rate of steam is 25 lbm/min. Calculate:

    (a) The change in entropy of the steam
    (b) The change in entropy of the reservoir
    (c) The change in entropy of the universe
    (d) The change in available energy of the reservoir for $T_0 = 70°F$
    (e) The steady-flow irreversibility for the steam for $T_0 = 70°F$

6-41 In a large power plant combustion gases ($c_p = 0.26$ Btu/lbm·°F) are cooled from 2300 to 1800°F at constant pressure while giving up heat to steam at 1200 psia operating between saturated liquid and 1400°F. For each pound of steam, calculate:

    (a) The mass flow of gas required
    (b) The change in entropy for both the gas and steam
    (c) The change in the availability function for each flow stream, using $T_0 = 70°F$
    (d) The irreversibility for the overall process

6-42 Calculate the change in available energy resulting from the process in Prob. 6-16.

6-43 Steam flows in a pipe at 100 psia and 500°F and is connected by a valve to a 30-ft³ tank initially containing steam at 20 psia and 300°F. The valve is opened and steam allowed to flow into the tank. The valve is closed when the pressure just reaches 100 psia. If the tank is perfectly insulated, calculate:

    (a) The mass added to the tank
    (b) The change in entropy of the contents of the tank
    (c) The change in entropy of the mass which *enters* the tank
    (d) The change in entropy of the universe resulting from the process

6-44 An inventor claims that he can store large quantities of energy for conversion to work by just storing high-pressure steam at 5000 psia and 1600°F. For a storage tank having a volume of 200 ft³ he claims that he could produce a work output of 8×10^8 J. Evaluate this claim by analyzing some ideal device which could be connected to the steam tank for producing work.

6-45 Ammonia vapor is compressed adiabatically from saturated vapor at 20°F to 240 psia in a steady-flow device. Because of frictional effects and other factors the actual work input is 17 percent greater than the work which would have been required in an isentropic process with the same discharge pressure. For each pound mass of ammonia compressed calculate:

    (a) The actual work required
    (b) The actual exit temperature from the compressor
    (c) The irreversibility for $T_0 = 70°F$

6-46 Freon 12 undergoes an adiabatic throttling process from saturated-liquid conditions at 100°F to a state where the final temperature is 40°F. For each pound mass of Freon calculate:

    (a) The change in entropy of the liquid
    (b) The irreversibility of the process with $T_0 = 70°F$

6-47 An inventor proposes a novel work-energy storage device which will operate by providing a work input to a Carnot refrigerator which adds heat to liquid water at 14.7 psia at the high temperature and extracts heat from liquid water at 14.7 psia at the low temperature. When the device is first turned on both water reservoirs are at 70°F and they each have equal masses of 500 lbm of liquid water. Calculate:

    (a) The work energy input to just produce solid ice at 32°F in the low-temperature reservoir

    (b) The final state of the high-temperature reservoir for this work input

    (c) The change in available energy of both reservoirs with $T_0 = 70°F$

6-48 For a certain refrigeration machine Freon 12 is compressed from saturated vapor at 10°F to 200 psia and 180°F in an adiabatic steady-flow process. Calculate the work input per unit mass of Freon, the change in entropy, and irreversibility for $T_0 = 70°F$.

PROBLEMS (Metric Units)

6-1M In a system which undergoes a reversible process the work done is 5 kJ and the heat rejected is 7 kJ. The change in entropy is:

    (a) Positive
    (b) Negative
    (c) Zero
    (d) Can't tell

6-2M In a system which undergoes an irreversible process the work done is 5 kJ and the heat rejected in 7 kJ. The change in entropy is:

    (a) Positive
    (b) Negative
    (c) Zero
    (d) Can't tell

6-3M In a system which undergoes an irreversible process the work done is 5 kJ and the heat added is 7 kJ. The change in entropy is:

    (a) Positive
    (b) Negative
    (c) Zero
    (d) Can't tell

6-4M A Carnot cycle receives 630 kJ of heat at 540°C and has an efficiency of 50 percent. Calculate the work output and temperature at which heat is rejected.

6-5M A Carnot cycle receives 1 MJ of heat at 800°C and rejects heat at 150°C. Calculate the work output.

6-6M A reversed Carnot cycle is to be used as a refrigerator operating between the temperature limits of −18 and 20°C. Calculate the horsepower input required to remove 3.5 kW from the low-temperature reservoir.

6-7M A certain heat pump for a house may be approximated by a reversed Carnot cycle operating between the outside air temperature of $-10°C$ and the inside air temperature of 30°C. Calculate the power input required to the cycle to deliver 30 kW to the high-temperature reservoir (the house).

6-8M   Prospective inventors frequently devise machines which are in violation of the second law. Suppose an inventor asks you to analyze his "new" device which will receive heat from boiling water at 100 kPa (99.6°C) and reject heat at 20°C. He claims the device could achieve an efficiency of 75 percent. How do you rate this claim? What is the maximum efficiency you would think is realistic?

6-9M   A Carnot cycle operates with air as the working fluid. The temperature limits of the cycle are 500 and 310 K, respectively, and the pressure at the start of heat addition is 350 kPa. Calculate the mass of air required to produce a work output of 1 kW if the cycle is repeated 5000 times per minute. What is the lowest pressure occurring in the cycle if the volume doubles in the heat addition?

6-10M   Steam is expanded reversibly and adiabatically in a nozzle from 6 MPa, 360°C, to 150 kPa. Calculate the exit velocity of the steam if the inlet velocity is very small.

6-11M   Steam is expanded in a reversible isothermal process from 700 kPa, 200°C, to 300 kPa in a piston-cylinder arrangement. Calculate the work and heat transfer for the process.

6-12M   Steam enters a turbine at 3 MPa, 260°C, and a velocity of 180 m/s. The process is isentropic and the exit velocity is 60 m/s. Calculate the work output of the turbine per kilogram of steam for discharge at 100 kPa.

6-13M   1 kg of saturated liquid at 700 kPa is mixed adiabatically with 1 kg of saturated vapor at 700 kPa. What is the change in entropy of the universe resulting from this process? Repeat, but for a liquid at 95°C, 700 kPa, and saturated vapor at 700 kPa. Assume steady flow.

6-14M   2 kg of water at 65°C are mixed with 5 kg of water at 40°C in an adiabatic process. Calculate the change in entropy of the universe.

6-15M   Two blocks of steel (c_p = 460 J/kg·°C), one at 700 K and the other at 420 K, are brought into contact inside an insulated container and allowed to reach thermal equilibrium. What is the change in entropy of the universe?

6-16M   Calculate the change in entropy of the universe resulting from the heat-exchange process of Prob. 4-13M. Assume that the air is heated at a constant pressure of 100 kPa.

6-17M   A 5-kg block slides along a horizontal surface with a velocity of 3 m/s. The coefficient of friction is 0.2 and the temperature of the surroundings is 20°C. Calculate the rate of entropy generation in this process. State assumptions.

6-18M   Steam enters a perfectly insulated pipe at 10.0 MPa and 500°C. After an extended distance in the pipe, frictional effects cause the pressure to drop to 8.0 MPa. Calculate the change in entropy for the steam and irreversibility with T_0 = 20°C.

6-19M   A tank of nitrogen at 4.0 MPa and 60°C is used to force liquid out of another tank. The two tanks are connected by a pressure-regulating valve which maintains the nitrogen-discharge pressure into the liquid tank at 700 kPa. A discharge valve on the liquid tank is opened, and 0.25 m³ of liquid are allowed to escape as a result of the nitrogen entry into the liquid tank. At the end of the process the pressure in the nitrogen tank has dropped to 700 kPa. Both tanks and the connecting valve are insulated. Calculate the volume of the nitrogen tank and the final nitrogen temperature in both tanks.

6-20M   Nitrogen in employed as a working fluid in a Carnot cycle. In the heat-addition process the volume *increases* from 1.0 to 1.44 liters, while in the heat-rejection process the volume *decreases* from 22.43 to 15.88 liters. Calculate the thermal efficiency of the cycle.

6-21M   Steam is throttled from 700 kPa, 75 percent quality, to 100 kPa. Calculate the increase in entropy.

6-22M   10 kg of air at 350 kPa and 420 K are mixed with 5 kg of air at 200 kPa and 530 K. The mixing takes place in an adiabatic chamber. Calculate the change in entropy if the final pressure is 170 kPa.

6-23M   Steam at 150 kPa and 160°C enters an insulated capillary. The pressure drops to 100 kPa as a result of fluid friction. Calculate the change in entropy per kilogram of steam.

6-24M   Air is expanded adiabatically in a turbine from 800 K and 1.4 MPa to 480 K. If the process had been reversible, the final temperature would have been 450 K for discharge at the same pressure. Calculate the change in entropy and the work produced per kilogram of air.

6-25M   A 1-kW electric heater is placed in an 8500-liter insulated room filled with air at 1 atm and 22°C. The heater is allowed to operate for 15 min. Calculate the change in entropy of the air.

6-26M   An electric heater is used to heat 110 liters of water in an insulated container. The temperature of the heater is constant at 370 K and the water is heated from 15 to 50°C. Calculate the change in entropy of the universe resulting from this heating process.

6-27M   A Carnot cycle receives heat from a constant-temperature reservoir at 700 K and rejects heat to 5 kg of water which is an initially saturated liquid at 100 kPa. As the engine operates, the water is heated at constant pressure until its temperature reaches 150°C. Calculate the work output of the Carnot engine.

6-28M   3.0 kg/s of water are to be heated from 30°C to 75°C in a steady-flow device. Two alternatives are to be considered: (a) electric heating and (b) heating with a constant-temperature source at 450°C. Calculate the overall change in entropy and irreversibility for both cases for $T_0 = 20°C$.

6-29M   A throttling calorimeter is connected to a steam line to sample steam at $p = 1.5$ MPa. After steam undergoes an adiabatic throttling process to $p = 100$ kPa, the temperature is measured as 105°C. Determine the quality of the steam at 1.5 MPa in the line.

6-30M   An inventor claims that he has designed an engine which will produce a power output of 170 kW while consuming 0.5 kg/min of fuel having an energy content (heating value) of 42 000 kJ/kg. He states that the upper and lower temperatures of the engine are 670 and 330 K, respectively. Do you believe his claim?

6-31M   Determine the change in entropy for the system in Problem 3-24M.

6-32M   Repeat Prob. 6-27M, assuming that the reservoir from which heat is supplied is 5 kg of steam initially at 700°K and 6 MPa, and that its pressure remains constant during the heat-removal process.

6-33M   A 1-kg piece of putty is dropped from a height of 3 m onto a flat surface. It immediately sticks to the surface. Calculate the change in entropy of the universe. State assumptions.

6-34M   A Carnot cycle is to be designed to operate with an efficiency of 30 percent with steam as the working fluid. Heat addition occurs at 260°C and is accomplished by changing the water from saturated liquid to saturated vapor. Calculate the temperature, entropy, and quality at each point in the cycle, and sketch the cycle on a T-S diagram. What is the work output per kilogram of steam?

6-35M   A reversed Carnot cycle is to be used as a refrigeration device with Freon 12 as the working fluid. The device is to operate between 50 and 120°F, and heat rejection (at the *upper* temperature) occurs as the fluid is changed from saturated vapor to saturated liquid. Calculate the entropy and quality at each point in the cycle, the refrigeration effect, and the work input per kilogram of Freon.

6-36M   Steam is contained in a perfectly insulated 280-liter tank at 360°C and 300 kPa. A valve on the side of the tank is opened and the steam allowed to discharge to the atmosphere. The valve is closed when the pressure in the tank just reaches 1 atm. Calculate the mass of steam which escapes in this process and the final temperature in the tank.

6-37M   A reversible process is executed with a closed system such that 40 kJ of work are delivered by the system while it receives 40 kJ of heat. The change in entropy of the system is
        (a) Positive
        (b) Negative
        (c) Zero
        (d) Can't tell

6-38M   1.0 MJ of heat is transferred from a high-temperature reservoir at 800 K to a low-temperature reservoir at 400 K. Calculate the overall change in entropy for this process and the loss in available energy for $T_0 = 20°C$.

6-39M   In a certain steam condenser, steam enters at 6 kPa and 120°C and leaves as a saturated liquid at the same pressure. Cooling water flows through the device to remove energy from the steam and effect the condensation. The cooling water enters at 25°C and leaves at 30°C, and the overall device is well insulated. Calculate the cooling water required to condense 28 kg/s of steam and the overall irreversibility for the device for $T_0 = 20°C$.

6-40M   A constant-temperature reservoir at 2000 K delivers 2.0 MJ of heat to a
        Carnot engine which receives the heat at 500 K and rejects heat at 300 K.
        Calculate the change in entropy of the universe, and the work output of the
        Carnot engine.

6-41M   A small spray can having a volume of 0.3 liter contains Freon 12 at 70°F, and
        30 percent of the volume is occupied by liquid. A small nozzle is attached to
        the top of the can which operates as a throttling device for discharge to the
        surroundings at 1 atm and 20°C. For a slow-discharge process the tempera-
        ture in the can remains constant at 70°F. Calculate the heat added to the
        can and change in entropy inside the can for a slow-discharge process in
        which half the initial mass is lost. Also calculate the change in entropy of the
        vapor which is lost.

6-42M   A 55 000-liter tank contains water in the saturated-vapor state at 2 MPa.
        Suppose the tank is to serve as a heat source, giving up energy as the steam is
        condensed to the liquid state. How much "available energy" does it have rel-
        ative to an environment temperature of 20°C?

6-43M   Water vapor is cooled at constant pressure of 1.5 MPa from the saturated-
        vapor state to the saturated-liquid state. How much has the available energy
        decreased for this system, assuming a lowest available temperature of 20°C?

6-44M   A tank of high-pressure steam having a volume of 5500 liters is available to
        drive a small turbine which may be assumed to operate reversibly and adia-
        batically. Initially, the steam is at 7 MPa and 540°C, and the tank is per-
        fectly insulated. Discharge from the turbine is at 700 kPa. A pressure-control
        valve on the discharge from the turbine throttles the steam to 1 atm. The
        turbine is allowed to run until the tank pressure reaches 2 MPa at which
        time the discharge valve is shut off. Calculate the work output of the turbine
        for this process.

6-45M   A Carnot cycle operates between temperature limits of 900 and 300 K. The
        heat supplied at the high temperature is 4.0 MJ. Calculate the change in en-
        tropy during the heat-addition and -rejection processes, the heat rejection,
        and the work output.

6-46M   An electric resistance heater is employed to heat a flow of air from 10 to
        50°C at a constant pressure of 1 atm. The power input to the heater is
        25 kW. Calculate the change in entropy of the air and the irreversibility for
        the process for $T_0 = 20°C$.

6-47M   Large power plants employ "economizers," devices which use the hot ex-
        haust gases from the boiler to preheat the water before entering the boiler.
        In one application, 68 kg/s of gases ($c_p = 1.05$ kJ/kg·°C) are used to heat
        50 kg/s of water from 120 to 200°C in a steady-flow device. Assuming that all
        the energy lost by the gas goes to heat the water, compute the change in
        availability of the gas, water, and overall system. What is the irreversibility
        for the overall system for $T_0 = 20°C$? T_{gas} at inlet $= 400°C$.

6-48M   Calculate the change in entropy for the steam in Prob. 5-38M.

6-49M   A heat exchanger is designed which will condense steam from the
        saturated-vapor to saturated-liquid state at 100 kPa. Air serves as the cooling

medium and enters and leaves the exchanger at 290 and 340 K, respectively. The device is insulated so that all of the heat lost by the steam is gained by the air. Steady-flow operation may be assumed. For each kilogram of steam condensed, calculate:

    (a) The mass flow of air required
    (b) The loss of availability of the steam
    (c) The gain of availability of the air (assume $T_0 = 290$ K)
    (d) The steady-flow irreversibility of the process
    (e) The change in entropy of the universe resulting from the process

6-50M   Steam at 150 kPa undergoes a constant-pressure steady-flow heating process from a quality of 50 percent to superheated vapor at 260°C. Heat for the process is obtained from a constant-temperature reservoir at 300°C, and the flow rate of steam is 10 kg/min. Calculate:

    (a) The change in entropy of the steam
    (b) The change in entropy of the reservoir
    (c) The change in entropy of the universe
    (d) The change in available energy of the reservoir for $T_0 = 20$°C
    (e) The steady-flow irreversibility for the steam for $T_0 = 20$°C

6-51M   In a large power plant, combustion gases ($c_p = 1.1$ kJ/kg·°C) are cooled from 1600 to 1300 K at constant pressure while giving up heat to steam at 8 MPa operating between saturated liquid and 740°C. For each kilogram of steam calculate:

    (a) The mass flow of gas required
    (b) The change in entropy for both the gas and steam
    (c) The change in the availability function for each flow system using $T_0 = 20$°C
    (d) The irreversibility for the overall process

6-52M   Steam flows in a pipe at 700 kPa and 260°C and is connected by a valve to an 850-liter tank initially containing steam at 150 kPa and 160°C. The valve is opened and steam allowed to flow into the tank. The valve is closed when the pressure just reaches 700 kPa. If the tank is perfectly insulated, calculate:

    (a) The mass added to the tank
    (b) The change in entropy of the contents of the tank
    (c) The change in entropy of the mass which *enters* the tank
    (d) The change in entropy of the universe resulting from the process

6-53M   An inventor claims he can store large quantities of energy for conversion to work by just storing high-pressure steam at 32 MPa and 900°C. For a storage tank having a volume of 5600 liters he claims he could produce a work output of 8×10^8 J. Evaluate this claim by analyzing some ideal device which could be connected to the steam tank for producing work.

6-54M   An inventor proposes a novel work-energy storage device which will operate by providing a work input to a Carnot refrigerator which adds heat to liquid water at 100 kPa at the high temperature and extracts heat from liquid water at 100 kPa at the low temperature. When the device is first turned on, both water reservoirs are at 20°C and they each have equal masses of 200 kg of liquid water. Calculate:

(a) The work energy input to just produce solid ice at 0°C in the low-temperature reservoir
(b) The final state of the high-temperature reservoir for this work input
(c) The change in available energy of both reservoirs with $T_0 = 20°C$

Assume that the processes for both reservoirs occur at constant pressure.

6-55M Saturated-water vapor at a temperature of 260°C is expanded isothermally until the pressure reaches 100 kPa. Using the Mollier diagram determine the change in enthalpy. Use the steam tables to also determine the change in volume.

6-56M 2.0 kg of steam are expanded isentropically from 3.0 MPa and 360°C to 40 kPa. Using the Mollier diagram determine the total change in enthalpy for the system. Also specify the final state.

6-57M Calculate the change in entropy for the air which enters the tank in Prob. 4-34M if the sphere has a diameter of 1.0 m.

6-58M An insulated tank contains steam at 8.0 MPa and 400°C. A valve is opened on the side and steam allowed to escape until the steam inside just reaches the saturated-vapor state. Calculate the fractional part of the steam removed when this state is attained.

6-59M A small cylinder having a volume of 15 cm³ contains nitrogen at 70 atm and 20°C. This cylinder is placed in an insulated chamber having a volume of 10 000 cm³ and punctured so that the nitrogen is allowed to fill the entire chamber. Assuming that the nitrogen behaves as an ideal gas, calculate:
(a) The change in entropy for the nitrogen
(b) The change in available energy of the nitrogen for $T_0 = 20°C$
(c) The work necessary to reversibly compress the nitrogen back to the initial state in a piston-cylinder arrangement

6-60M A novel reversible heat engine plots on the T-S diagram as a circle. The maximum and minimum temperatures are 1100 and 200 K, respectively, and the maximum entropy change in the cycle is 2000 J/K. Calculate the heat added to the cycle, the heat rejected, the net work output, and the thermal efficiency of the cycle.

6-61M A steady-flow compressor operates in an adiabatic manner to compress air from 1 atm, 20°C, to 5 atm, 210°C. Calculate:
(a) The work required per kilogram of air
(b) The change in entropy for the process
(c) The steady-flow irreversibility for the process assuming $T_0 = 20°C$

6-62M A large tank having a volume of 150 m³ contains air at 30 atm and 30°C. A valve is opened which allows the air to escape slowly to the atmosphere at 1 atm and 25°C. The tank is located outdoors in the sun, and the process occurs so slowly that the temperature in the tank remains constant at 30°C. The valve is closed when the pressure in the tank reaches 6 atm. Calculate the change in entropy of the universe resulting from this process.

REFERENCES

1 Callen, H. B.: "Thermodynamics," John Wiley & Sons, Inc., New York, 1960.

2 Lewis, G. N., and M. Randall: "Thermodynamics," 2d ed. (revised by K. S. Pitzer and L. Brewer), McGraw-Hill Book Company, New York, 1961.

3 Reynolds, W. C.: "Thermodynamics," 2d ed., McGraw-Hill Book Company, New York, 1968.

4 Sears, F. W.: "An Introduction to Thermodynamics, The Kinetic Theory of Gases, and Statistical Mechanics," Addison-Wesley Publishing Company, Inc., Reading, Mass., 1950.

5 Hatsopoulos, G. N., and J. H. Keenan: "Principles of General Thermodynamics," John Wiley & Sons, Inc., New York, 1965.

6 Van Wylen, G. J., and R. E. Sonntag: "Fundamentals of Classical Thermodynamics," John Wiley & Sons, Inc., New York, 1965.

7 Tribus, M.: "Thermostatics and Thermodynamics," D. Van Nostrand Company, Inc., Princeton, N.J., 1961.

chapter seven
Equations of state and general thermodynamic relations

7-1 INTRODUCTION

We now want to establish some general thermodynamic relations which may be applied to the development of equations of state for different substances. The property entropy will be quite useful in this development. No macroscopic thermodynamic analysis can be used to develop a realistic equation of state without the use of experimental data. The analysis may, however, be effectively utilized so that a minimum number of measurements are necessary to establish all properties of interest. We shall see which experimental measurements are necessary for this development.

7-2 SOME MATHEMATICAL PRELIMINARIES

It is worthwhile to review a few principles of the partial differential calculus to make most efficient use of later developments. Suppose that z is given as a function of the two independent variables x and y:

$$z = z(x,y) \tag{7-1}$$

Then, according to the calculus, we may write

$$dz = \frac{\partial z}{\partial x} dx + \frac{\partial z}{\partial y} dy \tag{7-2}$$

where dz is called an exact differential. Equation (7-1) is quite correct in a mathematical sense; however, the partial derivatives are usually written in the following way in thermodynamic presentations:

$$\frac{\partial z}{\partial x} = \left(\frac{\partial z}{\partial x}\right)_y = M \qquad \frac{\partial z}{\partial y} = \left(\frac{\partial z}{\partial y}\right)_x = N \tag{7-3}$$

Here the parentheses and subscripts are employed to show very clearly which variable is being held constant in the differentiation. Also, each one of these notations shows that x and y are the independent variables. The second mixed partial derivatives are equal, so that

$$\frac{\partial^2 z}{\partial x \, \partial y} = \frac{\partial^2 z}{\partial y \, \partial x}$$

or

$$\frac{\partial M}{\partial y} = \frac{\partial N}{\partial x} \tag{7-4}$$

where it is understood that Eq. (7-2) is written as

$$dz = M\, dx + N\, dy \qquad\qquad (7\text{-}5)$$

It is implied from Eq. (7-1) that, in principle, either x or y could be expressed explicitly in terms of the other two variables; i.e., there exist relations of the form

$$x = x(y,z) \qquad y = y(x,z) \qquad\qquad (7\text{-}6)$$

From these two relations we obtain

$$dx = \left(\frac{\partial x}{\partial y}\right)_z dy + \left(\frac{\partial x}{\partial z}\right)_y dz \qquad\qquad (7\text{-}7)$$

$$dy = \left(\frac{\partial y}{\partial x}\right)_z dx + \left(\frac{\partial y}{\partial z}\right)_x dz \qquad\qquad (7\text{-}8)$$

After some algebraic manipulation we find that

$$\left(\frac{\partial x}{\partial y}\right)_z \left(\frac{\partial y}{\partial z}\right)_x \left(\frac{\partial z}{\partial x}\right)_y = -1 \qquad\qquad (7\text{-}9)$$

We shall refer to Eq. (7-9) as a *cyclical* relation.

In thermodynamic relations we must be careful in handling products of partial derivatives. Suppose we have

$$r = r(x,y)$$
$$z = z(x,r)$$

We might form the derivatives

$$\frac{\partial r}{\partial x} = \left(\frac{\partial r}{\partial x}\right)_y$$

$$\frac{\partial z}{\partial x} = \left(\frac{\partial z}{\partial x}\right)_r$$

but it is important to note that

$$\frac{\partial r/\partial x}{\partial z/\partial x} \neq \frac{\partial r}{\partial z}$$

It is permissible, of course, to write

$$\left(\frac{\partial z}{\partial x}\right)_y = \frac{1}{(\partial x/\partial z)_y}$$

To avoid confusion, the parentheses and subscripts will *always* be employed in our discussions of partial derivatives. In this way, it will be possible to keep the independent variables identified throughout the differentiation processes.

7-3 HELMHOLTZ AND GIBBS FUNCTIONS

Let us define two new properties: the *Helmholtz function a*,

$$a = u - Ts \qquad (7\text{-}10)$$

and the *Gibbs function g*,

$$g = h - Ts \qquad (7\text{-}11)$$

At this point we shall not be concerned with the physical significance of these properties, except to say that they are of value in the analysis of system equilibrium.

7-4 THE ENERGY EQUATION

In Chap. 3 it was shown that the energy conservation principle for a pure substance having only a compression work mode could be written in differential form as

$$d'Q = du + p\,dv \qquad (7\text{-}12)$$

If we now assume that the system is in equilibrium during the differential variations described in Eq. (7-12), then the system is presumed to experience a quasistatic or reversible process. Accordingly, the heat transfer $d'Q$ may be written in terms of the entropy, so that

$$d'Q_{\text{rev}} = T\,ds = du + p\,dv \qquad (7\text{-}13)$$

7-5 THE MAXWELL RELATIONS

The definition of the enthalpy may be recalled as

$$h = u + pv \qquad (7\text{-}14)$$

Differentiating Eqs. (7-10), (7-11), and (7-14) and rearranging Eq. (7-13),

$$da = du - T\,ds - s\,dT$$
$$dg = dh - T\,ds - s\,dT$$
$$dh = du + p\,dv + v\,dp$$
$$du = T\,ds - p\,dv$$

or, making further use of Eq. (7-13),

$$da = -s\,dT - p\,dv \qquad\qquad\qquad (7\text{-}15a)$$

$$dg = -s\,dT + v\,dp \qquad\qquad\qquad (7\text{-}15b)$$

$$dh = T\,ds + v\,dp \qquad\qquad\qquad (7\text{-}15c)$$

$$du = T\,ds - p\,dv \qquad\qquad\qquad (7\text{-}15d)$$

Each of these equations is of the same form as Eq. (7-5). We may therefore equate mixed partial derivatives in accordance with Eq. (7-4) and define the coefficients of each differential in accordance with Eq. (7-3). The following results are obtained:

$$-s = \left(\frac{\partial a}{\partial T}\right)_v \qquad -p = \left(\frac{\partial a}{\partial v}\right)_T \qquad \left(\frac{\partial s}{\partial v}\right)_T = \left(\frac{\partial p}{\partial T}\right)_v \qquad (7\text{-}16a)$$

$$-s = \left(\frac{\partial g}{\partial T}\right)_p \qquad v = \left(\frac{\partial g}{\partial p}\right)_T \qquad \left(\frac{\partial s}{\partial p}\right)_T = -\left(\frac{\partial v}{\partial T}\right)_p \qquad (7\text{-}16b)$$

$$T = \left(\frac{\partial h}{\partial s}\right)_p \qquad v = \left(\frac{\partial h}{\partial p}\right)_s \qquad \left(\frac{\partial T}{\partial p}\right)_s = \left(\frac{\partial v}{\partial s}\right)_p \qquad (7\text{-}16c)$$

$$T = \left(\frac{\partial u}{\partial s}\right)_v \qquad -p = \left(\frac{\partial u}{\partial v}\right)_s \qquad \left(\frac{\partial T}{\partial v}\right)_s = -\left(\frac{\partial p}{\partial s}\right)_v \qquad (7\text{-}16d)$$

The four sets of relations between the partial derivatives as obtained from equating the second mixed partial derivatives are called the *Maxwell relations*.

Two other relations are of particular interest at this time:

$$T = \left(\frac{\partial u}{\partial s}\right)_v \qquad\qquad\qquad (7\text{-}17a)$$

$$T = \left(\frac{\partial h}{\partial s}\right)_p \qquad\qquad\qquad (7\text{-}17b)$$

These relations may be used to define temperature in terms of other thermodynamic properties, particularly the entropy. For example, if we can obtain expressions for u and s from microscopic considerations, then it should be possible to calculate the temperature of the system by using Eq. (7-17a). As it turns out, this relation serves as a crucial link between macroscopic and microscopic thermodynamic analysis and, in fact, provides the analytical facility for obtaining a definition of the temperature of a microscopic system. We shall examine the microscopic analysis further in the following chapter, but, for now, let us concentrate our attention on the macroscopic relations.

7-6 ENTHALPY, INTERNAL ENERGY, AND ENTROPY

Let us assume that the internal energy of a pure substance may be expressed as a function of temperature and volume (see Sec. 3-2). Then $u = u(T,v)$ and

$$du = \left(\frac{\partial u}{\partial T}\right)_v dT + \left(\frac{\partial u}{\partial v}\right)_T dv \qquad (7\text{-}18)$$

Recalling the definition

$$c_v = \left(\frac{\partial u}{\partial T}\right)_v$$

we have

$$du = c_v\, dT + \left(\frac{\partial u}{\partial v}\right)_T dv$$

From Eq. (7-13),

$$ds = \frac{du}{T} + \frac{p}{T}\, dv$$

$$= c_v \frac{dT}{T} + \left[\frac{1}{T}\left(\frac{\partial u}{\partial v}\right)_T + \frac{p}{T}\right] dv \qquad (7\text{-}19)$$

But we may write the entropy as a function of the independent variables T and v so that $s = s(T,v)$ and

$$ds = \left(\frac{\partial s}{\partial T}\right)_v dT + \left(\frac{\partial s}{\partial v}\right)_T dv \qquad (7\text{-}20)$$

Equating the coefficients of dT and dv in Eqs. (7-19) and (7-20) gives

$$\frac{c_v}{T} = \left(\frac{\partial s}{\partial T}\right)_v \qquad (7\text{-}21)$$

$$\left(\frac{\partial s}{\partial v}\right)_T = \frac{1}{T}\left[\left(\frac{\partial u}{\partial v}\right)_T + p\right] \tag{7-22}$$

Now the Maxwell relation of Eq. (7-16a) allows us to write

$$\left(\frac{\partial s}{\partial v}\right)_T = \left(\frac{\partial p}{\partial T}\right)_v$$

Thus Eq. (7-22) may be rewritten as

$$\left(\frac{\partial u}{\partial v}\right)_T = T\left(\frac{\partial p}{\partial T}\right)_v - p \tag{7-23}$$

and the relation for the internal energy [Eq. (7-18)] becomes

$$du = c_v\, dT + \left[T\left(\frac{\partial p}{\partial T}\right)_v - p\right] dv \tag{7-24}$$

Similarly, we may write for the entropy

$$ds = c_v\, \frac{dT}{T} + \left(\frac{\partial p}{\partial T}\right)_v dv \tag{7-25}$$

A similar development may be made for the enthalpy. The result is

$$dh = c_p\, dT + \left[v - T\left(\frac{\partial v}{\partial T}\right)_p\right] dp \tag{7-26}$$

where the definition of the specific heat at constant pressure will be recalled as

$$c_p = \left(\frac{\partial h}{\partial T}\right)_p \tag{7-27}$$

Taking the entropy as a function of pressure and temperature, and considering these independent variables instead of volume and temperature, results in

$$ds = c_p\, \frac{dT}{T} - \left(\frac{\partial v}{\partial T}\right)_p dp \tag{7-28}$$

The final equations for internal energy, Eq. (7-24), enthalpy, Eq. (7-26), and entropy, Eq. (7-25), involve five quantities: p, v, T, c_v, and c_p. We may develop a relationship between the specific heats in the following way. The two expressions for entropy, Eqs. (7-25) and (7-28), may be equated to give

$$c_v \frac{dT}{T} + \left(\frac{\partial p}{\partial T}\right)_v dv = c_p \frac{dT}{T} - \left(\frac{\partial v}{\partial T}\right)_p dp$$

Rearranging,

$$c_p - c_v = T \left[\left(\frac{\partial p}{\partial T}\right)_v \frac{dv}{dT} + \left(\frac{\partial v}{\partial T}\right)_p \frac{dp}{dT} \right] \tag{7-29}$$

For a variation at constant volume, $dv = 0$ and

$$(c_p - c_v)_v = T \left(\frac{\partial v}{\partial T}\right)_p \left(\frac{\partial p}{\partial T}\right)_v$$

or for a variation at constant pressure, $dp = 0$ and

$$(c_p - c_v)_p = T \left(\frac{\partial p}{\partial T}\right)_v \left(\frac{\partial v}{\partial T}\right)_p$$

These relations are identical. Using the cyclical relation of Eq. (7-9),

$$\left(\frac{\partial p}{\partial T}\right)_v = - \left(\frac{\partial v}{\partial T}\right)_p \left(\frac{\partial p}{\partial v}\right)_T$$

we obtain the final relation

$$c_p - c_v = -T \frac{[\partial v/\partial T]_p^2}{(\partial v/\partial p)_T} \tag{7-30}$$

By deriving Eq. (7-30) we have thus reduced the determination of u, h, and s to the specification of only four properties: p, v, T, and either c_v or c_p. Since the Gibbs and Helmholtz functions are expressed in terms of u, h, and s, we thus have also specified the information necessary for their determination.

The foregoing information is most important because it establishes the experimental measurements which must be performed to describe fully all the thermodynamic properties of a pure substance.

If we are concerned only with a single phase of the substance (i.e., liquid or vapor phase), we may presume that there exists an equation of state relating pressure, volume, and temperature:

$$v = v(T,p) \tag{7-31}$$

The p, v, T measurements may be employed to obtain an empirical relationship of the form of Eq. (7-31). This relation may then be combined with specific heat data to calculate other thermodynamic properties. We should recall at this point the utility of the Joule-Thomson porous-plug experiments in determining values of c_p from isothermal and adiabatic flow measurements. These experiments were discussed in Sec. 4-7.

7-7 SPECIFIC HEAT RELATIONS

Additional useful expressions relating specific heats may be developed in the following manner. Equation (7-28) is an exact differential equation of the form of Eq. (7-5). We may thus equate the second mixed partial derivatives to obtain

$$\left[\frac{\partial(c_p/T)}{\partial p}\right]_T = -\left[\frac{\partial}{\partial T}\left(\frac{\partial v}{\partial T}\right)_p\right]_p$$

or

$$\left(\frac{\partial c_p}{\partial p}\right)_T = -T\left(\frac{\partial^2 v}{\partial T^2}\right)_p \qquad (7\text{-}32)$$

Using Eq. (7-25) as a starting point, we obtain a similar expression for the variation of c_v with specific volume. Equating second mixed partial derivatives, there results

$$\left[\frac{\partial(c_v/T)}{\partial v}\right]_T = \left[\frac{\partial}{\partial T}\left(\frac{\partial p}{\partial T}\right)_v\right]_v$$

or

$$\left(\frac{\partial c_v}{\partial v}\right)_T = T\left(\frac{\partial^2 p}{\partial T^2}\right)_v \qquad (7\text{-}33)$$

Equations (7-32), (7-33), and (7-23) provide very useful results when applied to an ideal gas. We have

$$v = \frac{RT}{p} \qquad p = \frac{RT}{v}$$

Differentiating and inserting these expressions in Eqs. (7-32), (7-33), and (7-23),

$$\left(\frac{\partial c_p}{\partial p}\right)_T = -T(0) = 0 \qquad \left(\frac{\partial c_v}{\partial v}\right)_T = T(0) = 0$$

$$\left(\frac{\partial u}{\partial v}\right)_T = T\left(\frac{R}{v}\right) - p = 0$$

That is, the specific heats and internal energy of an ideal gas are functions of temperature alone. This result has already been used in Chap. 2 to perform calculations for processes involving ideal gases.

7-8 THE GAS TABLES

We have noted that the enthalpy and internal energy of a gas which obeys the equation of state $pv = RT$ are functions only of temperature and, thus, may be calculated with appropriate relations for specific heat as given in Table 2-3. Such calculations have been performed by Keenan and Kaye [3] and tabulated in the *gas tables*. For these tables the zero reference level for enthalpy and internal energy is taken at zero degrees absolute.

For an ideal gas we may write the change in entropy for a process as [see Eq. (6-31)]

$$s_2 - s_1 = \int_{T_1}^{T_2} c_p \frac{dT}{T} - R \ln \frac{p_2}{p_1} \tag{7-34}$$

Let us choose a reference state for entropy such that

$$s = 0 \quad \text{at } T = 0, \, p = 1 \text{ atm}$$

The entropy at any particular temperature and pressure is then

$$s = \int_0^T c_p \frac{dT}{T} - R \ln p \tag{7-35}$$

where, now, the pressure is expressed in units of atmospheres. The integral in this equation is a function of temperature alone and is tabulated in the *gas tables* as ϕ (see Tables A-17 and A-18),

$$\phi = \int_0^T c_p \frac{dT}{T} \tag{7-36}$$

The change in entropy between two given states, Eq. (7-34), may thus be written as

$$s_2 - s_1 = \phi_2 - \phi_1 - R \ln \frac{p_2}{p_1} \tag{7-37}$$

Now let us consider a reversible adiabatic process involving an ideal gas with variable specific heats. From Eq. (7-15c) we have

$$T \, ds = dh - v \, dp = 0$$

or

$$c_p \, dT = \frac{RT}{p} \, dp$$

Separating variables,

$$\frac{dp}{p} = \frac{c_p \, dT}{RT} \tag{7-38}$$

If this equation is now integrated between a reference state at T_0 and p_0, there results

$$\ln \frac{p}{p_0} = \frac{1}{R} \int_{T_0}^{T} \frac{c_p \, dT}{T} \tag{7-39}$$

Using the same reference temperature state as in the gas tables gives

$$\ln P_r = \frac{\phi}{R} \tag{7-40}$$

where the *relative pressure* is defined as

$$P_r = \frac{p}{p_0}$$

For an isentropic process it is apparent that the relative pressure is a function of temperature alone since ϕ is dependent only on temperature. Furthermore, for the isentropic process,

$$\frac{p_1}{p_2} = \left(\frac{P_{r_1}}{P_{r_2}} \right)_{s=\text{const.}} \tag{7-41}$$

A *relative specific volume* for the isentropic process may be developed in a similar manner such that

$$\frac{v_1}{v_2} = \left(\frac{v_{r_1}}{v_{r_2}} \right)_{s=\text{const.}} \tag{7-42}$$

v_r is also temperature dependent and is expressed by

$$\ln v_r = \ln \frac{v}{v_0} = -\frac{1}{R} \int_{T_0}^{T} \frac{c_v \, dT}{T} \tag{7-43}$$

Values of the relative pressure and relative specific volume are tabulated for air in Table A-17.

EXAMPLE 7-1 Isentropic expansion

Air expands from 100 psia, 700°F, to 30 psia in an isentropic process. Calculate the final temperature and change in enthalpy with the use of the gas tables.

SOLUTION   Using the gas tables, we have for the initial state

$T_1 = 1160°R$
$h_1 = 281.14$ Btu/lbm
$P_{r_1} = 21.18$

Using Eq. (7-41),

$$P_{r_2} = P_{r_1} \frac{p_2}{p_1}$$
$$= (21.18)\tfrac{30}{100} = 6.35$$

Entering the gas tables again with $P_{r_2} = 6.35$, we find

$T_2 = 833°R$
$h_2 = 199.45$ Btu/lbm   (463.9 kJ/kg)

and the change in enthalpy is

$$\Delta h = h_2 - h_1 = 199.45 - 281.14 = -81.69 \text{ Btu/lbm} \qquad (-190 \text{ kJ/kg})$$

EXAMPLE 7-1M Isentropic expansion

Air expands from 700 kPa, 650 K, to 200 kPa in an isentropic process. Calculate the final temperature and change in enthalpy with the use of the gas tables.

SOLUTION   Using the gas tables, we have for the initial state

$T_1 = 650$ K
$h_1 = 659.84$ kJ/kg
$P_{r_1} = 21.86$

Using Eq. (7-41),

$$P_{r_2} = P_{r_1} \frac{P_2}{P_1}$$
$$= (21.86)(\tfrac{200}{700}) = 6.246$$

Entering the gas tables again with $P_{r_2} = 6.246$, we find

$$T_2 = 460 \text{ K}$$
$$h_2 = 462.01 \text{ kJ/kg} \qquad (198.6 \text{ Btu/lbm})$$

and the change in enthalpy is

$$\Delta h = h_2 - h_1 = 462.01 - 659.84 = -197.83 \text{ kJ/kg} \qquad (-85 \text{ Btu/lbm})$$

EXAMPLE 7-2   Irreversible expansion

The expansion process of Example 7-1 occurs between the same pressure limits, but because of irreversible effects the final temperature is 400°F. Calculate the change in entropy for the process.

SOLUTION   The final temperature is given as $T_2 = 860°\text{R}$. We may, therefore, obtain the values of ϕ from the gas tables as

$$\phi_1 = 0.787\ 67$$
$$\phi_2 = 0.713\ 23$$

From Eq. (7-37) the change in entropy is

$$s_2 - s_1 = \phi_2 - \phi_1 - R \ln \frac{p_2}{p_1}$$

$$= 0.713\ 23 - 0.787\ 67 - \frac{53.35}{778} \ln \frac{30}{100}$$

$$= 0.008\ 12 \text{ Btu/lbm·°R} \qquad (0.03399 \text{ kJ/kg·K})$$

EXAMPLE 7-2M   Irreversible expansion

The expansion process of Example 7-1M occurs between the same pressure limits, but because of irreversible effects the final temperature is 480 K. Calculate the change in entropy for the process.

SOLUTION   The final temperature is given as $T_2 = 480$ K. We may, therefore, obtain the values of ϕ from the gas tables as

$$\phi_1 = 3.3069$$
$$\phi_2 = 2.9909$$

From Eq. (7-37) the change in entropy is

$$s_2 - s_1 = \phi_2 - \phi_1 - R \ln \frac{p_2}{p_1}$$

$$= 2.9909 - 3.3069 - \frac{287.1}{1000} \ln \frac{200}{700}$$

$$= 0.04367 \text{ kJ/kg·K} \qquad (0.01043 \text{ Btu/lbm·°R})$$

EXAMPLE 7-3

Using the air tables calculate the change in entropy for air under the following conditions:

(a) Compression from 1 atm, 300 K, to 10 atm, 550 K
(b) Constant-pressure heat addition at 340 kPa from 300 to 550 K
(c) Adiabatic throttling process from 1.4 MPa, 420 K, to 0.7 MPa
(d) 10 kJ heat addition to fixed mass inside 0.058-m³ rigid container from 300 K, 1 atm, to final state

SOLUTION

(a) The entropy change may be calculated directly from Eq. (7-37),

$$s_2 - s_1 = \phi_2 - \phi_1 - R \ln \frac{p_2}{p_1} \tag{a}$$

Using the air tables (Table A-17M),

$$T_1 = 300 \text{ K} \qquad \phi_1 = 2.5153 \text{ kJ/kg·K}$$
$$T_2 = 550 \text{ K} \qquad \phi_2 = 3.1314 \text{ kJ/kg·K}$$

From Eq. (a),

$$s_2 - s_1 = 3.1314 - 2.5153 - 0.287 \ln \tfrac{10}{1} \tag{b}$$
$$= -0.04474 \text{ kJ/kg·K}$$

(b) Equation (a) may still be used, but since $p_1 = p_2 = 340 \text{ kPa}$, the last term is zero, and

$$s_2 - s_1 = 3.1314 - 2.5153 = 0.6161 \text{ kJ/kg·K} \tag{c}$$

(c) For adiabatic throttling with negligible kinetic energies $h_1 = h_2$; therefore $T_1 = T_2$. This means that $\phi_1 = \phi_2$, and Eq. (a) becomes

$$s_2 - s_1 = 0 - 287 \ln \frac{0.7}{1.4}$$
$$= 198.9 \text{ J/kg·K} = 0.1989 \text{ kJ/kg·K} \tag{d}$$

(d) The mass of air is calculated from the initial state:

$$p_1 = 1 \text{ atm} = 101.32 \text{ kPa} \qquad T_1 = 300 \text{ K} \qquad V_1 = 0.058 \text{ m}^3$$

so that, from the ideal-gas law,

$$m = \frac{pV}{RT} = \frac{(1.0132 \times 10^5)(0.058)}{(287)(300)} = 0.068 \, 25 \text{ kg} \tag{e}$$

Because the container is rigid there is no work done on the air and the first law becomes

$$Q = U_2 - U_1 = m(u_2 - u_1) \tag{f}$$

Using the air tables (Table A-17M) at 300 K yields

$$u_1 = 214.09 \text{ kJ/kg} \qquad \phi_1 = 2.5153$$

Inserting the numerical values in Eq. (f) gives

$$10 \text{ kJ} = 0.068\ 25(u_2 - 214.09) \tag{g}$$
$$u_2 = 3.0358 \text{ kJ/kg·K}$$

From Table A-17M the value of T_2 corresponding to this value of u_2 is

$$T_2 = 511.5 \text{ K} \qquad \phi_2 = 3.0358 \text{ kJ/kg·K} \tag{h}$$

Because the volume remains constant the final pressure is calculated from

$$\frac{p_2}{p_1} = \frac{T_2}{T_1} \tag{i}$$

$$p_2 = (1.0) \left(\frac{511.5}{300}\right) = 1.705 \text{ atm}$$

Equation (a) is again used to calculate the change in entropy per unit mass,

$$s_2 - s_1 = 3.0358 - 2.5153 - \left(0.287 \ln\frac{1.705}{1.0}\right) = 0.3674 \tag{j}$$

The total entropy change is then

$$s_2 - s_1 = m(s_2 - s_1) = (0.068\ 25)(0.3674) \tag{k}$$
$$= 0.025\ 08 \text{ kJ/K}$$

EXAMPLE 7-4   Isentropic compression

Air is compressed isentropically in a certain steady-flow device from 80°F and 15 psia to a final temperature of 500°F. Calculate the work per pound mass of air, using the gas tables, and compare the result with the value which would be obtained if the air were assumed to have constant specific heats.

SOLUTION   We have

$$T_1 = 540°R \qquad T_2 = 960°R$$

For a steady-flow process in which kinetic energy effects are negligible, the adiabatic work done on the air is

$$W = h_2 - h_1$$

From the gas tables,

$$h_2 = 231.06 \text{ Btu/lbm}$$
$$h_1 = 129.06 \text{ Btu/lbm}$$

and

$$W = 231.06 - 129.06 = 102 \text{ Btu/lbm} \qquad (237 \text{ kJ/kg})$$

For a constant specific heat of $c_p = 0.24$ Btu/lbm·°F, the change in enthalpy would be

$$\begin{aligned} h_2 - h_1 &= c_p(T_2 - T_1) \\ &= (0.24)(500 - 80) \\ &= 100.8 \text{ Btu/lbm} \qquad (234.4 \text{ kJ/kg}) \end{aligned}$$

Thus, in this example, a constant specific heat assumption would be accurate within approximately 1 percent.

EXAMPLE 7-4M Isentropic compression

Air is compressed isentropically in a certain steady-flow device from 300 K and 100 kPa to a final temperature of 530 K. Calculate the work per kilogram of air, using the gas tables, and compare the result with the value which would be obtained if the air were assumed to have constant specific heats.

SOLUTION   We have

$$T_1 = 300 \text{ K} \qquad T_2 = 530 \text{ K}$$

For a steady-flow process in which kinetic energy effects are negligible, the adiabatic work done on the air is

$$W = h_2 - h_1$$

From the gas tables,

$$h_2 = 533.98 \text{ kJ/kg}$$
$$h_1 = 300.19 \text{ kJ/kg}$$

and

$$W = 533.98 - 300.19 = 233.8 \text{ kJ/kg} \qquad (100.5 \text{ Btu/lbm})$$

For a constant specific heat of $c_p = 1.005$ kJ/kg·°C, the change in enthalpy would be

$$h_2 - h_1 = c_p(T_2 - T_1)$$
$$= (1.005)(530 - 300)$$
$$= 231.2 \text{ kJ/kg} \qquad (99.4 \text{ Btu/lbm})$$

Thus, in this example, a constant specific heat assumption would be accurate within approximately 1 percent.

7-9 SPECIFIC HEATS AT ELEVATED PRESSURES

Equation (7-32) is particularly useful in the development of property data for various substances. We may integrate with respect to pressure to obtain

$$(c_p)_{T,p} - (c_p)_{T,p=0} = \int_{p=0}^{p} T \left(\frac{\partial^2 v}{\partial T^2} \right)_p dp_T \qquad (7\text{-}44)$$

Thus we may obtain the specific heat at any pressure and temperature once we know the specific heat at zero (very low) pressure and the (p,v,T) equation of state. Once the specific heats are determined, the enthalpy, entropy, and internal energies may be calculated. In summary, the properties of a single-phase pure substance may be calculated once empirical data are obtained which yield:

1 Specific heat behavior at zero pressure
2 (p,v,T) behavior over the property range of interest

Low-pressure specific heats of several substances were tabulated in Tables 2-2 and 2-3.

7-10 THE CLAUSIUS–CLAPEYRON EQUATION

The relations given in the foregoing obviously do not provide a basis for calculations of properties in a two-phase region. A useful relation for such calculations is the Clausius-Clapeyron equation which we shall now derive. Equation (7-16a) is rewritten for convenience:

$$\left(\frac{\partial p}{\partial T}\right)_v = \left(\frac{\partial s}{\partial v}\right)_T \tag{7-16a}$$

As a pure substance changes from a saturated-liquid state to a saturated-vapor state, it undergoes a process at constant temperature. Furthermore, the pressure and temperature are independent of volume in the saturation region. Thus the derivatives in Eq. (7-16a) may be written as

$$\left(\frac{\partial p}{\partial T}\right)_v = \frac{dp}{dT}$$

$$\left(\frac{\partial s}{\partial v}\right)_T = \frac{s_g - s_f}{v_g - v_f} = \frac{s_{fg}}{v_{fg}}$$

We also have the following relation for the heat added in the constant-pressure vaporization process:

$$\begin{aligned} Q &= \Delta u - W \\ &= u_g - u_f + p(v_g - v_f) \\ &= h_g - h_f = h_{fg} \end{aligned}$$

But $Q = Ts_{fg}$ in the constant-temperature process, so that

$$s_{fg} = \frac{h_{fg}}{T} \tag{7-45}$$

With the foregoing relations, Eq. (7-16a) becomes

$$\frac{dp}{dT} = \frac{h_{fg}}{Tv_{fg}} \tag{7-46}$$

Equation (7-46) is called the *Clausius-Clapeyron equation*. The derivative dp/dT represents the slope of the vapor-pressure curve, and the utility of the relation is clear. Presumably, saturated-liquid properties can be determined from some appropriate equation of state and an arbitrarily assigned reference state. For water, saturated liquid at 0°C is assigned a zero enthalpy and entropy. Experimental data may be used to establish an empirical equation for the vapor-pressure curve, so that the slope of this curve may be obtained from Eq. (7-46). Then v_{fg} is determined from the separate values of v_g and v_f, so that Eq. (7-46) may be used to calculate the value of h_{fg} and then the value of h_g. Once the saturated-vapor properties are established, the various equations of Sec. 7-6 may be employed to determine properties of the superheated vapor.

EXAMPLE 7-5   Solid-vapor equilibrium
Calculate the saturation pressure for an equilibrium mixture of solid-vapor water at $-80°F$ if $p_{sat} = 0.0019$ psia at $-40°F$ and h_{ig} is essentially constant at 1221.2 Btu/lbm.

SOLUTION   We shall employ the Clausius-Clapeyron equation for solution of this problem. At these low pressures the water vapor behaves very nearly as an ideal gas and

$$v_g = \frac{RT}{p} \tag{a}$$

Also $v_g \gg v_i$, so that $v_{ig} = v_g$ and Eq. (7-46) becomes, for the solid-vapor equilibrium line,

$$\frac{dp}{dT} = \frac{ph_{ig}}{RT^2} \tag{b}$$

where Eq. (*a*) has been substituted for v_g. Equation (*b*) may be integrated between two temperatures to give

$$\int_{p_1}^{p_2} \frac{dp}{p} = \int_{T_1}^{T_2} \frac{h_{ig}\, dT}{RT^2}$$

or

$$\ln \frac{p_2}{p_1} = \frac{h_{ig}}{R}\left(\frac{1}{T_1} - \frac{1}{T_2}\right) \tag{c}$$

Using the given data, $T_1 = 420°R$, $T_2 = 380°R$, and

$$\ln \frac{p_2}{p_1} = \frac{(1221.2)(778)}{85.6}\left(\frac{1}{420} - \frac{1}{380}\right)$$
$$= -2.78$$

or $p_2/p_1 = 0.062$ and $p_2 = (0.062)(0.0019) = 1.178 \times 10^{-4}$ psia.

7-11   DEVELOPMENT OF TABLES OF THERMODYNAMIC PROPERTIES

The overall procedure for developing tables of thermodynamic properties may now be summarized in outline form. The outline comments refer to the points and regions shown in Fig. 7-1.

A Experimental data required
  1 Low-pressure specific heat data for liquid and vapor
  2 p-v-T data for liquid, vapor, and supercritical-gas region
  3 Experimental data for p-T variation of vaporization curve
B Direct empirical relations to be obtained from data
  1 Low-pressure specific heat as a function of temperature
  2 (p,v,T) equations of state for liquid, vapor, and supercritical regions
  3 Empirical relation for vaporization curve
  4 Arbitrarily assigned reference state for enthalpy and entropy
C Property determination procedures
  1 Liquid properties, including saturated-liquid properties, are determined from (p,v,T) equation of state for liquid, liquid specific heats, and assigned reference state. See points 0, 3, and 4 in Fig. 7-1.
  2 p-v-T properties of vapor, including v_g, are obtained from (p,v,T) equation of state for vapor. See points 1, 2, and 5 in Fig. 7-1.
  3 Saturated-vapor enthalpies and entropies (h_g and s_g) are determined from saturated-liquid properties, v_g, and Eqs. (7-45) and (7-46) in conjunction with vaporization-curve data. See points 1 and 6 in Fig. 7-1.
  4 Vapor enthalpies and entropies are determined from h_g, s_g, (p,v,T) equation of state for vapor, and specific heat equation for vapor. See points 2 and 5 in Fig. 7-1.
  5 A separate equation of state may be required for the supercritical region.

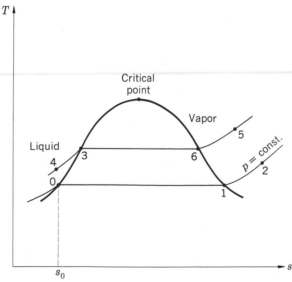

FIG. 7-1 *Schematic of regions for thermodynamic property determinations.*

It is well to note at this point that the development of tables of thermo-dynamic properties is a tedious task and requires much more physical insight than may be implied from the foregoing discussion. Experimental measure-ments of properties of substances can involve considerable uncertainties and the equation of state which is to be developed from these measurements must take the uncertainties into proper account. The purpose of this com-ment is not to alarm the reader, but rather to stress the fact that the rela-tively simple calculation procedure described must, in practice, be coupled with a vast amount of experimental data, physical insight, patience, and computer time to produce an accurate table of thermodynamic properties for a particular substance.

THE MOLLIER DIAGRAM

A diagram of practical utility was the Mollier chart or plot of enthalpy versus entropy, shown in schematic form in Fig. 7-2. An interesting feature of this diagram is that isotherms (lines of constant temperature) plot as straight lines in the two-phase region as a result of the relation given in Eq. (7-17b). The term $(\partial h/\partial s)_p$ is clearly the slope of a constant-pressure line on the Mol-lier diagram, but in the wet-mixture region a line of constant pressure is also a line of constant temperature so that Eq. (7-17b) predicts a constant slope in this region. As mentioned before, a Mollier diagram for steam is given in Fig. A-4 of the Appendix and will be very useful in conjunction with the steam-power-cycle analysis of Chap. 12.

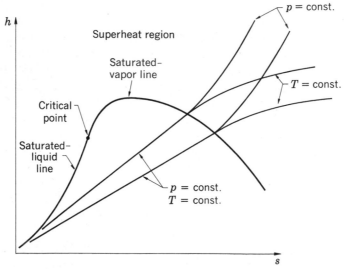

FIG. 7-2 *Schematic of Mollier diagram.*

7-12 EQUATIONS OF STATE

With the foregoing theoretical background in mind, it is now worthwhile to examine some particular forms of equation of state which are employed in practice.

THE VAN DER WAALS EQUATION

In Chaps. 8 and 9 we shall examine the microscopic behavior of an ideal gas in some detail. In accordance with the simple discussion of Sec. 1-14, we may consider an ideal gas as one in which:

1 The molecules are very widely spaced and behave as point masses.
2 The molecular force fields exert very little influence on collision processes because of the wide molecular spacing.
3 The volume occupied by the molecules is very small compared with the total volume occupied by the gas, again because of the wide molecular spacing.

When a gas is compressed, the molecules are forced into closer proximity of one another and nonideal-gas behavior will eventually develop. Finally, of course, there can be such strong force-field interaction that the gas is converted to a liquid. In 1879 van der Waals proposed an equation of state to account for nonideal-gas behavior. It has the form of

$$\left(p + \frac{a}{\bar{v}^2}\right)(\bar{v} - b) = \mathcal{R}T \tag{7-47}$$

where a is a constant which takes account of the molecular force-field interactions, and b is a constant which compensates for the volume occupied by the molecules. The reasoning used to obtain the $a/\bar{v}^2$ term is as follows. When a group of molecules strikes a surface, the pressure force exerted is proportional to the molecular density (molecules per unit volume, N/V), because it is this quantity that governs the number of molecular impacts. For molecule-molecule impacts, the interaction force is reasoned to be proportional to $(N/V)^2$, since the net effect results from two molecules striking each other. The molecular density is clearly proportional to the macroscopic gas density which, in turn, is proportional to $1/\bar{v}$. Thus the net interaction force is proportional to $1/\bar{v}^2$ and a is the proportionality constant. The constant b simply takes account of the effective volume occupied by the molecule.

To use the van der Waals equation, values of the constants a and b must be determined experimentally, and a summary of values for various gases is given in Table 7-1.

TABLE 7-1
Van der Waals constants

Gas	a		b		$Z_c = p_c v_c / R T_c$
	kN·m⁴/ (kg mol)²	atm·ft⁶/ (lbm mol)²	m³/ kg mol	ft³/ lbm mol	
Air	135.8	343.8	0.0365	0.585	0.284
O_2	138.0	349.5	0.0318	0.510	0.29
N_2	136.7	346	0.0386	0.618	0.291
H_2O	551.7	1397.1	0.0304	0.487	0.23
CH_4	228.6	578.9	0.0427	0.684	0.29
CO	147.9	374.5	0.0393	0.630	0.293
CO_2	365.6	926	0.0428	0.686	0.276
NH_3	424.9	1076	0.0373	0.598	0.242
H_2	24.8	62.9	0.0266	0.426	0.304
He	3.42	8.66	0.0235	0.376	0.30

A conventional technique used to estimate the van der Waals constants is based on the behavior of pure substances at the critical point. The isotherm passing through the critical point on a p-v diagram encounters a point of inflection at (T_c, p_c, v_c). Thus

$$\left(\frac{\partial p}{\partial v}\right)_{T_c} = 0 \qquad \text{at } T = T_c \tag{7-48}$$

$$\left(\frac{\partial^2 p}{\partial v^2}\right)_{T_c} = 0 \qquad \text{at } T = T_c \tag{7-49}$$

Solving Eq. (7-47) for p, we obtain

$$p = \frac{\mathscr{R}T}{\bar{v} - b} - \frac{a}{\bar{v}^2} \tag{7-50}$$

Applying Eqs. (7-48) and (7-49),

$$\frac{-\mathscr{R}T_c}{(\bar{v}_c - b)^2} + \frac{2a}{\bar{v}^3} = 0 \tag{7-51}$$

$$\frac{2\mathscr{R}T_c}{(\bar{v}_c - b)^3} - \frac{6a}{\bar{v}_c^4} = 0 \tag{7-52}$$

Equations (7-51) and (7-52) may now be solved for the constants a and b to yield

$$a = 3p_c \bar{v}_c^2 = \frac{9}{8} \mathscr{R}T_c \bar{v}_c = \frac{27}{64} \frac{\mathscr{R}^2 T_c^2}{p_c} \tag{7-53}$$

$$b = \frac{\bar{v}_c}{3} = \frac{\mathscr{R}T_c}{8p_c} \qquad (7\text{-}54)$$

Additionally, we obtain for a van der Waals gas

$$\frac{p_c\bar{v}_c}{\mathscr{R}T_c} = \frac{3}{8} \qquad (7\text{-}55)$$

The critical properties for several gases are given in Table 7-2 as determined from experimental measurements. The accuracy of the van der Waals equation at the critical point can be checked very quickly by using these data and Eq. (7-55). For water,

$$T_c = 1165.3°R \qquad p_c = 3208 \text{ psia} \qquad \bar{v}_c = 0.90 \text{ ft}^3/\text{lb mol}$$

$$\frac{p_c\bar{v}_c}{\mathscr{R}T_c} = \frac{(3208)(144)(0.90)}{(1545)(1165)} = 0.231$$

This value is in substantial disagreement with Eq. (7-55), and we would find, in general, that a similar calculation for many other substances would also produce disagreement. The reader should not be alarmed by this discrepancy, however. The simple proportionality relation for the force-field interaction $(a/\bar{v}^2)$ may only be expected to apply for modest deviations from ideal-gas behavior, i.e., for moderate- to long-range molecular spacings. For relatively close molecular spacings, as at the critical point, the interaction forces cannot be expressed by this simple relation, and we would not expect the van der Waals relation to give satisfactory results in this region.

A simpler equation of state is the *Clausius equation*, which is a modified form of the van der Waals equation that neglects interaction forces and is written as

TABLE 7-2
Some selected critical constants

Gas	T_c		p_c		$\bar{v}_c$	
	K	°R	MN/m²	atm	(m³/kg mol)10²	ft³/lbm mol
Air	132.41	238.34	3.774	37.25	8.30	1.33
O_2	154.78	278.6	5.080	50.14	7.43	1.19
N_2	126.2	227.16	3.398	33.54	8.99	1.44
H_2O	647.27	1165.3	22.105	218.167	5.68	0.91
CH_4	190.7	343.26	4.640	45.8	9.93	1.59
CO	132.91	239.24	3.496	34.53	9.30	1.49
CO_2	304.20	547.56	7.386	72.90	9.55	1.53
NH_3	405.4	729.72	11.277	111.3	7.24	1.16
H_2	33.24	59.83	1.297	12.8	6.49	1.04
He	5.19	9.34	0.229	2.26	5.81	0.93

$$p(\bar{v} - b) = \mathscr{R}T \tag{7-56}$$

BEATTIE-BRIDGEMAN EQUATION OF STATE

A widely used equation of state with good accuracy is the Beattie-Bridgeman equation:

$$p = \frac{\mathscr{R}T}{\bar{v}^2}(1 - e)(\bar{v} + B) - \frac{A}{\bar{v}^2} \tag{7-57}$$

where

$$A = A_0\left(1 - \frac{a}{\bar{v}}\right)$$

$$B = B_0\left(1 - \frac{b}{\bar{v}}\right)$$

$$e = \frac{c}{\bar{v}T^3}$$

There are five constants, a, A_0, b, B_0, and c, to be determined empirically. Table 7-3 lists values of these constants for several substances.

The Beattie-Bridgeman equation of state predicts properties accurately (within 2 percent) in regions where the density is less than 0.8 times the critical density.

Two other equations of state in use are the *Bertholet equation*:

TABLE 7-3
Constants for Beattie-Bridgeman equation

Gas	a (m³/kg mol)10²	ft³/lbm mol	A_0 kN·m⁴/(kg mol)²	atm·ft⁶/(lbm mol)²	b (m³/kg mol)10²
Air	1.93	0.309	131.9	334.1	−4.43
O_2	2.56	0.410	151.0	382.5	0.42
N_2	2.62	0.419	136.2	344.9	−0.692
CH_4	1.85	0.297	230.8	584.6	−1.58
C_3H_8	7.32	1.173	120.7	305.8	4.29
C_4H_{10}	12.16	1.948	180.2	456.5	9.43
CO	2.62	0.419	136.2	344.9	−0.069
CO_2	7.14	1.143	507.4	1 284.9	7.24
NH_3	17.04	2.729	242.4	613.9	19.12
H_2	−0.506	−0.0811	19.97	50.57	−4.35
He	5.98	0.958	2.19	5.55	0

$$p = \frac{\mathscr{R}T}{\bar{v} - b} - \frac{a}{T\bar{v}^2} \tag{7-58}$$

and the *Dieterici equation:*

$$p = \frac{\mathscr{R}T}{\bar{v} - b} e^{-a/\mathscr{R}T\bar{v}} \tag{7-59}$$

The Dieterici equation predicts properties very well in the neighbor-hood of the critical point and on the critical isotherm. It is much less satis-factory in other regions and may produce large errors when used away from the critical region. Following the same procedure as with van der Waals' equation the values of the constants are obtained as

$$a = \frac{4\mathscr{R}^2T_c^2}{p_ce^2} \qquad b = \frac{\mathscr{R}T_c}{p_ce^2}$$

In addition, we obtain

$$\frac{p_c\bar{v}_c}{\mathscr{R}T_c} = 0.271 = \frac{2}{e^2}$$

a value which is much closer to the experimental values for substances than the $\frac{3}{8}$ predicted by the van der Waals relation.

The Berthelot equation is similar to the van der Waals equation except for the inclusion of the temperature in the denominator of the second term. For low pressures it is accurate within 1 percent using the value of a given by Eq. (7-53) and $b = 9\mathscr{R}T_c/28p_c$.

TABLE 7-3 (Continued)

b		B_0		$10^{-6}C$	
ft³/ lbm mol	(m³/ kg mol)10²	ft³/ lbm mol	m³·K³/ kg mol	ft³·°R³/ lbm mol
−0.716	4.61	0.739	434	4.05
0.067 4	4.63	0.741	480	4.48
−0.111	5.04	0.808	419.3	3.917
−0.254	5.59	0.895	1 282	11.98
0.688	18.1	2.90	12 002	112.12
1.51	24.62	3.944	35 005	327.02
−0.011 05	5.04	0.808	419	3.92
1.159	10.48	1.678	6 599	61.65
3.062	3.41	0.547	47 698	445.6
−0.698	2.10	0.336	5.04	0.047 1
0	1.40	0.224	0.399	0.003 73

For high-density regions up to approximately two times the critical density a complicated empirical relation involving eight experimentally determined constants is the equation of Benedict-Webb-Rubin [10] (Table 7-4):

$$p = \frac{\mathscr{R}T}{\bar{v}} + \frac{\mathscr{R}TB_0 - A_0 - C_0/T^2}{\bar{v}^2}$$

$$+ \frac{\mathscr{R}Tb - a}{\bar{v}^3} + \frac{a\alpha}{\bar{v}^6} + \frac{c}{\bar{v}^3 T^2}\left(1 + \frac{\gamma}{\bar{v}^2}\right)e \qquad (7\text{-}60)$$

where A_0, B_0, C_0, a, b, c, α, and γ are the empirical constants.

Still another equation of state is that of Redlich and Kwong:

$$p = \frac{\mathscr{R}T}{\bar{v} - b} - \frac{a}{\bar{v}(\bar{v} + b)T^{1/2}} \qquad (7\text{-}61)$$

with the constants given in terms of critical state properties:

$$a = 0.427\ 48\ \frac{\mathscr{R}^2 T_c^{5/2}}{p_c} \qquad (7\text{-}62)$$

$$b = 0.086\ 64\ \frac{\mathscr{R}T_c}{p_c} \qquad (7\text{-}63)$$

VIRIAL EQUATION OF STATE

The virial equation of state represents an expansion of the $p\bar{v}$ product in infinite-series form:

$$p\bar{v} = \mathscr{R}T\left(1 + \frac{B}{\bar{v}} + \frac{C}{\bar{v}^2} + \frac{D}{\bar{v}^3} + \cdots\right) \qquad (7\text{-}64)$$

The constants B, C, D, ... are called *virial coefficients* and are temperature dependent. The word *virial* arises from the Latin word for force, and each term in the series provides a successive correction to the ideal-gas behavior, or no-interaction-force condition. In some cases it is possible to calculate values of the virial coefficients from statistical thermodynamic considerations together with empirical relations for molecular force fields. Although the calculation is quite complicated, a virial formulation does accurately predict p-v-T behavior over a wide range of densities.

It is sometimes more convenient to express the virial equation with pressure as the independent variable; i.e.,

$$p\bar{v} = \mathscr{R}T(1 + B'p + C'p^2 + D'p^3 + \cdots) \qquad (7\text{-}65)$$

There are many other equations of state which have been developed

TABLE 7-4
Constants for Benedict-Webb-Rubin equation of state (SI units)

Constant	N_2	CO	CO_2	CH_4	C_4H_{10} (n-butane)
a, $N \cdot m^7/(kg\ mol)^3$	2540	3710	13 860	5000	19 070
A_0, $N \cdot m^4/(kg\ mol)^2$	1.0673×10^5	13 587	27 730	18 790	10 220
b, $m^6/(kg\ mol)^3$	232.8	263.2	721	338	4000
B_0, $m^3/kg\ mol$	4074	5454	4991	4260	12 440
c, $N \cdot m^7 \cdot K^2/(kg\ mol)^3$	7.379×10^7	1.054×10^8	1.511×10^9	2.578×10^8	3.20×10^{10}
C_0, $N \cdot m^4 \cdot K^2/(kg\ mol)^2$	8.164×10^8	8.673×10^8	1.404×10^{10}	2.286×10^9	1.006×10^{11}
α, $m^9/(kg\ mol)^3$	1.272×10^{-4}	8.47×10^{-5}	1.35×10^{-4}	1.24×10^{-4}	1.101×10^{-3}
γ, $m^6/(kg\ mol)^2$	0.0053	0.0060	0.005 39	0.0060	0.0340

and tailored to match various types of substances. No single equation will be satisfactory for all ranges of temperatures and pressures, as the foregoing discussion has illustrated. The interested reader may wish to consult Refs. [2, 4, 6] for more detailed information on the applicability of the different equations of state.

7-13 THE GENERALIZED COMPRESSIBILITY FACTOR

Because the ideal-gas equation of state is so simple, it is not unnatural to seek a means of modifying it in order to match with nonideal-gas behavior. The technique employed is to define a factor Z, called the *compressibility factor*, such that

$$Z = \frac{pv}{RT} \tag{7-66}$$

Clearly, $Z = 1.0$ for an ideal gas, and the various virial coefficients in Eq. (7-64) simply provide the series corrections to the ideal-gas behavior. It is possible to give some generalized information for Z, as indicated in the following paragraphs.

Let us first consider a van der Waals gas and define new variables called the *reduced pressure*, *reduced volume*, and *reduced temperature*,

$$p_r = \frac{p}{p_c} = \text{reduced pressure}$$

$$v_r = \frac{v}{v_c} = \text{reduced specific volume}$$

$$T_r = \frac{T}{T_c} = \text{reduced temperature}$$

Some algebraic manipulation of the van der Waals equation, Eq. (7-47), along with the relations for the constants, Eqs. (7-53) and (7-54), gives

$$Z^3 - \left(\frac{p_r}{8T_r} + 1\right) Z^2 + \left(\frac{27 p_r}{64 T_r^2}\right) Z - \frac{27 p_r^2}{512 T_r^3} = 0 \tag{7-67}$$

where Z is the compressibility factor defined in Eq. (7-66).

Clearly, Eq. (7-67) can be solved for Z explicitly in terms of p_r and T_r and, consequently, we postulate the existence of a relation of the form

$$Z = f(p_r, T_r) \tag{7-68}$$

Equation (7-68) is called the *law of corresponding states*. Our previous discussions have shown that the van der Waals equation is inaccurate and, hence, at first glance, we would be tempted to question the validity of Eq. (7-68). However, experimental measurements show that a relation like Eq. (7-68) does indeed exist for a wide variety of substances, although the functional variation is quite different from that given by a solution to Eq. (7-67). Figure 7-3 presents a display of experimental data indicating the general existence of a relation like Eq. (7-68). Figures 7-4 to 7-6 present the Nelson-Obert calculation charts that have been derived from the various experimental data [2]. It turns out that it is convenient to plot a pseudo-reduced volume v_r', instead of v_r, which is defined by

$$v_r' = \text{pseudo-reduced volume} = \frac{v}{RT_c/p_c} \qquad (7\text{-}69)$$

which may be expressed in the alternate forms

$$\begin{aligned} v_r' &= \frac{v}{v_c} \frac{v_c}{RT_c/p_c} = v_r Z_c \\ &= \frac{pv}{RT} \frac{T/T_c}{p/p_c} = Z \frac{T_r}{p_r} \end{aligned} \qquad (7\text{-}70)$$

The generalized compressibility charts provide us with a more explicit basis for interpreting ideal-gas behavior than has been previously at our disposal. The ideal gas is one for which $Z = 1.0$. Consequently, we need only examine Fig. 7-5 to predict the ranges of pressures and temperatures in which a substance will behave like an ideal gas. The two general notions we obtain are:

Ideal-gas behavior exists when

1 p_r becomes small compared to 1.0.
2 T_r becomes large compared to 1.0.

As an example, we might try to anticipate the behavior of nitrogen at 70°F (530°R) and 14.7 psia. From the critical constants in Table 7-2, we obtain

$$p_r = \frac{14.7}{492} = 0.0299 \qquad T_r = \frac{530}{227.1} = 2.33$$

The reduced pressure is very small, so we would anticipate ideal-gas behavior, as we have assumed several times in previous chapters.

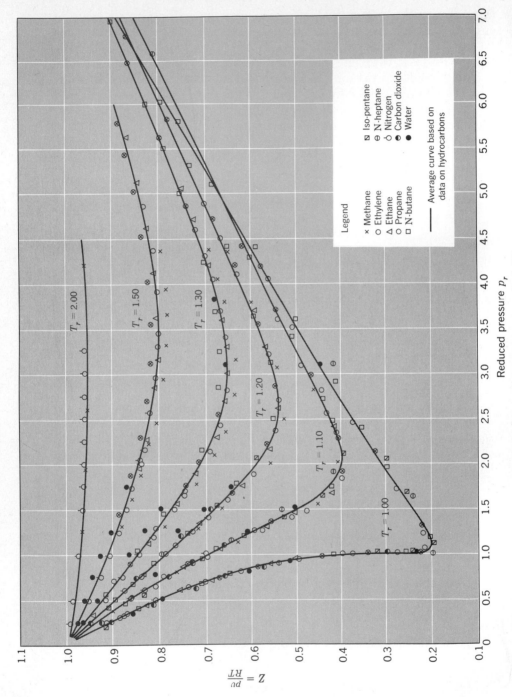

FIG. 7-3 *Experimental data showing generalized compressibility relation from Ref. [1].*

FIG. 7-4 *Nelson-Obert generalized compressibility chart according to Ref. [2].*

The chart axes and labels:

Reduced pressure p_r

$Z = \dfrac{pv}{RT}$

Compressibility factor $Z = \dfrac{pv}{RT}$

$T_r = 15.0$

5.00
7.00
10.00

$T_r = 3.50$
2.50
2.00
1.80
1.60
1.50
1.40
1.30
1.20
1.15
1.10
1.05
$T_r = 1.00$

$v'_r = 0$
0.25
0.30
0.35
0.40
$v'_r = 0.50$
0.60
0.70
0.80
0.90
1.00
1.20
1.40
1.60
2.00
2.50
3.00

Reduced pressure $p_r = \dfrac{p}{p_c}$

Reduced temperature $T_r = \dfrac{T}{T_c}$

Pseudo-reduced volume $v'_r = \dfrac{v}{RT_c/p_c}$

285

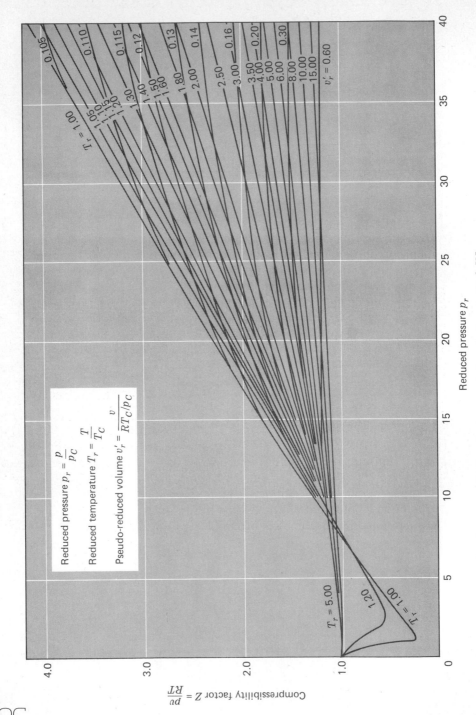

FIG. 7-5 *Nelson-Obert generalized compressibility chart, corrected according to Ref. [2].*

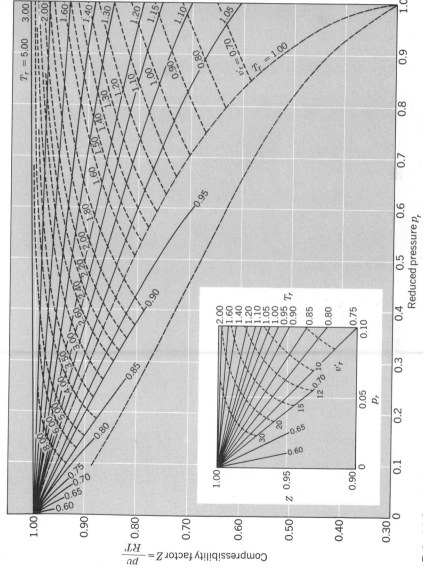

FIG. 7-6 *Nelson-Obert generalized compressibility chart, low pressure region, according to Ref. [2].*

287

EXAMPLE 7-6 Comparison of equations of state

A mass of CO_2 having a molal specific volume of 2.8 ft³/lbm mol is contained in a rigid vessel at 200°F. Estimate the pressure exerted by the gas using (a) the ideal-gas equation, (b) the van der Waals equation, and (c) the Beattie-Bridgeman equation.

SOLUTION We have $T = 200°F = 660°R$,

$$\bar{v} = 2.8 \text{ ft}^3/\text{lbm mol} \qquad (0.1748 \text{ m}^3/\text{kg mol})$$

(a) For the ideal-gas equation of state

$$p = \frac{\mathscr{R}T}{\bar{v}} = \frac{(1545)(660)}{(2.8)(144)} = 2529 \text{ psia} = 172.1 \text{ atm}$$

(b) From Table 7-1 the van der Waals constants for CO_2 are

$$a = 926 \text{ atm·ft}^6/(\text{lbm mol})^2$$
$$b = 0.686 \text{ ft}^3/\text{lbm mol}$$

Using Eq. (7-50) we then compute the pressure as

$$p = \frac{\mathscr{R}T}{\bar{v} - b} - \frac{a}{\bar{v}^2}$$

$$= \frac{(1545)(660)}{(2.8 - 0.686)(144)} - \frac{(926)(14.696)}{(2.8)^2}$$

$$= 1614 \text{ psia} = 109.8 \text{ atm}$$

(c) Finally, we use the Beattie-Bridgeman equation (7-57) with the constants obtained from Table 7-3:

$$a = 1.143 \text{ ft}^3/\text{lbm mol} \qquad A_0 = 1284.9 \text{ atm·(ft}^3/\text{lbm mol})^2$$
$$b = 1.159 \text{ ft}^3/\text{lbm mol} \qquad B_0 = 1.678 \text{ ft}^3/\text{lbm mol}$$
$$c = 61.65 \times 10^6 \text{ ft}^3 \cdot °R^3/\text{lbm mol}$$

Then,

$$A = A_0 \left(1 - \frac{a}{\bar{v}}\right) = 1284.9 \left(1 - \frac{1.143}{2.8}\right) = 760.4$$

$$B = B_0 \left(1 - \frac{b}{\bar{v}}\right) = 1.678 \left(1 - \frac{1.159}{2.8}\right) = 0.9384$$

$$e = \frac{c}{\bar{v}T^3} = \frac{61.56 \times 10^6}{(2.8)(660)^3} = 0.0766$$

Then, using Eq. (7-57),

$$p = \frac{\mathscr{R}T}{\bar{v}^2}(1 - e)(\bar{v} + B) - \frac{A}{\bar{v}^2}$$

$$= \frac{(1545)(660)(1 - 0.0766)(2.8 + 0.9384)}{(2.8)^2(144)} - \frac{(760.4)(14.696)}{(2.8)^2}$$

$$= 1692 \text{ psia} = 115.2 \text{ atm}$$

From these calculations we see that there are substantial differences in the answers obtained from the three equations of state. Based on our discussions we would anticipate that the Beattie-Bridgeman calculation is the most accurate and we will verify this fact shortly.

EXAMPLE 7-7   Corresponding states

Calculate the pressure exerted by the CO_2 in Example 7-6 using the principle of corresponding states and the generalized compressibility charts.

SOLUTION   From Table 7-2 the critical constants for CO_2 are

$$T_c = 548°R \qquad p_c = 72.9 \text{ atm} = 1071 \text{ psia}$$
$$\bar{v}_c = 1.53 \text{ ft}^3/\text{lbm mol}$$

We may then compute $\bar{v}_r$ and T_r for use with Fig. 7-4.

$$T_r = \frac{T}{T_c} = \frac{660}{548} = 1.204$$

$$\bar{v}_r = \frac{\bar{v}}{\mathscr{R}T_c/p_c} = \frac{(2.8)(1071)(144)}{(1545)(548)} = 0.510$$

Using these values we read from Fig. 7-4

$$p_r = 1.6 \qquad Z = 0.67$$

The pressure is then

$$p = p_r p_c = (1.6)(72.9) = 116.6 \text{ atm}$$

With this calculation we see that the Beattie-Bridgeman determination of Example 7-6 was indeed quite accurate, that is, 116.6 versus 115.2 atm. These numbers certainly agree within the accuracy with which one may read the charts. Checking back, we find the other two calculation methods in significant disagreement:

$$\text{Ideal-gas error} = \frac{172.1 - 116.6}{116.6} = +47.6\%$$

$$\text{van der Waals error} = \frac{109.8 - 116.6}{116.6} = -5.8\%$$

EXAMPLE 7-8 Properties from corresponding states
Estimate the density of nitrogen at $-150°\text{F}$ and 600 psia.

SOLUTION From Table 7-2 the critical constants are

$$p_c = 33.5 \text{ atm} = 482 \text{ psia}$$
$$T_c = 227°\text{R}$$
$$\bar{v}_c = 1.44 \text{ ft}^3/\text{lbm mol}$$

We compute the reduced pressure and temperatures as

$$p_r = \tfrac{600}{482} = 1.245$$
$$T_r = \tfrac{310}{227} = 1.365$$

Consulting Fig. 7-3 or 7-4, we find

$$Z = 0.83$$

The specific volume is now calculated from Eq. (7-66) as

$$v = \frac{ZRT}{p} = \frac{(0.83)(55.12)(310)}{(600)(144)} = 0.164 \text{ ft}^3/\text{lbm}$$

The density is the reciprocal of the specific volume, so

$$\rho = \frac{1}{0.164} = 6.1 \text{ lbm/ft}^3 \qquad (97.7 \text{ kg/m}^3)$$

EXAMPLE 7-8M Properties from corresponding states
Estimate the density of nitrogen at 170 K and 4 MPa.

SOLUTION From Table 7-2 the critical constants are

$$p_c = 3.398 \text{ MPa}$$
$$T_c = 126.2 \text{ K}$$
$$\bar{v}_c = 0.0899 \text{ m}^3/\text{kg mol}$$

We compute the reduced pressure and temperature as

$$p_r = \frac{4}{3.398} = 1.177$$

$$T_r = \frac{170}{126.2} = 1.347$$

Consulting Fig. 7-3 or 7-4, we find

$$Z = 0.85$$

The specific volume is now calculated from Eq. (7-66) as

$$v = \frac{ZRT}{p} = \frac{(0.85)(296.6)(170)}{(4 \times 10^6)} = 0.0107 \text{ m}^3/\text{kg}$$

The density is the reciprocal of the specific volume, so

$$\rho = \frac{1}{0.0107} = 93.3 \text{ kg/m}^3 \qquad (5.8 \text{ lbm/ft}^3)$$

EXAMPLE 7-9

Calculate the pressure at which $v = 0.06$ ft^3/lbm and $T = -150°F$ for nitrogen.

SOLUTION   The critical constants are given in Example 7-8 for nitrogen. With the foregoing data we have, for the reduced temperature and volume,

$$T_r = \tfrac{310}{227} = 1.365$$

$$v'_r = \frac{v}{RT_c/p_c} = \frac{0.06}{(55.1)(227)/(482)(144)} = 0.333$$

Consulting Fig. 7-4, we find

$$p_r = 2.85$$

The pressure is, therefore,

$$p = p_r p_c = (2.85)(482) = 1374 \text{ psia}$$

EXAMPLE 7-9M

Calculate the pressure at which $v = 0.004$ m^3/kg and $T = 170$ K for nitrogen.

SOLUTION   The critical constants are given in Example 7-8M for nitrogen. With the foregoing data we have, for the reduced temperature and volume,

$$T_r = \frac{170}{126.2} = 1.347$$

$$v_r' = \frac{v}{RT_c/p_c} = \frac{0.004}{(296.6)(126.2)/(3.398 \times 10^6)} = 0.363$$

Consulting Fig. 7-4, we find

$$p_r = 2.6$$

The pressure is, therefore,

$$p = p_r p_c = (2.6)(3.398) = 8.83 \text{ MPa}$$

7-14 ENERGY PROPERTIES OF REAL GASES

The law of corresponding states offers a means of calculating energy properties of substances in addition to the obvious information it provides on p-v-T properties. Consider the schematic chart for enthalpy behavior indicated in Fig. 7-7. At very low reduced pressures the enthalpy is a function of temperature alone since the gas obeys the $pv = RT$ equation of state. At higher pressures the enthalpy can be substantially dependent on pressure. The variation of enthalpy at low values of p_r can be obtained with the calculation

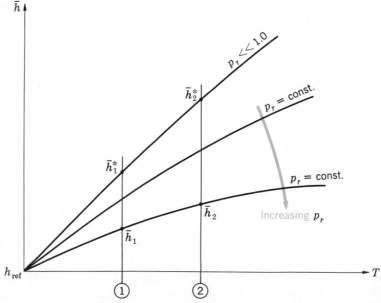

FIG. 7-7 *Schematic representation of enthalpy variation for pure substance.*

methods outlined in previous sections. We shall define the enthalpy of this low-pressure state as $\bar{h}^*$. Then, for some higher pressure, we may integrate Eq. (7-26) to obtain

$$(\bar{h} - \bar{h}^*)_T = \int_{p=0}^{p} \left[\bar{v} - T \left(\frac{\partial \bar{v}}{\partial T} \right)_p \right] dp_T \qquad (7\text{-}71)$$

But

$$\bar{v} = \frac{Z\mathscr{R}T}{p} \qquad (7\text{-}72)$$

and

$$\left(\frac{\partial \bar{v}}{\partial T} \right)_p = \frac{Z\mathscr{R}}{p} + \frac{\mathscr{R}T}{p} \left(\frac{\partial Z}{\partial T} \right)_p \qquad (7\text{-}73)$$

Introducing Eqs. (7-72) and (7-73) into Eq. (7-71), and expressing the variables in terms of reduced properties,

$$(\bar{h} - \bar{h}^*)_T = -\mathscr{R}T_c \int_0^{p_r} T_r^2 \left(\frac{\partial Z}{\partial T_r} \right)_p \frac{dp_r}{p_r} \qquad (7\text{-}74)$$

In a constant-temperature process $T_r = $ const., so that, finally,

$$\frac{(\bar{h}^* - \bar{h})}{T_c} = \mathscr{R}T_r^2 \int_0^{p_r} \left(\frac{\partial Z}{\partial T_r} \right)_p \frac{dp_r}{p_r} \qquad (7\text{-}75)$$

The right side of Eq. (7-75) is a function of T_r and p_r and may be obtained by a graphical integration of the generalized compressibility charts. The result of such a calculation is presented in Fig. 7-8.

A development similar to the previous one can be used to obtain generalized charts for entropy. The resulting equation is

$$\bar{s}_p^* - \bar{s}_p = \mathscr{R} \int_0^{p} \left[Z - 1 + T_r \left(\frac{\partial Z}{\partial T_r} \right)_{p_r} \right]_{T_r} \frac{dp_r}{p_r} \qquad (7\text{-}76)$$

where

$\bar{s}_p^*$ = entropy of ideal gas ($Z = 1.0$) at p and T in respect to some reference state

$\bar{s}_p$ = entropy of real gas at p and T in respect to same reference state

The generalized chart for entropy is presented in Fig. 7-9.

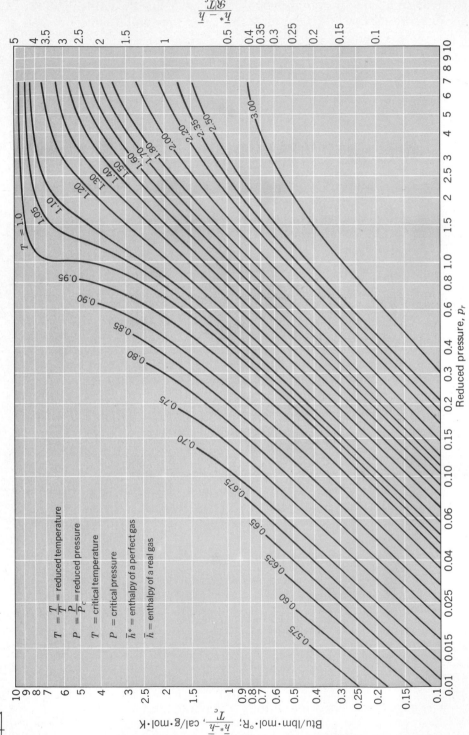

FIG. 7-8 *Generalized enthalpy chart according to Ref.* [4].

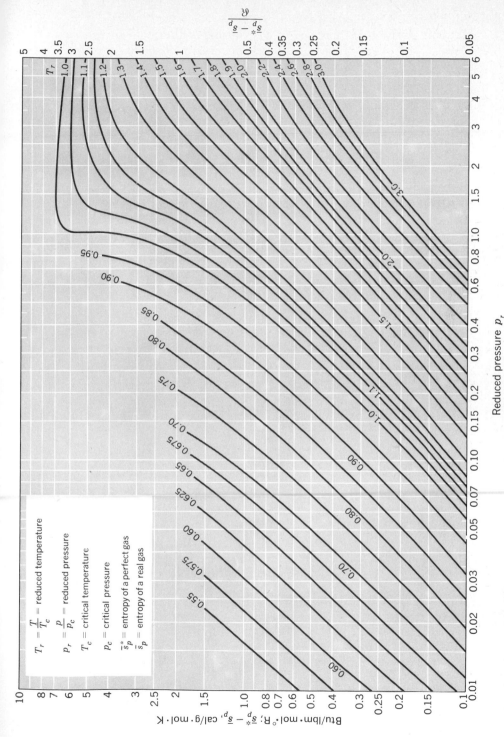

FIG. 7-9 Generalized chart for entropy according to Ref. [4].

EXAMPLE 7-10 Isothermal compression

Methane (CH_4) is compressed in a reversible isothermal process from 20 psia, 100°F, to 500 psia. The process occurs under steady-flow conditions. Calculate the work required per pound mass of methane.

SOLUTION If kinetic energies are negligible, the steady-flow energy equation for this process becomes

$$Q + W + h_1 = h_2 \tag{a}$$

where Q and W are the heat and work added, respectively. Since this is a reversible isothermal process, the heat transfer may be calculated from

$$Q = T(s_2 - s_1) \tag{b}$$

The work is, therefore,

$$W = h_2 - h_1 - T(s_2 - s_1) \tag{c}$$

We therefore need to obtain the values of the enthalpy and entropy from the generalized charts. The critical properties of methane are

$$T_c = 344°R \qquad (191.1 \text{ K})$$
$$p_c = 45.8 \text{ atm} = 673 \text{ psia}$$

The reduced properties of interest are

$$p_{r_1} = \tfrac{20}{673} = 0.0297$$
$$p_{r_2} = \tfrac{500}{673} = 0.743$$
$$T_r = \tfrac{560}{344} = 1.63$$

From Fig. 7-8 we have

$$\frac{\bar{h}_1^* - \bar{h}_1}{T_c} \approx 0.03 \qquad \text{by extrapolation}$$

$$\frac{\bar{h}_2^* - \bar{h}_2}{T_c} = 0.68 \text{ Btu/lbm·mol·°R}$$

Since $T_1 = T_2$, $\bar{h}_1^* = \bar{h}_2^*$ and we have

$$\bar{h}_2 - \bar{h}_1 = (0.03 - 0.68)T_c$$
$$= -(0.65)(344) = -233 \text{ Btu/lbm mol} \qquad (-519 \text{ kJ/kg mol})$$

From Fig. 7-9 we have

$\bar{s}_{p_1}^* - \bar{s}_{p_1} \approx 0.018$ by extrapolation

$\bar{s}_{p_2}^* - \bar{s}_{p_2} = 0.35$ Btu/lbm·mol·°R   (1.465 kJ/kg·mol·K)

In general, for the change in ideal-gas entropies we would write

$$\bar{s}_{p_2}^* - \bar{s}_{p_1}^* = \bar{\phi}_2 - \bar{\phi}_1 - \mathscr{R} \ln \frac{p_2}{p_1} \tag{d}$$

In this process $T_1 = T_2$ so that $\bar{\phi}_2 = \bar{\phi}_1$, and only the pressure ratio is retained. We have

$$\bar{s}_{p_2} - \bar{s}_{p_1} = (\bar{s}_{p_1}^* - \bar{s}_{p_1}) - (\bar{s}_{p_2}^* - \bar{s}_{p_2}) + (\bar{s}_{p_2}^* - \bar{s}_{p_1}^*)$$

$$= 0.018 - 0.35 - \mathscr{R} \ln \frac{p_2}{p_1}$$

$$= -0.332 - \tfrac{1545}{778} \ln \tfrac{500}{20}$$

$$= -6.71 \text{ Btu/lbm·mol·°R} \quad (-28.09 \text{ kJ/kg·mol·K})$$

We may now insert these properties back into Eq. (c) to obtain the work:

$$W = -223 - (560)(-6.71)$$
$$= 3537 \text{ Btu/lbm mol} \quad (8227 \text{ kJ/kg mol})$$

Since $M = 16$ for methane, the work per pound mass is

$$W = \tfrac{357}{16} = 221 \text{ Btu/lbm} \quad (514 \text{ kJ/kg})$$

EXAMPLE 7-10M   Isothermal compression

Methane (CH_4) is compressed in a reversible isothermal process from 140 kPa, 300 K, to 3.5 MPa. The process occurs under steady-flow conditions. Calculate the work required per kilogram of methane.

SOLUTION   If kinetic energies are negligible, the steady-flow energy equation for this process becomes

$$Q + W + h_1 = h_2 \tag{a}$$

where Q and W are the heat and work added, respectively. Since this is a reversible isothermal process, the heat transfer may be calculated from

$$Q = T(s_2 - s_1) \tag{b}$$

The work is, therefore,

$$W = h_2 - h_1 - T(s_2 - s_1) \tag{c}$$

We therefore need to obtain the values of the enthalpy and entropy from the generalized charts. The critical properties of methane are

$$T_c = 190.7 \text{ K} \qquad (343°\text{R})$$
$$P_c = 4.640 \text{ MPa} \qquad (673 \text{ psia})$$

The reduced properties of interest are

$$p_{r_1} = \tfrac{140}{4640} = 0.0302$$
$$p_{r_2} = \frac{3.5}{4.640} = 0.754$$
$$T_r = \frac{300}{190.7} = 1.57$$

From Fig. 7-8 we have

$$\frac{\bar{h}_1^* - \bar{h}_1}{\mathcal{R}T_c} \approx 0.016 \qquad \text{by extrapolation}$$
$$\frac{\bar{h}_2^* - \bar{h}_2}{\mathcal{R}T_c} = 0.39$$

Since $T_1 = T_2$, $\bar{h}_1^* = \bar{h}_2^*$ and we have

$$\bar{h}_2 - \bar{h}_1 = (0.016 - 0.39)\mathcal{R}T_c$$
$$= -(0.374)(8.314\ 41)(190.7)$$
$$= -593 \text{ kJ/kg mol} \qquad (-255 \text{ Btu/lbm mol})$$

From Fig. 7-9 we have

$$\frac{\bar{s}_{p_1}^* - \bar{s}_{p_1}}{\mathcal{R}} \approx 0.011 \qquad \text{by extrapolation}$$
$$\frac{\bar{s}_{p_1}^* - \bar{s}_{p_2}}{\mathcal{R}} = 0.13$$

In general, for the change in ideal-gas entropies we would write

$$\bar{s}_{p_2}^* - \bar{s}_{p_1}^* = \bar{\phi}_2 - \bar{\phi}_1 - \mathcal{R} \ln \frac{p_2}{p_1} \tag{d}$$

In this process $T_1 = T_2$, so that $\phi_2 = \phi_1$, and only the pressure ratio is retained. We have

$$\bar{s}_{p_2} - \bar{s}_{p_1} = (\bar{s}_{p_1}^* - \bar{s}_{p_1}) - (\bar{s}_{p_2}^* - \bar{s}_{p_2}) + (\bar{s}_{p_2}^* - \bar{s}_{p_1}^*)$$
$$= \mathcal{R}\left(0.011 - 0.13 - \ln \frac{p_2}{p_1}\right)$$

$$\bar{s}_{p_2} - \bar{s}_{p_1} = (8.31441)(-0.119 - \ln \tfrac{3500}{140})$$
$$= -27.75 \text{ kJ/kg·mol·K} \quad (-6.63 \text{ Btu/lbm·mol·°R})$$

We may now insert these properties back into Eq. (c) to obtain the work:

$$W = -593 - (300)(-27.75)$$
$$= 7732 \text{ kJ/kg mol} \quad (3324 \text{ Btu/lbm mol})$$

Since $M = 16$ for methane, the work per kilogram is

$$W = \tfrac{7732}{16} = 483.3 \text{ kJ/kg} \quad (208 \text{ Btu/lbm})$$

EXAMPLE 7-11   Constant-pressure heating

Carbon dioxide (CO_2) is heated at a constant pressure of 1000 psia from 100 to 800°F in a steady-flow process. Calculate the heating required per pound mass of CO_2.

SOLUTION   If kinetic energies are neglected, the heat transfer for this process is given by

$$Q = h_2 - h_1 \tag{a}$$

and the problem reduces to one of determining the change in enthalpy from the generalized charts. From Table 7-2 the critical constants for CO_2 are

$$p_c = 72.9 \text{ atm} = 1070 \text{ psia}$$
$$T_c = 548°R \quad (304 \text{ K})$$

The reduced properties of interest are, therefore,

$$p_{r_1} = p_{r_2} = \tfrac{1000}{1070} = 0.935$$
$$T_{r_1} = \tfrac{560}{548} = 1.02$$
$$T_{r_2} = \tfrac{1260}{548} = 2.30$$

From Fig. 7-8 we have

$$\frac{\bar{h}_1^* - \bar{h}_1}{T_c} = 3.5 \text{ Btu/lbm·mol·°R}$$

$$\frac{\bar{h}_2^* - \bar{h}_2}{T_c} = 0.48 \text{ Btu/lbm·°R}$$

We evaluate $\bar{h}_2^* - \bar{h}_1^*$ either by integrating the specific heat relation of Table 2-3 or by consulting the gas tables. We choose to consult the gas tables and obtain

$$\bar{h}_1^* = 4235.8 \text{ Btu/lbm mol}$$
$$\bar{h}_2^* = 11661.0 \text{ Btu/lbm mol}$$

We then have

$$\begin{aligned} \bar{h}_2 - \bar{h}_1 &= T_c(3.5 - 0.48) + \bar{h}_2^* - \bar{h}_1^* \\ &= (548)(3.5 - 0.48) + 11\,661.0 - 4235.8 \\ &= 9081 \text{ Btu/lbm mol} \end{aligned}$$

Since the molecular weight of CO_2 is 44, the heat transfer per pound mass is

$$Q = h_2 - h_1 = \tfrac{9081}{44} = 207 \text{ Btu/lbm} \qquad (481.5 \text{ kJ/kg})$$

It is interesting to compare this value with that which would be obtained by assuming ideal-gas behavior with constant specific heats. From Table 2-2,

$$c_p \text{ for } CO_2 = 0.202 \text{ Btu/lbm·°F} \qquad (0.846 \text{ kJ/kg·°C})$$

Assuming constant specific heats, we would calculate the heat transfer as

$$Q = c_p(T_2 - T_1) = (0.202)(800 - 100) = 141 \text{ Btu/lbm} \qquad (328 \text{ kJ/kg})$$

This value is in error by approximately 30 percent.

EXAMPLE 7-11M   Constant-pressure heating
Carbon dioxide (CO_2) is heated at a constant pressure of 7 MPa from 300 to 700 K in a steady-flow process. Calculate the heating required per kilogram of CO_2.

SOLUTION   If kinetic energies are neglected, the heat transfer for this process is given by

$$Q = h_2 - h_1 \qquad\qquad (a)$$

and the problem reduces to one of determining the change in enthalpy from the generalized charts. From Table 7-2 the critical constants for CO_2 are

$$p_c = 7.386 \text{ MPa}$$
$$T_c = 304.20 \text{ K} \qquad (548°R)$$

The reduced properties of interest are, therefore,

$$p_{r_1} = p_{r_2} = \frac{7}{7.386} = 0.948$$

$$T_{r_1} = \frac{300}{304.2} = 0.986$$

$$T_{r_2} = \frac{700}{304.2} = 2.30$$

From Fig. 7-8 we have

$$\frac{\bar{h}_1^* - \bar{h}_1}{\mathscr{R}T_c} = 2.35$$

$$\frac{\bar{h}_2^* - \bar{h}_2}{\mathscr{R}T_c} = 0.23$$

We evaluate $\bar{h}_2^* - \bar{h}_1^*$ either by integrating the specific heat relation of Table 7-3 or by consulting the gas tables. We choose to consult the gas tables and obtain

$$\begin{aligned}
\bar{h}_2^* - \bar{h}_1^* &= (\bar{h}_{700} - \bar{h}_{298}) - (\bar{h}_{300} - \bar{h}_{298}) \\
&= 17\ 761 - 67 = 17\ 694 \text{ kJ/kg mol}
\end{aligned}$$

We then have

$$\begin{aligned}
\bar{h}_2 - \bar{h}_1 &= \mathscr{R}T_c(2.35 - 0.23) + \bar{h}_2^* - \bar{h}_1^* \\
&= (8.314\ 41)(304.2)(2.12) + 17\ 694 \\
&= 23\ 056 \text{ kJ/kg mol}
\end{aligned}$$

Since the molecular weight of CO_2 is 44, the heat transfer per kilogram is

$$Q = h_2 - h_1 = \frac{23\ 056}{44} = 524 \text{ kJ/kg} \qquad (225 \text{ Btu/lbm})$$

It is interesting to compare this value with that which would be obtained by assuming ideal-gas behavior with constant specific heats. From Table 2-2,

$$c_p \text{ for } CO_2 = 0.846 \text{ kJ/kg·°C} \qquad (0.202 \text{ Btu/lbm·°F})$$

Assuming constant specific heats, we would calculate the heat transfer as

$$Q = c_p(T_2 - T_1) = (0.846)(700 - 300) = 338.4 \text{ kJ/kg} \qquad (145 \text{ Btu/lbm})$$

This value is in error by approximately 35 percent.

EXAMPLE 7-12 Adiabatic throttling

Nitrogen at 500 psia and $-100°F$ is throttled adiabatically to 50 psia in a Joule-Thomson porous-plug apparatus. Calculate the final temperature. Also estimate the Joule-Thomson coefficient in this region.

SOLUTION For this process we assume that there is no appreciable kinetic energy or external work, as described in Sec. 4-7. Therefore

$$h_1 = h_2$$

If the nitrogen behaved as an ideal gas this would mean that the initial and final temperatures would be the same because enthalpy is a function of temperature alone for an ideal gas. From Table 7-2 the critical constants are

$$p_c = 33.5 \text{ atm} = 492.5 \text{ psia}$$
$$T_c = 227°F$$

so that

$$p_{r_1} = \frac{500}{492.5} = 1.015$$

$$T_{r_1} = \frac{360}{227} = 1.586$$
$$p_{r_2} = \frac{50}{492.5} = 0.1015$$

In this temperature range we shall assume that the low pressure or ideal-gas enthalpies $\bar{h}^*$ may be calculated with a constant specific heat from Table 2-2. So, for nitrogen,

$$c_p = 0.248 \text{ Btu/lbm·°F} \qquad (1.038 \text{ kJ/kg·°C})$$
$$\bar{c}_p = 6.944 \text{ Btu/lbm·mol°F}$$

With the enthalpy reference taken as zero at 0°R,

$$\bar{h}^* = 6.944T$$

and

$$\bar{h}_1^* = (6.944)(360) = 2500 \text{ Btu/lbm mol}$$

Using the values of p_{r_1} and T_{r_1} we obtain from Fig. 7-8

$$\frac{\bar{h}_1^* - \bar{h}_1}{T_c} = 1.05$$

so that

$$\bar{h}_1 = 2500 - (1.05)(227) = 2262 \text{ Btu/lbm mol}$$

Unfortunately, there are no lines of constant enthalpy on Fig. 7-8 so we must adopt an iterative procedure which is directed toward obtaining T_2 so that $\bar{h}_1 = \bar{h}_2 = 2262$. The trials are as follows, using $p_{r_2} = 0.1015$ and Fig. 7-8.

T_2	$\bar{h}_2^*$	T_{r2}	$\dfrac{\bar{h}_2^* - \bar{h}_2}{T_c}$	$\bar{h}_2$
360	2500	1.586	0.11	2475
340.5	2364	1.5	0.12	2336
317.8	2206	1.4	0.13	2177
329.8	2290	1.453	0.13	2261

It is worthwhile to point out, of course, that at the low pressure of 50 psia the nitrogen is beginning to behave more nearly like an ideal gas. From this calculation we see that about a 30° temperature drop is experienced (360 − 329.8) in the throttling process and the final temperature is

$$T_2 = 329.8°R = -130.2°F \qquad (-90.1°C)$$

The Joule-Thomson coefficient is estimated from

$$\mu_J = \left(\frac{\partial T}{\partial p}\right)_h \approx \left(\frac{\Delta T}{\Delta p}\right)_h = \frac{329.8 - 360}{50 - 500}\left(\frac{14.7 \text{ psia}}{\text{atm}}\right)$$
$$= 0.987 \text{ F/atm} \qquad (0.548°C/\text{atm})$$

From this example we can see that graphical plots of the Joule-Thomson coefficient would be very useful for throttling calculations.

7-15 THERMODYNAMICS OF MAGNETISM

Our discussion of thermodynamic properties in Chap. 3 and in this chapter has been concerned primarily with the so-called *pure substance* or, more explicitly, the simple compressible substance. We now want to examine the behavior of a simple magnetic substance, or one in which only two properties are required to fix the thermodynamic state. We shall take the internal energy and the magnetic dipole moment M as the two independent variables. For this simple substance we assume that the $p\,dv$ work mode is negligible (constant volume), so that only two properties are necessary to establish the thermodynamic state of the system. The magnetic flux density B is given by

$$B = \mu_0(H + M) \tag{7-77}$$

where H is the magnetic intensity of the external magnetic field, M is the magnetic dipole moment or magnetization vector in the direction of the applied field, and μ_0 is the permeability of free space. In accordance with Eq. (2-11), we write for the magnetic work per unit mass,

$$d'W = vH \, dB$$
$$= \mu_0 vH \, dH + \mu_0 vH \, dM \tag{7-78}$$

The first term represents the work done on free space, and the second term expresses the work done by the magnetic substance. We may write the energy equation as

$$du = d'Q + d'W = T \, ds + \mu_0 \, vH \, dM \tag{7-79}$$

where we leave out of consideration the work done on free space, since we are analyzing the magnetic substance as the thermodynamic system. In Eq. (7-79) the internal energy is expressed in terms of the independent variables ds and dM, so we may write

$$du = \left(\frac{\partial u}{\partial s}\right)_M ds + \left(\frac{\partial u}{\partial M}\right)_s dM \tag{7-80}$$

Comparing Eq. (7-79) with (7-80) we obtain

$$\left(\frac{\partial u}{\partial s}\right)_M = T \tag{7-81}$$

$$\left(\frac{\partial u}{\partial M}\right)_s = \mu_0 vH \tag{7-82}$$

Equation (7-81) may be regarded as the definition of temperature for simple magnetic substances.

It should be noted at this point that the term $\mu_0 vH \, dM$ plays the same role as $p \, dv$ for a simple compressible substance; magnetic intensity H is analogous to pressure, and $\mu_0 v \, dM$ is analogous to a volume change.

Specific heats for the simple magnetic substance may be appropriately defined by rewriting Eq. (7-79) as

$$d'Q = du - \mu_0 vH \, dM \tag{7-83}$$

Let us consider the constant H and constant M process:

$$d'Q_H = c_H \, dT_H = du_H - \mu_0 vH \, dM_H \tag{7-84}$$

If we define a magnetic enthalpy h_m as

$$h_m = u - \mu_0 vHM \tag{7-85}$$

then Eq. (7-84) becomes

$$c_H \, dT_H = (dh_m)_H$$

or

$$c_H = \left(\frac{\partial h_m}{\partial T}\right)_H \tag{7-86}$$

For the process at constant M,

$$c_M \, dT_M = du_M$$

and

$$c_M = \left(\frac{\partial u}{\partial T}\right)_M \tag{7-87}$$

c_H and c_M are thus analogous to the specific heats at constant pressure and constant volume for a simple compressible substance.

Simple magnetic substances, like simple compressible substances, have equations of state and are expressed in the functional form

$$f(H, M, T) = 0$$

A *Curie substance* is one that has the very simple equation of state

$$M = C\frac{H}{T} \tag{7-88}$$

where C is called the *Curie constant*. Rewriting Eq. (7-79) and substituting $H = MT/C$,

$$ds = \frac{1}{T} \, du - \frac{\mu_0 v M}{C} \, dM \tag{7-89}$$

Since this is an exact differential equation, we may equate second mixed partial derivatives to obtain

$$\left[\frac{\partial (1/T)}{\partial M}\right]_u = \frac{-\mu_0 v}{C}\left(\frac{\partial M}{\partial u}\right)_M = 0 \tag{7-90}$$

We, therefore, have the result that the temperature is a function only of u, or $u = f(T)$. This result is analogous to the fact that the internal energy for an ideal gas is a function of temperature alone. The behavior of this particular type of magnetic substance is, therefore, quite simple just as the ideal gas is a very elementary type of compressible substance. Finally, we may write the following equations for a Curie substance in view of the result in Eq. (7-90):

$$du = c_M \, dT \tag{7-91}$$

$$ds = c_M \frac{dT}{T} - \frac{\mu_0 v}{C} M \, dM \tag{7-92}$$

7-16 MAGNETOCALORIC EFFECT

An expansion of the foregoing theory shows that we may express the entropy of a simple magnetic substance in terms of T and H with the following relation:

$$ds = \frac{c_H}{T} \, dT + \mu_0 v \left(\frac{\partial M}{\partial T} \right)_H dH \tag{7-93}$$

For paramagnetic materials, experiments indicate that $(\partial M/\partial T)_H$ is *always negative*. As a result, we see from Eq. (7-93) that:

1 A reversible isothermal increase in the magnetic field causes a rejection of heat; i.e., $T \, ds$ is negative when dH is positive.
2 A reversible isothermal decrease in the magnetic field causes an absorption of heat; i.e., $T \, ds$ is positive when dH is negative.

For a reversible adiabatic process it is clear from Eq. (7-93) that a decrease in the magnetic field causes a drop in temperature. This phenomenon is called the *magnetocaloric effect* and is widely used for achieving very low temperatures, of the order of 1 K. These low-temperature processes are fairly simple: a paramagnetic salt is cooled to a few degrees Kelvin with liquid helium and a strong magnetic field is applied. The energy from the field is subsequently dissipated in the liquid helium. After an equilibrium state is reached, the magnetic substance is isolated and the magnetic field is reduced to zero. As a result, the temperature of the substance drops to a very low value.

REVIEW QUESTIONS

1 Why is a thermodynamic property a point function?

2 Define the Helmholtz and Gibbs functions.

3 What minimum information is necessary to formulate a table of thermodynamic properties for a pure substance?

4 How is the Joule-Thomson experiment useful in determining thermodynamic properties?

5 What is known about the specific heat and internal-energy behavior for an ideal gas?

6 What basic assumptions are involved in formulating the gas tables?

7 What is the meaning of P_r, v_r, and ϕ as used in the gas tables?

8 What practical value is the Clausius-Clapeyron relation?

9 Why is a constant-pressure line a straight line in the wet-mixture region on a Mollier diagram?

10 Under what conditions, in terms of critical temperature and pressure, does one normally experience near-ideal-gas behavior?

11 What is the law of corresponding states?

12 Describe the simple magnetic substance. How does it differ from a simple compressible substance?

13 Describe the magnetocaloric effect.

14 What thermodynamic property relation is used to link microscopic and macroscopic thermodynamic analyses?

15 What experimental measurements are required to determine thermodynamic properties?

16 What is a van der Waals gas? What physical model is involved?

17 What is the generalized compressibility factor?

PROBLEMS (English units)

7-1 Air expands isentropically from a low velocity and 50 psia to atmospheric pressure. Calculate the final velocity for initial temperatures of (a) 1000°F, (b) 100°F. Compare with the final velocity which would be obtained if the air had constant specific heats.

7-2 Suppose the change in entropy for the two expansion processes in Prob. 7-1 is 0.008 Btu/lbm·°R. Calculate the exit velocity under these conditions, assuming that the process is still adiabatic, though irreversible.

7-3 Calculate the exit area of the nozzle for the conditions of Probs. 7-1 and 7-2, assuming a mass flow of 0.1 lbm/s.

7-4 Air is compressed from 10°F, 15 psia, to 500°F, 60 psia. Calculate the change in entropy.

7-5 Air expands in a turbine from 2000°F, 55 psia, to 20 psia. The initial-flow velocity is 200 ft/s and the exit velocity is 800 ft/s. Calculate the work output of the turbine if the process occurs isentropically.

7-6 How much energy would be required to heat 1 lbm of air from 200 to 1500°F at a constant pressure of 20 psia? What percent error would result if the calculation were performed assuming constant specific heats?

7-7 A rigid container having a volume of 10 ft³ is filled with air at 100°F and 15 psia. Heat is added until the temperature reaches 1800°F. What is the final pressure and the change in entropy? How much energy is added?

7-8 Calculate the energy which must be added to the air in Prob. 7-7 to raise the pressure to 100 psia. What is the final temperature?

7-9 Calculate the work per pound mass when air expands isentropically from 100 psia, 1500°F, to 30 psia in a cylinder behind a piston. What percent error would result from assuming constant specific heat behavior?

7-10 Using saturated-water-vapor properties at 40°F, calculate the vapor pressure at 35°F.

7-11 The enthalpy of fusion for water is 143.3 Btu/lbm at 32°F. Estimate the melting temperature of ice at a pressure of 3000 psia.

7-12 Using tabulated values of vapor pressure for water, calculate the enthalpy of vaporization for water at 100 psia. Compare this calculation with the value given in the steam tables.

7-13 Estimate the vapor pressure of Freon 12 at -150°F.

7-14 Estimate the vapor pressure of ammonia at -100 and 140°F.

7-15 Steam expands isentropically from 500 psia, 500°F, to 10 psia. With the use of the Mollier diagram obtain the change in enthalpy and the final temperature, if superheated, or the final quality, if saturated.

7-16 Repeat Prob. 7-15 for expansion from 100 psia, 1000°F, to 2 psia.

7-17 Steam is heated at constant pressure of 20 psia from saturated vapor to 500°F. With the use of the Mollier diagram calculate the heat required in a steady-flow process.

7-18 Show that the following relations hold for a van der Waals gas:

$$(\bar{h}_2 - \bar{h}_1)_T = p_2 \bar{v}_2 - p_1 \bar{v}_1 + a \left(\frac{1}{\bar{v}_1} - \frac{1}{\bar{v}_2} \right)$$

$$(\bar{s}_2 - \bar{s}_1)_T = \mathscr{R} \ln \frac{\bar{v}_2 - b}{\bar{v}_1 - b}$$

7-19 Derive the relation

$$\frac{c_p}{c_v} = \frac{(\partial p/\partial v)_s}{(\partial p/\partial v)_T}$$

7-20 Obtain an expression for $(\partial u/\partial p)_T$ that involves only p, v, and T.

7-21 Obtain an expression for $(\partial h/\partial v)_T$ that involves only p, v, and T.

7-22 Obtain an expression for $(u_2 - u_1)_T$ for a van der Waals gas. How does this answer differ from that obtained for an ideal gas? Can you offer an explanation for the difference on a physical basis?

7-23 Estimate the specific volume of oxygen and nitrogen at 500 atm and 1500°F.

7-24 Derive an expression for the Joule-Kelvin coefficient for a van der Waals gas.

7-25 Using the critical-state data given in the Appendix for ethane (C_2H_6), calculate the density at 300°F and 500 psia.

7-26 Calculate the work required to compress methane in a reversible isothermal process from 50 psia, 200°F, to 700 psia.

7-27 Nitrogen undergoes an adiabatic throttling process from 2000 psia and $-150°F$ to 20 psia. Calculate the final temperature in the process.

7-28 Carbon dioxide is compressed in an isentropic steady-flow process from 100 to 1500 psia. The initial temperature is 70°F. Calculate the work necessary to accomplish the compression and the final temperature. Compare the results with that obtained by assuming that CO_2 behaves like an ideal gas with constant specific heats.

7-29 A tank contains oxygen at 1000 psia and $-150°F$. Calculate the mass contained if the volume of the tank equals 1 gal (231 in³).

7-30 The tank in Prob. 7-29 is heated until its temperature reaches $-100°F$. Calculate the heat transfer.

7-31 How much heat must be added to the tank in Prob. 7-29 to increase the pressure to 1500 psia?

7-32 Steam at 1500 psia and 650°F is compressed isentropically in a steady-flow process to a pressure of 5500 psia. Calculate the work per pound mass, using the generalized charts, and compare with an answer obtained from the steam tables.

7-33 Show that the heat transfer per unit mass in an isothermal process of a Curie substance is

$$Q_T = - \frac{\mu_0 C_V}{2T} (H_2^2 - H_1^2)$$

where H_1 and H_2 are the initial and final magnetic intensities, respectively.

7-34 Show that the following relation applies to an isentropic change of magnetic field for a Curie substance:

$$T_2^2 = T_1^2 - \frac{2Q_{T_1}T_1}{C_H}$$

where T_1 and T_2 are the initial and final temperatures, respectively, and Q_{T_1} is the heat transfer for an isothermal process with the same change of field at temperature T_1. The value of Q_{T_1} is given in Prob. 7-33.

7-35   Show that, for a van der Waals gas,

$$\beta = \frac{\mathscr{R}\bar{v}^2(\bar{v} - b)}{\mathscr{R}T\bar{v}^3 - 2a(\bar{v} - b)^2}$$

$$\kappa = \frac{\bar{v}^2(\bar{v} - b)^2}{\mathscr{R}T\bar{v}^3 - 2a(\bar{v} - b)^2}$$

7-36   Show that for a van der Waals gas undergoing a reversible adiabatic process, the following relations apply:

$$T(\bar{v} - b)^{\mathscr{R}/\bar{c}_v} = \text{const.}$$

$$\left(p + \frac{a}{\bar{v}^2}\right)(\bar{v} - b)^{1+(\mathscr{R}/\bar{c})_v} = \text{const.}$$

7-37   Show that, for a van der Waals gas,

$$\bar{c}_p - \bar{c}_v = \frac{\mathscr{R}}{1 - 2a(v - b)^2/\mathscr{R}T\bar{v}^3}$$

7-38   Show that the Joule-Thomson coefficient for a van der Waals gas is

$$\mu_J = \frac{\bar{v}}{\bar{c}_p} \frac{2a(\bar{v} - b)^2 - \mathscr{R}Tb\bar{v}^2}{\mathscr{R}T\bar{v}^3 - 2a(\bar{v} - b)^2}$$

Subsequently show that the inversion curve on a p-v diagram could be expressed as

$$p = \frac{a}{b\bar{v}}\left(2 - \frac{3b}{\bar{v}}\right)$$

7-39   Show that the van der Waals equation can be expressed in the following generalized form:

$$p_r = \frac{T_r}{\bar{v}_r - \frac{1}{8}} - \frac{\frac{27}{64}}{\bar{v}_r^2}$$

From this show that the compressibility factor can be obtained as

$$Z = \frac{\bar{v}}{\bar{v}_r - \frac{1}{8}} - \frac{\frac{27}{64}}{T_r\bar{v}_r}$$

Subsequently, show that the $(\bar{v}_r - \frac{1}{8})^{-1}$ term can be expanded in a power series so that the van der Waals equation can be expressed in the virial form

$$Z = 1 + \left(\frac{1}{8} - \frac{\frac{27}{64}}{T_r}\right)\frac{1}{\bar{v}_r} + \frac{1}{64\bar{v}_r^2} + \frac{1}{512\bar{v}_r^3} + \cdots$$

7-40   Calculate the pressure of water vapor at 1000°F and $v = 0.1143$ ft³/lbm using (a) the ideal-gas law, (b) the van der Waals equation, (c) the Beattie-Bridgeman equation, and (d) the steam tables.

7-41   Compare the pressure predicted by the Dieterici equation with the steam tables for (a) $T = 700°F$, $v = 0.0287$ ft³/lbm and (b) $T = 1400°F$, $v = 1.0893$ ft³/lbm. How do the results of this comparison jibe with the comments in the text regarding the applicability of the Dieterici equation?

7-42   Compare the ideal-gas equation, the van der Waals equation, and the Berthelot equation with tabular values for ammonia at (a) 200°F, $v = 13.73$ ft³/lbm and (b) 360°F, $v = 2.265$ ft³/lbm.

7-43   Compare the ideal-gas equation and the Dieterici equation with tabular values for Freon 12 at 240°F and $v = 0.273\ 23$ ft³/lbm.

7-44   Verify that carbon dioxide and nitrogen may be considered to behave as ideal gases at room temperature and pressure (1 atm, 20°C).

7-45   Calculate the mass of hydrogen which would be contained in a 1000-cm³ vessel at 3700 psia and 60°C.

7-46   Using the compressibility charts calculate the specific volume of each of the following substances at the given conditions:
   (a) Air at 70°F and 500 psia
   (b) Butane (C_4H_{10}) at 100 psia and 900°F
   (c) Carbon dioxide at 500 K and 50 atm
   (d) Nitric oxide (NO) at 100 atm and 200 K
   (e) Water at 800 K and 50 atm

7-47   100 lbm of carbon dioxide is contained at 100°F and 1500 psia in a rigid container which may exchange heat with the surroundings. A valve on the side of the container is opened and the CO_2 allowed to escape very slowly until the pressure drops to 700 psia, at which time the valve is closed. At completion of the process the temperature in the tank is measured as 70°F. Calculate the mass of CO_2 while escapes from the tank and the heat transfer to the tank from the surroundings using (a) ideal-gas relations and (b) generalized compressibility charts.

7-48   How much mass would have escaped and what would have been the final temperature in the tank of Prob. 7-47 if it were perfectly insulated?

7-49   Ethane (C_2H_6) is throttled adiabatically from 850 psia, 750°F, to 100 psia. Calculate the final temperature and the change in entropy per unit mass. (Use Table A-5.)

7-50   Carbon dioxide at 1000 psia and 800°F is to be expanded isentropically in a nozzle to a pressure of 500 psia. The inlet velocity to the nozzle is low, and steady flow may be assumed. Calculate the exit temperature and exit velocity using (a) ideal gas with constant specific heats and (b) the generalized compressibility charts.

7-51   Suppose the carbon dioxide with the same initial conditions as in Prob. 7-50 is to be expanded to a velocity of 1200 ft/s in an isentropic nozzle. What exit flow area would be required for a mass flow of 0.4 lbm/s?

7-52 Propane (C_3H_8, Table A-5) is compressed adiabatically from 150°F, 25 psia, to 750°F, 500 psia. Assuming a steady-flow process, calculate the work input and change in entropy per pound mass using (a) ideal gas, constant specific heats and (b) the generalized compressibility charts.

7-53 Ammonia is contained in a 15-ft³ rigid vessel at 280 psia and 360°F. Calculate the heat which must be removed to lower the temperature to 180°F using (a) ideal-gas relations, (b) the generalized compressibility charts, and (c) the ammonia tables. What is the error in each determination?

7-54 Steam is expanded isentropically in a turbine from 5000 psia, 1600°F, to 500 psia in a steady-flow process. Calculate the work output and final temperature of the steam using (a) ideal-gas relations with constant specific heats, (b) the generalized compressibility charts, and (c) the steam tables.

PROBLEMS (Metric units)

7-1M Determine the density and compressibility factor for the following fluids at $p = 10.0$ MPa and $T = 650$ K:
    (a) Ammonia
    (b) Nitrogen
    (c) Oxygen
    (d) Heptane

7-2M Using the generalized compressibility charts, calculate the pressure at which density calculated with the ideal-gas equation will be in error by 5 percent for nitrogen at 45°C.

7-3M A volume flow of 4.5 m³/s of methane flows in a pipe at $p = 4.0$ MPa and $T = 30°C$. Determine the mass flow rate using both the ideal-gas equation and the compressibility factor.

7-4M 0.5 m³ of nitrogen are contained in a tank at $p = 60.0$ MPa and $T = 450$ K. Calculate the mass using (a) the ideal-gas equation, (b) the van der Waals equation, and (c) compressibility factors.

7-5M Calculate the enthalpy of nitrogen at 200 K and 7.0 MPa assuming the ideal-gas enthalpy is zero at 0 K.

7-6M Calculate the absolute entropy of nitrogen at 200 K and 7.0 MPa using the entropy deviation charts.

7-7M Nitrogen is throttled adiabatically from conditions of 20 MPa and 200 K to 1.5 MPa. Calculate the final temperature, change in entropy, and irreversibility for $T_0 = 20°C$.

7-8M Using the generalized compressibility charts, calculate the energy required to heat steam at a constant pressure of 20.0 MPa from the saturated-vapor state to a temperature of 640°C. Compare with appropriate values obtained from the steam tables.

7-9M Oxygen is cooled from 10.0 MPa, 250 to 160 K, in a steady-flow constant-pressure process. Using the generalized charts, calculate the cooling required, change in entropy, and final specific volume of the oxygen.

7-10M Show that the Joule-Thomson coefficient can be expressed as

$$\mu = \left(\frac{\partial T}{\partial p}\right)_h = (T\beta - 1)\frac{v}{c_p}$$

7-11M Using the relation given in Prob. 7-10M, determine the Joule-Thomson coefficient for (a) an ideal gas and (b) van der Waals gas.

7-12M Methane is contained in a rigid tank at 20 MPa and 0°C. Cooling is applied to the tank until the pressure drops to 10 MPa. Calculate the cooling required per unit mass of methane.

7-13M Oxygen undergoes a steady-flow adiabatic compression process from 5.08 MPa, 200 K, to 25.4 MPa, 360 K. Calculate the work required and irreversibility per unit mass for $T_0 = 20°C$.

7-14M Calculate the saturation pressure for an ice-water mixture in equilibrium at $-2°C$.

7-15M Using appropriate general relations show that the following results for a van der Waals gas:

$$d\bar{s} = \bar{c}_v \frac{dT}{T} + \frac{\mathcal{R}\, d\bar{v}}{\bar{v} - b}$$

$$d\bar{u} = \bar{c}_v\, dT + \frac{a\, d\bar{v}}{\bar{v}^2}$$

7-16M Ethylene (C_2H_4, Table A-5) is to be discharged into a 10-m³ tank which is initially evacuated. The discharge comes from a high-pressure line containing the gas at 125 atm and 325 K. The filling process occurs slowly and the tank is cooled so that the temperature of the gas inside remains constant at 300 K. Calculate the mass of gas in the tank when the pressure just reaches 100 atm and the heat which has been removed up to that point.

7-17M Air at 1 atm is heated from 300 to 800 K in a constant-pressure process. Calculate the heating required for a mass of 4.0 kg. Also determine the change in internal energy.

REFERENCES

1 Gouq-Jen Su: Modified Law of Corresponding States for Real Gases, *Ind. Engr. Chem.*, vol. 38, p. 803, 1946.

2 Obert, E. F.: "Concepts of Thermodynamics," McGraw-Hill Book Company, New York, 1960.

3 Keenan, J. H., and J. Kaye: "Gas Tables," John Wiley & Sons, Inc., New York, 1948.

4 Hougen, O. A., K. M. Watson, and R. A. Ragatz: "Chemical Process Principles," pt. II, John Wiley & Sons, Inc., New York, 1947.

5 Keenan, J. H., and F. G. Keyes: "Thermodynamic Properties of Steam," John Wiley & Sons, Inc., New York, 1936.

6 Lewis, G. N., and M. Randall: "Thermodynamics," 2d ed. (revised by K. S. Pitzer and L. Brewer), McGraw-Hill Book Company, New York, 1961.

7 Van Wylen, G. J., and R. E. Sonntag: "Fundamentals of Classical Thermodynamics," John Wiley & Sons, Inc., New York, 1965.

8 Hilsenrath, J., et al.: Tables of Thermal Properties of Gases, *Natl. Bur. Std. Circ.* 564, U.S. Government Printing Office, Washington, D.C., 1955.

9 Beattie, J. A., and O. C. Bridgeman: A New Equation of State for Fluids, *Proc. Am. Acad. Arts Sci.*, vol. 63, p. 229, 1928.

10 Benedict, M., G. Webb, and L. Rubin: An Empirical Equation for the Thermodynamic Properties of Light Hydrocarbons and their Mixtures, *J. Chem. Physics*, vol. 8, p. 334, 1940.

11 Nelson, L. C., and E. F. Obert: Generalized $pv.T$ Properties of Gases, *Trans. ASME*, p. 1057, October 1954.

chapter eight
Applications of statistical thermodynamics

8-1 INTRODUCTION

The basic statistical presentations of Chap. 5 along with macroscopic discussions of the second law of thermodynamics have indicated possible links between macroscopically observable thermodynamic phenomena and the underlying molecular behavior of matter. Statistical implications of this correspondence have already been noted. Our purpose in this chapter is to carry through with the analysis in order to develop calculation techniques for thermodynamic properties based on a microscopic model. One may raise a logical objection to the order of subject presentation: it may be argued that applications of statistical analysis should immediately follow the basic material of Chap. 5 without the intervening macroscopic development of the second law. It turns out that the macroscopic definition of entropy is needed to introduce the temperature concept into a microscopic analysis. This fact makes sense because temperature, as measured macroscopically, is a collective concept, that is, it represents the total effect of a collection of microscopic particles on some measuring device. It would be foolish then for us to speak of the temperature of a *single* molecule—we must consider a statistically large sample of particles. This means that we need a macroscopic concept of temperature to be able to link microscopic thermodynamic analysis with calculations of equilibrium thermodynamic properties.

By judicious use of the temperature concept, it will be possible for us to describe statistically several types of thermodynamic systems (gases, solids, thermal radiation) and calculate their properties. It is well to caution the reader at this point that the statistical approach followed here is not the only one available and, in many respects, is quite restrictive in scope. In a later section of this chapter we shall mention some features of the *ensemble* method, which furnishes powerful techniques for analysis of very complicated thermodynamic systems.

8-2 THE CONSTANTS β AND k

The development of general property relations in Chap. 7 furnishes an analytical basis for determination of the energy constant β that appears in all of the statistical distributions of Chap 5. We shall now establish the relationship between β and k, the entropy constant from

$$S = k \ln \Omega \tag{8-1}$$

In a later section we shall show that k is the Boltzmann constant.

Consider the Bose-Einstein (BE) energy distribution:

$$\frac{N_i}{g_i} = \frac{1}{Ae^{\beta\epsilon_i} - 1} \tag{8-2}$$

The logarithm of the thermodynamic probability for this distribution was given by

$$\ln \Omega = \Sigma[(N_i + g_i) \ln (N_i + g_i) - g_i \ln g_i - N_i \ln N_i] \tag{8-3}$$

when Stirling's approximation is employed. Rearranging terms and using Eq. (8-1), the entropy is written as

$$S = k \sum \left(g_i \ln \frac{N_i + g_i}{g_i} + N_i \ln \frac{N_i + g_i}{N_i} \right) \tag{8-4}$$

One of the intermediate steps in the derivation of the Maxwell relations [Eq. (7-16d)] yields

$$\frac{1}{T} = \left(\frac{\partial S}{\partial U} \right)_V \tag{8-5}$$

Equation (8-4) expresses the entropy for a fixed volume (isolated system), so that we easily perform the differentiation and obtain

$$\left(\frac{\partial S}{\partial U} \right)_V = k \sum g_i \frac{g_i}{N_i + g_i} \frac{1}{g_i} \left(\frac{\partial N_i}{\partial U} \right)_V + N_i \frac{N_i}{N_i + g_i} g_i \frac{-1}{N_i^2} \left(\frac{\partial N_i}{\partial U} \right)_V$$
$$+ \ln \frac{N_i + g_i}{N_i} \left(\frac{\partial N_i}{\partial U} \right)_V \tag{8-6}$$

In this differentiation g_i is considered constant since it does not vary with the total energy of the system. Canceling terms in Eq. (8-6) gives

$$\left(\frac{\partial S}{\partial U} \right)_V = k \sum \ln \frac{N_i + g_i}{N_i} \left(\frac{\partial N_i}{\partial U} \right)_V \tag{8-7}$$

From the equilibrium distribution of Eq. (8-2),

$$\ln \frac{N_i + g_i}{N_i} = \ln A + \beta \epsilon_i \tag{8-8}$$

so that

$$\left(\frac{\partial S}{\partial U} \right)_V = k \ln A \sum \left(\frac{\partial N_i}{\partial U} \right)_V + k\beta \sum \epsilon_i \left(\frac{\partial N_i}{\partial U} \right)_V \tag{8-9}$$

The constant total energy and number of particles restrictions may be differentiated to give

$$U = \Sigma \epsilon_i N_i$$

$$\left(\frac{\partial U}{\partial U}\right)_V = 1 = \sum \epsilon_i \left(\frac{\partial N_i}{\partial U}\right)_V \tag{8-10}$$

$$N = \Sigma N_i$$

$$\left(\frac{\partial N}{\partial U}\right)_V = 0 = \sum \left(\frac{\partial N_i}{\partial U}\right)_V \tag{8-11}$$

Inserting Eqs. (8-5), (8-10), and (8-11) into Eq. (8-9) gives

$$\frac{1}{T} = k\beta$$

or

$$\beta = \frac{1}{kT} \tag{8-12}$$

It may be shown that this same relation also applies for the Fermi-Dirac (FD) and Maxwell-Boltzmann (MB) distributions. As mentioned previously, we have yet to show that k is the Boltzmann constant.

8-3 PROPERTIES AND PARTITION FUNCTIONS

Let us assume that a collection of microscopic particles is available which obeys MB statistics or BE statistics in the limit where $g_i \gg N_i$. In this case we are able to define a partition function by

$$Z = \Sigma g_i e^{-\beta \epsilon_i} \tag{8-13}$$

and the equilibrium distribution becomes

$$N_i = \frac{N}{Z} g_i e^{-\beta \epsilon_i} \tag{8-14}$$

The entropy may be written as follows for $g_i \gg N_i$, using Eq. (8-4):

$$S = k \ln \Omega_{\max} = k \sum g_i \frac{N_i}{g_i} + N_i \ln \frac{g_i}{N_i}$$

$$= Nk + k \sum N_i \ln \frac{g_i}{N_i}$$

Making use of the equilibrium distribution, Eq. (8-14),

$$S = Nk + k \sum N_i \left(\ln \frac{Z}{N} + \beta \epsilon_i \right)$$

The total energy is $U = \Sigma \epsilon_i N_i$ and $\beta = 1/kT$, so that

$$S = Nk + Nk \ln \frac{Z}{N} + \frac{U}{T} \tag{8-15}$$

We thus have expressed the entropy in terms of the partition function and the internal energy of the system.

For an isolated system (constant total energy and number of particles), the partition function depends only on the temperature, since g_i and ϵ_i are established by quantum mechanical considerations. The other variable in the partition function is $\beta = 1/kT$, and the rate of change of the partition function with temperature is

$$\left(\frac{\partial Z}{\partial T} \right)_V = \sum - g_i \epsilon_i e^{-\beta \epsilon_i} \left(\frac{\partial \beta}{\partial T} \right)_V$$

$$= \frac{1}{kT^2} \sum g_i \epsilon_i e^{-\beta \epsilon_i} \tag{8-16}$$

The total internal energy may now be expressed as

$$U = \sum \epsilon_i N_i = \frac{N}{Z} \sum g_i \epsilon_i e^{-\beta \epsilon_i} \tag{8-17}$$

Comparing Eq. (8-16) with (8-17), we obtain

$$U = \frac{NkT^2}{Z} \left(\frac{\partial Z}{\partial T} \right)_V = NkT^2 \left(\frac{\partial \ln Z}{\partial T} \right)_V \tag{8-18}$$

Equation (8-15) for the entropy may now be rewritten as

$$S = Nk \left[1 + \ln \frac{Z}{N} + T \left(\frac{\partial \ln Z}{\partial T} \right)_V \right] \tag{8-19}$$

The differentiation is performed at constant volume because we are considering an isolated system with a constant number of particles.

We now have the important result that both the entropy and internal energy of the system for $g_i \gg N_i$ can be expressed in terms of a single property Z, the value of which may be calculated from microscopic considerations. Other thermodynamic properties may also be expressed in terms of the partition function, e.g., the Helmholtz function

$$A = U - TS = -NkT \left(1 + \ln \frac{Z}{N}\right) \qquad (8\text{-}20)$$

According to Eq. (7-16a), the pressure is expressible as

$$p = -\left(\frac{\partial A}{\partial V}\right)_T$$

Performing the indicated differentiation on Eq. (8-20) gives

$$p = NkT \left(\frac{\partial \ln Z}{\partial V}\right)_T \qquad (8\text{-}21)$$

The enthalpy and Gibbs function are now written immediately as

$$H = U + pV = NkT \left[T \left(\frac{\partial \ln Z}{\partial T}\right)_V + V \left(\frac{\partial \ln Z}{\partial V}\right)_T\right] \qquad (8\text{-}22)$$

$$G = A + pV - = -NkT \left[1 + \ln \frac{Z}{N} - V \left(\frac{\partial \ln Z}{\partial V}\right)_T\right] \qquad (8\text{-}23)$$

We now have related the partition function to all macroscopic thermodynamic properties of interest for simple substances.

The reader may find it interesting to muse over the developments relating the partition function to macroscopic thermodynamic properties. As the connecting link between these properties and microscopic analysis the partition function and its calculation form a substantial part of all statistical thermodynamic analysis. To aid in this reflection we may mention a few points of interest:

1   To link microscopic analysis with the macroscopic world, we must introduce the concept of temperature. This is done through Eq. (8-5), which is obtained from purely macroscopic considerations.
2   It is necessary to identify the behavior of the thermodynamic probability with the behavior of a quantity called the entropy. Otherwise the temperature definition could not be applied.
3   To define temperature, only systems in equilibrium may be considered. Otherwise it is difficult to attach a physical meaning to the temperature obtained from $\beta = 1/kT$.

8-4   THE IDEAL MONATOMIC GAS

As the first example of an application of statistical thermodynamic analysis, let us consider a simple monatomic gas whose molecules are so widely spaced that molecular force fields do not exert a significant influence on collision processes or distribution of energy. For this system we assume that the molecules represent point masses, so that their energy may be expressed as translational kinetic energy. The molecules are assumed to be indistinguishable and the translational kinetic energies are quantized. We, therefore, choose the BE statistical model,

$$\frac{N_i}{g_i} = \frac{1}{Ae^{\beta\epsilon_i} - 1} \tag{8-24}$$

If the gas is at a high temperature (room temperature is sufficiently high), there will be many more quantum states available than there are particles to fill them,[†] and the equilibrium distribution is given by

$$\frac{N_i}{N} = \frac{1}{Z} g_i e^{-\beta\epsilon_i} \tag{8-25}$$

Our problem is thus one of determining the values of g_i and ϵ_i so that the partition function may be calculated. Once the partition function is obtained, other thermodynamic properties may be evaluated.

The energies ϵ_i may be expressed by

$$\epsilon_i = \tfrac{1}{2}m(v_x^2 + v_y^2 + v_z^2) \tag{8-26}$$

where v_x, v_y, and v_z are the velocity components of the particles. Now consider an elemental volume in phase space, $dx\,dy\,dz\,dp_x\,dp_y\,dp_z$, where $p = mv$ is the particle momentum. According to the uncertainty principle, the accuracy of specification of momentum and displacement is limited by

$$\Delta x\,\Delta p_x \sim h$$

This means that the smallest discernible volume in phase space is h^3. The number of these small volumes in some larger volume element represents the degeneracy or the number of ways a certain energy level may occur. Thus we take

$$g_i = \frac{dx\,dy\,dz\,dp_x\,dp_y\,dp_z}{h^3} \tag{8-27}$$

In this analysis we assume that the volume element in phase space is large enough to contain a large number of particles so that the statistics will apply. Using Eqs. (8-25) and (8-27), the expression for the partition function is obtained as

[†] See discussion of Sec. 5-7.

$$Z = \sum \frac{m^3}{h^3} \exp\left[\frac{-m\beta}{2}(v_x^2 + v_y^2 + v_z^2)\right] dx\, dy\, dz\, dv_x\, dv_y\, dv_z \qquad (8\text{-}28)$$

where $p_x = mv_x$, etc., have been substituted for the momenta. The sum in Eq. (8-28) must be extended over six dimensions. If the energy states are sufficiently close spaced, and we assume they are for gases at high temperatures, the sum may be replaced by an integral. Thus

$$Z = \int\int\int\int\int\int \frac{m^3}{h^3} \exp\left[\frac{-m\beta}{2}(v_x^2 + v_y^2 + v_z^2)\right] dx\, dy\, dz\, dv_x\, dv_y\, dv_z$$

The triple integral over the three space coordinates is simply the total system volume V, and the exponential integrals involving the velocity components are all identical, so that the expression for the partition function becomes

$$Z = \frac{m^3V}{h^3}\left[\int_{-\infty}^{\infty} \exp\left(\frac{-\beta m v_x^2}{2}\, dv_x\right)\right]^3$$
$$= \frac{m^3V}{h^3}\left[2\int_{0}^{\infty} \exp\left(\frac{-\beta m v_x^2}{2}\, dv_x\right)\right]^3 \qquad (8\text{-}29)$$

The upper limit of this integral obviously has no physical meaning, but so few particles have very high velocities that no appreciable error is involved. In other words, the exponential function approaches zero so rapidly for large values of v_x that the integral is a suitable representation of the true physical situation.

The integral in Eq. (8-29) may be evaluated to give

$$Z = \frac{m^3V}{h^3}\left(\frac{2\pi}{m\beta}\right)^{3/2}$$

Letting $\beta = 1/kT$ gives

$$Z = V\left(\frac{2\pi mkT}{h^2}\right)^{3/2} \qquad (8\text{-}30)$$

Inserting Eqs. (8-30) and (8-27) into Eq. (8-24) results in the energy or velocity distribution

$$\frac{d^6N}{N} = \frac{m^3}{V}\left(\frac{1}{2\pi mkT}\right)^{3/2} e^{-mv^2/2kT}\, dx\, dy\, dz\, dv_x\, dv_y\, dv_z \qquad (8\text{-}31)$$

The symbol d^6N represents the number of particles in a six-dimensional differential element, as indicated in the equation. The symbol d^3N will be used to designate the number of particles in a three-dimensional differential element.

In this model of a monatomic gas, the gas is assumed to fill uniformly the system volume so that the molecular density (molecules per unit volume) is constant. Thus Eq. (8-31) may be integrated over the system volume to give

$$\frac{d^3N}{N} = \left(\frac{m}{2\pi kT}\right)^{3/2} e^{-mv^2/2kT} \, dv_x \, dv_y \, dv_z \tag{8-32}$$

Equation (8-32) is normally called the Maxwell-Boltzmann velocity distribution because the limiting case of BE statistics $(g_i \gg N_i)$ corresponds to that obtained from the MB model.

The internal energy of the monatomic gas may now be evaluated with Eq. (8-18). We have, from Eq. (8-30),

$$\ln Z = \ln \left[V \left(\frac{2\pi mk}{h^2}\right)^{3/2} \right] + \tfrac{3}{2} \ln T$$

$$\left(\frac{\partial \ln Z}{\partial T}\right)_V = \frac{3}{2} \frac{1}{T}$$

and

$$U = \tfrac{3}{2}NkT \tag{8-33}$$

Equation (8-33) is a very important result because it shows that the internal energy for this ideal monatomic gas is a function only of temperature.

We may derive the equation of state for the ideal monatomic gas very quickly by referring back to Eq. (8-21):

$$p = NkT \left(\frac{\partial \ln Z}{\partial V}\right)_T$$

or

$$p = NkT \frac{1}{V}$$

Thus

$$pV = NkT \tag{8-34}$$

This relation is identical to the perfect gas relation of Chap. 1 [Eq. (1-30)] if we identify

$$Nk = \eta \mathcal{R} \tag{8-35}$$

where η is the number of moles and $\mathscr{R}$ is the universal gas constant. The total number of molecules N is

$$N = \eta N_0$$

where N_0 is the number of molecules per mole, or Avogadro's number. Then from Eq. (8-35) we have the result that

$$k = \frac{\mathscr{R}}{N_0} \qquad (8\text{-}36)$$

where k is called the *gas constant per molecule*, or the *Boltzmann constant*, having the numerical value 1.3807×10^{-23} J/K·molecule.

Inserting the relations for internal energy and the partition function into Eq. (8-15) results in an expression for the entropy of the ideal monatomic gas:

$$S = Nk(\tfrac{5}{2} + \tfrac{3}{2}\ln 2\pi mk - \ln h^3 + \ln V + \tfrac{3}{2}\ln T - \ln N) \qquad (8\text{-}37)$$

Specific heats for the ideal monatomic gas may be obtained immediately with

$$C_V = \left(\frac{\partial U}{\partial T}\right)_V = \tfrac{3}{2}Nk = \tfrac{3}{2}\eta\mathscr{R}$$

or, on a molal basis,

$$\bar{c}_v = \tfrac{3}{2}\mathscr{R} \qquad (8\text{-}38)$$

Then

$$\bar{c}_p = \bar{c}_v + \mathscr{R} = \tfrac{5}{2}\mathscr{R} \qquad (8\text{-}39)$$

These specific heat relations are in excellent agreement with experiment for monatomic gases under pressure and temperature conditions which satisfy the original assumptions in the analysis, i.e., sufficiently low density so that the molecules are far apart and molecular force fields do not strongly influence molecular energy distributions.

8-5   DIATOMIC AND POLYATOMIC GASES

The major difference between monatomic and polyatomic gases is that the latter have rotational and vibrational energy states as well as translational energies based on the motion of the center of mass of the molecule. Still, if the molecules are sufficiently far apart, i.e., the gas density is low enough, the interaction force fields between molecules will not exert a significant influence on the energy distribution of the total system. As before, we assume $g_i \gg N_i$ and employ the limiting case of the BE statistics to calculate the thermodynamic properties. The problem reduces to that of determining the various energy levels and degeneracies to evaluate the system partition function. We assume that the total energy of a molecule may be expressed by the sum

$$\epsilon_{total} = \epsilon_{trans} + \epsilon_{rot} + \epsilon_{vib}$$

The partition function is then written as

$$Z = \Sigma(g_{trans}e^{-\beta\epsilon_{trans}})(g_{rot}e^{-\beta\epsilon_{rot}})(g_{vib}e^{-\beta\epsilon_{vib}})$$

if it is possible to assume that each energy mode is independent of the others. Then the sums may be taken separately and

$$Z = Z_{trans}Z_{rot}Z_{vib}$$

where

$$Z_{trans} = \Sigma g_{trans}e^{-\beta\epsilon_{trans}}$$
$$Z_{rot} = \Sigma g_{rot}e^{-\beta\epsilon_{rot}}$$
$$Z_{vib} = \Sigma g_{vib}e^{-\beta\epsilon_{vib}}$$

The assumption that various energy modes are independent is not strictly correct. Consider a diatomic dumbbell molecule, as shown in Fig. 8-1. The molecule has a translational energy expressed in terms of the motion of the center of mass, and a rotational energy which may be expressed in terms of the moment of inertia of the molecule about the center of mass. The vibrational energy results from vibration of the atoms in the force field acting between their centers. However, when these atoms vibrate, there can be a corresponding change in the moment of inertia of the molecule and, consequently, a change in the rotational energy. This is not usually an important consideration, except at high temperatures, and requires an analysis much too complicated for our study. At sufficiently high temperatures we should also consider changes in electron energy levels within the atoms.

For a diatomic dumbbell molecule like that shown in Fig. 8-1, a quantum mechanical analysis shows that the allowable rotational energy levels are given by

$$\epsilon_j = j(j + 1) \frac{h^2}{8\pi^2 I} \qquad j = 0, 1, 2, \ldots \tag{8-40}$$

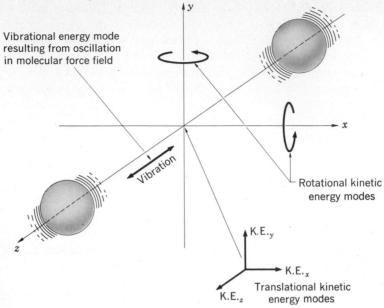

FIG. 8-1 *Energy modes of a diatomic dumbbell molecule.*

where I is the moment of inertia of the molecule about the center of mass. The degeneracy of the rotational energy levels can be shown to be

$$g_{rot} = 2j + 1 \tag{8-41}$$

We have used the index j to avoid confusion with the index i in the translational partition function. The rotational partition function is now written as

$$Z_{rot} = \sum_{j=0,1,2,...} (2j + 1) \exp\left[-j(j + 1)\frac{h^2}{8\pi^2 IkT}\right] \tag{8-42}$$

We now define the *characteristic temperature for rotation,* Θ_{rot}, as

$$\Theta_{rot} = \frac{h^2}{8\pi^2 Ik} \tag{8-43}$$

so that the expression for the partition function becomes

$$Z_{rot} = \sum_{j=0,1,2,...} (2j + 1) \exp\left[-j(j + 1)\frac{\Theta_{rot}}{T}\right] \tag{8-44}$$

Let us examine the behavior of Eq. (8-44) as a function of the ratio Θ_{rot}/T. For very low temperatures compared to Θ_{rot}, that is, $T \ll \Theta_{rot}$, the exponential function is very small and, hence, the value of Z_{rot} is very small. When T is of the same order as Θ_{rot}, the partition function undergoes rapid changes with T. For $T/\Theta_{rot} > 1$, Eq. (8-44) may be expanded to give

$$Z_{rot} = \frac{T}{\Theta_{rot}} \left[1 + \frac{1}{3} \left(\frac{\Theta_{rot}}{T} \right) + \frac{1}{15} \left(\frac{\Theta_{rot}}{T} \right)^2 + \cdots \right] \tag{8-45}$$

As the temperature becomes very large compared with Θ_{rot}, the higher-order terms become negligible and

$$Z_{rot} = \frac{T}{\Theta_{rot}} \qquad \text{for } \frac{T}{\Theta_{rot}} \gg 1 \tag{8-46}$$

We may now employ Eq. (8-46) to evaluate the rotational contribution to internal energy at high temperatures compared with Θ_{rot}. There results

$$U_{rot} = NkT \qquad \text{for } \frac{T}{\Theta_{rot}} \gg 1 \tag{8-47}$$

The molal specific heat for rotation then becomes

$$\bar{c}_{v,rot} = \mathcal{R} \qquad \text{for } \frac{T}{\Theta_{rot}} \gg 1 \tag{8-48}$$

To compute the specific heat for conditions other than the limiting one given by Eq. (8-48), it is necessary to evaluate the partition function from Eq. (8-44). The results of such a calculation are shown in Fig. 8-2. As expected, the rotational contributions to specific heat are small for $T/\Theta_{rot} \ll 1$ and approach a constant value of $\mathcal{R}$ at high temperatures compared with the characteristic temperature.

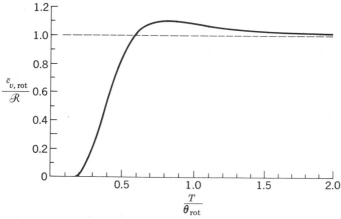

FIG. 8-2 *Rotational specific heat for a diatomic gas.*

To compute the vibrational partition function, an analysis similar to the previous one is performed. From quantum mechanics it is found that the allowable vibrational energy levels for a diatomic molecule are given by

$$\epsilon_{vib} = (n + \tfrac{1}{2})h\nu \qquad n = 0, 1, 2, \ldots \tag{8-49}$$

where ν is a characteristic frequency of vibration. It can be shown that each vibrational energy level may occur only one way so that the degeneracy is unity (or the vibrational energy states are nondegenerate). Equation (8-49) is obtained by assuming that the vibrating diatomic molecule behaves as a one-dimensional simple harmonic oscillator, and it is well to remark that this assumption may not be valid at very high temperatures.

The vibrational partition function is now computed with $g_{vib} = 1.0$:

$$Z_{vib} = \sum_{n=0}^{\infty} \exp\left[-\frac{(n + \tfrac{1}{2})h\nu}{kT}\right] \tag{8-50}$$

This sum may fortunately be expressed in closed form. Defining the *characteristic temperature for vibration*, Θ_{vib}, as

$$\Theta_{vib} = \frac{h\nu}{k} \tag{8-51}$$

the final relation for the partition function is

$$Z_{vib} = \frac{e^{-\Theta_{vib}/2T}}{1 - e^{-\Theta_{vib}/T}} \tag{8-52}$$

Using Eq. (8-18), the vibrational contribution to internal energy may be obtained as

$$U_{vib} = Nk\Theta_{vib}\left(\frac{1}{2} + \frac{1}{e^{\Theta_{vib}/T} - 1}\right) \tag{8-53}$$

The vibrational contribution to the molal specific heat then becomes

$$\bar{c}_{v,vib} = \left(\frac{\partial \bar{u}_{vib}}{\partial T}\right)_V = \mathcal{R}\left(\frac{\Theta_{vib}}{T}\right)^2 \frac{e^{\Theta_{vib}/T}}{(e^{\Theta_{vib}/T} - 1)^2} \tag{8-54}$$

Figure 8-3 gives a graphical display of the behavior of $\bar{u}_{vib}$ and $\bar{c}_{v,vib}$ as a function of the ratio T/Θ_{vib}. The behavior of specific heat is similar to that obtained for the rotational contribution. At very low temperatures compared with Θ_{vib}, the vibrational contribution to specific heat is negligible. At high temperatures it approaches a constant value of $\mathcal{R}$. The behavior of vibrational internal energy, however, is quite different from that of rotational

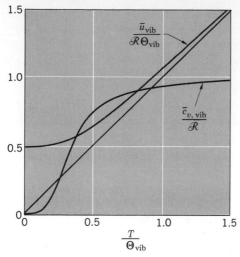

FIG. 8-3 *Vibrational specific heat and internal energy for a diatomic gas.*

internal energy. An inspection of Eq. (8-40) shows that the lowest rotational energy state is zero. This is not the case for vibration, as may be seen from Eq. (8-49). Therefore, as temperature is reduced, it is not surprising to find that the vibrational internal energy does not approach zero.

A summary of values of Θ_{rot} and Θ_{vib} for various gases is given in Table 8-1. It may be noted that the characteristic temperatures for rotation are substantially lower than those for vibration. This means that essentially

TABLE 8-1
Characteristic temperatures for rotation and vibration for various gases

Substance	Θ_{vib}, K	Θ_{rot}, K
H_2	6140	85.5
OH	5360	27.5
HCl	4300	15.3
CH	4100	20.7
N_2	3340	2.86
HBr	3700	12.1
HI	3200	9.0
CO	3120	2.77
NO	2740	2.47
O_2	2260	2.09
Cl_2	810	0.347
Br_2	470	0.117
Na_2	230	0.224
K_2	140	0.081

all of the rotational energy modes for a diatomic molecule will be excited before there is any appreciable change in the vibrational energy.

We are now in a position to predict the behavior of the total specific heat of a diatomic gas. Since the energies are additive, we may write

$$\bar{c}_v = \bar{c}_{v,\text{trans}} + \bar{c}_{v,\text{rot}} + \bar{c}_{v,\text{vib}} \qquad (8\text{-}55)$$

At low temperatures compared with Θ_{rot}, the essential contribution to specific heat results from translational energy modes, so that we would expect

$$\bar{c}_v = \tfrac{3}{2}\mathscr{R} \qquad T \ll \Theta_{\text{rot}}$$

As the temperature is raised, rotational energy modes are excited and the specific heat is increased corresponding to the behavior shown in Fig. 8-4. Once the temperature is raised substantially above Θ_{rot}, but still much below Θ_{vib}, the specific heat approaches

$$\bar{c}_v = \tfrac{3}{2}\mathscr{R} + \mathscr{R} = \tfrac{5}{2}\mathscr{R} \qquad \Theta_{\text{rot}} \ll T \ll \Theta_{\text{vib}}$$

Finally at much higher temperatures, vibrational energy states are excited and the specific heat is increased further until it approaches (see also Fig. 8-4)

$$\bar{c}_v = \tfrac{3}{2}\mathscr{R} + \mathscr{R} + \mathscr{R} = \tfrac{7}{2}\mathscr{R} \qquad \Theta_{\text{vib}} \ll T$$

Constant-pressure specific heats may be computed from

$$\bar{c}_p = \bar{c}_v + \mathscr{R} \qquad (8\text{-}56)$$

as long as ideal-gas behavior is observed.

This rather involved analysis may be put in quick perspective by examining the specific heat behavior of a familiar gas like N_2 at room temperature

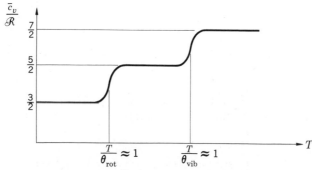

FIG. 8-4 *Approximate behavior of total specific heat of a diatomic gas.*

TABLE 8-2
Comparison of calculated specific heats of gases with experiment for $\Theta_{rot} \ll T \ll \Theta_{vib}$

Gas	γ Calc.	γ Exp.	$\frac{\bar{c}_p}{\mathcal{R}}$ Calc.	$\frac{\bar{c}_p}{\mathcal{R}}$ Exp.	$\frac{\bar{c}_v}{\mathcal{R}}$ Calc.	$\frac{\bar{c}_v}{\mathcal{R}}$ Exp.	$\frac{\bar{c}_p - \bar{c}_v}{\mathcal{R}}$ Calc.	$\frac{\bar{c}_p - \bar{c}_v}{\mathcal{R}}$ Exp.
A	1.667	1.67	2.500	2.52	1.500	1.509	1.000	1.008
He		1.659		2.52		1.519		1.001
CO	1.400	1.404	3.500	3.50	2.500	2.488	1.000	1.005
H_2		1.410		3.42		2.438		0.999
N_2		1.404		3.50		2.448		1.005
O_2		1.401		3.52		2.504		1.004

(~ 300 K). According to Table 8-1, the rotational energy levels are fully excited while there is essentially no excitation of vibrational energy states. Thus

$$\bar{c}_v = \tfrac{5}{2}\mathcal{R}$$
$$\bar{c}_p = \tfrac{7}{2}\mathcal{R}$$

The ratio of specific heats is

$$\gamma = \frac{\bar{c}_p}{\bar{c}_v} = \frac{7}{5} = 1.40$$

Table 8-2 gives a convenient comparison of specific heat values calculated with the foregoing relations and values obtained experimentally. The agreement is excellent.

For temperatures in the range of $T \approx \mathcal{Q}_{rot}$ hydrogen is an exception to the above analysis, because nuclear spin makes a contribution to its specific heat. A detailed analysis is beyond the scope of our discussion, but we may note that there are two types of hydrogen present in this temperature regime: *para*hydrogen with antisymmetrical nuclear spin states, and *ortho*hydrogen with symmetrical nuclear spin states. Experiments indicate that for $T \approx \mathcal{Q}_{rot}$ the actual hydrogen is a mixture of three parts ortho- to one part parahydrogen, and that rotational specific heats based on Fig. 8-2 will be too high. For further information the interested reader should consult Ref. [4].

EXAMPLE 8-1   Rotational energy levels
Calculate the values of the first three rotational energy levels for nitrogen.

SOLUTION   The rotational energy levels are given by Eq. (8-40):

$$\epsilon_j = j(j + 1) \frac{h^2}{8\pi^2 I} \qquad (a)$$

Using Table 8-1 the characteristic temperature for rotation is

$$\Theta_{rot} = 2.86 \text{ K} = \frac{h^2}{8\pi^2 I k} \qquad (b)$$

so that

$$\frac{h^2}{8\pi^2 I} = (2.86)(1.38 \times 10^{-23}) = 3.95 \times 10^{-23} \text{ J/molecule}$$

The first three nonzero energy levels are therefore obtained from (a) as

$$\epsilon_1 = (1)(1 + 1)(3.95 \times 10^{-23}) = 7.9 \times 10^{-23} \text{ J/molecule}$$
$$\epsilon_2 = (2)(2 + 1)(3.95 \times 10^{-23}) = 2.37 \times 10^{-22} \text{ J/molecule}$$
$$\epsilon_3 = (3)(3 + 1)(3.95 \times 10^{-23}) = 4.74 \times 10^{-22} \text{ J/molecule}$$

EXAMPLE 8-2

Calculate the specific heat of nitrogen (N_2) at 2000 K.

SOLUTION Consulting Table 8-1, we find that $T \gg \Theta_{rot}$ so that the rotational energy states are fully excited. The vibrational states are only partially excited, so we compute

$$\frac{T}{\Theta_{vib}} = \frac{2000}{3340} = 0.599$$

From Fig. 8-3 we have

$$\bar{c}_{v,vib} = 0.85 \mathscr{R}$$

and the total specific heat becomes

$$\bar{c}_{v,total} = \tfrac{3}{2}\mathscr{R} + \mathscr{R} + 0.85\mathscr{R}$$
$$= 3.35\mathscr{R}$$

Polyatomic molecules are much more complicated than the simple diatomic model considered in the foregoing, although the basic notion for evaluating partition functions and specific heats is the same. In particular, much more complicated rotational and vibrational energy modes are encountered and require a more tedious evaluation procedure for the partition function.

8-6 EQUIPARTITION OF ENERGY

Our analysis of monatomic and diatomic gases has yielded rather interesting information on the specific heat behavior of ideal gases. In this analysis, several energy modes were allowed, and we shall say that for each energy mode there is a *degree of freedom*. There are three translational energy modes, so there are three translational degrees of freedom. In the diatomic gas the molecule can rotate in three directions; however, the moment of inertia about the axis through the centers of the two atoms is negligibly small compared with the moments of inertia about the other two Cartesian axes passing through the center of mass. Thus we say that the molecule has two important rotational energy modes. The vibrational energy modes of a diatomic molecule result from oscillation along the axis joining the two atoms. To specify this vibrational energy, we would need to specify both the kinetic energy of the oscillating atoms and their potential energy resulting from position in the force field. Consequently, we say that there are two vibrational degrees of freedom.

A quick inspection of the preceding analysis shows that for each *fully excited* degree of freedom a contribution of $\frac{1}{2}Nk$ to the specific heat is observed. This is the essential content of the *equipartition of energy principle*. Note that this contribution to specific heat is independent of the mass of the particles.

The equipartition of energy principle is of limited value because we can never be sure when it applies without consulting tabular information such as shown in Table 8-1. A somewhat general analytical specification of the principle may be given in the following way. Suppose that the energy associated with a degree of freedom is expressible as a quadratic function of the coordinate specifying the particle position, velocity, etc.,

$$\epsilon_x = ax^2$$

Furthermore, let us assume the distribution among energy states is given by the MB exponential formula:

$$dN_x = e^{-\beta \epsilon_x}\, dx$$

where we are assuming a continuous distribution of energy states. The average energy associated with the degree of freedom is, therefore,

$$\bar{\epsilon}_x = \frac{\int \epsilon_x\, dN_x}{\int dN_x} = \frac{\displaystyle\int_0^\infty ax^2\, e^{-ax^2/kT}\, dx}{\displaystyle\int_0^\infty e^{-ax^2/kT}\, dx}$$

$$\bar{\epsilon}_x = \tfrac{1}{2}kT \tag{8-57}$$

This is the average energy per particle per degree of freedom. The total energy associated with the x mode would be $\frac{1}{2}NkT$, and the specific heat contribution would be $\frac{1}{2}Nk$.

There are several restrictive assumptions in the simple analysis of the preceding paragraphs. A quadratic energy function, a continuous distribution of energy, and the MB type of exponential distribution formula are all necessary to obtain the $\frac{1}{2}kT$ result. The translational energy modes for a monatomic gas would fulfill these conditions because of the v_x^2 energy dependence and an almost continuous distribution of energy states. If rotational energy modes behaved in a "classical" manner such that $\epsilon_{rot} \sim \frac{1}{2}I\omega^2$, where ω is the angular velocity, we should expect equipartition to hold in this case also. As already observed, however, this behavior is experienced only at temperatures which are high compared with Θ_{rot}.

8-7   THE SOLID STATE

Our previous discussions have considered the application of statistical thermodynamics to systems where the molecular spacing is very large compared with the size of the molecules. Because of this wide spacing, it was reasonable to assume that the molecules were free to move as independent entities, with their energy essentially unaffected by the force fields of neighbor molecules. The solid state represents an entirely different circumstance. In this instance the atoms are so closely spaced that their force fields essentially lock them in place, so that they are only free to oscillate about some equilibrium position under the influence of the force fields of neighboring atoms. One of the problems of a statistical analysis of the solid state is to be able to describe satisfactorily the contribution of this oscillatory motion to the internal energy and specific heat of the solid.

If the solid is an electric conductor, an appreciable number of electrons are relatively free to move about and transport electric charge. These free electrons are usually termed the *electron gas*. Since the electrons have motion and kinetic energy, they would be expected to contribute to the total internal energy and specific heat of the solid. An additional part of the statistical analysis of the solid state is, therefore, addressed to a specification of the energy distribution in the electron gas for a solid. This information is then combined with that concerning the lattice vibration energy to predict the overall energy behavior of the solid. For electric insulators it is expected that the electron-gas contribution to the internal energy of the solid would be small.

Let us first consider the energy contribution resulting from lattice vibrations. The first reasonably satisfactory analysis of this problem was performed by Einstein who assumed that the forces controlling the vibration of each atom produced a simple harmonic oscillator which could be analyzed with quantum mechanical methods to determine the allowable energy levels. The allowable energy levels are given by the same relation employed for analysis of vibrational energies in diatomic molecules:

$$\epsilon = (n + \tfrac{1}{2})h\nu \qquad n = 0, 1, 2, \ldots$$

where ν is the frequency of the oscillator. It is assumed that only one quantum state per energy level is allowed, so that the partition function is given as before by

$$Z = \sum_{n=0}^{\infty} \exp\left[- \left(n + \frac{1}{2}\right) \frac{h\nu}{kT}\right] = \frac{e^{-\Theta_E/2T}}{1 - e^{-\Theta_E/T}} \tag{8-58}$$

where, now, $\Theta_E = h\nu/K$ is called the *Einstein temperature* of the solid.

The total number of atoms in the solid is taken as N; however, it is expected that there will be three harmonic oscillators for each atom (one for each space coordinate), so that the total number of oscillators to be considered is $3N$. The internal energy is thus written as

$$U = 3NkT^2 \left(\frac{\partial \ln Z}{\partial T}\right)_V \tag{8-59}$$

Combining this relation with Eq. (8-58) gives

$$U = \frac{3}{2} NkT \frac{\Theta_E}{T} + 3NkT \frac{\Theta_E/T}{e^{\Theta_E/T} - 1} \tag{8-60}$$

The molal constant-volume specific heat is, thus,

$$\bar{c}_v = 3\mathscr{R} \left(\frac{\Theta_E}{T}\right)^2 \frac{e^{\Theta_E/T}}{(e^{\Theta_E/T} - 1)^2} \tag{8-61}$$

This relation for specific heat is plotted in Fig. 8-5 as a function of T/Θ_E. At very high temperatures the specific heat approaches a constant value of $3\mathscr{R}$. This value is exactly what would be expected from a simple equipartition of energy appraisal of the problem. The atoms have three directions of motion, and there are two vibrational energy modes for each coordinate, so that a total of six degrees of freedom are expected. Each degree of freedom contributes $\tfrac{1}{2}\mathscr{R}$ to the specific heat; thus the total of $3\mathscr{R}$ is quite reasonable. The shape of the curve in Fig. 8-5 is in general agreement with experimental data for specific heats of solids but predicts values for $\bar{c}_v$ which are too low at low temperatures.

The main fault in the Einstein analysis is that all atoms in a solid simply do not oscillate with the same frequency as is assumed for the calculation. This difficulty was alleviated by Debye, who assumed a continuous distribution of frequencies which could be computed from elastic-wave theory for

FIG. 8-5 *Specific heat of a solid according to Einstein theory compared with experimental values.*

isotropic media. We shall not present the rather cumbersome analysis but only state the result. For the internal energy, Debye obtained

$$U - U_0 = \frac{9NkT}{(\Theta_D/T)^3} \int_0^{\Theta_D/T} \frac{x^3}{e^x - 1} \, dx \qquad (8\text{-}62)$$

where U_0 is the internal energy of the lattice at $T = 0$. This expression may be differentiated to obtain the specific heat as

$$\bar{c}_v = 3\mathcal{R} \left[\frac{12}{(\Theta_D/T)^3} \int_0^{\Theta_D/T} \frac{x^3 \, dx}{e^x - 1} - \frac{3(\Theta_D/T)}{e^{\Theta_D/T} - 1} \right] \qquad (8\text{-}63)$$

In these expressions Θ_D is called the *Debye temperature* and is given by

$$\Theta_D = \frac{h\nu_m}{k} \qquad (8\text{-}64)$$

where ν_m is the maximum frequency encountered in the lattice vibrations. The integrals in Eqs. (8-62) and (8-63) must be evaluated numerically and results of such an evaluation are shown in Fig. 8-6. This curve is in excellent agreement with experiment, as indicated by the experimental data points. Values of the Debye temperature Θ_D for a number of solid substances are given in Table 8-3.

For high temperatures compared with Θ_D, the specific heat approaches

$$
\begin{array}{ll}
\times \ \text{Pb} & \Theta_D = 88 \\
\bullet \ \text{Ag} & 215 \\
\triangledown \ \text{KCl} & 230 \\
\square \ \text{Zn} & 235 \\
\diamond \ \text{NaCl} & 281 \\
\triangle \ \text{Cu} & 315 \\
+ \ \text{Al} & 398 \\
\circ \ \text{CaF}_2 & 474 \\
\blacktriangledown \ \text{C} & 1860
\end{array}
$$

FIG. 8-6 *Specific heat of a solid according to Debye theory and comparison with experimental values.*

$3\mathscr{R}$. For very low temperatures compared with Θ_D, the exponential term of the denominators in Eq. (8-63) dominates, and $\bar{c}_v$ is given approximately as

$$
\bar{c}_v \approx 3\mathscr{R} \left[\frac{12}{(\Theta_D/T)^3} \int_0^{\Theta_D/T} \frac{x^3 \, dx}{e^x} \right]
$$

$$
= \frac{12}{5} \mathscr{R} \frac{\pi^4}{(\Theta_D/T)^3} = 464.4 \left(\frac{T}{\Theta_D} \right)^3 \text{ cal/g·mol·K} \qquad T \ll \Theta_D \qquad (8\text{-}65)
$$

since $\mathscr{R} = 1.987$ cal/g·mol·K. Equation (8-65) is generally valid for $T < \Theta_D/10$.

TABLE 8-3
Debye temperatures for various solids

Substance	Range, K	Θ_D	Substance	Range, K	Θ_D
Lead	14–573	88	Silvium (KCl)	23–550	230
Thallium	23–301	96	Zinc	33–673	235
Mercury	31–232	97	Rock salt (NaCl)	25–664	281
Iodine	22–298	106	Copper	14–773	315
Cadmium	50–380	168	Aluminum	19–773	398
Sodium	50–240	172	Iron	32–95	453
Potassium bromide	79–417	177	Fluorspar (CaF$_2$)	17–328	474
Silver	35–873	215	Pyrites (FeS$_2$)	22–57	645
Calcium	22–62	226	Diamond	30–1169	1860

EXAMPLE 8-3
    Calculate the specific heat of aluminum at 200 K.

SOLUTION   From Table 8-3 we have $\Theta_D = 398$ K for aluminum. We compute

$$\frac{T}{\Theta_D} = \frac{200}{398} = 0.502$$

Consulting Fig. 8-6, we find

$$\bar{c}_v = 4.9 \text{ cal/g·mol·K} = 20.5 \text{ kJ/kg·mol·K}$$

EXAMPLE 8-4
    Calculate the specific heat of copper at 10 K.

SOLUTION   From Table 8-3, $\Theta_D = 315$ K for copper, so that $T \ll \Theta_D$. We may, therefore, use Eq. (8-65) to compute the specific heat:

$$\bar{c}_v = 464.4 \left(\frac{T}{\Theta_D}\right)^3$$

$$= 464.4(\tfrac{10}{315})^3$$
$$= 0.014\ 86 \text{ cal/g·mol·K} = 0.0622 \text{ kJ/kg·mol·K}$$

8-8   THE ELECTRON GAS

It will be recalled that our initial discussion of solids indicated that the free electrons in a metal, or electron gas, might contribute to the total energy of the solid and, hence, to the constant-volume specific heat. Assuming that these electrons do, indeed, behave like an ideal monatomic gas, then their contribution to the specific heat should be about $\tfrac{3}{2}\mathcal{R}$, as calculated in Sec. 8-4. But this apparently is not the case, because specific heats of both conductors and nonconductors are predicted quite nicely by the Debye theory alone. This theory, of course, does not consider the electrons in any way. In other words, the electrons apparently contribute nothing to the specific heat of the solid. This is a matter of serious concern, and we shall now see how the apparent anomaly is resolved.
    The electrons in a solid are not really "free" in the same sense as the atoms of a monatomic gas, even though they can move about. They are still bound to the atoms in the solid, even if in a very loose fashion. Their allowable energy levels are influenced by *all* the atoms in the solid, not by just one atom, as in the case of electrons attached to a gas atom. The basic statistical postulate concerning the electron gas is the *Pauli exclusion principle*, which

states that no more than two electrons may occupy the same quantum state. The factor of 2 arises because electrons may have a right- or left-handed spin. Note that the principle applies to quantum *state*, not energy *level*. It should be fairly easy to see that we are encountering a situation where the Fermi-Dirac statistics apply, i.e., a circumstance where only one particle may occupy each quantum state. (In Sec. 5-6 this meant no more than one ball per small box.) All we need to do is to introduce a factor of 2 to account for spin and proceed with the analysis. Still, we assume that the uncertainty principle applies, so that the degeneracy is given by

$$g_i = \frac{2}{h^3} \, dx \, dy \, dz \, dp_x \, dp_y \, dp_z \qquad (8\text{-}66)$$

The FD distribution was

$$\frac{N_i}{g_i} = \frac{1}{Ae^{\beta\epsilon} + 1} \qquad (8\text{-}67)$$

Let us now assume that there is a very large number of energy levels available to the electrons in a solid, so that the energy spectrum is almost continuous. Then, using Eqs. (8-66) and (8-67), we obtain

$$d^2N = \frac{2/h^3}{Ae^{\beta\epsilon} + 1} \, dx \, dy \, dz \, dp_x \, dp_y \, dp_z$$

Integrating over the system volume V gives

$$d^3N = \frac{2V/h^3}{Ae^{\beta\epsilon} + 1} \, dp_x \, dp_y \, dp_z \qquad (8\text{-}68)$$

We are now faced with the problem of evaluating the constant A. Let us make the substitution

$$A = e^{-\beta\mu_0} \qquad (8\text{-}69)$$

where μ_0 is a new constant. The FD distribution corresponding to Eq. (8-67) then becomes

$$\frac{N_i}{g_i} = \frac{1}{e^{(\epsilon_i - \mu_0)/kT} + 1} \qquad (8\text{-}70)$$

A plot of this relation is indicated in Fig. 8-7 for different temperatures. This plot reveals a rather remarkable result. At $T = 0$, all the energy states are completely filled up to an energy of μ_0. Above this energy none of the energy states is occupied. For higher temperatures (we have not yet said how

FIG. 8-7 *Fermi-Dirac distribution as a function of temperature.*

high), the energy distribution drops off, so that there are some energy states below μ_0 which are not completely filled and some above μ_0 which may be occupied. Still, the curves are approximately symmetric about $\epsilon = \mu_0$, so that for all these temperatures the total energy might be assumed to be approximately that observed at $T = 0$. The maximum energy of the electrons at $T = 0$ is μ_0, and this energy is called the *Fermi level*.

We now need to obtain an expression for μ_0. All of the electron energy is assumed to be translational kinetic energy, so that

$$\epsilon = \frac{1}{2} m_e v^2 = \frac{p^2}{2m_e} \tag{8-71}$$

where p is the momentum. The physical meaning of the plot in Fig. 8-7 is that energy space is uniformly populated at $T = 0$ up to an energy of μ_0. Above this energy there is no population. For this maximum energy there corresponds a maximum momentum

$$p_m^2 = 2m_e\mu_0 \tag{8-72}$$

Now the number of electrons per unit volume of momentum space at $T = 0$ is clearly given by Eq. (8-68) as

$$\rho_{eo} = \frac{2V}{h^3}$$

This number is a constant over all the applicable portion of momentum space, so to obtain the total number of electrons we multiply this electron density ρ_{eo} by the total volume of momentum space up to p_m. This volume is that of a sphere of radius p_m.

$$N_{eo} = \frac{2V}{h^3} (\tfrac{4}{3}\pi p_m^3) \tag{8-73}$$

Substituting Eq. (8-73) into Eq. (8-72), we obtain

$$\mu_0 = \frac{h^2}{8m_e} \left(\frac{3N_{e_0}}{\pi V} \right)^{2/3} \tag{8-74}$$

We now have an expression for the constant A at $T = 0$, and if we assume that this expression applies at higher temperatures, we may proceed to obtain more specific information about electron velocity distributions and contributions to specific heat. At temperatures above absolute zero it has been shown by Sommerfeld [9] that Eq. (8-69) may be used for the constant A if a parameter μ is substituted for μ_0, where

$$\mu = \mu_0 \left[1 - \frac{\pi^2}{12} \left(\frac{kT}{\mu_0} \right)^2 + \cdot \cdot \cdot \right] \tag{8-75}$$

We shall assume that all directions of electron motion are equally probable. The number density of electrons is, thus, uniform in a thin spherical shell of velocity space $4\pi v^2 \, dv$. This density is obtained from Eq. (8-68) as

$$\frac{d^3N}{dv_x \, dv_y \, dv_z} = \frac{2m^3V/h^3}{Ae^{\beta\epsilon} + 1}$$

Using $\epsilon = \frac{1}{2}mv^2$, the value for A, and multiplying by the volume of the spherical shell gives the velocity distribution as

$$dN_v = \frac{8\pi m^3 V}{h^3} = \frac{v^2 \, dv}{\exp \left[(\frac{1}{2}mv^2 - \mu)/kT \right] + 1} \tag{8-76}$$

This relation may be rearranged to give the energy distribution as

$$dN_\epsilon = 4\pi V \left(\frac{2m}{h^2} \right)^{3/2} \frac{\epsilon^{1/2} \, d\epsilon}{e^{(\epsilon - \mu)/kT} + 1} \tag{8-77}$$

In Eqs. (8-76) and (8-77) the factor

$$f_e = \frac{1}{e^{(\epsilon - \mu)/kT} + 1} \tag{8-78}$$

represents the fraction of available energy states which are occupied by electrons since, in the FD model, no more than one particle is allowed in each quantum state. We shall designate f_e as the *Fermi function*.

Figures 8-8 and 8-9 show plots of electron velocity and energy distributions as a function of temperature. At $T = 0$, ϵ is always less than μ_0 and the exponential term of Eq. (8-78) goes to zero, so that the energy distribution is given by

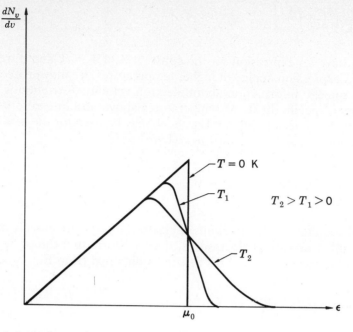

FIG. 8-8   *Velocity distribution for electron gas.*

$$dN_\epsilon = 4\pi V \left(\frac{2m}{h^2}\right)^{3/2} \epsilon^{1/2}\, d\epsilon \qquad \text{at } T = 0 \tag{8-79}$$

The *average* electron energy at $T = 0$ is

$$\bar{\epsilon}_0 = \frac{\displaystyle\int_0^{\mu_0} \epsilon\, dN_\epsilon}{\displaystyle\int_0^{\mu_0} dN_\epsilon} = \tfrac{3}{5}\,\mu_0 \tag{8-80}$$

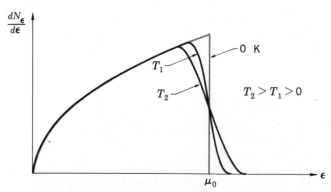

FIG. 8-9   *Energy distribution for electron gas.*

The upper limit for these integrals is μ_0 since this is the maximum electron energy at $T = 0$. At temperatures above absolute zero the average electron energy must be computed from

$$\bar{\epsilon} = \frac{\int_0^\infty \epsilon \, dN_\epsilon}{\int_0^\infty dN_\epsilon} \tag{8-81}$$

since, now, there is no upper limit on the electron energy levels. The integrations in Eq. (8-81) may be performed by substituting the energy distribution from Eq. (8-77) along with the Sommerfeld relation for μ from Eq. (8-75). The result is

$$\bar{\epsilon} = \tfrac{3}{5} \, \mu_0 \left[1 + \frac{5}{12} \, \pi^2 \left(\frac{kT}{\mu_0} \right)^2 + \cdots \right] \tag{8-82}$$

The internal energy is, then,

$$U = N\bar{\epsilon} = \tfrac{3}{5} \, N\mu_0 + \frac{1}{4} \frac{\pi^2 Nk^2}{\mu_0} T^2 \tag{8-83}$$

where N is the total number of electrons. The molal specific heat is then

$$\bar{c}_v = \left(\frac{\partial \bar{u}}{\partial T} \right)_v = \frac{\pi^2}{2} \frac{kT}{\mu_0} \mathcal{R} \tag{8-84}$$

If the electrons behaved as an ideal monatomic gas, we would obtain $\bar{c} = \tfrac{3}{2}\mathcal{R}$, a much larger value.

The foregoing analysis explains why it is unnecessary to consider the electron gas when computing specific heats of solids; their contribution is very small since $kT \ll \mu_0$ for all ordinary temperatures. Another way of stating this conclusion is that the temperature-dependent part of the electron internal energy in Eq. (8-83) is small compared with the total electron energy. This total energy, however, is quite large since $\mu_0 \gg kT$, as is illustrated in Example 8-5.

EXAMPLE 8-5 Fermi level for copper

Estimate the Fermi level for copper and compute the equivalent ideal-gas temperature for the free electrons based on their average energy at $T = 0$.

SOLUTION We shall assume that copper has one free electron per atom, so that

$$\frac{N_{eo}}{V} = \frac{N_0}{M} \frac{m}{V} = \frac{N_0}{M} \rho$$

where M is the molecular weight and ρ is the density. For copper, $M = 63.57$ and $\rho = 8.94$ g/cm^3, so that

$$\frac{N_{eo}}{V} = \frac{(6.025 \times 10^{23})(8.94)}{63.57}$$

$$= 8.47 \times 10^{22} \text{ electrons/cm}^3$$

We now use Eq. (8-74) to compute μ_0:

$$\mu_0 = \frac{h^2}{8m_e} \left(\frac{3N_{eo}}{\pi V}\right)^{2/3}$$

$$= \frac{(6.624 \times 10^{-27})^2}{(8)(9.1 \times 10^{-23})} \left[\frac{(3)(8.47 \times 10^{22})}{\pi}\right]^{2/3}$$

$$= 1.13 \times 10^{-11} \text{ erg} = 7.07 \text{ eV}$$

The average electron energy at $T = 0$ is given by Eq. (8-80) as

$$\bar{\epsilon}_0 = \tfrac{3}{5}\mu_0$$

$$= (0.6)(1.13 \times 10^{-11}) = 6.78 \times 10^{-12} \text{ erg}$$

$$= 6.78 \times 10^{-19} \text{ J}$$

To compute the equivalent ideal-gas temperature for the free electrons, we set

$$\bar{\epsilon}_0 = \tfrac{3}{2}kT_{equiv}$$

Solving for the equivalent temperature, we obtain

$$T_{equiv} = \frac{2\bar{\epsilon}_0}{3k}$$

$$= \frac{(2)(6.78 \times 10^{-19})}{(3)(1.38 \times 10^{-23})}$$

$$= 32\ 800 \text{ K}$$

EXAMPLE 8-6   Electron contribution to specific heat
 Calculate the electron contribution to specific heat of copper at 300 K (room temperature).

SOLUTION   This calculation may be performed with Eq. (8-84), and the value of μ_0 is obtained in Example 8-5. Thus,

$$\begin{aligned}
\bar{c}_{v,\text{elec}} &= \frac{\pi^2}{2} \frac{kT}{\mu_0} \mathcal{R} \\
&= \frac{\pi^2 (1.38 \times 10^{-23})(300)}{(2)(1.13 \times 10^{-18})} \mathcal{R} \\
&= 0.0181 \mathcal{R}
\end{aligned}$$

From this calculation we see that the electron contribution to specific heat is indeed very small.

Our brief discussion of the solid state has yielded several important bits of information.

1 Lattice vibration furnishes the primary contribution to specific heats of solids.
2 Specific heats of solids approach zero as the absolute temperature approaches zero.
3 The internal energy of solids does not approach zero as $T \rightarrow 0$. Not only is there potential energy stored in the force fields binding the atoms but also a considerable energy in electron kinetic energy. In fact, the kinetic energy of the electrons is much higher than that of an ordinary gas at room temperature.
4 Absolute zero of temperature does not represent a state where atomic and electronic motion stops.
5 The "free" electrons in a solid do not behave like a simple monatomic gas in that their specific heat may not be computed from the simple equipartition of energy principle.

8-9 CONDUCTORS AND SEMICONDUCTORS

Our analysis of the solid state up to this point has consisted of a determination of specific heats based on the Debye theory and on a demonstration that the electron gas makes no substantial contribution to specific heat. The analysis of the electron gas is based upon what might be called a *single-band model*; i.e., electron energies are primarily limited to the band between the zero energy level and the Fermi level. This is not an entirely realistic picture of a solid. A single atom may contain electrons in various energy levels which are governed by appropriate quantum mechanical relations. When a large number of atoms are forced into close proximity of one another, as in a solid, the electrons interact very strongly and occupy broadbands of allowable energy states instead of the nicely separated discrete values observed in a single atom. It can be shown by a rather involved quantum mechanical analysis that the overall result of the interaction process is to produce *allowable* and *forbidden energy bands* within the solid. An allowable energy band is one which may be occupied by electrons whereas a forbidden band is one which may not be occupied by electrons.

The lowest-level electron band is the *valence band* and results from interactions of the valence (outer shell) electrons in the single atoms. The next allowable band is called the *conduction band*. In general, the forbidden band separating these two allowable bands is very wide (i.e., $\gg kT$). There are other possible allowable bands, but these first two are of most interest.

The Fermi-Dirac distribution function gives the distribution of electrons among various *allowable* energy states in the solid. At $T = 0$, all electrons would occupy the valence band. The fraction of electrons which may be raised to the conduction band depends on T and the width of the forbidden band.

If the valence band is completely filled, the solid may not conduct unless sufficient energy is applied to move the electrons to make the jump and the material will act as an *insulator*.

If the forbidden band is narrow and the conduction band is empty, then it will be much easier for the electrons to cross from the valence band into the conduction band. In this instance the material is called a *semiconductor*.

If the conduction band is partially filled, then relatively little energy is required to raise the electrons to a slightly higher potential within the conduction band, i.e., to an unoccupied state at some higher energy level. In this case, electron motion is most easily influenced by an electric field, and we say that the material behaves as a *conductor*. Figures 8-10 and 8-11 illustrate the behavior described in the foregoing.

When an electron jumps the gap from the valence band to the conduction band, the vacant state created in the valence band is called a *hole*. The hole may remain vacant or be filled by another electron from some other energy site within the valence band. The number of electrons which move from the valence band to the conduction band increases with temperature and, since all electrons which occupy the conduction band come from the valence band, we assume that the number of electrons in the conduction band must equal the number of holes in the valence band.

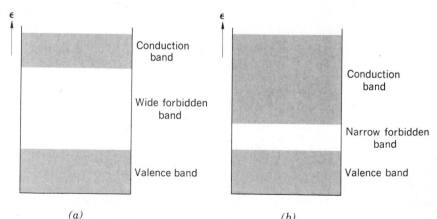

(a)   (b)

FIG. 8-10 *Band theory of (a) insulators and (b) semiconductors.*

FIG. 8-11 *Band theory of conductors.*

If conduction in the material results only from electron movement from valence to conduction band, it is called an *intrinsic semiconductor*. The performance characteristics of semiconductors may be changed appreciably by adding small amounts of impurities to the solid, which act as *acceptors* or *donors* of electrons in the forbidden gap, as illustrated in Fig. 8-12. An acceptor is an empty band at an energy level somewhat above the valence band. Since it provides additional "holes," such an impurity semiconductor is called a *p-type semiconductor*, where the *p* signifies *positive*. A donor is a filled band at an energy level somewhat below the conduction band. This type of impurity semiconductor provides additional negative charges (electrons), so it is called an *n-type semiconductor*.

The Fermi level may be related to the forbidden bandwidth in the following way. For convenience we take the energy level as zero at the top of the valence band and equal to ϵ_g at the bottom of the conduction band. It is assumed that every electron which makes the jump into the conduction band leaves a hole in the valence band. Thus

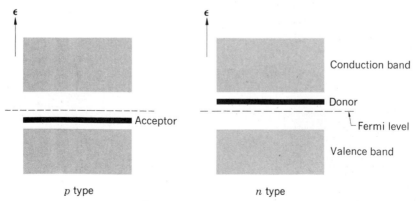

p type   n type

FIG. 8-12 *Impurity semiconductors.*

$$N_h \big|_{\text{valence}} = N_e \big|_{\text{conduction}} \tag{8-85}$$

where N_h represents the number of holes and N_e the number of electrons. The energy scale for this model is illustrated in Fig. 8-13. Because of the forbidden-band concept we are forced to require that the number of available quantum states for the electrons must be zero in this band. We then assume a FD distribution outside the forbidden band. This means that we must employ the following expression for g_i in the conduction band:

$$g_i \big|_{\text{conduction band}} = 4\pi V \left(\frac{2m_e}{h^2}\right)^{3/2} (\epsilon - \epsilon_g)^{1/2} \tag{8-86}$$

In other words, the allowable states start to build up from ϵ_g in the same way they built up from $\epsilon = 0$ for the single-band model. We still employ

$$f_e = \frac{1}{e^{(\epsilon - \mu)/kT} + 1} \tag{8-87}$$

for the FD energy distribution function.

The number of electrons in the conduction band is, thus,

$$N_e = 4\pi V \left(\frac{2m_e}{h^2}\right)^{3/2} \int_{\epsilon_g}^{\infty} \frac{(\epsilon - \epsilon_g)^{1/2} \, d\epsilon}{e^{(\epsilon - \mu)/kT} + 1} \tag{8-88}$$

Since the forbidden bandwidth is usually much greater than kT, it is expected that $(\epsilon - \mu) \gg kT$ in the conduction band, so that the one in the

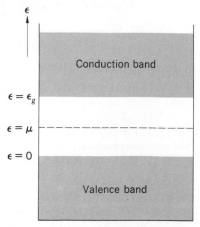

FIG. 8-13   *Energy scale for band analysis.*

denominator of Eq. (8-88) may be neglected and the integral may be evaluated with an appropriate variable substitution to give

$$N_e = V \left(\frac{2\pi m_e kT}{h^2}\right)^{3/2} e^{\mu/kT} e^{-\epsilon_g/kT} \tag{8-89}$$

The distribution function for the holes in the valence band is:

$$f_h = 1 - f_e = 1 - \frac{1}{e^{(\epsilon-\mu)/kT} + 1}$$

$$= \frac{e^{-(\mu-\epsilon)/kT}}{e^{-(\mu-\epsilon)/kT} + 1} \tag{8-90}$$

In the valence band it is expected that $(\mu - \epsilon) \gg kT$, so that the exponential in the denominator of Eq. (8-90) may be neglected. Then

$$f_h = e^{-(\mu-\epsilon)/kT}$$

The number of holes in the valence band is then obtained by forming the integral

$$N_h = 4\pi V \left(\frac{2m_h}{h^2}\right)^{3/2} \int_{-\infty}^{0} \epsilon^{1/2} e^{-(\mu-\epsilon)/kT} \, d\epsilon \tag{8-91}$$

where m_h is an effective mass of the hole. The integral may be evaluated to give

$$N_h = V \left(\frac{2\pi m_h kT}{h^2}\right)^{3/2} e^{-\mu/kT} \tag{8-92}$$

Equating Eqs. (8-89) and (8-92) gives

$$\mu = \tfrac{1}{2}\epsilon_g + \tfrac{3}{4}kT \ln\frac{m_h}{m_e} \tag{8-93}$$

or, if the masses of the electrons and holes are equal, we obtain

$$\mu = \tfrac{1}{2}\epsilon_g \tag{8-93a}$$

and the Fermi level lies exactly in the center of the forbidden band. This is an important result and is of practical value in explaining thermoelectric energy-conversion devices. Some typical bandwidths for semiconductors are given in Table 8-4.

TABLE 8-4
Typical forbidden bandwidths
for semiconductors

Material	ϵ_g, eV
Ge	0.78
Se	2.0
Si	1.21
Te	0.33
Cu_2O	1.0
InAs	0.5
InSb	2.3
UO_2	0.2
ZnS	3.6

8-10 THE PHOTON GAS (BLACKBODY RADIATION)

As an interesting application of quantum statistics, we consider a quantity of thermal radiation inside an insulated cavity and seek an expression for the energy distribution for such radiation. Such a system may be called a *photon gas* since it consists of energy quanta of magnitude

$$\epsilon_i = h\nu_i$$

where ν_i represents the frequency of the photons. The momentum of the photons is

$$p_i = \frac{h\nu_i}{c} = \frac{h}{\lambda_i} = \frac{\epsilon_i}{c} \tag{8-94}$$

The photons represent "particles" which occupy various quantum states, but the isolated collection does not consist of a constant number of particles. As with conventional particle systems, we seek the number density of quanta which occupy each energy state ϵ_i. The photons are indistinguishable and there is no restriction on the number of available quantum states, so we employ the BE statistical model.

$$\Omega_{BE} = \prod_i \frac{(g_i + N_i - 1)!}{N_i!(g_i - 1)!} \tag{8-95}$$

Forming the maximizing condition,

$$\delta(\ln \Omega) = 0$$

along with the constant energy condition,

$$\delta U = 0 = \Sigma \epsilon_i \delta N_i$$

gives

$$\frac{N_i}{g_i} = \frac{1}{e^{\beta \epsilon_i} - 1} \tag{8-96}$$

where it is to be noted that we have not imposed a constant number of par-ticles condition on the distribution, as in Sec. 8-4, since photons may be continuously created and destroyed.

Let us assume that the walls of the enclosure are at the temperature T and the total volume is V. The number of allowable quantum states g_i in each energy level ϵ_i is calculated by dividing the elemental volume in phase space having this energy by h^3, the smallest discernible volume in phase space. We cannot say "where" the photon is located in the Cartesian space of the container, except to say it is in the volume V. On the other hand, the energy level ϵ_i is linked to momentum through Eq. (8-94), so that we choose as our elemental volume in phase space a spherical shell:

$$V \times 4\pi p_i^2 \, dp_i$$

and calculate the number of quantum states in this volume by

$$g_i = \frac{4\pi V p_i^2 \, dp_i}{h^3} \tag{8-97}$$

This relation must be multiplied by a factor of 2 to account for the fact that light may be right or left circularly polarized. Combining Eq. (8-97) with Eq. (8-96) gives

$$N_i = \frac{8V\pi p_i^2 \, dp_i}{h^3(e^{\beta \epsilon_i} - 1)} \tag{8-98}$$

It is to be noted once again that N_i represents the number of photons in the volume V having the momentum between p_i and $p_i + dp_i$. This, in turn, is the number having an energy between ϵ_i and $\epsilon_i + d\epsilon$. Equation (8-98) may be expressed in terms of frequency by noting from Eq. (8-94) that

$$p_i^2 = \frac{h^2 \nu_i^2}{c^2} \qquad dp_i = \frac{h \, d\nu_i}{c}$$

so that

$$N_i = \frac{8V\pi\nu_i^2 \, d\nu_i}{c^3(e^{\beta h \nu_i} - 1)} \tag{8-98a}$$

The number of photons per unit volume and per unit frequency is, thus,

$$\frac{dN}{V \, d\nu} = \frac{N_i}{V \, d\nu_i} = \frac{8\pi\nu^2}{c^3(e^{\beta h \nu} - 1)}$$

where, now, we have assumed that an essentially continuous spectrum of frequencies is available so that ν_i is replaced by ν. The energy of each photon is $h\nu$, so that the energy per unit volume and per unit frequency is

$$u_\nu = \frac{8\pi h \nu^3}{c^3(e^{\beta h \nu} - 1)} \tag{8-99}$$

It turns out that $\beta = 1/kT$ when a condition like Eq. (8-5) is applied as before. Making use of $\nu = c/\lambda$ and $d\nu = -(c/\lambda^2) \, d\lambda$, Eq. (8-99) may then be written in terms of wavelength as

$$u_\lambda = \frac{8\pi h c \lambda^{-5}}{e^{hc/\lambda kT} - 1} \tag{8-99a}$$

where, now, u_λ is the energy per unit volume and per unit wavelength.

It can be shown that the radiant energy emitted from a surface in the enclosure per unit area per unit time and per unit wavelength $E_{b\lambda}$ is related to u_λ by

$$E_{b\lambda} = \frac{u_\lambda c}{4}$$

so that

$$E_{b\lambda} = \frac{2\pi h c^2 \lambda^{-5}}{e^{hc/\lambda kT} - 1}$$

$$E_{b\lambda} = \frac{c_1 \lambda^{-5}}{e^{c_2/\lambda T} - 1} \tag{8-100}$$

Equation (8-100) is called the *Planck blackbody radiation law* and forms a base for calculations of many radiation heat-transfer phenomena; a plot of this relation for several temperatures is given in Fig. 8-14. Integration of Eq. (8-100) over all possible wavelengths results in

$$E_b = \int_0^\infty E_{b\lambda}\, d\lambda = \int_0^\infty \frac{c_1\lambda^{-5}\, d\lambda}{e^{c_2/\lambda T} - 1} = \sigma T^4 \tag{8-101}$$

where σ is a new constant called the *Stefan-Boltzmann constant*. Equation (8-101) expresses the total blackbody radiation leaving a surface per unit area and per unit time. E_b is usually called the *emissive power* of an ideal radiator. It describes only the *thermal* radiation emitted by a body. Thermal radiation is distinguished from other types of electromagnetic radiation by the fact that it depends on temperature.

Some of the appropriate constants for use in the radiation formulas are:

$$\sigma = 5.669 \times 10^{-8}\ \text{W/m}^2\cdot\text{K}^4$$
$$= 1.714 \times 10^{-9}\ \text{Btu/h}\cdot\text{ft}^2\cdot{}^\circ\text{R}^4$$
$$c_1 = 3.743 \times 10^{-8}\ \text{W}\cdot\mu\text{m}^4/\text{m}^2$$
$$= 1.187 \times 10^8\ \text{Btu}\cdot\mu\text{m}^4/\text{ft}^2\cdot\text{h}$$
$$c_2 = 1.4387 \times 10^4\ \mu\text{m}\cdot\text{K}$$
$$= 2.5896 \times 10^4\ \mu\text{m}{}^\circ\text{R}$$

where μm designates the unit of micrometers (formerly microns; $1\mu\text{m} = 10^{-6}$ m).

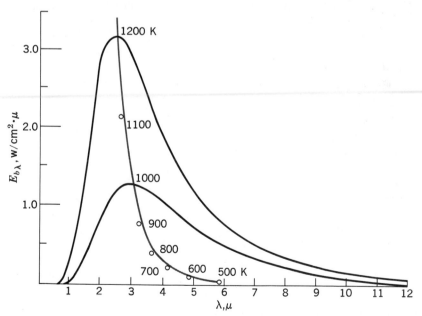

FIG. 8-14 *Planck blackbody radiation distribution.*

EXAMPLE 8-7

Calculate the total energy emitted by an ideal radiator at 1000 and 2000 K.

SOLUTION   This calculation is made immediately with Eq. (8-101). At $T = 1000$ K,

$$E_b = \sigma T^4 = (5.669 \times 10^{-8})(1000)^4$$
$$= 56.69 \text{ kW/m}^2$$
$$= 5.669 \text{ W/cm}^2$$

At $T = 2000$ K,

$$E_b = (5.669 \times 10^{-8})(2000)^4$$
$$= 907 \text{ kW/m}^2$$

8-11   ENSEMBLES AND INTERACTING SYSTEMS

In the basic statistical presentations of Chap. 5 and in the applications of these statistics throughout this chapter we have been concerned with particles which do not interact with one another. To be sure, the vibration of atoms in a solid are dependent on the behavior of their neighbors. But the *statistical* portion of the analysis still considered the particles as independent, and a quantum mechanical specification of allowable energy states was relied upon to take interactions into account. For the systems considered so far, an isolated assembly of particles has been assumed. For the isolated assembly the total energy is constant as well as the total number of particles.

In the real world there are many situations where particle interaction exerts a significant influence on the distribution of particles among available energy states. Another way of saying this is that the number of particles in a particular element of volume in phase space is not unlimited but may depend on the interaction forces between particles as well as on the size of the particles. The solid state is an obvious example. High-pressure gases and liquids are others. In these instances we are forced to take the interactions into account. These interesting and complicated effects are normally treated by *ensemble theory*, a brief outline of which we shall give in this section.

A *canonical ensemble* is depicted in Fig. 8-15. The large system is supposed to be a collection of many smaller subsystems, each having a fixed number of particles N and a fixed volume V. The ensemble is the collection of all these η subsystems. The total energy of the ensemble is assumed to be constant, but the energy of each subsystem may vary as the particles comprising this subsystem move about in the total system. Each of the η subsystems is assumed to be distinguishable. Note that the subsystems may exchange energy with one another, even though the total energy of the en-

FIG. 8-15 *Canonical ensemble.*

semble remains constant. To specify the allowable energy states of the sub-systems, we would need to consider particle interactions as well as other factors. Presumably, these allowable energy states are calculable. If the number of subsystems having the energy U_i is η_i, then the total average energy of the ensemble is

$$U = \sum_i \eta_i U_i \tag{8-102}$$

The total number of subsystems is η so that

$$\eta = \sum_i \eta_i \tag{8-103}$$

We are interested in the number of ways of arranging the subsystems among the various η_i states. This was calculated in Eq. (5-10) as

$$\Omega = \frac{\eta!}{\Pi \eta_i!} \tag{8-104}$$

Of course, the relations in Eqs. (8-102) to 8-104) bear a striking resemblance to the basic MB relations of Chap. 5. The point to remember is that the ensemble artifice allows an inclusion of interaction energies in this simplified distribution.

A *grand canonical ensemble* is depicted in Fig. 8-16. In this type of ensemble we remove the restriction of a constant number of particles in each subsystem. Now we have η subsystems of volume V, all of which are distinguishable. The total energy and number of particles of the ensemble are assumed constant, but the energy and number of particles in each subsystem may vary. In this way we are able to study systems of particles which may exchange both energy and mass. The inclusion of a variable number of particles in the analysis complicates things considerably; in particular, multiple

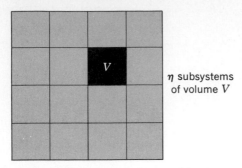

η subsystems
of volume V

FIG. 8-16 *Grand canonical ensemble.*

summations must be employed. This complication is justified, however, by the power supplied for analysis of more involved physical systems.

A *microcanonical ensemble* is one which consists of a group of subsystems which have constant N, V, *and* energy.

We shall not be able to expand our discussion of ensembles further except to say that a detailed study of ensemble methods is essential to a more advanced study of statistical thermodynamics. The interested reader may consult the references at the end of the chapter for more information on this subject.

REVIEW QUESTIONS

1 What is a partition function?

2 How is the macroscopic concept of temperature connected with microscopic thermodynamic models?

3 What is meant by an ideal gas in a physical sense?

4 How does rotational energy contribute to the specific heats of ideal gases?

5 What is the equipartition of energy principle?

6 Discuss the relative contributions of lattice vibrations and electron energies to the specific heats of solids.

7 In many secondary school science programs (and some colleges) there is a strong implication that absolute zero of temperature means zero energy. Discuss this concept.

8 What is a "hole"?

9 Distinguish between conductors, semiconductors, and electrical insulators in a microscopic sense.

10 What is blackbody radiation?

11 What is entropy?

12 What is degeneracy?

13 How is entropy related to the partition function?

14 What is meant by characteristic rotational and vibrational temperatures?

15 What is a "degree of freedom"?

16 Distinguish between the Einstein and Debye models of a solid.

17 What are free electrons?

18 What is the Fermi level?

19 What is a photon?

20 What is an ensemble?

21 Distinguish between the MB, BE, and FD models of energy distributions.

22 How do electrons contribute to specific heat of a solid?

23 Discuss the effect of impurities on the behavior of semiconductors.

24 What are meant by p- and n-type semiconductors?

25 What is a photon gas?

26 What is meant by radiation energy density.

27 What is the Stefan-Boltzmann law?

PROBLEMS

8-1 Estimate the specific heats of the following gases:
    (a) CO at 50 K, 100 K, 300 K, 3000 K
    (b) O_2 at 100 K, 300 K, 3000 K
    (c) N_2 at 100 K, 300 K, 3000 K

8-2 Estimate the specific heats of the following solids:
    (a) Sodium at 40 K, 100 K, 200 K
    (b) Silver at 50 K, 100 K, 300 K, 500 K
    (c) Copper at 50 K, 100 K, 300 K, 700 K
    (d) Lead at 20 K, 100 K, 300 K, 500 K

8-3 Compute the first four terms of the series of Eq. (8-44), and calculate the corresponding value of rotational specific heat for:

(a) $\dfrac{T}{\Theta_{rot}} = 0.2$

(b) $\dfrac{T}{\Theta_{rot}} = 0.5$

(c) $\dfrac{T}{\Theta_{rot}} = 0.8$

(d) $\dfrac{T}{\Theta_{rot}} = 1.0$

Compare the results of these first four terms of the series with the curve in Fig. 8-2.

8-4 Estimate the Fermi level for silver and for tungsten.

8-5 Plot the Fermi function versus $\epsilon - \mu$ for:
 (a) $T = 0$ K
 (b) $T = 50$ K
 (c) $T = 100$ K
 (d) $T = 300$ K
 (e) $T = 500$ K
Discuss this plot.

8-6 Plot the Fermi function versus ϵ/μ for the valence and conduction bands of a semiconductor in which $\mu = \frac{1}{2}\epsilon_g$ for the following temperatures:
 (a) $T = 0$ K
 (b) $T = 50$ K
 (c) $T = 100$ K
 (d) $T = 300$ K
 (e) $T = 500$ K
Make the plots over the range $-5 < \epsilon/\mu < +5$, where the zero level is taken at the top of the valence band. For this calculation take the forbidden bandwidth as 1.0 eV. Discuss the plots.

8-7 Why is it that free electrons do not contribute substantially to the specific heat of solids?

8-8 Four indistinguishable particles are to be placed in two energy levels, each of which has a degeneracy of three. Evaluate the thermodynamic probability for *all* possible arrangements. What is the most probable arrangement?

8-9 Determine the entropy for H_2 at 50, 300, and 3000 K.

8-10 Derive an expression for enthalpy in terms of the partition function.

8-11 Calculate the relative populations in the first three rotational energy levels for N_2 at 30 and 2000 K.

8-12 Calculate the relative populations of the first three vibrational energy levels for N_2 at 2000 and 3500 K.

8-13 Construct plots of u and $\bar{c}_v$ for N_2 versus temperature from 50 to 7000 K.

8-14 Calculate the specific heat of diamond at 30 and 1000 K.

8-15 For what temperature range(s) will the equipartition of energy principle apply for (a) N_2, (b) O_2, (c) H_2, with an error of less than ± 10 percent for specific heat?

8-16 Estimate the moment of inertia for (a) O_2, (b) N_2, (c) H_2 molecules.

8-17 Calculate the maximum free-electron velocity in copper at $T = 0$.

8-18 Calculate, for silver at 35 K, the percent contribution to specific heat by the electron gas.

8-19   Compare the total energy of the electron gas in silver at 0 K with the energy required to raise the temperature of silver from 35 to 55 K.

8-20   Estimate the characteristic maximum vibrational frequency for the lattice of (a) copper, (b) lead, (c) diamond.

8-21   Calculate the radiant emission from an ideal radiator at (a) 100 K, (b) 1000 K, (c) 5000 K.

8-22   Calculate the energy density of radiation per unit wavelength at 1 μm and 5 μm for each of the cases in Prob. 8-21.

8-23   Show, by a suitable differentiation process, that a maximum point exists in the Planck blackbody radiation formula, and that this point is given by

$$(\lambda T)_{max} = 5215.6 \ \mu\text{m}\cdot°\text{R}$$

This relation is called *Wien's displacement law*. Repeat Prob. 8-22, using the value of λ obtained from Wien's law for each temperature.

REFERENCES

1   Allis, W. P., and M. A. Herlin: "Thermodynamics and Statistical Mechanics," McGraw-Hill Book Company, New York, 1952.

2   Crawford, F. H.: "Heat, Thermodynamics and Statistical Physics," Harcourt, Brace & World, New York, 1963.

3   Hill, L.: "Statistical Mechanics," McGraw-Hill Book Company, New York, 1956.

4   Lee, J. F., F. W. Sears, and D. L. Turcotte: "Statistical Thermodynamics," Addison-Wesley Publishing Company, Inc., Reading, Mass., 1963.

5   Mayer, J. E., and M. G. Mayer: "Statistical Mechanics," John Wiley & Sons, Inc., New York, 1940.

6   Reynolds, W. C.: "Thermodynamics," McGraw-Hill Book Company, New York, 1965.

7   Schrödinger, E.: "Statistical Thermodynamics," Cambridge University Press, Cambridge, 1952.

8   Sears, F. W.: "An Introduction to Thermodynamics, The Kinetic Theory of Gases, and Statistical Mechanics," Addison-Wesley Publishing Company, Inc., Reading, Mass., 1950.

9   Sommerfeld, A.: "Thermodynamics and Statistical Mechanics," Academic Press, Inc., New York, 1956.

10   Tolman, C.: "The Principles of Statistical Mechanics," Oxford University Press, London, 1938.

11 Tribus, M.: "Thermostatics and Thermodynamics," D. Van Nostrand Company, Inc., Princeton, N.J., 1961.

12 Hatsopoulos, G. N., and J. H. Keenan: "Principles of General Thermodynamics," John Wiley & Sons, Inc., New York, 1965.

13 Fay, J. A.: "Molecular Thermodynamics," Addison-Wesley Publishing Company, Inc., Princeton, N.J., 1961.

14 Knuth, E. S.: "Introduction to Statistical Thermodynamics," McGraw-Hill Book Company, New York, 1966.

chapter nine
Kinetic theory and transport phenomena

9-1 INTRODUCTION

Our discussion in Chap. 8 has indicated the application of statistical thermo-dynamics to calculation of thermodynamic properties of gases and solids. We now want to extend the analysis of gases a few steps further to study transport phenomena. By transport phenomena we mean the observable physical effects which result from the *transport* of mass, energy, mo-mentum, and electric charge by molecular motion. These phenomena are distinctly different from the physical behavior which are characterized by thermodynamic properties. A thermodynamic property like internal energy, for example, tells us how energy is distributed among available molecular energy states, but it does not give any indication of the *rate* at which we may add energy to the system; i.e., it does not tell us how long it will take the mol-ecules to transport energy from a high-temperature region of a gas to a low-temperature region. The study of transport processes is a very important part of physical science, and it is to our advantage to develop a basic under-standing of the physical mechanism for these processes. The discussion will be restricted to gaseous systems since they are relatively simple to analyze and illustrate the important physical concepts. Liquid and solid systems are much more complicated and are beyond the scope of our discussion.

We have already seen that the energies of simple monatomic molecules may be computed in terms of translational kinetic energies. The energy dis-tribution thus gives the distribution of molecular velocities. In principle, then, we have the basic information necessary to calculate the rates of molecular movement and, subsequently, to calculate various mass, energy, and momentum transport rates.

9-2 THE MAXWELL-BOLTZMANN VELOCITY DISTRIBUTION

We have already used the MB statistics (as a limiting case of BE statistics) to find the energy distribution in a gas. Let us consider a monatomic gas whose molecules behave like point masses and have only translational kinetic en-ergy, so that

$$\epsilon_i = \tfrac{1}{2}mv_i^2$$

For such a gas the MB velocity distribution was derived in Sec. 8-4 as

$$\frac{d^3N}{N} = \left(\frac{m}{2\pi kT}\right)^{3/2} e^{-mv^2/2kT} \, dv_x \, dv_y \, dv_z \tag{9-1}$$

where

$$v^2 = v_x^2 + v_y^2 + v_z^2$$

In this derivation it was assumed that all directions of molecular motion are equally probable; i.e., the fraction of molecules having a certain velocity v is dependent only on the magnitude of the velocity and not on the direction. This means that the velocity vectors must be uniformly distributed in the spherical velocity shell between v and $v + dv$, as shown in Fig. 9-1. Equation (9-1) may be rearranged to give the density of particles in velocity space as

$$\rho_v = \frac{d^3N}{dv_x \, dv_y \, dv_z} = N \left(\frac{m}{2\pi kT} \right)^{3/2} e^{-mv^2/2kT} \tag{9-2}$$

This density is dependent only on the magnitude of the total velocity. The number of particles in the spherical shell shown in Fig. 9-1 will be given as a product of the volume of the shell and the density function of Eq. (9-2). Thus

$$dN_v = 4\pi v^2 \, dv \, N \left(\frac{m}{2\pi kT} \right)^{3/2} e^{-mv^2/2kT} \tag{9-3}$$

or, written another way,

$$\frac{dN_v}{N} = \left(\frac{2}{\pi} \right)^{1/2} \left(\frac{m}{kT} \right)^{3/2} v^2 e^{-mv^2/2kT}$$

$$= f(v) \, dv \tag{9-3a}$$

The function $f(v) \, dv$ represents the fraction of molecules having speeds between v and $v + dv$.

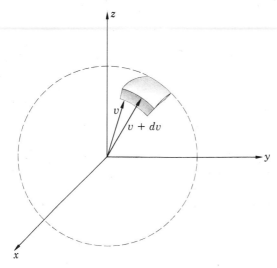

FIG. 9-1 *Velocity space for Maxwell-Boltzmann distribution.*

Two average molecular speeds are of interest. First, the arithmetic average speed $\bar{v}$ is calculated from

$$\bar{v} = \frac{\int_0^\infty vNf(v)\,dv}{N} = \int_0^\infty \left(\frac{2}{\pi}\right)^{1/2} \left(\frac{m}{kT}\right)^{3/2} v^3 e^{-mv^2/2kT}\,dv \tag{9-4}$$

where N is the number of molecules. The integral may be evaluated with the aid of Table 9-1, so that

$$\bar{v} = \left(\frac{8kT}{\pi m}\right)^{1/2} \tag{9-5}$$

The root-mean-square speed v_{rms} is defined by

$$v_{rms} = (\overline{v^2})^{1/2}$$
$$= \left[\frac{1}{N}\int_0^\infty Nv^2 f(v)\,dv\right]^{1/2} \tag{9-6}$$

Evaluating the integral in Eq. (9-6) gives

$$v_{rms} = \left(\frac{3kT}{m}\right)^{1/2} \tag{9-7}$$

The square of the rms speed is a measure of the average kinetic energy of the molecules.

The most probable speed v_m is that speed for which $f(v)$ is a maximum; it is the speed which the largest number of molecules have, on the average. We evaluate v_m by setting

TABLE 9-1
Values of the integral
$$I(n) = \int_0^\infty x^n e^{-ax^2}\,dx$$

n	$I(n)$
0	$\frac{1}{2}\sqrt{\pi/a}$
1	$\frac{1}{2}a$
2	$\frac{1}{4}\sqrt{\pi/a^3}$
3	$\frac{1}{2}a^2$
4	$\frac{3}{8}\sqrt{\pi/a^5}$
5	$1/a^3$
6	$\frac{15}{16}\sqrt{\pi/a^7}$
7	$3/a^4$

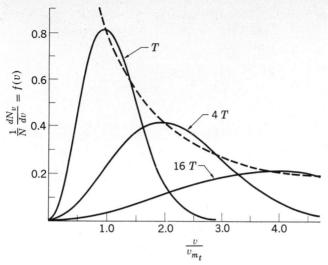

FIG. 9-2 *Maxwell-Boltzmann velocity distribution. Dashed line represents locus of most probable velocities.* v_{m_t} *represents the most probable velocity at temperature T. Function f(v) is that defined by Eq. (9-3a).*

$$\frac{d}{dv}[f(v)] = 0 \qquad \text{at } v = v_m$$

resulting in

$$v_m^2 e^{-mv_m^2/2kT}\left(-\frac{mv_m}{kT}\right) + 2v_m e^{-mv_m^2/2kT} = 0$$

and

$$v_m = \left(\frac{2kT}{m}\right)^{1/2} \tag{9-8}$$

Figure 9-2 illustrates the MB velocity distribution and the relative effects of temperature changes on the distribution.

EXAMPLE 9-1   Mean velocities
Calculate the value of $\bar{v}$, v_{rms}, and v_m for oxygen (O_2) at 300 K.

SOLUTION   Equations (9-5), (9-7), and (9-8) apply. We have:

$$m = \frac{M}{N_0} = \frac{32}{6.022 \times 10^{26}} = 5.31 \times 10^{-26} \text{ kg}$$

Then,

$$\bar{v} = \left(\frac{8kT}{\pi m}\right)^{1/2} = \left[\frac{(8)(1.38 \times 10^{-23})(300)}{\pi(5.31 \times 10^{-26})}\right]^{1/2}$$

$$= 446 \text{ m/s} = 1461 \text{ ft/s}$$

$$v_{rms} = \left(\frac{3kT}{m}\right)^{1/2} = \left[\frac{(3)(1.38 \times 10^{-23})(300)}{(5.31 \times 10^{-26})}\right]^{1/2}$$

$$= 484 \text{ m/s} = 1587 \text{ ft/s}$$

$$v_m = \left(\frac{2kT}{m}\right)^{1/2} = 395 \text{ m/s} = 1296 \text{ ft/s}$$

EXAMPLE 9-2   Molecular speed distribution

Compare the fraction of oxygen molecules at 300 K which have speeds within ± 1 percent of (a) the most probable speed and (b) the rms speed.

SOLUTION   We may write Eq. (9-3a) in finite-difference form as

$$\frac{\Delta N_v}{N} = f(v) \, \Delta v \tag{a}$$

as long as Δv is small compared with v. In this problem

$$\Delta v = 0.01v - (-0.01v)$$
$$= 0.02v$$

and the finite-difference approximation is justified. We now evaluate Eq. (a) for $v = v_m$ and $v = v_{rms}$. From Example 9-1,

$$v_m = 395 \text{ m/s} \qquad m = 5.31 \times 10^{-26} \text{ kg}$$
$$v_{rms} = 484 \text{ m/s}$$

At $v = v_m$,

$$\frac{\Delta N_v}{N} = \left(\frac{2}{\pi}\right)^{1/2} \left[\frac{5.31 \times 10^{-26}}{(1.38 \times 10^{-23})(300)}\right]^{3/2} (395)^2 e^{-1}(0.02)(395)$$

$$= 0.0166$$

At $v = v_{rms}$,

$$\frac{\Delta N_v}{N} = \left(\frac{2}{\pi}\right)^{1/2} \left[\frac{5.31 \times 10^{-26}}{(1.38 \times 10^{-23})(300)}\right]^{3/2} (484)^2 e^{-3/2}(0.02)(484)$$

$$= 0.0185$$

9-3 MOLECULAR FLUX

Let us now consider an imaginary plane, as shown in Fig. 9-3. We want to calculate the flux of molecules through the element of area dA located as shown. By flux we mean the number of molecules striking this area, per unit area, and per unit time. It is assumed that the molecular speed distribution is given by the MB relation, Eq. (9-3), and, furthermore, that the *direction* of the incoming molecules which strike dA is random, so that there is no preferred direction of molecular motion. For the moment we shall assume that there are no collisions between molecules.

The number of molecules per unit volume with speeds between v and $v + dv$ is

$$f(v)n \, dv \tag{9-9}$$

where n is the molecular density. Of this number only those directed toward dA are to be counted. Since there is no preferred direction of motion, the fraction of molecules which hit the element dA can be calculated with the aid of the projection in Fig. 9-4. The fraction of the molecules arriving at the area element in the specified direction (θ, ϕ) is

$$\frac{r \sin \phi \, d\theta \, r \, d\phi}{4\pi r^2} = \frac{\sin \phi \, d\phi \, d\theta}{4\pi} \tag{9-10}$$

In time dt, only the following fraction of molecules will strike the element of area dA:

$$(dA \cos \phi)v \, dt \, dn_v \tag{9-11}$$

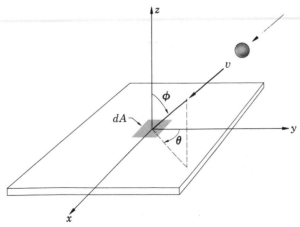

FIG. 9-3 *Model for molecular flux calculation.*

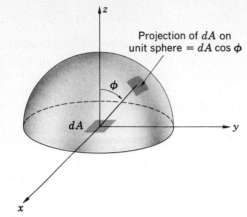

Projection of dA on
unit sphere $= dA \cos \phi$

FIG. 9-4

where, now, we use the symbol dn_v to designate the number of molecules per unit volume and per unit velocity and direction. The quantity $v\,dt$ is the distance moved in time dt, and $dA \cos \phi$ is the projected area in the direction of the velocity v. Thus the number of molecules contained in the volume $dA \cos \phi\, v\, dt$ is the number which strikes dA. dn_v is the number density of molecules in this imaginary volume and is given as the product of Eqs. (9-9) and (9-10). Finally, the number striking the area dA in time dt is

$$f(v)n\,dv(dA \cos \phi)v\,dt\,\frac{\sin \phi\,d\phi\,d\theta}{4\pi} \qquad (9\text{-}12)$$

The molecular flux per unit area and time is obtained by integrating this expression over all directions and velocities:

$$J_N = \int_0^{2\pi} \int_0^{\pi/2} \int_0^{\infty} \frac{n}{4\pi}\, vf(v)\,dv \sin \phi \cos \phi\, d\phi\, d\theta \qquad (9\text{-}13)$$

The integration results in

$$J_N = \frac{n\bar{v}}{4} \qquad (9\text{-}14)$$

where J_N designates the molecular flux.

It is important to realize that the foregoing analysis does not consider possible effects of collisions between molecules on the molecular flux. We shall examine these effects in a later section.

EXAMPLE 9-3   Molecular flux

A tank contains oxygen at 300 K and 1 atm of pressure. The tank is placed in a highly evacuated chamber and a small hole is pierced in the tank. The area of the hole is 0.001 mm². Calculate the number of molecules escaping from the tank per unit time.

SOLUTION We have $p = 1$ atm $= 1.013 \times 10^5$ N/m² (101.3 kPa), and the molecular density may be computed from

$$p = nkT$$

$$n = \frac{(1.013 \times 10^5)}{(1.38 \times 10^{-23})(300)} = 2.45 \times 10^{25} \text{ molecules/m}^3$$

From Example 9-1, $\bar{v} = 446$ m/s, so the molecular flux from Eq. (9-14) is

$$J_N = \frac{n\bar{v}}{4} = \frac{(2.45 \times 10^{25})(446)}{4}$$

$$= 2.73 \times 10^{27} \text{ molecules/m}^2 \cdot \text{s}$$

This flux must be multiplied by the area of the hole to calculate the number of molecules escaping:

$$\text{No. escaping} = J_N A$$
$$= (2.73 \times 10^{27})(0.001 \times 10^{-6})$$
$$= 2.73 \times 10^{18} \text{ molecules/s}$$

9-4 MEAN FREE PATH

No transport can take place at the molecular level without collisions, and to describe transport processes properly the influence of collisions must be taken into account. For this analysis we use as the gas the monatomic, point-mass model discussed before. We want to compute the *average* distance a molecule travels before it collides with another molecule. To simplify the analysis, let us examine an imaginary system consisting of many molecules *at rest* through which moves a single molecule of speed v and diameter d. The moving molecule will strike a stationary molecule when the distance between the centers of the molecules is less than or equal to the diameter, as depicted in Fig. 9-5. Thus the number of collisions in time dt is equal to the number of molecules contained in the volume $\pi d^2 v \, dt$, as shown in Fig. 9-5, or $n\pi d^2 v \, dt$. The quantity πd^2 is called the *collision cross section* σ. The number of collisions per unit time is called the *collision frequency* Θ, and is given by

$$\Theta = \sigma n v$$

The average speed of a molecule is $\bar{v}$; therefore, on the average, we would expect to have a collision frequency of

FIG. 9-5 *Molecular collision cross section.*

$$\Theta = \sigma n \bar{v} \qquad (9\text{-}15)$$

The distance traveled in time t is $\bar{v}t$, and the number of collisions in this time is

$$\Theta t = \sigma n \bar{v} t$$

Thus the average distance traveled between collisions is

$$\lambda = \frac{\bar{v}t}{\sigma n \bar{v} t} = \frac{1}{\sigma n} \qquad (9\text{-}16)$$

where λ is called the *mean free path*.

The foregoing analysis is oversimplified in that it does not take into account the fact that *all* of the molecules are in motion and that the collision frequency must be governed by the relative velocity between molecules. To analyze the situation properly, we would have to use the MB velocity distribution and take the relative velocities of the molecules into account. The result of such an analysis is

$$\lambda = \frac{1}{\sqrt{2}\,\sigma n} \qquad (9\text{-}17)$$

It is to be noted that the foregoing equations give expressions for the *mean free path*. We have yet to establish a relation for the *distribution* of free paths. Once again, let us consider a large number of molecules, and assume that a stream of N_0 molecules impinges upon the large group. As some of the molecules have collisions, they are removed from the stream, as depicted in Fig. 9-6a. The number removed should be proportional to the total number in the stream at that time and distance traveled dx. Thus we write

$$dN = -\alpha N\, dx \qquad (9\text{-}18)$$

Impinging molecular
stream, N_0

Molecular field

(a)

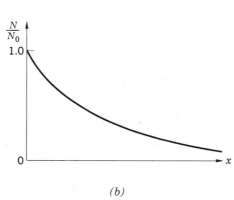

(b)

FIG. 9-6 (a) *Molecular impingement;* (b) *free-path distribution.*

where dN represents the number lost from the stream owing to collisions, and the proportionality constant α is called the *collision probability*. We have the initial condition that

$$N = N_0 \quad \text{at } x = 0 \tag{9-19}$$

so that the solution to Eq. (9-18) becomes

$$\frac{N}{N_0} = e^{-\alpha x} \tag{9-20}$$

In this equation N/N_0 represents the fraction of molecules which travel a distance x before having a collision. The constant α may be evaluated by observing that the mean free path must be given by

$$\lambda = \frac{\int_0^\infty x \, dN}{N_0} = \frac{-\int_0^\infty \alpha N_0 x e^{-\alpha x} \, dx}{N_0} = \frac{1}{\alpha} \tag{9-21}$$

Finally, the distribution of free paths, Eq. (9-20), becomes

$$\frac{N}{N_0} = e^{-x/\lambda} \tag{9-22}$$

This distribution is shown in Fig. 9-6b.

EXAMPLE 9-4 Mean free path

Calculate the mean free path for oxygen molecules at 300 K and 1-atm pressure. What fraction of molecules have free paths greater than 2λ? The effective diameter of the oxygen molecule may be taken as 3.5×10^{-10} m.

SOLUTION   The collision cross section is calculated as

$$\sigma = \pi d^2 = \pi(3.5 \times 10^{-10})^2 = 3.85 \times 10^{-19} \text{ m}^2$$

From Example 9-3, $n = 2.45 \times 10^{25}$ molecules/m³, so that the mean free path may be calculated from Eq. (9-17):

$$\lambda = \frac{1}{\sqrt{2}\sigma n}$$

$$= \frac{1}{\sqrt{2}(3.85 \times 10^{-19})(2.45 \times 10^{25})}$$

$$= 7.50 \times 10^{-8} \text{ m}$$

Equation (9-22) expresses the fraction of molecules which have free paths of distance x or greater. Thus, for $x = 2\lambda$,

$$\frac{N}{N_0} = e^{-2\lambda/\lambda} = 0.135$$

or 13.5 percent of the molecules have free paths greater than twice the mean free path.

9-5 INFLUENCE OF COLLISIONS ON MOLECULAR FLUX

We now reconsider the analysis in Sec. 9-3, taking into account the influence of collisions. The physical situation is shown in Fig. 9-7. In the volume dV there are $n \, dV$ molecules, and of these

$$\Theta n \, dV \, dt$$

will have a collision in the time dt. A number of those molecules which have just had a collision will be directed toward dA. This number is

FIG. 9-7 *Model for analysis of influence of collisions on molecular flux.*

$$\frac{dA \cos \phi}{4\pi r^2} \Theta n \, dV \, dt$$

We want to count the number of molecules reaching dA before having another collision. This number is

$$dN = \frac{dA \cos \phi}{4\pi r^2} \Theta n \, dV \, dt \, e^{-r/\lambda} \qquad (9\text{-}23)$$

We have $dV = r^2 \sin \phi \, d\phi \, d\theta \, dr$, so that the molecular flux on dA becomes

$$J_N = \int_0^\infty \int_0^{\pi/2} \int_0^{2\pi} \frac{n}{4\pi} \Theta \sin \phi \cos \phi \, e^{-r/\lambda} \, d\theta \, d\phi \, dr \qquad (9\text{-}24)$$

The result is the same as in Sec. 9-3.

$$J_N = \tfrac{1}{4}\Theta \, n = \tfrac{1}{4}n\bar{v} \qquad (9\text{-}25)$$

9-6 AVERAGE COLLISION DISTANCE

We now want to calculate a quantity which we choose to call *average collision distance*. In Fig. 9-7 it is the average height $\bar{z}$ above the xy plane at which the molecules had their last collision before crossing dA. The average distance away from dA, $\bar{r}$, at which the molecules had a collision is clearly the mean free path λ, but the average z coordinate is not equal to λ. To calculate this average distance, we evaluate

$$\bar{z} = \frac{\int z \, dN}{\int dN} \tag{9-26}$$

where the dN is the number crossing dA as given by Eq. (9-23) and $\bar{z} = r \cos\phi$. Evaluating Eq. (9-26), we obtain

$$\bar{z} = \frac{\frac{1}{6}\Theta \, n\lambda^2 \, dA \, dt}{\frac{1}{4}\Theta \, n\lambda \, dA \, dt} = \frac{2}{3}\lambda \tag{9-27}$$

9-7 GENERALIZED TRANSPORT COEFFICIENT

The relations derived in the foregoing may now be applied to calculate molecular transport rates. Imagine a field of molecules, as shown in Fig. 9-8, which is uniform in density (n = const.), but in which there may be a gradient in some molecular property Γ in the z direction. Γ might be energy, momentum, concentration, electric charge, etc. The transport of the quantity Γ across the area dA is governed by the variation of Γ in the z direction. We may write

$$\Gamma = \Gamma|_{z=0} + \frac{d\Gamma}{dz}\bigg|_{z=0} dz \tag{9-28}$$

for the immediate neighborhood of the plane dA. For pressure and temperature conditions where λ is small, the relation in Eq. (9-28) may be assumed to hold within a few mean free paths of $z = 0$. The downward transport of Γ through dA is calculated by multiplying the molecular flux through dA by the value of the property Γ at its last collision before hitting dA. Thus

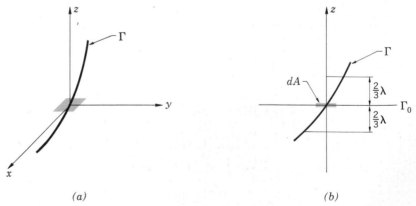

(a)   (b)

FIG. 9-8 *Molecular property gradient.*

$$(J_\Gamma)_{-z} = -\tfrac{1}{4}n\bar{v}\left[\Gamma_0 + \left(\frac{d\Gamma}{dz}\right)_0 \frac{2}{3}\lambda\right]$$

Similarly, the upward transport is

$$(J_\Gamma)_{+z} = +\tfrac{1}{4}n\bar{v}\left[\Gamma_0 - \left(\frac{d\Gamma}{dz}\right)_0 \frac{2}{3}\lambda\right]$$

The net transport rate in the positive z direction is the algebraic sum of these two relations, or

$$J_\Gamma = -\tfrac{1}{3}n\bar{v}\lambda\frac{d\Gamma}{dz} \tag{9-29}$$

We shall call the quantity $\tfrac{1}{3}n\bar{v}\lambda$ a *generalized transport coefficient*. Its units are $(\text{cm}\cdot\text{s})^{-1}$ in the cgs system.

9-8 THERMAL CONDUCTIVITY

Thermal conduction is defined by the Fourier relation

$$J_Q = \frac{q_z}{A} = -K\frac{\partial T}{\partial z} \tag{9-30}$$

that is, heat flux per unit area is proportional to temperature gradient. The proportionality constant K is defined as the *thermal conductivity*. In this process the quantity being transported is molecular energy. Energy is transported from a high-temperature region to a low-temperature region by the molecular motion, and the general process is described by Eq. (9-30). Let us assume that our molecular system is such that the average molecular energy can be expressed in terms of the number of degrees of freedom f and the temperature

$$\epsilon_i = \frac{f}{2}kT \tag{9-31}$$

The energy transport rate is then calculated with Eq. (9-29):

$$J_Q = \frac{q}{A} = -\tfrac{1}{3}n\bar{v}\lambda\frac{d}{dz}\left(\frac{f}{2}kT\right) \tag{9-32}$$

Comparing Eq. (9-30) with Eq. (9-32), we find

$$K = \tfrac{1}{3} n \bar{v} \lambda \left(\frac{f}{2} k \right) \tag{9-33}$$

This result could also be expressed as

$$K = \frac{1}{3} \frac{n \bar{v} \lambda \bar{c}_v}{N_0} = \frac{1}{3} \frac{\bar{v} \bar{c}_v}{N_0 \sigma} \tag{9-34}$$

where N_0 is Avogadro's number. From this equation we find that the thermal energy transport rate is directly proportional to the mean molecular speed. A typical set of units for thermal conductivity may be obtained from Eq. (9-34) as kilojoules per second-meter-degree centigrade.

9-9   DYNAMIC VISCOSITY

Viscous action in a fluid is evidenced as a shear stress between fluid layers, and the shear stress is assumed to be proportional to the normal velocity gradient, as shown in Fig. 9-9.

$$\tau = \mu \frac{\partial u}{\partial z} \tag{9-35}$$

where τ is the shear stress, and the proportionality constant μ is defined as the dynamic viscosity. According to this definition, a typical set of units for the dynamic viscosity would be newtons-second per square meter in the SI system. The physical mechanism of viscosity is one of momentum exchange. Molecules transport momentum from regions of high *macroscopic* velocity to regions of low velocity. This transport tends to pull the slower

FIG. 9-9   *Relation of dynamic viscosity to fluid shear.*

moving fluid layers along, and the "drag" or shear force is the quantity τ. The momentum of a molecule *due to the macroscopic velocity u* is

$$p = mu$$

The net momentum transport through an area element, as shown in Fig. 9-9, may be written with the aid of Eq. (9-29):

$$J_p = \tau = -\tfrac{1}{3}n\bar{v}\lambda \frac{\partial}{\partial z}(mu) = -\tfrac{1}{3}n\bar{v}\lambda m \frac{\partial u}{\partial z} \tag{9-36}$$

The net momentum transport is negative when $\partial u/\partial z$ is positive. This indicates a net transport *downward* through dA and a shear stress acting to the right. Thus, comparing Eq. (9-35) with Eq. (9-36), we obtain

$$\mu = \tfrac{1}{3}nm\bar{v}\lambda = \frac{1}{3}\frac{m\bar{v}}{\sigma} \tag{9-37}$$

The viscosity, like thermal conductivity, is directly proportional to the mean molecular speed.

EXAMPLE 9-5   Thermal conductivity

Calculate the thermal conductivity for oxygen at 1 atm and 300 K. Express in units of Btu/h·ft·°F.

SOLUTION   At $T = 300$ K, the molal specific heat for oxygen is (see Chap. 8):

$$\bar{c}_v = \tfrac{5}{2}\mathscr{R} = 20.79 \text{ kJ/kg·mol·K}$$

The mean molecular speed was calculated in Example 9-1 as

$$\bar{v} = 446 \text{ m/s} = 1461 \text{ ft/s}$$

From Example 9-4 the collision cross section was

$$\sigma = 3.85 \times 10^{-19} \text{ m}^2$$

The thermal conductivity may now be calculated from Eq. (9-34) as

$$K = \frac{1}{3}\frac{\bar{v}\bar{c}_v}{N_0\sigma}$$

$$= \frac{(20.79 \times 10^3)(446)}{(3)(6.022 \times 10^{26})(3.85 \times 10^{-19})}$$

$$= 0.0133 \text{ W/m·K}$$

$$= 7.72 \times 10^{-3} \text{ Btu/h·ft·°F}$$

EXAMPLE 9-6   Dynamic viscosity

Calculate the dynamic viscosity of oxygen at 300 K and 1-atm pressure.

SOLUTION   Equation (9-37) is appropriate for this problem. From previous examples we have:

$$m = 5.32 \times 10^{-26} \text{ kg}$$
$$\bar{v} = 446 \text{ m/s}$$
$$\sigma = 3.85 \times 10^{-19} \text{ m}^2$$

Thus

$$\mu = \frac{1}{3}\frac{m\bar{v}}{\sigma}$$
$$= \frac{(5.32 \times 10^{-26})(446)}{(3)(3.85 \times 10^{-19})}$$
$$= 2.05 \times 10^{-5} \text{ N·s/m}^2$$

9-10   MUTUAL DIFFUSION OF LIKE MOLECULES

Consider the container shown in Fig. 9-10, where there are light and dark molecules, all of the same size, distributed uniformly throughout the container. Let us change this situation by placing a reservoir of light molecules

"Dark molecule" reservoir   "Light molecule" reservoir

FIG. 9-10   *Illustration of mutual diffusion of like molecules. (Top) Uniform distribution of "light" and "dark" molecules; (bottom) nonuniform distribution resulting from mutual diffusion.*

at one end and a reservoir of dark molecules at the other end of the container. Under these circumstances we have impressed a molecular concentration gradient on the system. The total number of molecules in the system remains the same, but they are no longer uniformly distributed throughout the system volume. There is a net flow of light molecules from the region of high concentration to the region of low concentration; this flow results from the light molecule density gradient and the random molecular motion. Of course, to maintain a flow of molecules through the system, we would have to provide a mechanism which continuously supplies the reservoir at one end and extracts molecules at the other end.

The physical situation described in the preceding paragraph is called mutual diffusion, and the property we are transporting is molecular *identity*, in this instance, light or dark color. If the like molecules did not have different colors, we could not tell them apart, and the calculation of transport rates would not be of much value. Let the number of light molecules per unit volume at any location z be n_1. The *concentration* of light molecules is, then,

$$c_1 = \frac{n_1}{n} \tag{9-38}$$

where n is still taken as the total number of molecules per unit volume and is assumed constant. The transport of light molecules per unit area and per unit time is, thus,

$$J_{N_1} = -\tfrac{1}{3} n \bar{v} \lambda \frac{\partial}{\partial z} \left(\frac{n_1}{n} \right) \tag{9-39}$$

The diffusion coefficient D is defined by

$$J_{N_1} = -D \frac{\partial n_1}{\partial z} \tag{9-40}$$

Thus

$$D = \tfrac{1}{3} \bar{v} \lambda \tag{9-41}$$

A few additional comments concerning concentration and molecular identity are in order at this point. The concentration parameter is the one we chose to measure the fraction of molecules which have a certain identity. Since we have assumed the total molecular density as constant, the total net flux of molecules across some plane is zero while the flux of tagged (light or dark) molecules is not zero. The rate of change of the concentration is, therefore, the quantity governing the net transport rate of tagged molecules.

9-11 MUTUAL DIFFUSION OF UNLIKE MOLECULES

The preceding analysis considered diffusion of like molecules, i.e., those of equal mass and shape. Insofar as collisions of these molecules are concerned, the momentum and energy exchange is fairly easy to calculate because they all have the same mass and collision cross section. Suppose, however, that we had a system consisting of two different types of molecules, as depicted in Fig. 9-11. One molecule, m_1, is substantially heavier than the other. Now we assume that the molecules are no longer distinguished by different colors but rather by their difference in mass and size. Imagine the plane shown in Fig. 9-12 and consider the molecular collisions which may occur in the immediate neighborhood of the plane. These are the collisions which determine the transport rate across the plane. There are only three types of collisions that can occur: (a) collisions between two type 1 molecules; (b) collisions between two type 2 molecules; and (c) collisions between a type 1 and a type 2 molecule.

It is fairly easy to see that collisions between like molecules do not influence the transport rate of type 1 molecules, because an observer does not care whether a type 1 molecule crossed plane P without making a collision or whether it was hit by a type 1 molecule and pushed across the plane. The net result is the same because the same mass transport has occurred in either circumstance. But an unlike-molecule collision makes quite a difference. If, instead of crossing plane P, a type 1 molecule hits a type 2 molecule and drives it across the plane, a completely different mass transfer rate is observed.

The purpose of this brief discussion is to show that the collisions between *unlike* molecules exert significant influence on the mass transport rate. It is a rather complicated process to calculate the diffusion coefficient for the system of unlike molecules, and we present only the result:

$$D_{12} = \frac{3}{8}\left(\frac{\pi kT}{2\overline{m}}\right)^{1/2}\frac{1}{n\pi d_{12}^2} \tag{9-42}$$

where

$$m = \frac{m_1 m_2}{m_1 + m_2}$$

$$d_{12} = \frac{d_1 + d_2}{2}$$

FIG. 9-11 *Mutual diffusion of unlike molecules.*

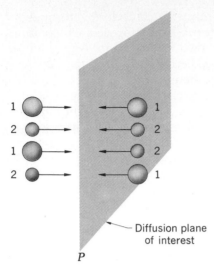

FIG. 9-12 *Collisions of unlike molecules.*

and d_1 and d_2 are the diameters of the two molecules which are assumed to behave as perfectly elastic spheres. The quantity D_{12} designates the diffusion coefficient for type 1 molecules diffusing through type 2 molecules.

9-12 THERMAL DIFFUSION

Consider the system shown in Fig. 9-13a. The container is filled with a mixture of type 1 and type 2 molecules having different masses and sizes. As

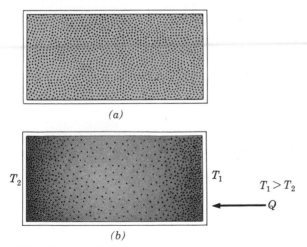

FIG. 9-13 *Thermal diffusion of unlike molecules. (a) Uniform distribution with no temperature gradients; (b) concentration gradients resulting from temperature gradients.*

shown, the two types of molecules are distributed uniformly throughout the volume when the temperature is uniform. Now suppose a temperature gradient is imposed on the system, as shown in Fig. 9-13b. As a result of this temperature gradient, it is observed experimentally that the lighter molecules tend to concentrate in the higher-temperature region and the heavier molecules tend to concentrate in the lower-temperature region.

The phenomenon just described is called *thermal diffusion*, because the diffusion results from a thermal (temperature) gradient instead of a concentration gradient. A calculation of thermal diffusion rates is rather complicated and beyond the scope of our discussion; however, it is possible to give a simple qualitative explanation of the physical mechanism of the effect.

In Sec. 9-11 we established the fact that collisions between *unlike* molecules appreciably influence the mass transport rate. Let us use this principle to explain thermal diffusion. Consider the typical collision shown in Fig. 9-14. The rate of transport will be governed by the speeds at which the molecules leave the collision, which, in turn, are dependent on the *relative velocity* of the molecules before impact. For Fig. 9-14, molecule 1 gives up momentum to molecule 2, thereby pushing it to the right. The higher the relative velocity at impact the faster molecule 2 is pushed to the right.

The foregoing collision process may be related to the thermal diffusion system in the following way. At any one z location in the gas mixture all molecules have the same average kinetic energy (since T is uniform at any z location), but the lighter molecules have a higher average velocity than the heavier molecules. We have seen that the greatest momentum exchange occurs in a collision between unlike molecules having greatest relative velocity. This situation occurs when a light molecule from a warm region collides with a heavy molecule from a cool region. After the collision the light molecule is pushed back toward the cool region. The basic thermal diffusion effect is, thus, a result of the fact that the unlike molecule collisions having the greatest momentum exchange will be those tending to force the light molecules into the hot region.

We hasten to emphasize that this model of thermal diffusion is considerably simplified, and that molecular size as well as molecular mass can influence the diffusion process.

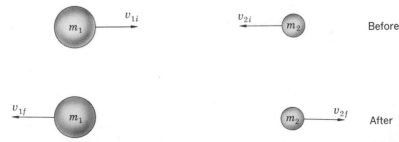

FIG. 9-14 *Typical collision for analysis of thermal diffusion.*

9-13   ELECTRON TRANSPORT IN GASES

Electric conduction in gases arises as a result of motion of the free electrons present in the gas. These electrons are liberated as a result of collisions between the few high-energy molecules in the system. Most gases at room temperature do not have many such high-energy molecules and, thus, have very few free electrons. At elevated temperatures, however, an appreciable number of electrons may be liberated and the gases may become highly conductive. If there are no external electric fields impressed on the gas, the electrons will be distributed uniformly throughout the gas volume. In this instance there is no net electron transport at a particular location in the gas, which, of course, is what we could expect. As an electric field is impressed on the gas, the charged particles are accelerated according to

$$F_e = q_e E = m_e a_e = m_e \frac{dv_e}{dt} \qquad v_e, \ x_e = 0 \text{ at } t = 0 \qquad (9\text{-}43)$$

where q_e is the charge on the electron and E is the electric field strength. Integrating Eq. (9-43),

$$v_e = \frac{dx_e}{dt} = \frac{q_e}{m_e} E t \qquad (9\text{-}44)$$

The electron velocity v_e is randomly oriented in the case of a zero electric field, so that there is no preferred direction of motion. The motion described by Eq. (9-44) then expresses the directed velocity the electron would attain t seconds after the electric field is turned on, assuming it had no collisions.

To take the effects of collisions into account, we assume that the electrons are brought to rest after each collision, reaccelerated by the electric field, and then experience another collision. The distance traveled between each collision is obtained by integrating Eq. (9-44):

$$x_e = \frac{q_e}{2m_e} E t^2$$

Solving Eq. (9-44) for t and inserting the expression for x_e gives

$$x_e = \frac{m_e v_e^2}{2E q_e} \qquad (9\text{-}45)$$

Differentiating,

$$dx_e = \frac{m_e v_e dv_e}{E q_e}$$

The number of electrons having collisions in the distance dx_e is given by the distribution of free paths [Eq. (9-18)]:

$$dN = \frac{N}{\lambda_e} e^{-x_e/\lambda_e} \, dx_e \tag{9-46}$$

where λ_e is the mean free path for the electrons moving among the ions. The average electron velocity between collisions is, therefore,

$$\bar{v}_e = \frac{1}{N} \int v_e \, dN = \frac{1}{N} \int_0^\infty \frac{v_e N}{\lambda_e} e^{-x_e/\lambda_e} \, dx_e$$

$$= \frac{1}{N} \int_0^\infty \frac{v_e^2 N m_e}{\lambda_e E q_e} e^{-m_e v_e^2/2\lambda_e E q_e} \, dv_e$$

$$= \left(\frac{\pi \lambda_e E q_e}{2 m_e}\right)^{1/2} \tag{9-47}$$

The quantity $\bar{v}_e$ is called the *electron drift velocity* and represents the average velocity at which the electrons move through the gas as a result of the impressed electric field. The net flux of charge is, therefore,

$$J_e = n_e q_e \bar{v}_{e_d} \tag{9-48}$$

The flux of charge is called the *current density*. We define the *electric conductivity* K_e by

$$J_e = K_e E \tag{9-49}$$

Using Eqs. (9-48) and (9-47), we obtain

$$K_e = n_e q_e \left(\frac{\pi \lambda_e q_e}{2 m_e E}\right)^{1/2} \tag{9-50}$$

The quantity n_e represents the number density of electrons in the gas. The *electron mobility* μ_e is defined by

$$\bar{v}_{e_d} = \mu_e E \tag{9-51}$$

or

$$\mu_e = \left(\frac{\pi \lambda_e q_e}{2 m_e E}\right)^{1/2} \tag{9-52}$$

To calculate the electron mean free path, the collision cross section may be estimated, as shown in Fig. 9-15. Since the electron is essentially a point

$$\sigma_e = \pi r^2$$

FIG. 9-15 *Collision cross section for electrons.*

compared to the normal molecules with which it collides, the collision cross section is given as

$$\sigma_e = \pi r^2 \tag{9-53}$$

This results in

$$\frac{\lambda_e}{\lambda} = 4$$

or, since the collision cross section for an electron is only one-fourth that of the larger molecules, the electrons must travel, on the average, four times as far between collisions.

In the foregoing analysis of electric conductivity we considered only electron motion. If the overall charge of the gas is neutral, there must be a number of positive ions equal to the number of free electrons to balance the charge. The ions are also influenced by the electric field and have some drift velocity which contributes to the electric conduction. However, the mass of the ions is much larger than that of the electrons. If thermal equilibrium is assumed so that both ions and electrons have the same average energy, the electrons will have a much higher velocity than the ions. In addition, the larger mass of the ions means that they will not be accelerated as much by the electric field. Thus it is expected that the transport of charge by electron motion is much larger than the transport due to ion motion, and the foregoing analysis should furnish a reasonable estimate for electric conductivity, even though it neglects the transport of charge by the ions.

9-14 INFLUENCE OF MOLECULAR FORCE FIELDS ON TRANSPORT PROCESSES

The preceding analyses assumed a physical model of "billiard-ball" molecules with no surrounding force fields. In Chap. 8 we have already noted that molecular force fields can play a significant role in determining molecular energy levels and interactions. This is particularly true for collision processes. Consider the typical molecular collision shown in Fig. 9-16. The

FIG. 9-16   *Typical molecular collision.*

molecular force field between the two molecules behaves approximately as shown in Fig. 9-17. At large distances the force field is mildly attractive. As the molecules come closer together, the force becomes more attractive and then strongly repulsive. In some instances the strong repulsive force prevents the molecules from ever coming into actual contact. A "collision," therefore, must be described in a new way. In the example shown in Fig. 9-16 we expect molecule 1 to be diverted from its path when it is appreciably influenced by the force field of molecule 2. The molecular collision cross section is, therefore, described in terms of the parameter b, which is called the *impact parameter*. Then

where b_m is the maximum value of b for which molecule 1 will be appreciably altered in its course.

It is fairly easy to see that the molecular deflection angle is dependent on the value of b and the relative velocity for the molecules as well as on the

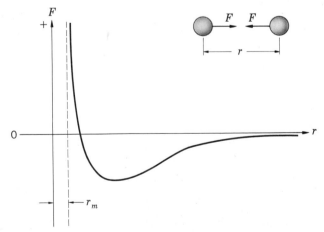

FIG. 9-17   *Typical molecular force field.*

strength of the molecular force fields. A complete analysis of molecular collisions must take all these factors into account in order to calculate transport properties. This analysis is quite beyond the scope of our discussion, but it is well to alert the reader to the mechanism of the effects of molecular force fields on these transport properties.

Some additional words on collision cross sections are in order. In all of the previous discussions the cross section has been described as the area presented to an incoming molecule which will cause a collision. If the incoming molecule enters this area, it will have a collision. The various transport properties described in Sec. 9-7 were calculated with the assumption that there is complete transfer of a property (momentum, energy, identity) in each collision. In other words, we assumed that when one molecule hits another, it transfers momentum, energy, and molecular identity all in the same proportion. This assumption, however, can be quite wrong. Consider the very simple exchange process shown in Fig. 9-18. The incoming molecule hits the stationary wall and experiences a perfectly elastic collision. Thus it bounces off the wall with the same velocity magnitude v, and we assume that $\phi_1 = \phi_2$. Even in this perfectly elastic collision there has been a momentum transfer *to the wall* in the amount

$$2mv \cos \phi$$

Note, however, that the *energy* transfer to the wall is *zero*, since there is no change in kinetic energy of the particle.

The point of these remarks is that the detailed interaction process between molecules can influence energy and momentum transfer in different ways. It is frequently necessary, therefore, to use different collision cross sections for energy and momentum exchange. These collision cross sections will be strongly dependent on the type of gas molecule (monatomic, diatomic, etc.) and on the molecular force fields.

An important parameter employed in many practical heat-transfer problems involving gas flows is the Prandtl number Pr:

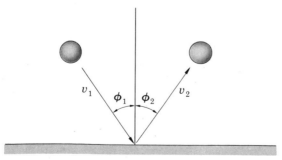

FIG. 9-18   *Elastic collision with a wall.*

$$Pr = \frac{c_p \mu}{K} = \gamma \frac{c_v \mu}{K} \qquad (9\text{-}55)$$

where $\gamma = c_p/c_v$. Assuming that the gas is in a temperature region such that equipartition of energy applies, we have

$$\bar{c}_v = \frac{f}{2} \mathscr{R}$$

$$\bar{c}_p = \frac{f + 2}{2} \mathscr{R}$$

$$\gamma = \frac{f + 2}{f}$$

$$K = \frac{1}{3} \frac{\bar{v} \bar{c}_v}{N_0 \sigma}$$

$$\mu = \frac{1}{3} \frac{m \bar{v}}{\sigma}$$

Combining these relations, we have

$$Pr = \frac{f + 2}{f} N_0 m = \frac{f + 2}{f} M \qquad \bar{c}_v \text{ on molal basis}$$

where $M = N_0 m$ is the molecular weight of the gas. If the specific heat is expressed on a mass basis, this relation is simply divided by the molecular weight to obtain

$$Pr = \frac{f + 2}{f} \qquad (9\text{-}56)$$

It will be recalled that $f = 3$ for a monatomic gas and $f = 5$ for a diatomic gas at a temperature sufficiently low to ensure that vibrational energy states are not excited. Thus we anticipate:

$$Pr = \tfrac{5}{3} = 1.667 \qquad \text{ideal monatomic gas}$$
$$Pr = \tfrac{7}{5} = 1.40 \qquad \text{ideal diatomic gas}$$

Table 9-2 gives a summary of experimental values of the Prandtl number for several gases. As can be seen, the estimated values are not in agreement with experiment. The reason for the disagreement is the basically different character of energy and momentum exchange in molecular collisions previously discussed and the inadequacy of the simple billiard-ball molecular model.

Even though some disagreement between theory and experiment has been encountered, this simple presentation of transport phenomena has developed quite a number of important concepts which are useful in acquiring a basic physical understanding of the processes involved.

TABLE 9-2
Prandtl numbers for various gases

Gas	Prandtl number, $Pr = c_p \mu / K$
He	0.69
Ne	0.68
A	0.69
H_2	0.71
N_2	0.72
O_2	0.71

REVIEW QUESTIONS

1 Why is it possible to use the MB distribution to calculate transport rates?

2 Define *mean free path*.

3 What is meant by the term *average collision distance?*

4 Distinguish between the transport processes associated with thermal conductivity and viscosity.

5 Describe the phenomenon of mutual diffusion.

6 Describe the phenomenon of thermal diffusion.

7 How does electron transport in gases differ from that associated with ordinary diffusion of gas molecules?

8 What is a property?

9 When does MB distribution apply?

10 How are mean molecular speeds related to temperature?

11 What is a collision crossection?

12 What is *collision frequency?*

13 How are free paths distributed?

14 What is meant by *molecular flux?*

15 What is Fourier's law?

16 When does equipartition of energy apply?

17 How is the diffusion coefficient defined?

18 What is meant by "electron drift velocity"?

19 What is meant by "electron mobility"?

20 What is the Prandtl number?

21 Distinguish between a *velocity* distribution and a *speed* distribution.

PROBLEMS

9-1 A chamber contains O_2 at 300 K and 10^{-8}-atm pressure (1 atm = 1.013 × 10^5 N/m²). The effective diameter of the oxygen molecule may be taken as 3.5 × 10^{-10} m. The chamber is placed in outer space and a small hole is pierced in the side having an area of 0.01 mm². Estimate the number of molecules which escape from the container per unit time.

9-2 Calculate the mean free path of the molecules in Prob. 9-1.

9-3 The mean free path of the molecules of a certain gas at a temperature of 20°C is 3.00 × 10^{-5} m. If the radius of the molecule is 3.00 × 10^{-10} m, find the pressure of the gas and calculate the number of collisions per meter of path made by a molecule.

9-4 Using simple kinetic theory, estimate the dynamic viscosity of helium at 400 K.

9-5 The viscosity of N_2 at 0°C and 1-atm pressure is 16.6 × 10^{-6} N·s/m². Estimate the effective molecular diameter from this information.

9-6 We might reasonably assume that the heat of vaporization for a fluid would be the energy necessary to increase the velocity of the molecules from an essentially zero value in the liquid state to the rms value representative of the mean energy in the vapor state. Check the validity of this assumption for vaporization of water at 1-atm pressure.

9-7 What fraction of molecules have speeds (a) greater than v_m, (b) less than v_m, (c) greater than some arbitrary velocity v?

9-8 What fraction of molecules have velocities terminating in the solid angle for which (a) $0 < \phi < 20°$, $0 < \theta < 2\pi$, (b) $0 < \phi < 10°$, (c) $0 < \phi < 30°$?

9-9 A cathode-ray tube must be evacuated to low pressures to enable the electrons emitted by the cathode to reach the screen without having a collision. Assuming that the gas remaining in the tube has the properties of oxygen, estimate the required absolute pressure to ensure that (a) 90 percent and (b) 98 percent of the electrons reach the screen which is located a distance of 0.3 m away from the cathode.

9-10 Consider a reversible adiabatic expansion of an ideal gas in a cylinder. The piston is assumed to move with some velocity u such that $u \ll \bar{v}$. For a molecule impinging on the piston at an incident angle ϕ, show that the decrease in kinetic energy of the molecule is

$$2mvu \cos \phi$$

where v is the velocity at impact. By a suitable integration obtain an expression for the kinetic energy lost by *all* molecules, and show that this energy is equal to the work done on the piston by the expanding gas.

9-11 A container has two identical chambers separated by an appropriate membrane. One chamber is filled with a gas and the other chamber is evacuated. A small pinhole of area A is punched in the membrane. Derive an expression for the time required for the pressure to drop to three-fourths of its original value.

9-12   Derive an expression for the rms free path in terms of the mean free path. What is the most probable free path?

9-13   Obtain an expression for the distribution law of *free times* defined as the time required to execute a free path. Also obtain an expression for the mean free time.

9-14   Calculate v_{rms} for H_2, He, Hg, and N_2 molecules at 0°C. Repeat the calculations for free electrons.

9-15   Positive ions of N_2 are subjected to an electric field of 10^6 V/m. The ions move through N_2 at 1 atm and 300 K. Calculate the average drift velocity of the ions and compare this velocity with the rms velocity of the gas. Also calculate the distance an ion moves in the direction of the field in a time equal to that required to traverse one mean free path.

REFERENCES

1   Lee, J. F., F. W. Sears, and D. L. Turcotte: "Statistical Thermodynamics," Addison-Wesley Publishing Company, Inc., Reading, Mass., 1963.

2   Chapman, S., and T. G. Cowling: "The Mathematical Theory of Non-Uniform Gases," Cambridge University Press, Cambridge, 1953.

3   Present, R. D.: "Kinetic Theory of Gases," McGraw-Hill Book Company, New York, 1958.

4   Kennard, E. H.: "Kinetic Theory of Gases," McGraw-Hill Book Company, New York, 1938.

5   Jeans, J. H.: "The Dynamical Theory of Gases," Dover Publications, Inc., New York, 1954.

6   Cobine, J. D.: "Gaseous Conductors," Dover Publications, Inc., New York, 1958.

chapter ten
Gaseous mixtures

10-1 INTRODUCTION

Previous chapters have been concerned with the calculation and use of thermodynamic properties for substances which require only two independent properties to define the thermodynamic state. In this chapter we shall examine the behavior of mixtures of ideal gases and ideal-gas–vapor mixtures which require additional information pertaining to composition in order to establish the thermodynamic state. This presentation will then form a basis for studies of chemical equilibrium and combustion reactions to be developed in Chap. 11. Calculations involving gas-vapor mixtures are particularly applicable to air conditioning processes and, thus, will merit examination in this chapter. Following the presentation of ideal-gas mixtures we shall discuss some methods for calculating properties of real-gas mixtures and relate the procedures to the equation of state material in Chap. 7.

10-2 THE GIBBS–DALTON LAW

Let us begin our discussion by considering some basic definitions applicable to a mixture of ideal gases. The total mass of the mixture is clearly the sum of the masses of each component, so that

$$m = m_1 + m_2 + \cdots + m_j = \sum_{i=1}^{j} m_i \qquad (10\text{-}1)$$

The *mole fraction* of a component x_i is defined by

$$x_i = \frac{n_i}{n} \qquad (10\text{-}2)$$

where n_i is the number of moles of the ith component and n is the total number of moles of the mixture. The sum of the fractions is equal to the whole, or

$$\Sigma x_i = 1.0 \qquad (10\text{-}3)$$

The mass of each component can be calculated from

$$m_i = n_i M_i \qquad (10\text{-}4)$$

where M_i is the molecular weight of the ith component. Now, if Eq. (10-4) is inserted into Eq. (10-1), we obtain

$$m = \sum_i n_i M_i \tag{10-5}$$

An equivalent molecular weight for the particular mixture may be defined as

$$M = \frac{m}{n} = \sum \frac{n_i}{n} M_i$$

or

$$M = \sum x_i M_i \tag{10-6}$$

An equivalent gas constant for the particular mixture may then be calculated by

$$R = \frac{\mathscr{R}}{M} \tag{10-7}$$

As long as ideal-gas behavior is studied, the assumption is made that each gaseous component of the mixture behaves as if it existed at the temperature of the mixture and filled the entire volume occupied by the mixture. This assumption is a statement of the *Gibbs-Dalton law*. From the discussions in Chaps. 8 and 9 it is clear that an ideal gas is one in which there is negligible influence of molecular force fields; hence, in a mixture of ideal gases, we would not expect appreciable interference between the different molecular components.

The *partial pressure* of the ith component is defined as

$$p_i = \frac{n_i \mathscr{R} T}{V} \tag{10-8}$$

In other words, this is the pressure the component would exert if it occupied the entire volume of the mixture V at the temperature of the mixture T. The *total pressure* of the mixture is defined as

$$p = \frac{n \mathscr{R} T}{V} \tag{10-9}$$

Dividing Eq. (10-8) by Eq. (10-9),

$$\frac{p_i}{p} = \frac{n_i}{n} = x_i \tag{10-10}$$

Applying Eq. (10-3),

$$\Sigma p_i = \Sigma x_i p = p \qquad (10\text{-}11)$$

or the sum of the partial pressures equals the total pressure.

Now suppose that we separate a mixture of ideal gases into its individual components and place each component in a container such that it is maintained at the pressure and temperature of the total mixture. This separation process is indicated in Fig. 10-1. The volume occupied by each component in this case would be

$$V_i = \frac{n_i \mathscr{R} T}{p} \qquad (10\text{-}12)$$

The total volume of the mixture at conditions p and T is

$$V = \frac{n \mathscr{R} T}{p} \qquad (10\text{-}13)$$

Dividing Eq. (10-12) by Eq. (10-13),

$$\frac{V_i}{V} = \frac{n_i}{n} = x_i \qquad (10\text{-}14)$$

Applying Eq. (10-3),

$$\Sigma V_i = \Sigma x_i V = V \qquad (10\text{-}15)$$

Equation (10-15) is called the *Amagat-Leduc law of additive volumes*, or the total volume of a mixture of ideal gases is the sum of the volumes each com-

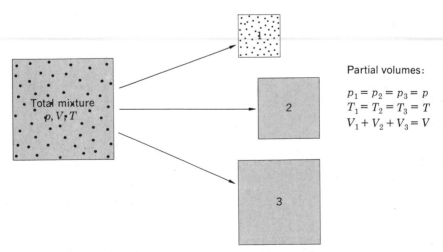

Partial volumes:

$$p_1 = p_2 = p_3 = p$$
$$T_1 = T_2 = T_3 = T$$
$$V_1 + V_2 + V_3 = V$$

FIG. 10-1 *Illustration of Amagat-Leduc law of additive volumes.*

ponent gas would occupy if contained at the pressure and temperature of the mixture. V_i is usually called the *partial volume* of the ith component.

The concentrations of the constituents in a mixture may be expressed on either a mass or molal basis. Using a molal basis the partial pressures or partial volumes are obtained very easily from either Eq. (10-10) or (10-14). Sometimes the concentrations in a mixture are expressed on a "volume" basis. The meaning of this term is that the concentrations are expressed in terms of the partial volumes. So, the term *volume basis*, or *volume fraction*, means the same thing as molal basis, or mole fraction, in accordance with Eq. (10-14).

EXAMPLE 10-1

A rigid tank contains 2 lbm of nitrogen at 300 psia and 100°F. A sufficient quantity of oxygen is added to increase the pressure to 400 psia while the temperature remains constant at 100°F. Calculate the mass of oxygen added.

SOLUTION   In this process the number of moles of nitrogen remains constant at

$$n_{N_2} = \frac{m}{M} = \frac{2}{28} = 0.0714$$

After the oxygen is added, the partial pressure of the nitrogen will be 300 psia since the volume does not change. We, thus, have

$$
\begin{aligned}
p_{O_2} &= p_{total} - p_{N_2} \\
&= 400 - 300 \\
&= 100 \text{ psia} \quad (6.89 \times 10^5 \text{ N/m}^2)
\end{aligned}
$$

From Eq. (10-8) we can write, using a ratio,

$$\frac{n_{O_2}}{n_{N_2}} = \frac{p_{O_2}}{p_{N_2}}$$

Using the numerical values given in the foregoing,

$$n_{O_2} = \frac{(0.0714)(100)}{(300)} = 0.0238$$

Finally, the mass of oxygen is calculated as

$$
\begin{aligned}
m_{O_2} &= M_{O_2} n_{O_2} = (32)(0.0238) \\
&= 0.761 \text{ lbm} \quad (0.345 \text{ kg})
\end{aligned}
$$

EXAMPLE 10-2

Calculate the volumes which would be occupied by the oxygen and nitrogen in Example 10-1 if each gas were contained separately at the temperature and pressure of the mixture, i.e., 100°F and 400 psia.

SOLUTION   The volume of nitrogen is calculated from the ideal-gas law (Eq. 10-8):

$$V_{N_2} = \frac{n\mathscr{R}T}{p} = \frac{(0.0714)(1545)(560)}{(400)(144)}$$
$$= 1.071 \text{ ft}^3 \qquad (0.0303 \text{ m}^3)$$

For the oxygen,

$$V_{O_2} = \frac{(0.0238)(1545)(560)}{(400)(144)}$$
$$= 0.357 \text{ ft}^3 \qquad (0.0101 \text{ m}^3)$$

The total volume of the mixture would be

$$V = 1.071 + 0.357$$
$$= 1.428 \text{ ft}^3 \qquad (0.0404 \text{ m}^3)$$

It should be observed that the volume of each constituent is in direct proportion to the number of moles of that constituent.

10-3   ENERGY PROPERTIES OF MIXTURES

The principles and definitions of the preceding section may be easily extended to facilitate the calculation of energy properties of mixtures. Consider the internal energy. We may write

$$U = U_1 + U_2 + U_3 + \cdots = \Sigma U_i = \Sigma m_i u_i \qquad (10\text{-}16)$$

or, on a molal basis,

$$U = \Sigma n_i \bar{u}_i \qquad (10\text{-}17)$$

where $\bar{u}_i$ designates the internal energy per mole for the ith component. The specific internal energy for the mixture is then calculated with

$$u = \frac{1}{m} \sum m_i u_i \tag{10-18}$$

or the molal specific internal energy is

$$\bar{u} = \frac{1}{n} \sum n_i \bar{u}_i = \sum x_i \bar{u}_i \tag{10-19}$$

Relations for the enthalpy and entropy of the mixture are developed in a similar manner with the following results:

$$H = \Sigma H_i = \Sigma m_i h_i \tag{10-20a}$$

$$H = \Sigma n_i \bar{h}_i \tag{10-20b}$$

$$\bar{h} = \Sigma x_i \bar{h}_i \tag{10-21}$$

$$S = \Sigma S_i = \Sigma m_i s_i \tag{10-22a}$$

$$S = \Sigma n_i \bar{s}_i \tag{10-22b}$$

$$\bar{s} = \Sigma x_i \bar{s}_i \tag{10-23}$$

Specific heats for the mixture are obtained by applying the definitions

$$c_v = \left(\frac{\partial u}{\partial T}\right)_v \quad \text{and} \quad c_p = \left(\frac{\partial h}{\partial T}\right)_p$$

to the foregoing relations for internal energy and enthalpy. For the specific heat at constant volume, we have

$$c_v = \left[\frac{\partial}{\partial T}\left(\frac{1}{m}\sum m_i u_i\right)\right]_v$$

$$= \frac{1}{m}\sum m_i \left(\frac{\partial u_i}{\partial T}\right)_v$$

$$c_v = \frac{1}{m}\sum m_i c_v \tag{10-24}$$

On a molal basis,

$$\bar{c}_v = \left(\frac{\partial \bar{u}}{\partial T}\right)_v = \sum x_i \left(\frac{\partial \bar{u}_i}{\partial T}\right)_{v_i}$$

$$\bar{c}_v = \Sigma x_i \bar{c}_{v_i} \tag{10-25}$$

The specific heat at constant pressure is developed in a similar fashion with the results

$$c_p = \frac{1}{m} \sum m_i c_{pi} \tag{10-26}$$

$$\bar{c}_p = \sum x_i \bar{c}_{p_i} \tag{10-27}$$

In the study of thermodynamic *processes* involving ideal-gas mixtures, it is of interest to calculate *changes* in the energy properties of the mixture. Such a calculation proceeds on the basic notion incorporated in the Gibbs-Dalton law that each component of the mixture occupies the entire volume of the mixture at the temperature of the mixture. The change of internal energy of the mixture would be calculated by

$$\Delta U = \sum \Delta U_i = \sum m_i \, \Delta u_i = \sum m_i c_{v_i} \, \Delta T \tag{10-28}$$

or, making use of Eq. (10-24),

$$\Delta U = mc_v \, \Delta T \tag{10-29}$$

where now c_v is the mass specific heat of the mixture. ΔU could also be calculated from

$$\Delta U = n\bar{c}_v \, \Delta T \tag{10-30}$$

Similarly, the change in enthalpy for the mixture could be written as

$$\Delta H = \sum \Delta H_i = \sum m_i \, \Delta h_i = \sum m_i c_{p_i} \, \Delta T \tag{10-31}$$

$$\Delta H = mc_p \, \Delta T \tag{10-32}$$

$$\Delta H = n\bar{c}_p \, \Delta T \tag{10-33}$$

The computation of the change in entropy of a mixture is a bit more complicated since the entropy is not a function of temperature alone. From Eq. (6-32) the change in entropy for an ideal gas with constant specific heats is

$$s_2 - s_1 = c_p \ln \frac{T_2}{T_1} - R \ln \frac{p_2}{p_1} \tag{10-34}$$

where the subscripts 1 and 2 pertain to the initial and final states, respectively. The change in entropy for a mixture of ideal gases is calculated by applying Eq. (10-34) to each component and summing the results. Thus

$$\Delta S = S_2 - S_1 = \sum m_i c_{pi} \ln \frac{T_2}{T_1} - \sum m_i R_i \ln \frac{p_{i_2}}{p_{i_1}} \tag{10-35}$$

where p_{i_1} and p_{i_2} are the initial and final *partial pressures* of the ith component, respectively. An alternate expression for the entropy change may be written on a molal basis. Thus

$$\Delta S = S_2 - S_1 = \sum n_i \bar{c}_{pi} \ln \frac{T_2}{T_1} - \sum n_i \mathscr{R} \ln \frac{p_{i_2}}{p_{i_1}} \qquad (10\text{-}36)$$

The following example illustrates the use of these relations.

EXAMPLE 10-3

Two lbm of CO_2 at 100°F and 20 psia are mixed with 5 lbm of N_2 at 300°F and 15 psia to form a mixture at a final pressure of 10 psia. The process occurs adiabatically in a steady-flow apparatus. Calculate the final temperature of the mixture and the change in entropy.

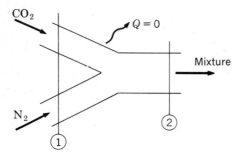

FIG. EXAMPLE 10-3

SOLUTION   The process is shown schematically in the accompanying figure. The steady-flow energy balance is

$$(mh_1)_{N_2} + (mh_1)_{CO_2} = (mh_2)_{mixture}$$
$$= (mh_2)_{N_2} + (mh_2)_{CO_2}$$

Assuming ideal-gas behavior with constant specific heats, this equation becomes

$$[mc_p(T_1 - T_2)]_{N_2} + [mc_p(T_1 - T_2)]_{CO_2} = 0$$

Inserting the appropriate numerical values,

$$(5)(0.248)(300 - T_2) + (2)(0.203)(100 - T_2) = 0$$

since the final temperature for both constituents is the same, i.e., the temperature of the mixture. This equation yields

$$T_2 = 250°F = 710°R \qquad (394\ K)$$

The change in entropy may be calculated with either Eq. (10-35) or (10-36). For either equation the final partial pressures of the constituents must be computed. The molal quantities are:

$$n_{CO_2} = \tfrac{2}{44} = 0.0455$$
$$n_{N_2} = \tfrac{5}{28} = 0.1785$$
$$n_{total} = 0.1785 + 0.0455$$
$$= 0.224$$

The final partial pressures may now be calculated from Eq. (10-10):

$$(p_2)_{CO_2} = \frac{(10)(0.0455)}{0.224} = 2.03 \text{ psia} \qquad (14 \text{ kPa})$$

$$(p_2)_{N_2} = \frac{(10)(0.1785)}{0.224} = 7.97 \text{ psia} \qquad (55 \text{ kPa})$$

The change in entropy is now calculated with Eq. (10-35):

$$S_2 - S_1 = \left(mc_p \ln \frac{T_2}{T_1} - mR \ln \frac{p_2}{p_1} \right)_{CO_2} + \left(mc_p \ln \frac{T_2}{T_1} - mR \ln \frac{p_2}{p_1} \right)_{N_2}$$

Inserting the appropriate numerical values,

$$S_2 - S_1 = (2) \left(0.203 \ln \frac{710}{560} - \frac{35.1}{778} \ln \frac{2.03}{20} \right)$$
$$+ (5) \left(0.248 \ln \frac{710}{760} - \frac{55.2}{778} \ln \frac{7.97}{15} \right)$$
$$= 0.443 \text{ Btu/°R} \qquad (841 \text{ J/K})$$

10-4 MIXTURES OF AN IDEAL GAS AND A VAPOR

Let us now consider an important type of gaseous mixture, that of an ideal gas and a condensable vapor. This class of mixture is encountered daily in the air we breathe and in the heating and air conditioning processes associated therewith. The formation of dew on the grass on a still night, the drying and humidification of gases for various applications, and all air conditioning calculations depend on an understanding of the behavior of gas-vapor mixtures.

We shall be concerned only with mixtures involving a single condensable vapor, and shall focus our attention on air–water-vapor mixtures because of the wide range of practical circumstances in which they are encountered.

The *dew point* of a mixture is the temperature at which the vapor starts to condense when the mixture is cooled at constant pressure. The *dry-bulb temperature* of a mixture is the temperature indicated by an ordinary thermometer placed in the mixture. The *wet-bulb temperature* is the temperature indicated by a thermometer covered with a wicklike material saturated with liquid, after the arrangement has been allowed to reach evaporation equilibrium with the mixture. These temperatures are indicated schematically in Fig. 10-2.

A gas-vapor mixture is said to be *saturated* when a reduction in temperature would cause part of the vapor to condense. Therefore, a saturated mixture does exist at the dew point, and the partial pressure of the vapor in such a mixture is the saturation pressure corresponding to the temperature of the mixture.

The *relative humidity* ϕ is defined as the ratio of the actual mass of vapor to the mass of vapor required to produce a saturated mixture at the same temperature. If the vapor behaves like an ideal gas, we can write

$$\phi = \frac{m_v}{m_{sat}} = \frac{p_v V / R_v T}{p_g V / R_v T} = \frac{p_v}{p_g} \tag{10-37}$$

where p_v is the actual partial pressure of the vapor and p_g is the saturation pressure of the vapor at the temperature of the mixture.

The *specific humidity* or humidity ratio ω is refined as the ratio of the mass of vapor to the mass of noncondensable gas. Our primary concern in this chapter will be with air–water-vapor mixtures, so that for these cases

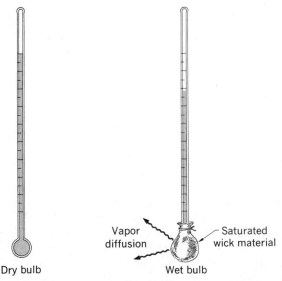

Dry bulb   Vapor diffusion   Wet bulb   Saturated wick material

FIG. 10-2 *Measurement of dry-bulb and wet-bulb temperature.*

$$\omega = \frac{m_v}{m_a} \qquad (10\text{-}38)$$

where m_a designates the mass of air. If the assumption is made that the vapor behaves like an ideal gas, we can write

$$\omega = \frac{p_v V M_v / \mathscr{R} T}{p_a V M_a / \mathscr{R} T} = \frac{M_v}{M_a} \frac{p_v}{p_a} \qquad (10\text{-}39)$$

For air–water-vapor mixtures this expression reduces to

$$\omega = 0.622 \frac{p_v}{p_a} = 0.622 \frac{p_v}{p - p_v} \qquad (10\text{-}40)$$

Equation (10-37) may be introduced to obtain

$$\phi = \frac{\omega p_a}{0.622 p_g} \qquad (10\text{-}41)$$

Another quantity of interest is the *degree of saturation* μ, defined as the ratio of the actual specific humidity to the specific humidity of saturated air at the dry-bulb temperature:

$$\mu = \left. \frac{\omega}{\omega_{\text{sat}}} \right|_T \qquad (10\text{-}42)$$

For ideal-gas behavior this may be expressed as

$$\mu = \frac{0.622 p_v / (p - p_v)}{0.622 p_g / (p - p_g)} = \frac{p_v(p - p_g)}{p_g(p - p_v)} = \phi \frac{p - p_g}{p - p_v} \qquad (10\text{-}43)$$

Because both p_v and p_g are small compared to the total pressure p, the degree of saturation is very nearly equal to the relative humidity for air–water-vapor mixtures at normal room temperatures and pressures.

In accordance with our discussion so far we see that a process which does not add or remove moisture from the mixture occurs at constant specific humidity. If the process occurs at constant total pressure also, then from Eq. (10-40) the partial pressure of the vapor must also remain constant. From these facts we can interpret the dew point of the mixture in a specific way. Recalling that the dew point is the temperature at which vapor will just start to condense when cooled at constant pressure, the vapor partial pressure at the dew point must be the saturation pressure at that temperature. But this is equal to the actual vapor pressure before the mixture is cooled, so

$$p_v = p_g \qquad \text{(evaluated at dew-point temperature)}$$

When the mixture is fully saturated (ϕ = 100 percent) the dry-bulb, wet-bulb, and dew-point temperatures are all the same.

EXAMPLE 10-4   Air–water-vapor mixture

An air–water-vapor mixture exists at 25 psia, 140°F, and 50 percent relative humidity. Calculate the mass fraction of water vapor in the mixture.

SOLUTION   The specific humidity is the quantity of interest in this problem. The saturation pressure at 140°F is

$$p_g = 2.892 \text{ psia} \qquad (1.994 \times 10^4 \text{ Pa})$$

From Eq. (10-37) the actual vapor pressure is

$$p_v = \phi p_g$$
$$= (0.5)(2.892) = 1.446 \text{ psia} \qquad (9970 \text{ Pa})$$

The partial pressure of the air is, therefore,

$$p_a = 25 - 1.446 = 23.554 \text{ psia} \qquad (1.624 \times 10^5 \text{ Pa})$$

The specific humidity may now be calculated with Eq. (10-40):

$$\omega = \frac{(0.622)(1.446)}{23.554}$$
$$= 0.0382 \text{ lbm vapor/lbm air}$$

The mass fraction of water vapor is

$$\frac{m_v}{m_{\text{total}}} = \frac{m_v}{m_a + m_v} = \frac{\omega}{1 + \omega}$$
$$= 0.0368 \text{ lbm vapor/lbm mixture}$$

10-5   ADIABATIC SATURATION

Because so many air–water-vapor processes involve the concepts of humidity and saturation, it is important to examine the energy balances encountered in a simple adiabatic saturation process. The system to be analyzed is shown schematically in Fig. 10-3.

An air–water-vapor mixture is blown over a pool of water until it leaves the container in a saturated state. Fresh liquid is continuously supplied at the exit temperature to compensate for water picked up by the air in the sat-

FIG. 10-3   *Adiabatic saturation process.*

uration process. The container is insulated so that no heat exchange with the surroundings occurs during the process.

A steady-flow energy balance may now be applied to the process. We have

> Enthalpy of mixture in + enthalpy of liquid supplied
>> = enthalpy of mixture out

or,

$$m_a h_{a_1} + m_{v_1} h_{v_1} + (m_{v_2} - m_{v_1}) h_{f_2} = m_a h_{a_2} + m_{v_2} h_{v_2}$$

Dividing by the air flow rate m_a,

$$h_{a_1} + \omega_1 h_{v_1} + (\omega_2 - \omega_1) h_{f_2} = h_{a_2} + \omega_2 h_{v_2} \qquad (10\text{-}44)$$

If ideal-gas behavior is assumed,

$$h_{a_1} - h_{a_2} = c_{pa}(T_1 - T_2)$$

Furthermore,

$$h_{v_2} - h_{f_2} = h_{fg_2}$$

so that Eq. (10-42) may be solved for ω_1 to yield

$$\omega_1 = \frac{c_{pa}(T_2 - T_1) + \omega_2 h_{fg_2}}{h_{v_1} - h_{f_2}} \qquad (10\text{-}45)$$

Equation (10-45) is useful in that the specific humidity of a mixture may be determined through a measurement of T_1 and T_2 and a calculation with Eq. (10-45). The wet-bulb temperature of a mixture is very nearly equal to the adiabatic saturation temperature. It differs slightly because of the mass diffusion processes which occur at the surface of the "wick" and the temperature gradients associated therewith.

The foregoing discussion leads to the conclusion that a specification of the wet-bulb and dry-bulb temperatures of a mixture serves to define the humidity state of the mixture. This information may then be used to calculate the energy properties of the mixture.

For most psychrometric mixtures the partial pressure of the water vapor is low, and simplified relations may be used to calculate thermodynamic properties. For water vapor at low pressures almost ideal-gas behavior is experienced and the enthalpy and internal energy are essentially functions of temperature alone. Using the same reference levels as for the steam tables (0°C, saturated liquid), it is possible to give a summary of low-pressure water properties as shown in Table 10-1. Some of these relations have already been given in Chap. 3 for the solid-vapor saturation region.

Below freezing temperatures it is possible, of course, to have frost or ice formation on cooling coils, and an energy balance on the system must take proper account of the energy required to effect the change of phase from water vapor to solid ice. It is interesting to note that in the sublimation region between -40 and $0°C$ the enthalpy of sublimation is very nearly constant at

TABLE 10-1
Some approximate formulas for properties of water at low pressures

Property	Units	Temp. range	Equation number
$h_f = u_f = (T - 32)$ Btu/lbm	T in °F	32 to 100°F	10-46a
$h_f = u_f = 4.025T$ kJ/kg	T in °C	0 to 40°C	10-46b
$h_g = (1061 + 0.445T)$ Btu/lbm	T in °F	-40 to 100°F	10-47a
$h_g = (2501 + 1.863T)$ kJ/kg	T in °C	-40 to 40°C	10-47b
$u_g = (1010.3 + 0.335T)$ Btu/lbm	T in °F	-40 to 100°F	10-48a
$u_g = (2374.9 + 1.403T)$ kJ/kg	T in °C	-40 to 40°C	10-48b
$u_i = h_i = (-158.9 + 0.467T)$ Btu/lbm	T in °F	-40 to 32°F	10-49a
$u_i = h_i = (-334.6 + 1.96T)$ kJ/kg	T in °C	-40 to 0°C	10-49b
$p_g = 5.103 \exp\left(18.42 - \dfrac{11\ 059}{T}\right)$ psia	T in °R	-40 to 32°F	10-50a
$\ln\left(\dfrac{p_g}{0.3390}\right) = 12\ 221\left(\dfrac{1}{527.67} - \dfrac{1}{T}\right)$ $-\ 5.031 \ln\left(\dfrac{T}{527.67}\right)$	T in °R p_g in psia	32 to 100°F	10-51a
$p_g = 35.18 \exp\left(18.42 - \dfrac{6144}{T}\right)$ kN/m²	T in K	-40 to 0°C	10-50b
$\ln\left(\dfrac{p_g}{2337}\right) = 6789\left(\dfrac{1}{293.15} - \dfrac{1}{T}\right)$ $-\ 5.031 \ln\left(\dfrac{T}{293.15}\right)$	T in K p_g in N/m²	0 to 40°C	10-51b

$$h_g - h_i = h_{ig} = 2838 \text{ kJ/kg} = 1220 \text{ Btu/lbm} \tag{10-52}$$

Using a relation like Eq. (10-47), Eq. (10-45) may be modified to enable a calculation of actual vapor pressure directly from dry-bulb and wet-bulb temperatures. The resulting relation is called *Carrier's equation* [6] and is given as

$$p_v = p_{g_w} - \frac{(p - p_{g_w})(T_{DB} - T_{WB})}{K_w - T_{WB}} \tag{10-53}$$

where
p_v = actual vapor pressure
p_{g_w} = saturation pressure corresponding to wet-bulb temperature
p = total pressure of mixture
T_{DB} = dry-bulb temperature, °F or °C
T_{WB} = wet-bulb temperature, °F or °C
K_w = 2800°F or 1537.8°C

For wet-bulb temperatures below 32°F, Doolittle [1] suggests that the following relation be employed:

$$p_v = p_{g_w} - \frac{p(T_{DB} - T_{WB})}{K_L} \tag{10-54}$$

K_L = 3160°F or 1756°C

EXAMPLE 10-5 Closed-system cooling of air–water-vapor mixture
An air–water-vapor mixture is contained in a 4000-ft³ tank at 30 psia and 100°F. The relative humidity is 80 percent. How much cooling is required to lower the temperature to 40°F, and what is the final pressure of the air–water-vapor mixture?

SOLUTION At 100°F, $p_g = 0.9503$ psia, so the initial vapor pressure is obtained from

$$\phi_1 = \frac{p_{v_1}}{p_{g_1}} \qquad p_{v_1} = (0.9503)(0.8) = 0.7602 \text{ psia} \qquad (5241 \text{ Pa})$$

At 40°F, $p_g = 0.1217$ psia, so the final state must be saturated. The initial partial pressure of the air is

$$p_{a_1} = p_1 - p_{v_1}$$
$$= 20 - 0.7602 = 19.2398 \text{ psia} \qquad (1.327 \times 10^5 \text{ Pa})$$

During the cooling process the volume remains constant, so the final partial pressure of the air may be calculated from

$$\frac{p_{a_2}}{p_{a_1}} = \frac{T_2}{T_1} \qquad p_{a_2} = \frac{(19.2398)(460+40)}{(460+100)}$$

$$= 17.178 \text{ psia} \qquad (1.184 \times 10^5 \text{ Pa})$$

The final total pressure is therefore

$$p_2 = p_{a_2} + p_{v_2}$$
$$= 17.178 + 0.1217 = 17.3 \text{ psia} \qquad (1.193 \times 10^5 \text{ Pa})$$

The mass of air is calculated from

$$m_a = \frac{p_{a_1} V}{R T_1}$$

$$= \frac{(19.2398)(144)(4000)}{(53.35)(560)} = 371 \text{ lbm} \qquad (168 \text{ kg}) \qquad (a)$$

and the specific humidities are

$$\omega = 0.622 \frac{p_v}{p_a} \qquad\qquad (b)$$

$$\omega_1 = \frac{(0.622)(0.7602)}{19.2398} = 0.0246$$

$$\omega_2 = \frac{(0.622)(0.1217)}{17.178} = 0.004\,41$$

Because the tank is a closed system the cooling is calculated from the following:

$$Q = U_1 - U_2$$
$$= m_a(u_{a_1} + \omega_1 u_{v_1}) - m_a(u_{a_2} + \omega_2 u_{v_2}) - m_a(\omega_1 - \omega_2)u_{f_2} \qquad (c)$$

The value of u_{f_2} is obtained from Table A-7:

$$u_{f_2} = 8.02 \text{ Btu/lbm} \qquad (\text{saturated liquid at } 40°F)$$

and the vapor internal energies from Eq. (10-48):

$$u_{v_1} = 1010.3 + 0.335(100) = 1043.8 \text{ Btu/lbm}$$
$$u_{v_2} = 1010.3 + 0.335(40) = 1023.7 \text{ Btu/lbm}$$

We now insert the numerical values in Eq. (c) and obtain

$$Q = (371)[(0.1715)(100 - 40) + (0.0246)(1043.8) - (0.004\,41)(1023.7)$$
$$- (0.0246 - 0.004\,41)(8.02)]$$
$$= (371)(10.29 + 25.68 - 4.51 - 0.16)$$
$$= 11\,612 \text{ Btu} \qquad (12\,251 \text{ kJ})$$

EXAMPLE 10-6

Calculate the relative humidity of an air stream having $T_{DB} = 100°F$, $T_{WB} = 80°F$, and a total pressure of 14.696 psia.

SOLUTION   We shall employ Eq. (10-53) to calculate the vapor partial pressure and then determine ϕ. The quantities of interest are:

$$p_{gw} = 0.5073 \text{ psia} \quad (3498 \text{ Pa})$$
$$p = 14.696 \text{ psia} \quad (1.0132 \times 10^5 \text{ Pa})$$

Then, from Eq. (10-53)

$$p_v = 0.5073 - \frac{(14.696 - 0.5073)(100 - 80)}{2800 - 80}$$

$$Pa = 0.4030 \text{ psia} \quad (2778 \text{ Pa})$$

The saturation pressure at 100°F is

$$p_g = 0.9503 \text{ psia} \quad (6552 \text{ Pa})$$

so that, from Eq. (10-37),

$$\phi = \frac{p_v}{p_g} = \frac{0.4030}{0.9503}$$

$$= 42.4 \text{ percent}$$

EXAMPLE 10-7   Helium–water-vapor mixture

A mixture of helium and water vapor at a total pressure of 1 atm and a dry-bulb temperature of 35°C has a dew point of 20°C. Calculate the relative humidity and specific humidity of the mixture.

SOLUTION   Assuming ideal-gas behavior, the actual vapor pressure will be the saturation pressure evaluated at the dew point or

$$p_v = p_g \text{ at } 20°C = 2.339 \text{ kPa} \qquad (a)$$

The saturation pressure at the dry-bulb temperature is

$$p_g = 5.628 \text{ kPa} \quad \text{at } 35°C$$

so the relative humidity is

$$\phi = \frac{p_v}{p_g} = \frac{2.339}{5.628} = 41.6 \text{ percent} \qquad (b)$$

The specific humidity may be calculated from Eq. (10-39) by replacing the molecular weight and partial pressure for air with those of helium.

$$\omega = \frac{M_v}{M_{He}} \frac{p_v}{p_{He}} \qquad (c)$$

The helium partial pressure is

$$p_{He} = p - p_v = 101.32 - 2.339 = 98.981 \text{ kPa}$$

Using $M_{He} = 4.0$ and inserting the values in (c) gives

$$\omega = \frac{(18)(2.339)}{(4)(98.981)} = 0.106 \text{ kg vapor/kg dry helium}$$

EXAMPLE 10-8 Iterative calculation

An air–water-vapor mixture at a total pressure of 150 kPa has a dry-bulb temperature of 35°C and a relative humidity of 60 percent. Calculate the wet-bulb temperature and dew point.

SOLUTION At 35°C the saturation pressure is

$$p_g = 5.628 \text{ kPa}$$

so, from Eq. (10-37) we may calculate the actual vapor pressure as

$$p_v = \phi p_g = (0.6)(5.628) = 3.3768 \text{ kPa}$$

The dew point is the saturation temperature corresponding to this pressure or

$$T_{DP} = 25.96°C$$

To determine the wet-bulb temperature we must employ the Carrier relation of Eq. (10-53) and an iterative procedure. Rewriting Eq. (10-53) in the form

$$p_v - p_{g_w} + \frac{(p - p_{gw})(T_{DB} - T_{WB})}{1537.8 - T_{WB}} = 0 = f(T_{WB})$$

we see that with p_v and T_{DB} known, the equation involves only one unknown, T_{WB}, because p_{g_w} is a function of T_{WB}. This is a nonlinear function so we must use tabular values and iterate until we obtain $f(T_{WB}) = 0$. The trials are shown below.

T_{WB}, °C	p_{g_w}, kPa	$f(T_{WB})$
20	2.339	2497.1
25	3.169	1178.4
30	4.246	−385.9
28.77		0

Thus, we find $T_{WB} = 28.77°C$.

10-6 THE PSYCHROMETRIC CHART

Although the equations of the foregoing sections provide an adequate basis for calculations involving air–water-vapor mixtures, it is desirable to have a calculation chart which may be used in those cases where only modest accuracy is required but with the obvious utility of a speedier computation procedure. The display of interest is called a *psychrometric chart* and is shown schematically in Fig. 10-4.

Three large charts drawn to scale are given in the Appendix:

1 English units for dry-bulb temperatures of 20 to 110°F
2 English units for dry-bulb temperatures of −20 to +50°F
3 SI units for dry-bulb temperatures of −10 to +55°C

All three of these charts are for a mixture total pressure of 1 atm. On the English unit charts the specific humidity is given in both pound mass of vapor per pound mass of dry air and *grains* of vapor per pound mass of dry air. The conversion factor is

1 lbm = 7000 grains

The volume lines on the chart give the volume of air–water-vapor *mixture* per unit mass of dry air in the mixture. We shall designate this quantity with the symbol v_a

$$v_a = \frac{\text{volume mixture}}{\text{mass dry air in mixture}} = \frac{V}{m_a} \tag{10-55}$$

The true specific volume of the mixture may be obtained by manipulating Eq. (10-55) along with

FIG. 10-4 *Schematic of psychrometric chart. 7000 grains = 1 lbm.*

$$m = m_a + m_v = m_a(1 + \omega) \tag{10-56}$$

The true specific volume is $v = V/m$ so that

$$v = \frac{m_a v_a}{m_a(1 + \omega)} = \frac{v_a}{1 + \omega} \tag{10-57}$$

We should note that in most air–water-vapor processes m_a is the invariant quantity, and thus v_a is of greatest practical utility in solving problems. Therefore, it is not surprising that it is the volumetric quantity displayed on the psychrometric chart.

Values of the mixture enthalpies per unit mass of dry air in the mixture are also shown in the chart with the following zero base levels assumed:

> *English units*
> Enthalpy of dry air = 0 at 0°F
> Enthalpy of saturated-liquid water = 0 at 32°F
>
> *SI units*
> Enthalpy of dry air = 0 at 0°C
> Enthalpy of saturated-liquid water – 0 at 0°C

The mixture enthalpy per unit mass of dry air in the mixture is defined as

$$h_{\text{mix}} = h_a + \omega h_v \tag{10-58}$$

It turns out that, to a good approximation, lines of constant wet-bulb temperature are very nearly lines of constant mixture enthalpy and are so indicated on the chart. The error in this approximation is usually less than 1 percent for mixtures at 1 atm.

For those readers wishing to avoid the approximation, the charts in the Appendix have a scale for mixture enthalpies at saturation, i.e., at 100 percent relative humidity, and dashed lines indicating deviations from these enthalpies at lower humidities and the same wet-bulb temperature. As an example, consider a mixture at 90°F dry bulb and 30 percent relative humidity. The mixture enthalpy at saturation is 31.8 Btu/lbm dry air while the deviation at 30 percent humidity is −0.19 Btu/lbm dry air, or a deviation of only 0.6 percent.

For a given total pressure of the mixture the partial pressure of the water vapor is a direct function, through Eq. (10-40), of the specific humidity, and so a line of constant ω is also a line of constant p_v and constant dew point. The dew point is found by first locating the point in question and then following a line of constant ω (horizontal line) until 100 percent relative humidity is encountered. The temperature when saturation is reached is the dew point.

It is very important to note that a given psychrometric chart is applicable only for the total-pressure conditions indicated; for the charts in the Appendix a total pressure of 1 atm applies.

10-7 AIR CONDITIONING PROCESSES

To illustrate the application of the principles of the foregoing sections, we now shall consider some typical air conditioning processes and show how they may be represented on the psychrometric diagram.

DEHUMIDIFICATION BY COOLING

Consider first the simple cooling process illustrated in Fig. 10-5. The humid air stream is passed across cooling coils which lower the mixture temperature below the dew point. As a result of the cooling process, a portion of the vapor is condensed on the coils and removed as liquid. The process is shown schematically on the psychrometric chart of Fig. 10-6. For constant-pressure cooling the stream is first cooled at constant specific humidity from point 1 to 2', at which point the water vapor starts to condense. Further cooling reduces the specific humidity to point 2. An energy balance on the system yields

$$q = m_a(h_{a_1} - h_{a_2} + \omega_1 h_{v_1} - \omega_2 h_{v_2}) - m_a(\omega_1 - \omega_2)h_f \qquad (10\text{-}59)$$

where q is the heat removed by the cooling coils. The dry air may be reheated back to the original dry-bulb temperature if desired. Such a reheat process would proceed to point 3 in Fig. 10-6.

EVAPORATIVE COOLING

A cooling technique which finds considerable application in desert climates is that of evaporative cooling, as shown in Fig. 10-7. Relatively dry air is delivered to the chamber at point 1 and is either sprayed with water or allowed to pass across a soaked pad which is continuously replenished with water. The process is illustrated on the psychrometric chart of Fig. 10-8, and is assumed to occur adiabatically so that a drop in temperature results from the evaporation process. The energy balance for the system is

FIG. 10-5 *Simple cooling and reheat process for air–water-vapor mixture.*

FIG. 10-6 *Psychrometric diagram for the process of Fig. 10-5.*

$$m_a(h_{a_1} + \omega_1 h_{v_1}) + m_a(\omega_2 - \omega_1)h_f = m_a(h_{a_2} + \omega_2 h_{v_2}) \qquad (10\text{-}60)$$

There is, of course, some minimum temperature which may be achieved by this process, designated by point 2′ in Fig. 10-8. It is fairly easy to see that the evaporative cooling process is an adiabatic saturation process which is carried more or less to completion and, therefore, follows a line of constant wet-bulb temperature. The amount of makeup water which must be supplied is very small and its temperature does not exert a significant influence on the process.

FIG. 10-7 *Adiabatic humidification or evaporative cooling process.*

FIG. 10-8 *Psychrometric diagram for the process of Fig. 10-7.*

HUMIDIFICATION WITH HEATING

When cool air is heated at constant specific humidity, the relative humidity is reduced and the effect is to produce rather "dry" air if the heating process occurs over a substantial temperature range. This circumstance is observed in heating applications in extremely cold weather. Very dry air can cause discomfort to building occupants such as chapping of lips, etc., so that many heating systems include provisions for humidification. The heating system for a hospital operating room or maternity ward is a particular case in point. A typical humidification process is shown in Fig. 10-9. The additional water vapor is supplied either by a water or steam spray—the steam having the advantage that it provides heating as well as an increased vapor content and thereby reduces the quantity of external heat which must be supplied to the process. A disadvantage of steam humidification is that an objectionable odor is experienced unless the purity of the steam is carefully controlled. The heating-humidification process is illustrated on the psychrometric diagram of Fig. 10-10.

ADIABATIC MIXING OF TWO STREAMS

Consider the arrangement shown in Fig. 10-11. Two humid air streams 1 and 2 are mixed adiabatically in the chamber to produce the outlet stream 3. The air and water-vapor mass balances for this system yield the following relations:

$$m_{a_1} + m_{a_2} = m_{a_3} \tag{10-61}$$

FIG. 10-9 *Humidification-heating process.*

$$m_{a_1}\omega_1 + m_{a_2}\omega_2 = m_{a_3}\omega_3 \qquad\qquad (10\text{-}62)$$

For adiabatic mixing the energy balance for steady flow is

$$m_{a_1}(h_{a_1} + \omega_1 h_{v_1}) + m_{a_2}(h_{a_2} + \omega_2 h_{v_2}) = m_{a_3}(h_{a_3} + \omega_3 h_{v_3}) \qquad\qquad (10\text{-}63)$$

Introducing Eq. (10-61) into Eq. (10-62) gives

$$\frac{m_{a_1}}{m_{a_2}} = \frac{\omega_3 - \omega_2}{\omega_1 - \omega_3} \qquad\qquad (10\text{-}64)$$

If Eq. (10-61) is applied to Eq. (10-63), the result is

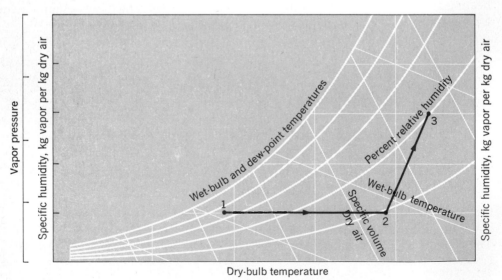

FIG. 10-10 *Psychrometric diagram for the process of Fig. 10-9.*

FIG. 10-11 *Adiabatic mixing of two air–water-vapor systems.*

$$\frac{m_{a1}}{m_{a2}} = \frac{(h_{a3} + \omega_3 h_{v3}) - (h_{a2} + \omega_2 h_{v2})}{(h_{a1} + \omega_1 h_{v1}) - (h_{a3} + \omega_3 h_{v3})} = \frac{h_3 - h_2}{h_1 - h_3} \qquad (10\text{-}65)$$

Equations (10-64) and (10-65) may be combined to yield

$$\frac{m_{a1}}{m_{a2}} = \frac{h_3 - h_2}{h_1 - h_3} = \frac{\omega_3 - \omega_2}{\omega_1 - \omega_3} \qquad (10\text{-}66)$$

We may now examine the different physical conditions which can persist when two streams are mixed. In deriving the foregoing relations the tacit assumption has been made that no vapor will be condensed. If vapor is condensed, we would need to add another exit stream for the liquid condensate. The possibility exists that the relative humidity of the exit stream may exceed 100 percent and that the excess water will be carried along as small droplets in a foglike flow. If we assume that the total enthalpy is a function primarily of wet-bulb temperature, then the mixing process may be represented on a psychrometric chart as shown in Fig. 10-12, where three possibilities are indicated:

1 $\phi_3 < 1$ no vapor condensed
2 $\phi_3 = 1$ outlet air stream is saturated
3 $\phi_3 > 1$ vapor condenses, producing foglike flow

The foregoing theory for adiabatic mixing is useful in the design of air conditioning systems which receive and mix air from multiple sources.

Table 10-2 gives a convenient summary of the different types of air conditioning and psychrometric mixture problems discussed in the foregoing sections.

(c)

FIG. 10-12   *Psychrometric diagrams for the process of Fig. 10-11. (a) $\phi_3 < 1$, no vapor condensed; (b) $\phi_3 = 1$, outlet air stream is saturated; (c) $\phi_3 > 1$, vapor condenses producing a foglike flow.*

EXAMPLE 10-9   Cooling-reheat

The air of Example 10-6 enters a cooler-reheat section like that shown in Fig. 10-5. The outlet air stream is to be at 75°F and 50 percent relative humidity. Calculate the heat which must be removed in the cooling coils and the heat which must be added in the heating section.

SOLUTION   We shall employ the psychrometric chart for solution of this problem. The processes are shown in schematic form in Fig. 10-6. From the chart we have:

$$\omega_1 = \omega_2' = 123 \text{ grains} = 0.0176 \text{ lbm} \quad (7.98 \text{ g})$$
$$\omega_2 = \omega_3 = 65 \text{ grains} = 0.0093 \text{ lbm} \quad (4.22 \text{ g})$$
$$T_2' = 73°F \quad (22.8°C)$$
$$T_2 = 55°F \quad (12.8°C)$$

The energy balance for the cooling section is given in Eq. (10-59), and the vapor enthalpy can be taken from Eq. (10-47) or, as a good approximation,

TABLE 10-2
Process relations for air–water-vapor mixtures

Process or property	Mixture properties				
	Dry-bulb temp., T_{DB}	Wet-bulb temp., T_{WB}	Dew point temp., T_{DP}	Specific humidity, ω	Mixture enthalpy, $h_m = h_a + \omega h_v$
Simple heating, no moisture addition	Increases	Increases	Constant	Constant	Increases
Cooling, no dehumidification	Decreases	Decreases	Constant	Constant	Decreases
Cooling, with dehumidification	Decreases	Decreases	Constant to 100% R.H., then decreases	Constant to 100% R.H., then decreases	Decreases
Evaporative cooling	Decreases	Approx. constant	Increases	Increases	Approx. constant
Heating, with humidification	Increases	Increases	Increases	Increases	Increases
Cooling, dehumidification, reheat	Variable, detailed mass and energy balances required				
Mixture enthalpy per mass of dry air	Primarily function of wet-bulb temperature				
Constant dew point	Same as constant p_v and ω				
Wet-bulb temperature	Approximately same as adiabatic saturation temperature				
Dry-bulb temperature	*Actual* temperature of mixture				

saturated vapor enthalpies from the steam tables at the given temperatures may be used. We shall use Eq. (10-47):

$$h_{v_1} = 1061 + (0.445)(100) = 1106 \text{ Btu/lbm} \quad (2572 \text{ J/g})$$
$$h_{v_2} = 1061 + (0.445)(55) = 1085 \text{ Btu/lbm} \quad (2523 \text{ J/g})$$
$$h_f \text{ at } 55°F = 23.06 \text{ Btu/lbm} \quad (53.63 \text{ kJ/kg})$$

Now inserting these values in Eq. (10-59) gives

$$q = m_a[(0.24)(100 - 55) + (0.0176)(1106) - (0.0093)(1085)]$$
$$- m_a(0.0176 - 0.0093)(23.06)$$

or

$$q = 20 \text{ Btu/lbm dry air} \quad (46.5 \text{ kJ/kg})$$

In the heating section there is no water lost or gained, and the energy balance yields

$$q = m_a[h_{a_3} - h_{a_2} + \omega_3(h_{v_3} - h_{v_2})]$$

where q is the heat added. With the proper numerical values added, this relation becomes

$$q = m_a[(0.24)(75 - 55) + (0.0093)(0.445)(75 - 55)]$$

or

$$q = 4.88 \text{ Btu/lbm dry air} \quad (11.35 \text{ kJ/kg})$$

EXAMPLE 10-10 Evaporative cooling

Determine the minimum temperature which may be achieved by using an evaporative cooling process with desert air at 115°F and 5 percent relative humidity.

SOLUTION   We could obtain this minimum temperature either by using a psychrometric chart as illustrated in Fig. 10-8 or by calculating the adiabatic saturation temperature. From the psychrometric chart we read, immediately, for $T_{DB} = 115°F$, $\phi = 5$ percent,

$$T_{WB} = 67°F$$

The usefulness of evaporative coolers in desert climates is quite obvious.

EXAMPLE 10-11   Adiabatic mixing

Two pounds of air at 100°F and 70 percent humidity are mixed adiabatically with 3 lbm of air at 70°F and 50 percent humidity. Calculate the temperature of the mixture if the process occurs under steady-flow conditions at a constant pressure of 14.696 psia. Also determine the relative humidity of the mixture.

SOLUTION   From the psychrometric chart we have:

$$T_1 = 100°F \qquad \omega_1 = 206 \text{ grains}$$
$$T_2 = 70°F \qquad \omega_2 = 55 \text{ grains}$$
$$m_{a_1} = 2$$
$$m_{a_2} = 3$$

We may now calculate the exit specific humidity from Eq. (10-64):

$$\frac{m_{a_1}}{m_{a_2}} = \frac{\omega_3 - \omega_2}{\omega_1 - \omega_3}$$

$$\frac{2}{3} = \frac{\omega_3 - 55}{206 - \omega_3}$$

$$\omega_3 = 115 \text{ grains}$$

A line is now drawn on the psychrometric chart shown schematically in Fig. 10-12a. The final state is found at the intersection of this line and $\omega_3 = 115$ gr. We obtain:

$$T_{DB_3} = 82°F$$
$$T_{WB_3} = 74.5°F$$
$$\phi_3 = 70 \text{ percent}$$

EXAMPLE 10-12   Low-temperature cooling

A special insulated room is designed to store meat at $-20°F$ and it is estimated that the leakage of outside air through the entry doors amounts to 500 ft³/min at 95°F and 40 percent relative humidity. How much cooling is required to compensate for this leakage assuming that the air in the room is saturated at $-20°F$?

SOLUTION   The total energy removed from the air is

$$Q = \dot{m}_a(h_1 - h_2) \qquad (a)$$

where h_1 is the enthalpy of the outside air–water-vapor mixture and h_2 is the enthalpy of the air–water-vapor–frost mixture inside the room. For the outside air we can use the psychrometric chart to obtain

$$\omega_1 = 101 \text{ grains} = 0.014\ 43 \text{ lbm/lbm dry air}$$
$$T_{DB_1} = 95°F$$
$$v_1 = 14.3 \text{ ft}^3/\text{lbm dry air} \qquad (0.893 \text{ m}^3/\text{kg})$$

The mass flow of dry air is therefore

$$\dot{m}_a = \frac{500 \text{ ft}^3/\text{min}}{14.3 \text{ ft}^3/\text{lbm}} = 34.96 \text{ lbm/min} \qquad (15.86 \text{ kg/min}) \qquad (b)$$

The vapor pressure at $-20°F$ (440°R) is calculated from Eq. (10-50a)

$$p_{g_2} = 5.103 \exp\left(18.42 - \frac{11\ 059}{440}\right) = 0.006\ 19 \text{ psia}$$
$$(42.67 \text{ Pa}) \qquad (c)$$

so the specific humidity in the room is

$$\omega_2 = 0.622 \frac{p_{g_2}}{p_{a_2}} = \frac{(0.622)(0.006\ 19)}{14.696 - 0.006\ 19} = 2.62 \times 10^{-4} \qquad (d)$$

The mass of inlet vapor converted to frost in the room is $\dot{m}_a(\omega_1 - \omega_2)$ so the enthalpy of the room mixture is

$$h_2 = h_{a_2} + \omega_2 h_{v_2} + (\omega_1 - \omega_2)h_{i_2} \qquad (e)$$

while the inlet enthalpy is

$$h_1 = h_{a_1} + \omega_1 h_{v_1} \qquad (f)$$

The vapor and frost enthalpies are now determined as

$$h_{v_1} = 1061 + (0.445)(95) = 1103.3 \text{ Btu/lbm}$$
$$h_{v_2} = 1061 + (0.445)(-20) = 1052.1 \text{ Btu/lbm}$$
$$h_{i_2} = -158.9 + (0.467)(-20) = -168.24 \text{ Btu/lbm}$$

We may now insert the numerical values in Eq. (a) to determine the heat removal

$$\begin{aligned}
Q &= 34.96[(0.24)(95 + 20) + (0.014\ 43)(1103.3) \\
&\quad - (2.62 \times 10^{-4})(1052.1) \\
&\quad - (0.014\ 43 - 2.62 \times 10^{-4})(-168.24)] \\
&= (34.96)(27.6 + 15.92 - 0.276 + 2.384) \\
&= 1595 \text{ Btu/min} = 95\ 709 \text{ Btu/h} \qquad (28.04 \text{ kW})
\end{aligned}$$

This represents a cooling requirement of 7.98 tons of refrigeration.

EXAMPLE 10-13 Air conditioning of office space

In a small office space it is determined that a heat removal of 10 kW will suffice for air conditioning. The room is to be maintained at 22°C, 50 percent relative humidity, and 25 percent of the moisture is picked up in the room. The air is completely recirculated and the cooling system must remove the moisture added in the room as indicated in the schematic diagram. Calculate the volume of air flow required under the condition that the air is cooled just enough to remove the moisture.

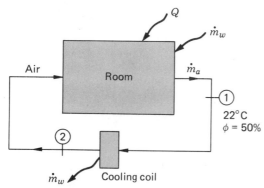

FIG. EXAMPLE 10-13

SOLUTION From the problem statement

$$T_1 = 22°C \qquad \phi_1 = 50\% \qquad Q = 10 \text{ kW}$$
$$\dot{m}_w = 0.25\dot{m}_{v_1} \qquad \phi_2 = 100\%$$

The water mass balance thus requires that

$$\dot{m}_{v_2} = \dot{m}_{v_1} - \dot{m}_w = 0.75\dot{m}_{v_1} \qquad\qquad (a)$$

From the psychrometric chart

$$\omega_1 = 0.008\ 20 \text{ kg/kg dry air} \qquad\qquad (b)$$

State 2 is saturated according to the problem statement and has

$$\omega_2 = 0.75\omega_1 = 0.006\ 15 \qquad\qquad (c)$$

The temperature at point 2 is then obtained from the psychrometric chart as

$$T_2 = T_{DB_2} = T_{WB_2} = 7°C \qquad\qquad (d)$$

The energy balance on the room is now written

$$\dot{m}_a(h_2 + \omega_2 h_{v_2}) + \dot{m}_w h_w + Q = \dot{m}_a(h_1 + \omega_1 h_{v_1}) \tag{e}$$

The vapor enthalpies are calculated from Eq. (10-47b)

$$h_v = 2501 + 1.863T$$
$$h_{v_1} = 2501 + (1.863)(22) = 2542 \text{ kJ/kg}$$
$$h_{v_2} = 2501 + (1.863)(7) = 2514 \text{ kJ/kg}$$

A reasonable estimate for h_w is

$$h_w = h_f \text{ at } 22°C = 92.33 \text{ kJ/kg} \tag{f}$$

Noting that $\dot{m}_w = \dot{m}_a(\omega_1 - \omega_2)$, Eq. (e) becomes

$$\dot{m}_a[(1.005)(7 - 22) + (0.006\ 15)(2514) - (0.008\ 20)(2542) \\ + (0.008\ 20 - 0.006\ 15)(92.33)] + 10 = 0 \tag{g}$$

which may be solved to give

$$\dot{m}_a = 0.4934 \text{ kg/s} = 29.6 \text{ kg/min} \tag{h}$$

From the psychrometric chart the specific volume leaving the cooling coils is

$$v_2 = 0.802 \text{ m}^3/\text{kg dry air} \tag{i}$$

so the volume flow rate at that point would be

$$\dot{V}_2 = \dot{m}_a v_2 = (29.6)(0.802) = 23.7 \text{ m}^3/\text{min} \tag{j}$$

At the inlet conditions to the cooling coil the specific volume is

$$v_1 = 0.847 \text{ m}^3/\text{kg dry air}$$

so the volume flow at that point would be

$$\dot{V}_1 = \dot{m}_a v_1 = (29.6)(0.847) = 25.1 \text{ m}^3/\text{min} \tag{k}$$

EXAMPLE 10-14

It is common practice in many air conditioning systems to mix a small amount of fresh outside air with the cool air coming off the refrigerant coils. In a Texas application the air from the cooling coils is at 10°C and 100 percent humidity, and it is to be mixed with 20 percent outside air at 38°C and 30 percent humidity. Calculate the final temperature and relative humidity of the mixture.

SOLUTION For this problem we consult the psychrometric chart and read

$$T_1 = 10°C \qquad \phi_1 = 100\% \qquad \omega_1 = 0.0077 \text{ kg/kg dry air}$$
$$T_2 = 38°C \qquad \phi_2 = 30\% \qquad \omega_2 = 0.0123 \text{ kg/kg dry air}$$

From Eq. (10-64)

$$\frac{\dot{m}_{a_1}}{\dot{m}_{a_2}} = \frac{\omega_3 - \omega_2}{\omega_1 - \omega_3} \qquad\qquad (a)$$

The term *20 percent outside air* means $\dot{m}_{a_1}/\dot{m}_{a_2} = 1/0.2 = 5$, so that

$$5 = \frac{\omega_3 - 0.0123}{0.0077 - \omega_3} \qquad\qquad (b)$$

and

$$\omega_3 = 0.0085$$

We now draw a line between points 1 and 2 on the psychrometric chart, and determine T_3 at the intersection of $\omega = 0.0085$. Thus,

$$T_3 = 14.5°C \qquad \phi_3 = 83\%$$

10-8 COOLING TOWERS

As we shall see in Chap. 12, large power and refrigeration cycles require the dissipation of considerable quantities of heat to their atmospheric surroundings. Typically, water is the cooling medium for the power or refrigeration cycle, and a *wet cooling tower* is employed to transfer the energy to the atmosphere, as shown schematically in Fig. 10-13.

We assume that there is little heat transferred from the surroundings. The inlet warm water is sprayed into the upward flow of relatively dry air. As the air is saturated the temperature drops, similar to the evaporative cooler, and the cooler liquid is collected in the bottom of the cooling tower. Because of the water removal with the air, make-up water must be supplied in order for the cool-water return to equal the warm-water inlet. Thus, we assume

$$\dot{m}_{w_1} = \dot{m}_{w_2} \qquad\qquad (10\text{-}67)$$

We may now make the appropriate mass and energy balances on the system. Clearly, the dry air at inlet is equal to that at exit. The mass balance on vapor and liquid water is

FIG. 10-13   *Schematic of cooling tower.*

$$\dot{m}_{w_1} + \dot{m}_{v_1} + \dot{m}_{w_3} = \dot{m}_{w_2} + \dot{m}_{v_2} \tag{10-68}$$

Using Eq. (10-67) this reduces to

$$\dot{m}_{w_3} = \dot{m}_{v_2} - \dot{m}_{v_1} \tag{10-69}$$

and, using specific humidities, we obtain

$$\dot{m}_{w_3} = \dot{m}_a(\omega_2 - \omega_1) \tag{10-70}$$

The energy balance is then

$$\dot{m}_a(h_{a_1} + \omega_1 h_{v_1}) + \dot{m}_{w_1} h_{w_1} + \dot{m}_{w_3} h_{w_3}$$
$$= \dot{m}_a(h_{a_2} + \omega_2 h_{v_2}) + \dot{m}_{w_2} h_{w_2} \tag{10-71}$$

or, using mixture enthalpies and taking the water enthalpies as those of saturated liquid,

$$\dot{m}_a[h_{m_1} - h_{m_2} + h_{f_3}(\omega_2 - \omega_1)] = \dot{m}_{w_1}(h_{f_2} - h_{f_1}) \tag{10-72}$$

An example illustrates the use of these concepts.

EXAMPLE 10-15   Cooling tower for a power plant

A power plant requires that 12 000 kg/s of water be cooled from 40 to 30°C. Inlet air for a cooling tower is available at 25°C and 35 percent relative humidity and may be assumed to leave the tower at 35°C and 90 percent relative humidity. Make-up water is available at 20°C. Calculate the mass flow rates of air and make-up water required.

SOLUTION   We first compute the vapor pressures from Eq. (10-37) and the saturation pressures

$$
\begin{aligned}
p_{g_1} &= 3.169 \text{ kPa} \qquad \text{at } T_{a_1} = 25°C \\
p_{g_2} &= 6.268 \text{ kPa} \qquad \text{at } T_{a_2} = 35°C
\end{aligned}
\tag{a}
$$

$$
\begin{aligned}
p_{v_1} &= \phi_1 p_{g_1} = (3.169)(0.35) = 1.109 \text{ kPa} \\
p_{v_2} &= \phi_2 p_{g_2} = (5.268)(0.90) = 4.741 \text{ kPa}
\end{aligned}
\tag{b}
$$

Then, the specific humidities are

$$
\omega_1 = 0.622 \frac{p_{v_1}}{p_{a_1}} = \frac{(0.622)(1.109)}{101.32 - 1.109} = 6.88 \times 10^{-3}
$$

$$
\omega_2 = \frac{(0.622)(4.741)}{101.32 - 4.741} = 0.0305 \text{ kg/kg dry air}
\tag{c}
$$

The liquid enthalpies are obtained from the steam tables as

$$
\begin{aligned}
h_{f_1} &= 167.57 \text{ kJ/kg} \qquad \text{at } 40°C \\
h_{f_2} &= 125.79 \text{ kJ/kg} \qquad \text{at } 30°C \\
h_{f_3} &= 83.96 \text{ kJ/kg} \qquad \text{at } 20°C
\end{aligned}
\tag{d}
$$

The vapor enthalpies are calculated with Eq. (10-47b):

$$
\begin{aligned}
h_{v_1} &= 2501 + (1.863)(25) = 2548 \text{ kJ/kg} \\
h_{v_2} &= 2501 + (1.863)(35) = 2566 \text{ kJ/kg}
\end{aligned}
\tag{e}
$$

The difference in air enthalpies is calculated as $c_{pa}(T_1 - T_2)_a$ with $c_{pa} = 1.005 \text{ kJ/kg·°C}$.

All of the properties may now be inserted in Eq. (10-72) to give

$$
\begin{aligned}
\dot{m}_a[(1.005)(25 - 35) &+ (6.88 \times 10^{-3})(2548) - (0.0305)(2566) \\
&+ (0.0305 - 6.88 \times 10^{-3})(83.96)] \\
&= (12\ 000 \text{ kg/s})(125.79 - 167.57)
\end{aligned}
\tag{f}
$$

The dry air flow required is thus obtained as

$$\dot{m}_a = 7287 \text{ kg/s} \tag{g}$$

The make-up water required is thus

$$
\begin{aligned}
\dot{m}_{w3} &= \dot{m}_a(\omega_2 - \omega_1) \\
&= (7287)(0.0305 - 6.88 \times 10^{-3}) \\
&= 172.1 \text{ kg/s}
\end{aligned}
\tag{h}
$$

We should note that the total air and water-vapor flow at inlet is

$$
\begin{aligned}
\dot{m}_{tot} &= \dot{m}_a + \dot{m}_{v_1} = \dot{m}_a(1 + \omega_1) \\
&= (7287)(1 + 6.88 \times 10^{-3}) \\
&= 7337 \text{ kg/s}
\end{aligned}
\tag{i}
$$

10-9   REAL–GAS MIXTURES

All of the previous discussions in this chapter have been concerned with the behavior of gaseous mixtures in which each constituent behaves as an ideal gas, independent of its composition in the mixture. Physically, this behavior will be observed when the constituent gases exist at low enough pressures that their molecules are spaced far apart and molecular force fields do not influence the pressure. In terms of the generalized compressibility factors of Chap. 7 this means that the gaseous constituents will exist at $T_r \gg 1.0$ and $p_r \ll 1.0$. Our concern in this section is to develop a technique for calculating the behavior of gaseous mixtures in temperature and pressure regions where the ideal-gas law no longer applies. Regretfully, such calculations can be rather complicated.

Let us first consider two simplified approaches to the problem. We might assume that the total pressure of the mixture could be calculated as a linear sum of the pressures exerted by each constituent at the temperature and total volume of the mixture. Thus,

$$p_m = \Sigma p_i \tag{10-73}$$

where p_m is the total pressure of the mixture. This is called *Dalton's rule of additive pressures*, and differs from the *law* of partial pressures in that we now must calculate the p_i from some real-gas equation of state. A similar relation is the *Amagat rule of additive volumes*, which assumes that the total volume of the mixture can be calculated as a sum of constituent volumes, each evaluated at the temperature and pressure of the mixture:

$$V_m = \Sigma V_i \tag{10-74}$$

The V_i must be evaluated with a real-gas equation of state. The reader must note carefully that the p_i and V_i are *not* the partial pressure and volume discussed in Sec. 10-2 because they are not evaluated with the ideal-gas equation of state.

If we use the generalized compressibility equation of state for the constituents

$$p_i = \frac{Z_i n_i \mathscr{R} T}{V_m} \tag{10-75}$$

and

$$V_i = \frac{Z_i n_i \mathscr{R} T}{p_m} \tag{10-76}$$

where p_m and V_m are the total pressure and volume of the mixture, respectively, which we assume are related to a Z value for the mixture through

$$p_m V_m = n_m Z_m \mathscr{R} T \tag{10-77}$$

Inserting these compressibility relations back into either Eq. (10-75) or (10-76) gives

$$n_m Z_m = \Sigma n_i Z_i$$

or

$$Z_m = \Sigma x_i Z_i$$

However, we must go back and note that the conditions for evaluating the Z_i are different, depending on whether the additive-pressure or additive-volume rule is employed. For the additive-pressure rule

$$Z_m = \Sigma (x_i Z_i)_{T,v} \tag{10-78}$$

while for the additive-volume rule

$$Z_m = \Sigma (x_i Z_i)_{T,p} \tag{10-79}$$

For strict ideal-gas behavior $Z_i = 1.0$ and the two rules would be equivalent, as we already know.

We can anticipate some errors in the use of the above relations by noting that the Dalton rule neglects interaction effects between the constituent molecules because it forces a calculation at the total volume of the mixture; consequently, we anticipate that it would be most useful for low gas densities. The Amagat rule creates an error in just the opposite direction because it does allow for a molecular force interaction effect by forcing a calculation at the total pressure of the mixture. Consequently, we expect that the Amagat rule might overcompensate for molecular interactions when the mixture exists at low densities.

There are many relations which have been developed for the calculation of real-gas mixtures in addition to the simple additive rules given above. A simple procedure was developed by Kay [5] which is accurate within 10 percent over a wide range of temperatures and pressures. The method assumes that the mixture may be treated as a pseudo-pure substance with some empirical artifice used to determine the pseudo-critical constants for the mixture. Kay uses a simple linear combination of the critical properties so that

$$p_{cm} = \Sigma x_i p_{ci} \tag{10-80}$$

$$T_{cm} = \Sigma x_i T_{ci} \tag{10-81}$$

where p_{cm} and T_{cm} are the pseudo-critical pressure and temperature of the mixture. Once these pseudo-critical constants are determined the calculation can then proceed, based on the generalized compressibility charts.

EXAMPLE 10-16 High-pressure air–water-vapor mixture

1 lbm of air is saturated with water vapor and contained in a rigid tank at 320°F and 150 psia. Calculate the cooling required to lower the temperature to 180°F. How much moisture is condensed in this process?

SOLUTION   In this problem we may anticipate that the air will behave as an ideal gas while the water vapor will exist at a pressure high enough to deviate from the ideal-gas equation of state. However, the pressure is low enough compared to the critical pressure for water to anticipate that an additive-pressure rule might be employed. For the saturated mixture the vapor pressure must be

$$p_v = p_g \quad \text{at } 320°F$$
$$= 89.60 \text{ psia (Table A-7)} \quad v_g = 4.919 \text{ ft}^3/\text{lbm}$$

The pressure of the air is then

$$p_a = p_m - p_v$$
$$= 150 - 89.60 = 60.40 \text{ psia} \quad (4.16 \times 10^5 \text{ Pa})$$

and the mixture volume is calculated from

$$V_m = \frac{m_a R_a T}{p_a} = \frac{(1)(53.35)(780)}{(60.40)(144)} = 4.784 \text{ ft}^3 \quad (0.1355 \text{ m}^3)$$

Carrying through with the assumption of no interaction between the gases, the mass of vapor at the total mixture volume is calculated as

$$m_v = \frac{V_m}{v_g} = \frac{4.784}{4.919} = 0.973 \text{ lbm} \qquad (441 \text{ g})$$

At the 180°F condition,

$$p_g = 7.515 \text{ psia} \qquad v_g = 50.20 \text{ ft}^3/\text{lbm}$$

For the same container volume, the mass of vapor at this condition is

$$m_v = \frac{4.784}{50.20} = 0.0953 \text{ lbm} \qquad (43.2 \text{ g})$$

For the constant-volume process the final air pressure can be calculated from

$$p_{a2} = p_{a1}\frac{T_2}{T_1} = \frac{(60.40)(640)}{780} = 49.56 \text{ psia} \qquad (3.417 \times 10^5 \text{ Pa})$$

and the final mixture pressure is

$$p_{m2} = p_{a2} + p_{v2} = 49.56 + 7.515 = 57.08 \text{ psia} \qquad (3.935 \times 10^5 \text{ Pa})$$

For the process at constant volume there is no work and the heat transfer is calculated from

$$Q = U_2 - U_1 = (U_2 - U_1)_{\text{air}} + (U_2 - U_1)_{\text{vapor}} \qquad (a)$$

We may evaluate the change in internal energy separately for the air and vapor. For the air

$$\begin{aligned} U_2 - U_1 &= mc_v(T_2 - T_1) = (1)(0.1715)(180 - 320) \\ &= -24.01 \text{ Btu} \qquad (-25.33 \text{ kJ}) \end{aligned}$$

The internal energy for the water vapor is obtained from the steam tables:

$$u_{v_1} = 1104.2 \text{ Btu/lbm} \qquad (2568 \text{ kJ/kg}) \qquad \text{sat. vapor at 320°F}$$

The mass of water condensed is

$$m_w = m_{v_1} - m_{v_2} = 0.973 - 0.0953 = 0.8777 \text{ lbm} \qquad (b)$$

The final vapor and liquid internal energies are

$$u_{v_2} = 1068.3 \text{ Btu/lbm} \qquad (2484 \text{ kJ/kg}) \qquad \text{sat. vapor at } 180°F$$
$$u_{w_2} = 147.97 \text{ Btu/lbm} \qquad (344 \text{ kJ/kg}) \qquad \text{sat. liquid at } 180°F$$

The total change in internal energy of the water is

$$
\begin{aligned}
U_2 - U_1 &= m_{v_2} u_{v_2} + m_{w_2} u_{w_2} - m_{v_1} u_{v_1} \\
&= (0.0953)(1068.3) + (0.8777)(147.97) - (0.973)(1104.2) \\
&= -842.7 \text{ Btu} \qquad (-889.1 \text{ kJ}) \tag{c}
\end{aligned}
$$

Adding together the changes in internal energy for air and vapor in Eq. (a) gives the heat transfer as

$$Q = -24.01 - 842.7 = -866.71 \text{ Btu} \qquad (-914 \text{ kJ})$$

The negative sign means that heat is removed from the system during the cooling process.

It is interesting to see what would happen in the calculation if we applied an additive-volume rule instead of the additive-pressure rule. The calculation of the volume of air at the temperature and pressure of the mixture is an easy matter:

$$V_a = \frac{m_a R_a T}{p} = \frac{(1)(53.35)(780)}{(150)(144)} = 1.927 \text{ ft}^3 \qquad (0.0546 \text{ m}^3)$$

But if one evaluates the volume of the water at 150 psia and 320°F we find that a compressed liquid state is encountered, which of course makes no sense in this problem. So, as discussed earlier, a use of the additive-volume concept at low pressures may result in substantial errors.

EXAMPLE 10-16M High-pressure air–water-vapor mixture

0.5 kg of air is saturated with water vapor and contained in a rigid tank at 160°C and 1 MPa. Calculate the cooling required to lower the temperature to 80°C. How much moisture is condensed in this process?

SOLUTION In this process we may anticipate that the air will behave as an ideal gas while the water vapor will exist at a pressure high enough to deviate from the ideal-gas equation of state. However, the pressure is low enough compared to the critical pressure for water to anticipate that an additive-pressure rule might be employed. For the saturated mixture the vapor pressure must be

$$
\begin{aligned}
p_v &= p_g \qquad \text{at } 160°C \\
&= 617.8 \text{ kPa (Table A-7M)} \qquad v_g = 0.3071 \text{ m}^3/\text{kg}
\end{aligned}
$$

The pressure of the air is then

$$p_a = p_m - p_v = 1000 - 617.8 = 382.2 \text{ kPa} \qquad (55.4 \text{ psia})$$

and the mixture volume is calculated from

$$V_m = \frac{m_a R_a T}{p_a} = \frac{(0.5)(287.1)(433)}{382.2 \times 10^3} = 0.1626 \text{ m}^3 \qquad (5.743 \text{ ft}^3)$$

Carrying through with the assumption of no interaction between the gases, the mass of vapor at the total mixture volume is calculated as

$$m_v = \frac{V_m}{v_g} = \frac{0.1626}{0.3071} = 0.529 \text{ kg} \qquad (1.167 \text{ lbm})$$

At the 80°C condition,

$$p_g = 47.39 \text{ kPa} \qquad v_g = 3.407 \text{ m}^3/\text{kg}$$

For the same container volume, the mass of vapor at this condition is

$$m_v = \frac{0.1626}{3.407} = 0.0477 \text{ kg} \qquad (0.105 \text{ lbm})$$

For the constant-volume process the final air pressure can be calculated from

$$p_{a2} = p_{a1} \frac{T_2}{T_1} = \frac{(382.2)(353)}{433} = 311.6 \text{ kPa} \qquad (45.19 \text{ psia})$$

and the final mixture pressure is

$$p_{m2} = p_{a2} + p_{v2} = 311.6 + 47.39 = 359.0 \text{ kPa} \qquad (52.07 \text{ psia})$$

For the process at constant volume there is no work and the heat transfer is calculated from

$$Q = U_2 - U_1 = (U_2 - U_1)_{\text{air}} + (U_2 - U_1)_{\text{vapor}} \qquad (a)$$

We may evaluate the change in internal energy separately for the air and vapor. For the air

$$U_2 - U_1 = mc_v(T_2 - T_1) = (0.5)(0.718)(80 - 160)$$
$$= -28.72 \text{ kJ} \qquad (-27.22 \text{ Btu})$$

The internal energy for the water vapor is obtained from the steam tables:

$$u_{v_1} = 2568.4 \text{ kJ/kg} \quad (1104.2 \text{ Btu/lbm}) \quad \text{sat. vapor at } 160°C$$

The mass of water condensed is

$$m_w = m_{v_1} - m_{v_2} = 0.529 - 0.0477 = 0.4813 \text{ kg} \qquad (b)$$

The final vapor and liquid internal energies are

$$\begin{aligned} u_{v_2} &= 2482.2 \text{ kJ/kg} \quad (1067 \text{ Btu/lbm}) \quad \text{sat. vapor at } 80°C \\ u_{w_2} &= 334.86 \text{ kJ/kg} \quad (144 \text{ Btu/lbm}) \quad \text{sat. liquid at } 80°C \end{aligned}$$

The total change in internal energy of the water is

$$\begin{aligned} U_2 - U_1 &= m_{v_2} u_{v_2} + m_{w_2} u_{w_2} - m_{v_1} u_{v_1} \\ &= (0.0477)(2482.2) + (0.4813)(334.86) - (0.529)(2568.4) \\ &= -1079.1 \text{ kJ} \quad (-1023 \text{ Btu}) \qquad (c) \end{aligned}$$

Adding together the changes in internal energy for air and vapor in Eq. (a) gives the heat transfer as

$$Q = -28.72 - 1079.1 = -1107.8 \text{ kJ} \quad (-1050 \text{ Btu})$$

The negative sign means that heat is removed from the system during the cooling process.

It is interesting to see what would happen in the calculation if we applied an additive-volume rule instead of the additive-pressure rule. The calculation of the volume of air at the temperature and pressure of the mixture is an easy matter:

$$V_a = \frac{m_a R_a T}{p} = \frac{(0.5)(287.1)(433)}{1 \times 10^6} = 0.06216 \text{ m}^3 \quad (2.195 \text{ ft}^3)$$

But if one evaluates the volume of the water at 1 MPa and 160°C, we find that a compressed-liquid state is encountered, which of course makes no sense in this problem. So, as discussed earlier, a use of the additive-volume concept at low pressures may result in substantial errors.

EXAMPLE 10-17 Comparison of calculation techniques

A mixture containing 1.0 lbm mol and equal molal parts of CO_2 and N_2 is contained at 100°F in a vessel having a volume of 1.5 ft³. Calculate the total pressure of the mixture using (a) the ideal-gas law, (b) the rule of additive pressures and the compressibility charts, (c) the rule of additive volumes and the generalized compressibility charts, and (d) Kay's pseudo-critical method.

SOLUTION We have 1 mol of total mixture so

$$x_{N_2} = 0.5 \qquad \text{and} \qquad x_{CO_2} = 0.5$$

Also, $T_m = 100°F = 560°R$, $V_m = 1.5$ ft³, and $n_m = 1.0$ lbm mol.

(a) For ideal-gas behavior the total pressure is calculated immediately from

$$p_m = \frac{n_m \mathscr{R} T_m}{V_m} = \frac{(1.0)(1545)(560)}{(1.5)(144)} = 4005 \text{ psia}$$

(b) The critical properties for CO_2 and N_2 are obtained from Table 7-2 as

$$T_{c,N_2} = 227°R \qquad T_{c,CO_2} = 548°R$$
$$p_{c,N_2} = 33.5 \text{ atm} \qquad p_{c,CO_2} = 72.9 \text{ atm}$$

For the rule of additive pressures we must evaluate the pressure of each constituent at the temperature and volume of the total mixture. Because we have 0.5 mol of each constituent the molal specific volume of each is

$$\bar{v} = \frac{V_m}{n} = \frac{1.5}{0.5} = 3.0 \text{ ft}^3/\text{lbm mol}$$

The pseudo-reduced volume for each constituent is thus

$$\bar{v}'_r = \frac{\bar{v}}{\mathscr{R} T_c / p_c}$$

$$\bar{v}'_{r,CO_2} = \frac{3.0}{(1545)(548)/(72.9)(14.7)(144)} = 0.547$$

$$\bar{v}'_{r,N_2} = \frac{3.0}{(1545)(227)/(33.5)(14.7)(144)} = 0.607$$

and the reduced temperatures are

$$T_{r,CO_2} = \frac{560}{548} = 1.022$$

$$T_{r,N_2} = \frac{560}{227} = 2.467$$

We may consult the generalized compressibility charts of Chap. 7 to obtain the reduced pressures as

$$p_{r,CO_2} = 1.0 \qquad p_{r,N_2} = 4.25$$

The pressures to be used in the additive-pressure rule are therefore

$$p_{CO_2} = (1.0)(72.9) = 72.9 \text{ atm} = 1071 \text{ psia}$$
$$p_{N_2} = (4.25)(33.5) = 142.4 \text{ atm} = 2092 \text{ psia}$$

The total pressure of the mixture is then

$$p_m = p_{N_2} + p_{CO_2} = 2092 + 1071 = 3163 \text{ psia}$$

This value is substantially lower than that predicted by the ideal-gas law.
(c) To employ the additive-volume rule an iterative procedure is required because we do not know the total mixture pressure for evaluation of the volumes. The easiest way to proceed is to assume a value for the total pressure, calculate the Z_i's for insertion in Eq. (10-79), compute Z_m, and then calculate the total pressure from this. The procedure continues until agreement is reached between the assumed and calculated values. For these calculations the values of T_r are the same as in part (b). As a first trial we might take

$$p_m = 3500 \text{ psia} = 238.2 \text{ atm}$$

Then

$$p_{r,CO_2} = \frac{238.2}{72.9} = 3.27$$

$$p_{r,N_2} = \frac{238.2}{33.5} = 7.11$$

Consulting the generalized charts,

$$Z_{CO_2} = 0.48 \qquad Z_{N_2} = 1.09$$

Then, from Eq. (10-69),

$$Z_m = \Sigma x_i Z_i = (0.5)(0.48) + (0.5)(1.09) = 0.785$$

and the mixture pressure is calculated from

$$p_m = \frac{Z_m n_m \mathcal{R} T_m}{V_m} = \frac{(0.785)(1.0)(1545)(560)}{(1.5)(144)} = 3144 \text{ psia}$$

This differs substantially from our 3500 psia assumption so we try

$$p_m = 3100 \text{ psia} = 210.9 \text{ atm}$$

$$p_{r,CO_2} = \frac{210.9}{72.9} = 2.89$$

$$p_{r,N_2} = \frac{210.9}{33.5} = 6.30$$

Consulting the generalized charts again,

$$Z_{CO_2} = 0.43 \qquad Z_{N_2} = 1.075$$

and

$$Z_m = \Sigma x_i Z_i = (0.5)(0.43) + (0.5)(1.075) = 0.753$$

Then

$$p_m = \frac{Z_m n_m \mathcal{R} T_m}{V_m} = \frac{(0.753)(1.0)(1545)(560)}{(1.5)(144)} = 3014 \text{ psia}$$

Further iteration would give agreement between the assumed and calculated values at

$$p_m = 2970 \text{ psia} = 202.1 \text{ atm}$$

This value is also in substantial disagreement with the ideal-gas calculation.

(d) Kay's rule requires that we calculate pseudo-critical properties for the mixture using Eqs. (10-80) and (10-81):

$$p_{cm} = \Sigma x_i p_{ci} = (0.5)(33.5) + (0.5)(72.9) = 53.2 \text{ atm}$$
$$T_{cm} = \Sigma x_i T_{ci} = (0.5)(227) + (0.5)(548) = 387.5°R$$

The mixture-reduced properties for use with the compressibility charts are then calculated as

$$T_{rm} = \frac{T_m}{T_{cm}} = \frac{560}{387.5} = 1.445$$

$$\bar{v}'_{rm} = \frac{\bar{v}}{\mathcal{R} T_{cm}/p_{cm}} = \frac{1.5}{(1545)(387.5)/(53.2)(14.7)(144)}$$
$$= 0.282$$

Consulting Fig. 7-4 we obtain

$$Z_m = 0.775$$

so that

$$p_m = \frac{Z_m n_m \mathcal{R} T_m}{V_m} = \frac{(0.775)(1.0)(1545)(560)}{(1.5)(144)}$$
$$= 3104 \text{ psia} = 211.12 \text{ atm}$$

The true value may only be obtained from experimental data, but we can compare the answers obtained by the different methods taking the result obtained from Kay's rule as a reference point. The deviations from this value are then

Ideal-gas law: 29 percent high
Additive-pressure rule: 1.9 percent high
Additive-volume rule: 4.3 percent low

In this particular problem the additive rules and Kay's rule are in fairly close agreement.

EXAMPLE 10-17M   Comparison of calculation techniques

A mixture containing 1.0 kg mol and equal molal parts of CO_2 and N_2 is contained at 40°C in a vessel having a volume of 85 liters. Calculate the total pressure of the mixture using (a) the ideal-gas law, (b) the rule of additive pressures and the compressibility charts, (c) the rule of additive volumes and the generalized compressibility charts, and (d) Kay's pseudo-critical method.

SOLUTION   We have 1 mol of total mixture so that

$$x_{N_2} = 0.5 \quad \text{and} \quad x_{CO_2} = 0.5$$

Also

$$T_m = 40°C = 313 \text{ K} \qquad V_m = 0.085 \text{ m}^3 \qquad n_m = 1.0 \text{ kg mol}$$

(a) For ideal-gas behavior the total pressure is calculated immediately from

$$p_m = \frac{n_m \mathscr{R} T_m}{V_m} = \frac{(1.0)(8314.41)(313)}{0.085} = 30.62 \text{ MPa}$$

(b) The critical properties for CO_2 and N_2 are obtained from Table 7-2 as

$$T_{c,N_2} = 126.2 \text{ K} \qquad T_{c,CO_2} = 304.20 \text{ K}$$
$$p_{c,N_2} = 3.398 \text{ MPa} \qquad p_{c,CO_2} = 7.386 \text{ MPa}$$

For the rule of additive pressures we must evaluate the pressure of each constituent at the temperature and volume of the total mixture. Because we have 0.5 mol of each constituent, the molal specific volume of each is

$$v = \frac{V_m}{n} = \frac{0.085}{0.5} = 0.170 \text{ m}^3/\text{kg mol}$$

The pseudo-reduced volume for each constituent is thus

$$\bar{v}_r' = \frac{\bar{v}}{\mathcal{R}T_c/P_c}$$

$$\bar{v}_{r,CO_2}' = \frac{0.170}{(8314.41)(304.20)/(7.386 \times 10^6)} = 0.496$$

$$\bar{v}_{r,N_2}' = \frac{0.170}{(8314.41)(126.2)/(3.398 \times 10^6)} = 0.551$$

and the reduced temperatures are

$$T_{r,CO_2} = \frac{313}{304.2} = 1.029$$

$$T_{r,N_2} = \frac{313}{126.2} = 2.480$$

We may consult the generalized compressibility charts of Chap. 7 to obtain the reduced pressures as

$$p_{r,CO_2} = 1.05 \qquad p_{r,N_2} = 4.78$$

The pressures to be used in the additive-pressure rule are therefore

$$p_{CO_2} = (1.05)(7.386) = 7.755 \text{ MPa}$$
$$p_{N_2} = (4.78)(3.398) = 16.24 \text{ MPa}$$

The total pressure of the mixture is then

$$p_m = p_{N_2} + p_{CO_2} = 7.755 + 16.24 = 24.00 \text{ MPa}$$

This value is substantially lower than that predicted by the ideal-gas law.

(c) To employ the additive-volume rule an iterative procedure is required because we do not know the total mixture pressure for evaluation of the volumes. The easiest way to proceed is to assume a value for the total pressure, calculate the Z_i's for insertion in Eq. (10-79), compute Z_m, and then calculate the total pressure from this. The procedure continues until agreement is reached between the assumed and calculated values. For these calculations the values of T_r are the same as in part (b). As a first trial we might take

$$p_m = 30.0 \text{ MPa}$$

Then

$$p_{r,CO_2} = \frac{30}{7.386} = 4.06$$

$$p_{r,N_2} = \frac{30}{3.398} = 8.83$$

Consulting the generalized charts,

$$Z_{CO_2} = 0.57 \qquad Z_{N_2} = 1.15$$

Then, from Eq. (10-78),

$$Z_m = \Sigma x_i Z_i = (0.5)(0.57) + (0.5)(1.15) = 0.860$$

and the mixture pressure is calculated from

$$p_m = \frac{Z_m n_m \mathcal{R} T_m}{V_m} = \frac{(0.860)(1.0)(8314.41)(313)}{0.085} = 26.33 \text{ MPa}$$

This differs substantially from our 30.0 MPa assumption, and so we try

$$p_m = 25.0 \text{ MPa}$$

$$p_{r,CO_2} = \frac{25}{7.386} = 3.38$$

$$p_{r,N_2} = \frac{25}{3.398} = 7.36$$

Consulting the generalized charts again,

$$Z_{CO_2} = 0.49 \qquad Z_{N_2} = 1.10$$

and

$$Z_m = \Sigma x_i Z_i = (0.5)(0.49) + (0.5)(1.10) = 0.795$$

Then

$$p_m = \frac{Z_m n_m \mathcal{R} T_m}{V_m} = \frac{(0.795)(1.0)(8314.41)(313)}{0.085} = 24.34 \text{ MPa}$$

Further iteration would give agreement between the assumed and calculated values at

$$p_m = 24.22 \text{ MPa}$$

This value is also in substantial disagreement with the ideal-gas calculation.

(d) Kay's rule requires that we calculate pseudo-critical properties for the mixture using Eqs. (10-80) and (10-81):

$$p_{cm} = \Sigma x_i p_{ci} = (0.5)(3.398) + (0.5)(7.386) = 5.392 \text{ MPa}$$
$$T_{cm} = \Sigma x_i T_{ci} = (0.5)(126.2) + (0.5)(304.2) = 215.2 \text{ K}$$

The mixture-reduced properties for use with the compressibility charts are then calculated as

$$T_{rm} = \frac{T_m}{T_{cm}} = \frac{313}{215.2} = 1.454$$

$$\bar{v}'_{rm} = \frac{\bar{v}}{\mathcal{R} T_{cm}/p_{cm}} = \frac{0.085}{(8314.41)(215.2)/(5.392 \times 10^6)} = 0.256$$

Consulting Fig. 7-4, we obtain

$$Z_m = 0.78$$

so that

$$p_m = \frac{Z_m n_m \mathcal{R} T_m}{V_m} = \frac{(0.78)(1.0)(8314.41)(313)}{0.085} = 23.88 \text{ MPa}$$

The true value can only be obtained from experimental data, but we can compare the answers obtained by the different methods taking the result obtained from Kay's rule as a reference point. The deviations from this value are then

Ideal-gas law: 28 percent high
Additive-pressure rule: 0.5 percent high
Additive-volume rule: 1.4 percent high

In this particular problem the additive rules and Kay's rule are in fairly close agreement.

REVIEW QUESTIONS

1 State the Gibbs-Dalton law. Why is it applicable to ideal gases?

2 What is the Amagat-Leduc law?

3 It is sometimes stated that the composition of a gaseous mixture on a volume basis is the same as a mole basis. Explain.

4   How is the term *partial pressure* defined?

5   Show that the entropy change of a mixture in a process at constant total pressure is a function of only the initial and final temperatures and the mixture specific heat.

6   Define the terms *dew point, dry- and wet-bulb temperature, relative humidity, specific humidity,* and *degree of saturation.*

7   Why is it possible to assume that water vapor behaves as an ideal gas for purposes of making psychrometric mixture calculations?

8   Why cannot the psychrometric chart in the Appendix be used for air–water-vapor mixtures at 25 psia?

9   What is meant by the term *evaporative cooling?*

10 What is a *grain?*

11 How is wet-bulb temperature measured?

12 What is a cooling tower and how does it work?

13 What is the rule of additive pressures? When does it apply?

14 What is the rule of additive volumes? When does it apply?

15 What is Kay's rule for real-gas mixtures? When does it apply?

16 How is specific humidity related to dew point?

17 How is mixture enthalpy related to wet-bulb temperature for an air–water-vapor mixture?

18 How do you know when to expect that moisture will be condensed from an air–water-vapor stream in an air conditioning process?

19 What is meant by the term *mole fraction?*

20 What is an ideal gas?

21 What is a *saturated mixture* in air–water-vapor problems?

PROBLEMS (English units)

10-1   A certain mixture has the following composition on a mass basis:
N_2 = 50 percent
He = 10 percent
CO = 12 percent
O_2 = 15 percent
CO_2 = 13 percent
Calculate the specific heats for the mixture and the gas constant. Also calculate the mole fraction of each constituent.

10-2   The mixture of Prob. 10-1 is compressed from 15 psia, 70°F, to 50 psia, 200°F, in a reversible nonflow process. Calculate the work per pound mass of mixture and the heat transfer.

10-3 The mixture of Prob. 10-1 is compressed isentropically from 15 psia, 70°F, to 50 psia. Calculate the work done and the change in entropy for each constituent.

10-4 A certain mixture has the following composition on a *volumetric* basis:

CO = 10 percent
CO_2 = 25 percent
O_2 = 10 percent
N_2 = 55 percent

Calculate the mass fractions of each constituent, the gas constant, apparent molecular weight, and specific heats.

10-5 1 lbm of N_2 at 100°F and 4 psia is mixed with 2 lbm of H_2 at 200°F and 5 psia in a steady-flow process. The mixture leaves at 3 psia and the process is adiabatic. Calculate the change in entropy.

10-6 A rigid tank contains 1 lbm of H_2 at 200 psia and 50°F. A tap is connected to a high-pressure nitrogen line at 150°F, and a sufficient quantity is drawn off until the total pressure in the tank reaches 400 psia. The process is adiabatic. Calculate the final temperature in the tank, the mass of nitrogen added, and the change in entropy.

10-7 Repeat Prob. 10-6 for a *slow* filling process, where the temperature of the gas in the tank remains constant at 50°F.

10-8 Two 1.0-ft³ tanks are connected by a pipe and valve. One tank is filled with N_2 at 15 psia and 100°F, and the other tank is filled with argon at 100°F. The valve is opened and the two gases are allowed to mix adiabatically. Calculate the change in entropy when the initial argon pressure is (a) 15 psia, (b) 30 psia.

10-9 Complete the following table for an air–water-vapor mixture at atmospheric pressure.

T_{DB}, °F	T_{WB}, °F	ϕ, percent	Dew point, °F	p_v
100		30		
80	60			
		50	55	
	80		60	
90			60	
	70	40		

10-10 Repeat Prob. 10-9 for an air–water-vapor mixture at a total pressure of 30 psia.

10-11 The partial pressure of water vapor in a certain psychrometric mixture is 0.3 psia and the total pressure of the mixture is 14.696 psia. Calculate the relative humidity when the dry-bulb temperature is 80°F, 90°F, and 100°F.

10-12 An air–water-vapor mixture is compressed isentropically from 100°F, 14.7 psia, and 50 percent relative humidity to 100 psia. Calculate the temperature at the end of compression. If the mixture is then cooled at constant pressure calculate the amount of moisture removed when the high-pressure gas is cooled to (a) 80°F, (b) 100°F.

10-13 A cold-water pipe at 55°F passes through a room with atmospheric air at 80°F. What is the maximum relative humidity the air may have such that condensation does not occur on the pipe?

10-14 Meats and other food products will dry out when exposed directly to the very-low-temperature air in cold-storage vaults. Explain this phenomenon on the basis of information contained in this chapter. For calculation purposes, assume the air temperature is at 0°F.

10-15 In a certain northern city a heating system is designed to heat outside air at −10°F and 100 percent relative humidity to 80°F. Calculate the heat which must be supplied under these circumstances per pound mass of air. What is the final relative humidity of the mixture?

10-16 The heating system in Prob. 10-15 is to be modified so that the final humidity of the mixture will be 50 percent. For this purpose a quantity of steam at 20 psia, 80 percent quality, is mixed with the air during the heating process. How much steam must be supplied per pound mass of air and what are the heating requirements under these new conditions?

10-17 The process in Prob. 10-16 is to be accomplished by spraying the heated air with liquid water at 180°F. What quantity of water will be required per pound mass of air, and what are the total heating requirements?

10-18 A small cooling system is designed solely for dehumidification purposes. The system is designed to remove 1 gal of moisture per hour from air at 80°F and 80 percent relative humidity. What air flow must the system handle, and what must its cooling capacity be? (*Note:* There are several correct answers; specify two suitable choices.)

10-19 An evaporative cooler is used to deliver 5000 ft³/min of air, operating in a desert climate where $T_{DB} = 110°F$, $\phi = 10$ percent. Calculate the quantity of water which must be supplied to the unit if the air leaves the unit at 90 percent relative humidity.

10-20 In infant intensive-care wards a *supersaturated* air condition is frequently maintained. Suppose air is available from the regular heating system at 80°F and 40 percent relative humidity. The air-flow rate for the ward is 1200 ft³/min. How much water must be added to the air per hour, in fine droplet form, to produce an effective humidity of 105 percent at 75°F? How much additional heat must be supplied?

10-21 A certain precision-gage room must be maintained at low-humidity conditions to ensure proper accuracy of the instruments. A cooling-reheat air conditioning process is to be employed. The incoming air is at 90°F and 50 percent relative humidity and is to be delivered to the room at 65°F and 20 percent humidity. 1000 ft³/min of air are required. Calculate the total cooling and heating requirements.

10-22 1000 ft³/min of air at 110°F and 15 percent relative humidity are mixed adiabatically with 500 ft³/min of air at 65°F and 90 percent humidity. Calculate the final temperature and humidity.

10-23 The output air from the evaporative cooler of Prob. 10-19 is mixed with 1000 ft³/min of the outside air. What is the resultant temperature and humidity?

10-24 1000 ft³/min of air at 100°F and 80 percent relative humidity are mixed with an air flow at 55°F. What flow rate of the cooler air and humidity is necessary just to produce a mixture at 100 percent relative humidity? What is the final mixture temperature?

10-25 4 lbm of air at 25 psia, 40°F, and 80 percent relative humidity are mixed with 8 lbm of air at 25 psia, 100°F, and 20 percent relative humidity in a steady-flow adiabatic process. Calculate the final mixture temperature and humidity for a final pressure of (a) 25 psia, (b) 20 psia.

10-26 One way to produce very dry air is to cool it to very low temperatures and then reheat to the desired operating temperature. Suppose a mixture is available at 1 atm, 75°F, and 80 percent relative humidity. What final relative humidity would result if the mixture is cooled to $-30°F$ and then reheated to 75°F? How much cooling and heating would be required for a flow rate of initial mixture of 1700 ft³/min?

10-27 The "chiller" section in Prob. 10-26 must be defrosted when 50 lbm of frost or ice has accumulated. Assuming that the water removed from the air at temperatures above 35°F is able to run off freely, calculate the time required to accumulate 50 lbm of ice. Repeat the calculation for no water runoff.

10-28 A special vaporizer device for use with small children in cold dry weather is designed to mix one part of a supersaturated mixture at 60°F and 110 percent effective relative humidity with two parts of room air at 75°F and 15 percent relative humidity. What are the humidity and temperature of the outlet mixed stream if the extra 10 percent humidity in the supersaturated stream is in the form of small liquid droplets? Assume the mixture parts are on a dry-air basis.

10-29 An inexpensive humidifier for home heating systems operates by atomizing liquid water and spraying it directly into the air duct system. One particular device is to be designed to handle the following air flows:

Input to heater-humidifier section: 300 ft³/min at 20°F, $\phi = 80\%$
          plus   1200 ft³/min at 60°F, $\phi = 80\%$
          plus   liquid water at 60°F

Output from heater-humidifier section: $T_{DB} = 85°F$, $\phi = 35$ percent

Calculate the amount of heating required and the amount of liquid water which must be supplied to the humidifier.

10-30 A small building in a desert climate requires comfort cooling of 120 000 Btu/h, and an evaporative cooler is to be designed to perform the task. Outside air is available at 110°F and $\phi = 5$ percent. What inlet air-flow rate is required, and how much water must be supplied to the cooler?

10-31 A certain home has 3500 ft² of floor space with an 8-ft ceiling height. How much total water vapor is contained in the house at 75°F and 50 percent relative humidity?

10-32 Calculate the mass specific heats for a mixture of 1 kg of helium and 1 kg of nitrogen. If such a mixture is heated and compressed from 1 atm, 20°C, to 5 atm, 150°C, in a steady-flow device, calculate the heat added, the shaft work input, and change in entropy of the mixture. Repeat for heating and compressing in a closed system.

10-33 Repeat Prob. 10-32 for the same quantities of helium and nitrogen but with enough water vapor added so that the initial relative humidity is 20 percent.

10-34 A large cooling tower for a power plant is designed on the evaporation cooling concept. In one installation the heat which must be absorbed in the cooling tower is 1000 MW, and this is to be accomplished by spraying air at 90°F, $\phi = 30$ percent, with water at 80°F. The cooling-tower design assumes that the air leaves at 70°F saturated. What inlet air and water flows are needed to accomplish the required cooling rate?

10-35 Calculate the mixture enthalpies for air–water-vapor mixtures at 1 atm and (a) $T_{DB} = 100°F$, $T_{WB} = 60°F$, (b) $T_{DB} = 70°F$, $T_{WB} = 60°F$. What percent error would result from assuming that the mixture enthalpy is a function only of wet-bulb temperature?

10-36 How much cooling is required to cool air at 90°F and 60 percent relative humidity to $-30°F$ in a constant-pressure process at (a) 1 atm and (b) 2 atm? How much moisture will be removed in these processes?

10-37 A mixture of one part nitrogen and two parts carbon dioxide by mass contains enough water vapor to produce 50 percent relative humidity at 100°F and 1 atm total pressure. How much cooling is required to cool 1 lbm of mixture to 40°F at constant pressure? How much water is removed in the process?

10-38 A mixture containing 1 lbm oxygen and 1 lbm helium has enough water vapor to produce 10 percent relative humidity at 105°F. The mixture is to be cooled in an adiabatic saturation process. What is the lowest temperature which can be achieved in this manner, and how much water must be supplied per lbm of initial mixture? What is the mass specific heat of the mixture before and after the saturation process?

10-39 A mixture of methane (CH_4) and air is to be produced which has mass proportions of 1 part methane to 15 parts dry air. The mixing is accomplished in a steady-flow process with both methane and air entering at 80°F. The inlet air has $\phi = 70$ percent and the gases are heated during the mixing process to produce an outlet temperature of 200°F. The process occurs at constant pressure of 1 atm. How much heating must be supplied for an inlet volume flow of methane of 1000 ft³/h, and what is the relative humidity of the outlet mixture?

10-40 A tank having a volume of 20 ft³ contains an air–water-vapor mixture at 14.7 psia, 80°F, and $\phi = 50$ percent. A high-pressure line with dry air at 120°F and 70 psia is used to fill the tank with additional air. The filling process occurs slowly so that the temperature inside the tank remains constant at 80°F. How much air must be added to raise the tank pressure to 50 psia, what will be the relative humidity under the new conditions, and how much heat will have been lost by the tank in the process?

10-41 A mixture of equal molal parts of air and water vapor at 14.7 psia and 300°F is heated to 500°F in a constant-pressure process. How much heat is required per pound mass of mixture?

10-42 Estimate the temperature at which the vapor in the mixture of Prob. 10-41 will start to condense if the mixture is cooled at constant pressure.

10-43 A mixture of air and water vapor is saturated at 200 psia and 300°F. Calculate the cooling required to cool the mixture to 150°F in a steady-flow process at constant pressure.

10-44 Assuming that air may be considered as a mixture of 78 percent nitrogen and 22 percent oxygen on a molal basis, calculate the critical constants using Kay's rule and compare with tabulated values.

10-45 1 lbm mol of a mixture of 50 percent CO_2 and 50 percent CH_4 on a molal basis is contained in a rigid vessel at 300°F and 500 psia. Calculate the volume of the mixture using (a) the ideal-gas law, (b) the rule of additive volumes, and (c) the Kay pseudo-critical method.

10-46 Calculate the energy required to heat the mixture of Prob. 10-45 to 500°F in a constant-pressure steady-flow process using (a) the ideal-gas law and (b) the pseudo-critical method.

10-47 A mixture of 60 percent methane (CH_4) and 40 percent ethane (C_2H_6) on a molal basis is to be compressed from 150°F, 100 psia, to 350 psia, in a steady-flow device. The mass flow rate is 10 lbm/min. Assuming that the process is reversible and adiabatic calculate the work input using (a) ideal-gas relations and (b) the pseudo-critical method.

10-48 A mixture containing 30 percent CO_2, 10 percent CO, 40 percent N_2, and 20 percent CH_4 on a molal basis is contained at 4000 psia and 200°F in a 5-ft³ rigid vessel. Calculate the heat which must be removed to lower the temperature to 100°F using (a) the ideal-gas law and (b) Kay's rule.

10-49 Oxygen is contained in a rigid 1-ft³ vessel at 50 atm and 100°F. Nitrogen from a high-pressure line at 4000 psia and 100°F is bled slowly into the tank until the mixture proportions in the tank approximate that for air, i.e., 78% N_2 and 22% O_2. During the filling process cooling is supplied to the tank to maintain the temperature at 100°F. Calculate the heat removed from the tank and the mass of nitrogen added. Use Kay's rule.

10-50 A mixture of equal molal parts of methane (CH_4) and ethane (C_2H_6) is compressed reversibly and isothermally from 150°F, 100 psia, to 700 psia, in a steady-flow process. Calculate the work required per pound mass of mixture.

10-51 A mixture of 40 percent argon and 60 percent nitrogen on a molal basis is to be heated from $-100°F$ and 700 psia to 200°F in a steady-flow heat exchanger device. Because of fluid friction the pressure at exit of the device drops to 650 psia. Flow velocities are low enough to neglect kinetic energies. Calculate the heating which must be supplied per pound mass of mixture.

10-52 An air–water-vapor mixture has 25 percent water vapor on a molal basis and is contained at 500 psia and 800°F. Calculate the specific volume of the mixture using (a) the ideal-gas law, (b) the rule of additive pressures, (c) the rule of additive volumes, and (d) Kay's pseudo-critical rule.

10-53 How much heat per pound mass of mixture would be required to heat the mixture of Prob. 10-52 to 1000°F in a constant-pressure steady-flow process?

PROBLEMS (Metric units)

10-1M The mixture of Prob. 10-1 is compressed from 100 kPa, 295 K, to 350 kPa, 370 K, in a reversible nonflow process. Calculate the work per kilogram of mixture and the heat transfer.

10-2M The mixture of Prob. 10-1 is compressed isentropically from 100 kPa, 295 K, to 350 kPa. Calculate the work done and the change in entropy for each constituent.

10-3M 1 kg of N_2 at 310 K and 28 kPa is mixed with 2 kg of H_2 at 370 K and 35 kPa in a steady-flow process. The mixture leaves at 20 kPa and the process is adiabatic. Calculate the change in entropy.

10-4M A rigid tank contains 1 kg of H_2 at 1.4 MPa and 280 K. A tap is connected to a high-pressure nitrogen line at 340 K, and a sufficient quantity is drawn off until the total pressure in the tank reaches 2.8 MPa. The process is adiabatic. Calculate the final temperature in the tank, the mass of nitrogen added, and the change in entropy.

10-5M Repeat Prob. 10-4M for a *slow* filling process, where the temperature of the gas in the tank remains constant at 280 K.

10-6M Two 28-liter tanks are connected by a pipe and valve. One tank is filled with N_2 at 100 kPa and 310 K, and the other tank is filled with argon at 310 K. The valve is opened and the two gases are allowed to mix adiabatically. Calculate the change in entropy when the initial argon pressure is (a) 100 kPa and (b) 200 kPa.

10-7M Complete the following table for an air–water-vapor mixture at atmospheric pressure.

T_{DB}, °C	T_{WB}, °C	ϕ, percent	Dew point, °C	p_v
38		30		
27	15			
		50	13	
	27		15	
32			15	
	20	40		

10-8M   Repeat Prob. 10-7M for an air–water-vapor mixture at a total pressure of 200 kPa.

10-9M   The partial pressure of water vapor in a certain psychrometric mixture is 2 kPa, and the total pressure of the mixture is 100 kPa. Calculate the relative humidity when the dry-bulb temperature is 30, 35, and 40°C.

10-10M An air–water-vapor mixture is compressed isentropically from 40°C, 1 atm, and 50 percent relative humidity to 700 kPa. Calculate the temperature at the end of the compression. If the mixture is then cooled at constant pressure, calculate the amount of moisture removed when the high-pressure gas is cooled to (a) 30°C and (b) 40°C.

10-11M A cold water pipe at 13°C passes through a room with atmospheric air at 27°C. What is the maximum relative humidity the air may have such that condensation does not occur on the pipe?

10-12M Meats and other food products will dry out when exposed directly to the very-low-temperature air in cold-storage vaults. Explain this phenomenon on the basis of information contained in this chapter. For calculation purposes, assume the air temperature is at −18°C.

10-13M In a certain northern city a heating system is designed to heat outside air at −25°C and 100 percent relative humidity, to 30°C. Calculate the heat which must be supplied under these circumstances per kilogram of air. What is the final relative humidity of the mixture?

10-14M The heating system in Prob. 10-13M is to be modified so that the final humidity of the mixture will be 50 percent. For this purpose a quantity of steam at 140 kPa, 80 percent quality, is mixed with the air during the heating process. How much steam must be supplied per kilogram of air, and what are the heating requirements under these new conditions?

10-15M The process in Prob. 10-14M is to be accomplished by spraying the heated air with liquid water at 80°C. What quantity of water will be required per kilogram of air, and what are the total heating requirements?

10-16M A small cooling system is designed solely for dehumidification purposes. The system is designed to remove 4 liters of moisture per hour from air at 27°C and 80 percent relative humidity. What air flow must the system handle, and what must its cooling capacity be? (*Note*: There are several correct answers; specify two suitable choices.)

10-17M An evaporative cooler is used to deliver 140 m³/min of air, operating in a desert climate where $T_{DB} = 43°C$, $\phi = 10$ percent. Calculate the quantity of water which must be supplied to the unit if the air leaves the unit at 90 percent relative humidity.

10-18M In infant intensive-care wards a *supersaturated* air condition is frequently maintained. Suppose air is available from the regular heating system at 27°C and 40 percent relative humidity. The air-flow rate for the ward is 34 m³/min. How much water must be added to the air per hour, in fine droplet form, to produce an effective humidity of 105 percent at 24°C? How much additional heat must be supplied?

10-19M A certain precision-gage room must be maintained at low-humidity conditions to ensure proper accuracy of the instruments. A cooling-reheat air
conditioning process is to be employed. The incoming air is at 32°C and 50
percent relative humidity and is to be delivered to the room at 18°C and 20
percent humidity. 28 m³/min of air are required. Calculate the total
cooling and heating requirements.

10-20M 28 m³/min of air at 43°C and 15 percent relative humidity are mixed adiabatically with 14 m³/min of air at 18°C and 90 percent humidity. Calculate
the final temperature and humidity.

10-21M The output air from the evaporative cooler of Prob. 10-17M is mixed with
28 m³/min of the outside air. What is the resultant temperature and humidity?

10-22M 30 m³/min of air at 33°C and 90 percent relative humidity are mixed with
an air flow at 15°C. What flow rate of the cooler air and humidity is necessary just to produce a mixture at 100 percent relative humidity? What is the
final mixture temperature?

10-23M 2 kg of air at 170 kPa, 5°C, and 80 percent relative humidity are mixed with
4 kg of air at 170 kPa, 40°C, and 20 percent relative humidity in a
steady-flow adiabatic process. Calculate the final mixture temperature and
humidity for a final pressure of (a) 170 kPa and (b) 140 kPa.

10-24M One way to produce very dry air is to cool it to very low temperatures and
then reheat to the desired operating temperature. Suppose a mixture is
available at 1 atm, 25°C, and 80 percent relative humidity. What final relative humidity would result if the mixture is cooled to −35°C and then reheated to 25°C? How much cooling and heating would be required for a
flow rate of initial mixture of 48 m³/min?

10-25M The "chiller" section in Prob. 10-24M must be defrosted when 25 kg of frost
or ice has accumulated. Assuming that the water removed from the air at
temperatures above 2°C is able to run off freely, calculate the time required
to accumulate 25 kg of ice. Repeat the calculation for no water runoff.

10-26M A special vaporizer device for use with small children in cold dry weather is
designed to mix one part of a supersaturated mixture at 15°C and 110 percent effective relative humidity with two parts of room air at 24°C and 15
percent relative humidity. What are the humidity and temperature of the
outlet mixed stream if the extra 10 percent humidity in the supersaturated
stream is in the form of small liquid droplets? Assume the mixture parts are
on a dry-air basis.

10-27M An inexpensive humidifier for home heating systems operates by atomizing
liquid water and spraying it directly into the air duct system. One particular
device is to be designed to handle the following air flows:

Input to heater-humidifier section: 8.5 m³/min at −7°C, $\phi = 80\%$
plus   34 m³/min at 15°C, $\phi = 80\%$
plus   liquid water at 15°C

Output from heater-humidifier section: $T_{DB} = 30°C$, $\phi = 35$ percent

Calculate the amount of heating required and the amount of liquid water which must be supplied to the humidifier.

10-28M   A small building in a desert climate requires comfort cooling of 35 kW and an evaporative cooler is to be designed to perform the task. Outside air is available at 45°C and $\phi = 5$ percent. What inlet air-flow rate is required, and how much water must be supplied to the cooler?

10-29M   A certain home has 325 m² of floor space with a 2.5-m ceiling height. How much total water vapor is contained in the house at 24°C and 50 percent relative humidity?

10-30M   Calculate the mass specific heats for a mixture of 1 kg of helium and 1 kg of nitrogen. If such a mixture is heated and compressed from 1 atm, 20°C, to 5 atm, 150°C, in a steady-flow device, calculate the heat added, the shaft work input, and change in entropy of the mixture. Repeat for heating and compressing in a closed system.

10-31M   A large cooling tower for a power plant is designed on the evaporation cooling concept. In one installation the heat which must be absorbed in the cooling tower is 1000 MW, and this is to be accomplished by spraying air at 32°C, $\phi = 30$ percent, with water at 27°C. The cooling-tower design assumes that the air leaves at 20°C saturated. What inlet air and water flows are needed to accomplish the required cooling rate?

10-32M   Calculate the mixture enthalpies for air–water-vapor mixtures at 1 atm and (a) $T_{DB} = 40°C$, $T_{WB} = 16°C$ and (b) $T_{DB} = 20°C$, $T_{WB} = 16°C$. What percent error would result from assuming that the mixture enthalpy is a function only of wet-bulb temperature?

10-33M   How much cooling is required to cool air at 32°C and 60 percent relative humidity to $-35°C$ in a constant-pressure process at (a) 1 atm and (b) 2 atm? How much moisture will be removed in these processes?

10-34M   A mixture of one part nitrogen and two parts carbon dioxide by mass contains enough water vapor to produce 50 percent relative humidity at 40°C and 1 atm total pressure. How much cooling is required to cool 1 kg of mixture to 5°C at constant pressure? How much water is removed in the process?

10-35M   A mixture containing 1 kg of oxygen and 1 kg of helium has enough water vapor to produce 10 percent relative humidity at 40°C. The mixture is to be cooled in an adiabatic saturation process. What is the lowest temperature which can be achieved in this manner, and how much water must be supplied per kilogram of initial mixture? What is the mass specific heat of the mixture before and after the saturation process?

10-36M   A mixture of methane (CH_4) and air is to be produced which has mass proportions of 1 part methane to 15 parts dry air. The mixing is accomplished in a steady-flow process with both methane and air entering at 27°C. The inlet air has $\phi = 70$ percent, and the gases are heated during the mixing process to produce an outlet temperature of 93°C. The process occurs at constant pressure of 1 atm. How much heating must be supplied for an inlet volume flow of methane of 28 m³/min, and what is the relative humidity of the outlet mixture?

10-37M A tank having a volume of 570 liters contains an air–water-vapor mixture at 1 atm, 27°C, and ϕ = 50 percent. A high-pressure line with dry air at 50°C and 480 kPa is used to fill the tank with additional air. The filling process occurs slowly so that the temperature inside the tank remains constant at 27°C. How much air must be added to raise the tank pressure to 350 kPa, what will the relative humidity be under the new conditions, and how much heat will have been lost by the tank in the process?

10-38M A mixture of equal molal parts of air and water vapor at 1 atm and 150°C is heated to 260°C in a constant-pressure process. How much heat is required per kilogram of mixture?

10-39M Estimate the temperature at which the vapor in the mixture of Prob. 10-38M will start to condense if the mixture is cooled at constant pressure.

10-40M A mixture of air and water vapor is saturated at 1.4 MPa and 150°C. Calculate the cooling required to cool the mixture to 65°C in a steady-flow process at constant pressure.

10-41M 1 kg of a mixture of 50 percent CO_2 and 50 percent CH_4 on a molal basis is contained in a rigid vessel at 150°C and 3.5 MPa. Calculate the volume of the mixture using (a) the ideal-gas law, (b) the rule of additive volumes, and (c) the Kay pseudo-critical method.

10-42M Calculate the energy required to heat the mixture of Prob. 10-41M to 260°C in a constant-pressure steady-flow process using (a) the ideal-gas law and (b) the pseudo-critical method.

10-43M A mixture of 60 percent methane (CH_4) and 40 percent ethane (C_2H_6) on a molal basis is to be compressed from 65°C, 700 kPa, to 2.4 MPa, in a steady-flow device. The mass flow rate is 5 kg/min. Assuming that the process is reversible and adiabatic, calculate the work input using (a) ideal-gas relations and (b) the pseudo-critical method.

10-44M A mixture containing 30 percent CO_2, 10 percent CO, 40 percent N_2, and 20 percent CH_4 on a molal basis is contained at 28 MPa and 90°C in a 140-liter rigid vessel. Calculate the heat which must be removed to lower the temperature to 40°C using (a) the ideal-gas law and (b) Kay's rule.

10-45M Oxygen is contained in a rigid 30-liter vessel at 50 atm and 40°C. Nitrogen from a high-pressure line at 28 MPa and 40°C is bled slowly into the tank until the mixture proportions in the tank approximate that for air, i.e., 78 percent N_2 and 22 percent O_2. During the filling process cooling is supplied to the tank to maintain the temperature at 40°C. Calculate the heat removed from the tank and the mass of nitrogen added. Use Kay's rule.

10-46M A mixture of equal molal parts of methane (CH_4) and ethane (C_2H_6) is compressed reversibly and isothermally from 65°C, 700 kPa, to 5 MPa, in a steady-flow process. Calculate the work required per kilogram of mixture.

10-47M   A mixture of 40 percent argon and 60 percent nitrogen on a molal basis is to be heated from -70°C, 4.8 MPa, to 90°C in a steady-flow heat-exchanger device. Because of fluid friction the pressure at exit of the device drops to 4.5 MPa. Flow velocities are low enough to neglect kinetic energies. Calculate the heating which must be supplied per kilogram of mixture.

10-48M   A mixture of equal molal parts of argon and CO_2 is contained in a rigid vessel at 20 atm and 25°C. How much heat must be supplied per kilogram to raise the temperature to 1500°C.

10-49M   An air–water-vapor mixture has 25 percent water vapor on a molal basis and is contained at 3.5 MPa and 430°C. Calculate the specific volume of the mixture using (a) the ideal-gas law, (b) the rule of additive pressures, (c) the rule of additive volumes, and (d) Kay's pseudo-critical rule.

10-50M   How much heat per kilogram of mixture would be required to heat the mixture of Prob. 10-49M to 540°C in a constant-pressure steady-flow process?

10-51M   4000 liters/s of air at 10°C, 100 percent relative humidity, are mixed in a steady-flow adiabatic process with 3000 liters/s of air at 33°C, 60 percent relative humidity, at a pressure of 1 atm. Calculate the properties for the final mixture.

10-52M   800 liters/s of air at 1 atm, 5°C, and 20 percent relative humidity are delivered to a heating chamber where the air is also sprayed with water at 25°C. The air discharges from the chamber at 25°C, 50 percent relative humidity. Calculate the amount of heat which must be supplied and the mass of water spray needed.

10-53M   Hot air at 1 atm, 35°C, and 60 percent relative humidity is cooled by spraying with liquid water at 3°C. The air leaves the spray chamber saturated with a specific humidity of 0.009 kg/kg dry air, and the water leaves the chamber at 9°C. The inlet air flow is 2800 liters/s. Calculate the flow rate of spray water required assuming the chamber is adiabatic.

10-54M   A mixture of CO_2 and N_2 has equal proportions by volume. Determine the mass fraction of each constituent and the gas constant for the mixture.

10-55M   1 kg of O_2 is mixed with 4 kg of another ideal gas, and the total mixture occupies a volume of 0.6 m³ at a total pressure of 320 kPa and 50°C. Determine the gas constant of the other gas and the mole fraction of each constituent.

10-56M   500 liters/s of air at 1 atm, 20°C, and 50 percent relative humidity are cooled to 5°C in a steady-flow process at constant pressure. Calculate the cooling required and amount of moisture removed.

10-57M   5.0 kg of an air-steam mixture at 5.0 atm and 35°C are expanded isothermally until the partial pressure of the water vapor is 1.228 kPa. The mixture is saturated at the initial condition. Find the total pressure for the final state and the heat transfer.

10-58M 500 liters of an air–water-vapor mixture at 1 atm, 35°C, and 90 percent relative humidity are cooled at constant volume to 5°C. Calculate the cooling required, amount of water vapor condensed, and temperature at which condensation starts.

10-59M 2300 liters/s of air at 1 atm, 30°C dry-bulb, and 35 percent relative humidity enter an adiabatic chamber where water is added to the flow in the amount of 8 g for each kilogram of dry air in the mixture. During the process the wet-bulb temperature remains constant. Calculate the dry-bulb temperature and relative humidity of the exit flow. How much total water is added per hour?

10-60M 700 liters/s of air at 20°C and 100 percent relative humidity are heated to 35°C in a steady-flow process. Determine the final relative humidity and the heating required.

10-61M 700 liters/s of atmospheric air at 32°C and 60 percent relative humidity is to undergo a cooling process to remove moisture and then a heat-addition process to bring the mixture to 24°C, 40 percent relative humidity. Calculate the amount of cooling and heating required and the moisture removed.

10-62M For a sample of air at 100 kPa the dry-bulb temperature is 34°C and the wet-bulb temperature is 20°C. Determine the specific and relative humidities, dew point, mole fractions of water vapor and dry air, and volume of mixture per kilogram of dry air.

10-63M Air and water vapor are contained in a rigid tank at 30°C. The water vapor is in the saturated-vapor state, and there is only 0.1 kg of dry air per kilogram of vapor. Calculate the specific humidity and total pressure of the mixture. Also calculate the mole fractions of air and water vapor.

10-64M 8.0 kg/s of dry air at 105°C and 1 atm are mixed with 0.5 kg/s of steam at 1 atm and 90 percent quality in an adiabatic steady-flow device at constant pressure. Calculate the absolute humidity of the final mixture and the irreversibility for the process for $T_0 = 20$°C.

10-65M Equal mass fractions of CH_4 and CO_2 are contained at 700 kPa and 30°C in a tank having a volume of 0.2 m³. 1.0 kg of O_2 is added to the mixture while maintaining the total volume and temperature constant. Calculate the molal analysis of the new mixture and the final pressure.

10-66M A certain mixture has the following molal analysis: 40 percent CH_4, 35 percent CO_2 and 25 percent N_2. Determine the mass fractions of each constituent and the gas constant for the mixture.

10-67M 3.5 kg of a mixture of CO_2 and N_2 are contained in a rigid vessel at 300 kPa and 70°C. The mole fraction of each gas is 50 percent. 1.0 kg of N_2 is added to the vessel while maintaining the temperature constant at 70°C. Determine the mass analysis of the new mixture, its molecular weight, and total pressure.

10-68M   In an air conditioning application for a building, 4.7 MW of energy must be dissipated to the surroundings using a wet-cooling tower. Inlet air is at 35°C and 30 percent relative humidity, and the air leaves at 38°C and 90 percent relative humidity. Make-up water enters the cooling tower at 23°C. Cooling water from the air conditioning unit enters at 40°C and is returned to the unit at 30°C. Calculate the mass flow of make-up water and inlet volume flow of air required.

10-69M   A power plant operates in the winter with outside air at 0°C, 80 percent relative humidity. 1000 MW of energy are to be dissipated by a wet-cooling tower which cools liquid water from 30°C to 20°C. Make-up water is available at 10°C, and the exit air from the tower is at 20°C, 80 percent relative humidity. Calculate the mass flow of make-up water and inlet volume flow of air required.

REFERENCES

1   Doolittle, J. S.: "Thermodynamics for Engineers," 2d ed., International Textbook Company, Scranton, Pa., 1964.

2   Obert, E. F.: "Concepts of Thermodynamics," McGraw-Hill Book Company, New York, 1960.

3   Van Wylen, G. J., and R. E. Sonntag: "Fundamentals of Classical Thermodynamics," John Wiley & Sons, Inc., New York, 1965.

4   Hall, N. A., and W. E. Ibele: "Engineering Thermodynamics," Prentice-Hall, Inc., Englewood Cliffs, N.J., 1960.

5   Kay, W. B.: Density of Hydrocarbon Gases and Vapors, *Ind. Eng. Chem.*, vol. 28, p. 1014, 1936.

6   Carrier, W. H.: Rational Psychrometric Formulae, *Trans. ASME*, vol. 33, p. 1005, 1911.

chapter eleven
Chemical thermodynamics and equilibrium

11-1 INTRODUCTION

The thermodynamics of chemical processes is an important part of engineering practice in a number of industries. This chapter presents an introduction to such processes. The exposition begins with a discussion of simple combustion reactions from the standpoint of mass balances. Next, techniques for describing chemical energy properties are examined, followed by application to combustion reactions. A chemical reaction is always somewhat incomplete in that reactants are never fully converted to products. The degree to which a reaction is completed depends on temperature, pressure, and other factors, and falls under the general subject of thermodynamic equilibrium. In the latter sections of this chapter we shall develop some general relations for thermodynamic equilibrium and apply them to chemical processes. Finally, we shall state the third law of thermodynamics and note its relation to the concepts of chemical equilibrium and statistical thermodynamics.

11-2 COMBUSTION REACTIONS

A combustion process involves the burning of a fuel with oxygen or a substance containing oxygen such as air. The fuel and oxidizer are called the *reactants*, and the constituents resulting from the combustion process are called the *products*. The combustion of carbon, for example, involves the simple reaction

$$C + O_2 \rightarrow CO_2 \tag{11-1}$$

When the combustion process involves the burning of a hydrocarbon fuel, the products will normally contain carbon dioxide and water. Thus, for methane, CH_4, we have the reaction

$$CH_4 + 2O_2 \rightarrow CO_2 + 2H_2O \tag{11-2}$$

The numbers that precede the chemical symbols in these reactions are called the *stoichiometric coefficients*. When a fuel is burned with air, all constituents in normal atmospheric air, other than oxygen, are assumed to be inert insofar as the reaction is concerned, unless very high temperatures are encountered such that chemical dissociation takes place. For normal air the composition is approximately 78 percent nitrogen, 21 percent oxygen, and 1 percent argon. For calculation purposes the argon is usually ignored and the air is assumed to be 79 percent nitrogen and 21 percent oxygen. Consequently, there are $\frac{79}{21} = 3.76$ mol of nitrogen for each mole of oxygen in the air. For the burning of methane with air we would have

$$CH_4 + 2O_2 + (2)(3.76)N_2 \rightarrow CO_2 + 2H_2O + (2)(3.76)N_2 \qquad (11\text{-}3)$$

The quantity of air necessary to balance this equation is called the *stoichiometric air*. It is not uncommon, however, to have a reaction where too much air is supplied for the quantity of fuel involved. In such cases we say that *excess air* is used, and the excess is usually expressed as a percentage of the stoichiometric air. The amount of excess air simply carries through the reaction. Using the combustion of mechane as an example again, 10 percent excess air would mean that $(2.0)(1.1)$ mol of oxygen were involved in the reactants. In the products $(0.1)(2.0)$ mol of oxygen would be left unused, and the chemical equation would be written as

$$CH_4 + (1.1)(2)O_2 + (1.1)(2)(3.76)N_2 \rightarrow$$
$$CO_2 + 2H_2O + (0.1)(2)O_2 + (1.1)(2)(3.76)N_2 \qquad (11\text{-}4)$$

The term *theoretical air* is sometimes used instead of stoichiometric air.

A combustion reaction for a general hydrocarbon C_nH_y burning with the theoretical proportion of air may be obtained as follows

$$C_nH_y + xO_2 + 3.76xN_2 \rightarrow aCO_2 + bH_2O + 3.76xN_2 \qquad (11\text{-}5)$$

From the carbon balance

$$n = a$$

so that the hydrogen balance yields

$$y = 2b$$

Finally, the oxygen balance is

$$2x = 2n + \frac{y}{2}$$
$$x = n + \frac{y}{4}$$

The final balance is then

$$C_nH_y + \left(n + \frac{y}{4}\right)O_2 + 3.76\left(n + \frac{y}{4}\right)N_2 \rightarrow$$
$$nCO_2 + \left(\frac{y}{2}\right)H_2O + 3.76\left(n + \frac{y}{4}\right)N_2 \qquad (11\text{-}6)$$

If less than the stoichiometric air is supplied, the reaction does not go to completion; i.e., all of the carbon does not burn and appear in the form of

CO_2 in the products. Instead a portion of the carbon will usually appear in the form of carbon monoxide, CO. For methane an incomplete reaction might appear as

$$CH_4 + xO_2 + 3.76xN_2 \rightarrow aCO_2 + bCO + 2H_2O + 3.76xN_2 \qquad (11\text{-}7)$$

In this case we would expect that $x < 2.0$. The oxygen balance on the equation would yield

$$2x = 2a + b + 2 \qquad (11\text{-}8)$$

Further information concerning the temperature of the reaction and chemical equilibrium considerations would be necessary to predict the exact split between CO_2 and CO, i.e., the exact values of the coefficients a and b. If enough O_2 were added to Eq. (11-7) to bring it up to stoichiometric proportions, the CO would then react with the oxygen to form additional CO_2 in accordance with the reaction equation:

$$CO + \tfrac{1}{2}O_2 \rightarrow CO_2 \qquad (11\text{-}9)$$

Even when excess air is supplied, real reactions do not usually go to completion because of various factors involved in the combustion process. In a practical sense, therefore, it is usually necessary to make experimental measurements of the composition of the products of combustion to balance the equation.

Figure 11-1 illustrates a simple *Orsat* apparatus which is used to analyze products of combustion. It consists of a measuring burette and three reagent pipettes which may be used to successively absorb carbon dioxide, oxygen, and carbon monoxide in the mixture. First, a sample of the flue gas is drawn into the measuring burette. Next, the sampling manifold is shut off from the flue gas and the sample forced into the first reagent pipette where the carbon dioxide is absorbed. The sample is then brought back into the measuring

FIG. 11-1 *Orsat apparatus.*

burette and the reduction in volume noted. The procedure is then repeated for the other two pipettes which absorb the O_2 and CO successively. In the measuring process the combustion products are saturated with water vapor, and the procedure is conducted so that the volumetric proportions of the products are obtained on a so-called "dry basis," i.e., exclusive of the water vapor which is present.

It turns out that even when excess air is supplied for the combustion process both CO and CO_2 may be obtained in the products because of chemical reactions occurring at the high temperatures of combustion. In actual practice then, the excess air reaction of Eq. (11-4) could contain both CO and O_2 in the products. This matter will be clarified when we discuss chemical equilibrium in later sections of this chapter.

A parameter of interest in practical combustion problems is the *air/fuel ratio*, which is the ratio of air to fuel in the reactants expressed on either a mass or molal basis.

EXAMPLE 11-1   Burning of octane

Octane (C_8H_{18}) is burned with 150 percent theorctical air. Calculate the air/fuel ratio and the molal analysis of the products of combustion. What is the dew point of the products for a total pressure of 14.696 psia?

SOLUTION   For theoretical air the combustion equation is

$$C_8H_{18} + 12.5O_2 + (3.76)(12.5)N_2 \rightarrow 8CO_2 + 9H_2O + (3.76)(12.5)N_2$$

For 150 percent theoretical air the corresponding reaction would be

$$C_8H_{18} + (1.5)(12.5)O_2 + (1.5)(3.76)(12.5)N_2 \rightarrow$$
$$8CO_2 + 9H_2O + (0.5)(12.5)O_2 + (1.5)(3.76)(12.5)N_2$$

We therefore have

$$m_f = (1)(114) = 114 \text{ lbm/mol fuel}$$
$$m_a = (1.5)(12.5)(1 + 3.76)(28.96) = 2580 \text{ lbm/mol fuel}$$

and the air/fuel ratio is

$$AF = \frac{m_a}{m_f} = \frac{2580}{114} = 22.6 \text{ lbm air/lbm fuel}$$

The total number of moles of products is

$$n_p = 8 + 9 + (0.5)(12.5) + (1.5)(3.76)(12.5)$$
$$= 93.75 \text{ mol/mol fuel}$$

so that the mole fractions of the constituents are

$$x_{CO_2} = \frac{8}{93.75} = 8.53 \text{ percent}$$

$$x_{H_2O} = \frac{9}{93.75} = 9.60 \text{ percent}$$

$$x_{O_2} = \frac{6.25}{93.75} = 6.66 \text{ percent}$$

$$x_{N_2} = \frac{70.5}{93.75} = 75.21 \text{ percent}$$

The dew point of the products is the saturation temperature corresponding to the partial pressure of the water vapor. We have

$$\begin{aligned}
p_{H_2O} &= x_{H_2O}\, p_{total} \\
&= (0.096)(14.696) \\
&= 1.41 \text{ psia} \qquad (9.72 \text{ kPa})
\end{aligned}$$

The saturation temperature corresponding to 1.41 psia is

$$T_{\text{dew point}} = 113.5°F \qquad (45.3°C)$$

Thus moisture would condense if the products were cooled below 113°F. This explains, of course, why water drips from the tailpipe of an automobile on a cold morning before the exhaust system has warmed up.

11-3 ENTHALPY OF FORMATION

In the following sections we shall be interested in performing energy balances on chemical reactions in order to determine the quantity of heat which may be liberated in a particular combustion process, the highest temperature attainable for a given fuel when burning with air, etc. For these energy balances we require a knowledge of the enthalpies of the substances involved in the reaction. Tables are available for various substances which give the properties as a function of temperature. To formulate such tables, however, it is necessary to establish a reference state for the properties which takes proper account of the internal chemical energy of the substance.

The base (zero level) energy level is chosen so that the energy of all *elements* is zero at 25°C (77°F) and 1-atm pressure. The energy of a compound is not zero at this reference state, but is defined in accordance with the following reasoning as applied to CO_2 at 25°C and 1 atm. To form CO_2 in a steady-flow process, we would have

$$C + O_2 \rightarrow CO_2 \tag{11-10}$$

and the energy balance

$$H_R + Q = H_P \qquad\qquad\qquad\qquad (11\text{-}11)$$

where H_R is the total enthalpy of the reactants and H_P is the total enthalpy of the products. Q is the heat transfer necessary to make the reaction go. Using the above reference state, $H_R = 0$ since all the reactants are elements and $H_P = Q$. The value of Q may be determined experimentally and has the following value for CO_2:

$$Q = H_p = -169\ 297 \text{ Btu/lbm mol} \qquad (-393\ 766 \text{ kJ/kg mol})$$

In other words, the reaction in Eq. (11-10) *liberates* 169 297 Btu of energy for each mole of carbon burned. The enthalpy of CO_2 at 25°C and 1 atm is therefore less than the reference state by this amount. This quantity is called the *enthalpy of formation* and has the symbol $\bar{h}_f^\circ$, where the bar designates per mole of substance. Values for typical substances are given in Table A-6 of the Appendix. A reaction in which heat is liberated is called *exothermic*, with the term *endothermic* being employed to describe reactions which absorb heat.

It is now fairly easy to see how we may calculate the enthalpies of substances with respect to the zero element state of 298 K and 1 atm. We employ the tables in the Appendix to compute the change from the reference state and add this to the enthalpy of formation. Thus the total molal enthalpy at temperature T is

$$\bar{h}(T)_{\text{total}} = \bar{h}_f^\circ + \bar{h}_T - \bar{h}_{298} \qquad\qquad\qquad (11\text{-}12)$$

11-4   REFERENCE LEVELS FOR TABLES

As we have seen before, it is possible to choose different zero levels, or reference states, for tabulations of thermodynamic properties. For the steam tables we chose the saturated-liquid state at 0°C, while for the air tables we chose 0 K and a pressure of 1 atm. For use with Eq. (11-12) tables are employed using a reference temperature of 0 K as well as the standard state temperature of 298 K (77°F). The 0 K state is perhaps more common for English units (as in Table A-18) while the 298 K reference is employed for SI tabulations (as in Table A-18M). The modern approach is also to use a reference pressure of 0.1 MPa instead of 1 atm (0.101 32 MPa). This small difference has essentially no effect on the enthalpy or internal energies.

11-5   ENTROPY OF FORMATION

As with the enthalpy, it is possible to determine absolute entropies of formation $\bar{s}_f^\circ$ for substances as given in Table A-6. This table is based on the standard state of 298 K and 1 atm. For those substances which follow ideal-gas behavior the entropy variation with pressure is given through an application of Eq. (7-37), written in terms of molal properties

$$\bar{s}_2 - \bar{s}_1 = \bar{\phi}_2 - \bar{\phi}_1 - \mathscr{R} \ln \left(\frac{p_2}{p_1}\right) \tag{11-13}$$

The absolute entropy can thus be written in terms of the entropy of formation as

$$\bar{s}°(p,T) = \bar{s}_f° + \bar{\phi}_T - \bar{\phi}_{298} - \mathscr{R} \ln \left(\frac{p}{p_0}\right) \tag{11-14}$$

where p_0 is the reference pressure of 1 atm. Table A-18M gives the values of $\bar{s}°$ for gases at various temperatures and a pressure of 0.1 MPa. Absolute values for entropy are required for certain second-law analyses of combustion reactions as we shall see in later sections.

The important point to remember is that reference states must be carefully specified when dealing with combustion reactions. The tables in the Appendix specify these states and indicate some appropriate conversion equations.

EXAMPLE 11-2

An analysis is made of the products of combustion when methane (CH_4) is burned with atmospheric air. The mole fractions of the *dry* products (those other than water) are as follows:

CO_2 = 9.00 percent
CO = 1.20 percent
O_2 = 1.50 percent
N_2 = 88.30 percent

Calculate the percent theoretical air which is used in the combustion process.

SOLUTION   We shall assume 100 mol of dry products and balance the combustion equation on this basis. We have

$$a CH_4 + x O_2 + 3.76 x N_2 \rightarrow$$
$$9.00 CO_2 + 1.2 CO + 1.5 O_2 + b H_2O + 88.3 N_2$$

Carbon balance:

$$a = 9.00 + 1.2 = 10.2$$

Hydrogen balance:

$$4a = 2b$$
$$b = 2a = 20.4$$

Oxygen balance:

$$2x = (2)(9.00) + 1.2 + (2)(1.5) + 20.4$$
$$x = 21.3$$

Nitrogen balance:

$$3.76x = 88.3$$
$$x = 23.5$$

Note that the two determinations of x do *not* agree. In actual practice, the nitrogen content of the products is usually determined by subtracting the other constituents from the total; i.e., it is not measured directly. For this reason, the value of x determined from the nitrogen balance is not believed to be as reliable as the other calculation. Using $x = 21.3$ gives

$$\frac{\text{Moles } O_2}{\text{Moles fuel}} = \frac{21.3}{10.2} = 2.09$$

According to Eq. (11-3) there are 2.00 mol of O_2 required for each mole of fuel when the theoretical air is burned. The desired result is, therefore,

$$\text{Percent theoretical air} = \frac{2.09}{2.00}(100) = 104.5 \text{ percent}$$

EXAMPLE 11-3
Calculate the molal enthalpy of CO_2 at 440°F.

SOLUTION We employ Eq. (11-12) for this calculation along with Table A-6 and the gas tables (Table A-18).

$$\bar{h}^\circ_{CO_2} = -169\ 290 \text{ Btu/lbm mol}$$
$$\bar{h}_{440} = 7597.6 \text{ Btu/lbm mol} \qquad (17\ 671 \text{ kJ/kg mol})$$
$$\bar{h}_{77} = 4030.2 \text{ Btu/lbm mol} \qquad (9373.8 \text{ kJ/kg mol})$$

Thus

$$\bar{h}_{\text{total}} = -169\ 290 + 7597.6 - 4030.2$$
$$= -165\ 722 \text{ Btu/lbm mol}$$
$$= -385\ 451 \text{ kJ/kg mol}$$

EXAMPLE 11-4

Ethane (C_2H_6) is burned with the theoretical amount of oxygen. Determine the heat transfer for reactants and products, both at 298 K and 1 atm.

SOLUTION   The combustion equation is

$$C_2H_6 + 3.5O_2 \rightarrow 2CO_2 + 3H_2O(l)$$

From Eq. (11-11) the heat transfer is

$$Q = H_P - H_R$$

Also,

$$H_R = \bar{h}^\circ_{C_2H_6} = -36\ 420 \text{ Btu}$$
$$H_P = 2\bar{h}^\circ_{CO_2} + 3\bar{h}^\circ_{H_2O}$$
$$= (2)(-169\ 290) + (3)(-122\ 970)$$
$$= -707\ 490 \text{ Btu} \qquad (-7.464 \times 10^5 \text{ kJ})$$

The heat transfer is, therefore,

$$Q = -707\ 490 - (-36\ 420)$$
$$= -671\ 070 \text{ Btu/mol fuel} \qquad (-1.56 \times 10^6 \text{ kJ/kg mol})$$

The negative sign indicates that heat is liberated during the process.

11-6   HEAT OF REACTION

The energy liberated in a chemical reaction is given by the following relations. For a constant-volume reaction,

$$Q_V = (\Delta U)_v = (U_P - U_R)_v = \bar{u}_{RP} \qquad (11\text{-}15)$$

and, at constant pressure,

$$Q_P = (\Delta H)_P = (H_P - H_R)_P = \bar{h}_{RP} \qquad (11\text{-}16)$$

The quantities given in Eqs. (11-15) and (11-16) are called the *heats of reaction* or the *internal energy and enthalpy of combustion*. If the heats are expressed per mole of fuel, then the symbols $\bar{u}_{RP}$ and $\bar{h}_{RP}$ are appropriately used to designate the energies indicated in the equation.

The term *heating value* for a combustion process is in wide use and is a

TABLE 11-1
Higher heating values for fuels with combustion at 25°C (77°F) and 1 atm according to Refs. [1, 2, 3]. Liquid water is in the products of combustion

Fuel	Formula	HHV, Btu/lbm mol	kJ/kg mol	Vaporization enthalpy $\bar{h}_{fg}$ for fuel Btu/lbm mol	kJ/kg mol
Hydrogen	$H_2(g)$	−122 970	−285 840		
Carbon	$C(s)$	−169 290	−393 520		
Carbon monoxide	$CO(g)$	−121 750	−282 990		
Methane	$CH_4(g)$	−383 040	−890 360		
Acetylene	$C_2H_2(g)$	−559 120	−1 299 600		
Ethylene	$C_2H_4(g)$	−607 010	−1 410 970		
Ethane	$C_2H_6(g)$	−671 080	−1 559 900		
Propylene	$C_3H_6(g)$	−885 580	−2 058 500		
Propane	$C_3H_8(g)$	−955 070	−2 220 000	6 480	15 060
n-butane	$C_4H_{10}(g)$	−1 237 800	−2 877 100	9 060	21 060
n-pentane	$C_5H_{12}(g)$	−1 521 300	−3 536 100	11 360	26 410
n-hexane	$C_6H_{14}(g)$	−1 804 600	−4 194 800	13 563	31 530
n-heptane	$C_7H_{16}(g)$	−2 088 000	−4 853 500	15 713	36 520
n-octane	$C_8H_{18}(g)$	−2 371 400	−5 512 200	17 835	41 460
Benzene	$C_6H_6(g)$	−1 420 300	−3 301 500	14 552	33 830
Toluene	$C_7H_8(g)$	−1 698 400	−3 947 900	17 176	39 920
Methyl alcohol	$CH_3OH(g)$	−328 700	−764 540	16 092	37 900
Ethyl alcohol	$C_2H_5OH(g)$	−606 280	−1 409 300	18 216	42 340

synonym for heat of reaction. When a hydrocarbon fuel is burned in a combustion process, water will appear in the products. The maximum energy release will be obtained when all the water in the products *due to combustion*
is in the liquid state; in such cases the heat of reaction is called a *higher
heating value* (HHV). Similarly, a lower heat of reaction will be experienced
when all the water in the products due to combustion is in the vapor state; in
this case we say that a *lower heating value* (LHV) is obtained. In reality, both
liquid and vapor will be present in many combustion processes, so that a
heating value between the HHV and LHV can be observed. A tabulation of
the heating values of several common fuels is given in Table 11-1. The examples illustrate the use of these values.

Table 11-1 gives the heating values of fuels for combustion at the convenient reference temperature of 25°C (77°F). Most combustion processes certainly do not occur at this temperature, and it is of interest to show how
these tabular values may be used to calculate heats of combustion at other
temperatures. The combustion process at some arbitrary temperature T can
be expressed in terms of $\bar{h}_{RP}$ for the standard state by means of the simple
sketch shown in Fig. 11-2. The reactants are first cooled to 25°C, combustion is allowed to proceed at 25°C, and the products are then heated back to
the temperature T. In terms of the diagram, the heat of reaction may be expressed as

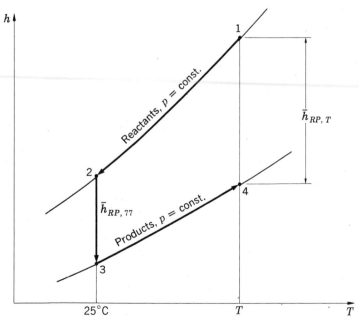

FIG. 11-2 *Calculation of heating values.*

$$\bar{h}_{RP_T} = H_4 - H_1 = (H_4 - H_3) + (H_3 - H_2) + (H_2 - H_1)$$

This relation may be rewritten as

$$\bar{h}_{RP_T} = (H_T - H_{25})_P + \bar{h}_{RP_{25}} + (H_{25} - H_T)_R \tag{11-17}$$

If property data for the various constituents of the reactants and products are available, the value of $\bar{h}_{RP_T}$ is easily calculated.

Of course, it is not necessary to use the concept of heating values for energy analysis of combustion reactions. One might use the enthalpies of formation directly to carry out the desired calculations.

EXAMPLE 11-5

Calculate the enthalpy of combustion for gaseous methane at 440°F.

SOLUTION We assume that the water in the products is in the vapor phase. The higher heating value for methane at 25°C is given in Table 11-1 as

$$\bar{h}_{RP_{25}} = -383\,040 \text{ Btu/lbm mol}$$

The combustion equation is

$$CH_4 + 2O_2 \rightarrow CO_2 + 2H_2O$$

For the modest temperature range involved we may assume the specific heat for CH_4 as a constant of 0.532 Btu/lbm·°F (from Table 2-2). The enthalpies of other constituents in the reactants and products are determined from the gas tables. We shall employ Eq. (11-17) for the calculation of enthalpy of combustion. The quantities of interest are:

$$
\begin{aligned}
(H_{77} - H_T)_R &= n_{CH_4}(\bar{h}_{77} - \bar{h}_T)_{CH_4} + n_{O_2}(\bar{h}_{77} - \bar{h}_T)_{O_2} \\
&= (1)(16)(0.532)(77 - 440) + (2)(3725.1 - 6337.9) \\
&= -8315 \text{ Btu} \\
(H_T - H_{77})_P &= n_{CO_2}(\bar{h}_T - \bar{h}_{77})_{CO_2} + n_{H_2O}[\bar{h}_T(g) - \bar{h}_{77}(l)]_{H_2O} \\
&= (1)(7597.6 - 4030.2) + (2)(18)(1260.3 - 45.02) \\
&= 47\,317 \text{ Btu} \quad (49\,922 \text{ kJ})
\end{aligned}
$$

Note that the enthalpy of water at 77°F is taken as the value for saturated liquid, since we will be using the higher heating value for $\bar{h}_{RP_{77}}$.

The foregoing values may now be inserted in Eq. (11-17) to yield:

$$
\begin{aligned}
\bar{h}_{RP_T} &= 47\,317 + (-383\,040) + (-8315) \\
&= -344\,038 \text{ Btu/lbm mol fuel}
\end{aligned}
$$

EXAMPLE 11-5M

Calculate the enthalpy of combustion for gaseous methane at 500 K.

SOLUTION   We assume that the water in the products is in the vapor phase. The higher heating value for methane at 298 K is given in Table 11-1 as

$$\bar{h}_{RP_{298}} = -890\ 360 \text{ kJ/kg mol}$$

The combustion equation is

$$CH_4 + 2O_2 \rightarrow CO_2 + 2H_2O$$

For the modest temperature range involved we may assume the specific heat for CH_4 as a constant of 2.227 kJ/kg·°C (from Table 2-2). The enthalpies of other constituents in the reactants and products are determined from the gas tables. We shall employ Eq. (11-17) for the calculation of enthalpy of combustion. The quantities of interest are:

$$
\begin{aligned}
(H_{298} - H_T)_R &= n_{CH_4}(\bar{h}_{298} - \bar{h}_T)_{CH_4} + n_{O_2}(\bar{h}_{298} - \bar{h}_T)_{O_2} \\
&= (1)(16)(2.227)(298 - 500) + (2)(-6088) \\
&= -19\ 374 \text{ kJ/kg mol fuel} \\
(H_T - H_{298})_P &= n_{CO_2}(\bar{h}_T - \bar{h}_{298})_{CO_2} + n_{H_2O}[\bar{h}_T(g) - \bar{h}_{298}(g) + \bar{h}_{fg298}]_{H_2O} \\
&= (1)(8314) + (2)[6920 + (18)(2442.3)] \\
&= 110\ 077 \text{ kJ/kg mol} \qquad (47\ 325 \text{ Btu/lbm mol})
\end{aligned}
$$

Note that the enthalpy of water at 298 K is taken as the value for saturated liquid, since we will be using the higher heating value for $\bar{h}_{RP_{298}}$. The foregoing values may now be inserted in Eq. (11-17) to yield:

$$
\begin{aligned}
\bar{h}_{RPT} &= 110\ 077 + (-890\ 360) + (-19\ 374) \\
&= -799\ 657 \text{ kJ/kg mol fuel} \qquad (-343\ 791 \text{ Btu/lbm mol})
\end{aligned}
$$

EXAMPLE 11-6

A certain gaseous fuel consists of 35 percent methane (CH_4), 25 percent propane (C_3H_8), and 40 percent pentane (C_5H_{12}), where the percentages are on a molal basis. Calculate the mass fraction of each constituent and the gas constant for the mixture. Also calculate the higher heating value for this fuel when burned with the theoretical air at 25°C.

SOLUTION   For 1 mol of mixture the mass and mass fraction of each constituent are:

Mass fraction

$$m(CH_4) = (0.35)(16) = 5.6 \qquad 0.123$$
$$m(C_3H_8) = (0.25)(44) = 11 \qquad 0.242$$
$$m(C_5H_{12}) = (0.40)(72) = \underline{28.8} \qquad \underline{0.635}$$
$$\text{Total} \qquad 45.4 \qquad 1.000$$

The molecular weight of the mixture is

$$M = \frac{45.4}{1.0} = 45.4 \tag{a}$$

so that the gas constant of the mixture is

$$R = \frac{\mathscr{R}}{M} = \frac{8314.41}{45.4} = 183.14 \text{ J/kg·K} \tag{b}$$

To calculate the heating value we first construct the combustion equation. Considering 1 mol of fuel, the equation would be

$$0.35CH_4 + 0.25C_3H_8 + 0.40C_5H_{12} + xO_2 + 3.76xN_2 \rightarrow$$
$$aCO_2 + bH_2O + 3.76xN_2 \tag{c}$$

The carbon balance then gives

$$0.35 + (0.25)(3) + (0.40)(5) = a = 3.10 \tag{d}$$

From the hydrogen balance we obtain

$$(0.35)(4) + (0.25)(8) + (0.40)(12) = 2b$$
$$b = 4.1 \tag{e}$$

Finally, the oxygen balance yields

$$2x = 2a + b = (2)(3.10) + 4.1 \tag{f}$$
$$x = 5.15$$

The heating value is now obtained from

$$Q_p = H_p - H_R \tag{g}$$

where all of the enthalpies are evaluated at 25°C. This means that we can use the formation enthalpies obtained from Table A-6.

$$CH_4: \overline{h_f^\circ} = -74\ 850 \text{ kJ/kg mol}$$
$$C_3H_8: \overline{h_f^\circ} = -103\ 850 \text{ kJ/kg mol}$$
$$C_5H_{12}: \overline{h_f^\circ} = -146\ 440 \text{ kJ/kg mol}$$

$$O_2: \overline{h_f^\circ} = 0$$
$$N_2: \overline{h_f^\circ} = 0$$
$$CO_2: \overline{h_f^\circ} = -393\ 520\ kJ/kg\ mol$$
$$H_2O\ (l): \overline{h_f^\circ} = -285\ 830\ kJ/kg\ mol$$

We then obtain for the enthalpies of products and reactants

$$H_R = (0.35)(-74\ 850) + (0.25)(-103\ 850) + (0.40)(-146\ 440)$$
$$= -110\ 736\ kJ/mol\ fuel \tag{h}$$
$$H_P = (3.10)(-393\ 520) + (4.10)(-285\ 830)$$
$$= -2\ 391\ 815\ kJ/mol\ fuel \tag{i}$$

The heating value is obtaining by inserting the values into Eq. (g).

$$Q = -2\ 391\ 815 - (-110\ 736)$$
$$= -2\ 281\ 079\ kJ/mol\ fuel \tag{j}$$

11-7 ADIABATIC FLAME TEMPERATURE

Consider the combustion process shown in Fig. 11-3. If the process is allowed to proceed adiabatically, the final temperature which the products will attain is called the *adiabatic flame temperature*. In this instance, all of the heat of reaction is converted to internal energy and is manifested in the form of a high temperature for the products of combustion. For the temperatures involved in most reactions, the combustion process usually will not go to completion, and equations of chemical equilibrium must be employed to determine the final temperature.

To compute the adiabatic flame temperature an iterative procedure must be employed because of the complex relation between the products temperature and their constituent enthalpies. The following example illustrates such a procedure.

EXAMPLE 11-7

Calculate the adiabatic flame temperature for liquid octane burning with 200 percent theoretical air at 77°F.

FIG. 11-3 *Adiabatic combustion process.*

SOLUTION   The combustion equation for this reaction is

$$C_8H_{18}(l) + (2)(12.50)O_2 + (2)(12.5)(3.76)N_2 \rightarrow$$
$$8CO_2 + 9H_2O + (2)(12.5)(3.76)N_2 + (12.5)O_2$$

Because the process is adiabatic, we must have

$$H_R = H_P \qquad\qquad (a)$$

The final temperature of the products is obtained by an iterative technique outlined in the following. The total enthalpy of the reactants must be that of liquid octane at 77°F. For octane in the gaseous state the enthalpy of formation is given in Table A-6 as $-89\ 680$ Btu/lbm mol. The enthalpy of vaporization is given in Table 11-1 as 17 835 Btu/lbm mol so the enthalpy in the liquid state would be

$$H_R = \bar{h}^\circ_{C_8H_{18}} = -89\ 680 - 17\ 835 = -107\ 515 \text{ Btu/mol fuel}$$

and the relation for the enthalpy of the products is

$$\begin{aligned}
H_P = {}& 8(\bar{h}^\circ + \bar{h}_T - \bar{h}_{537})_{CO_2} \\
& + 9(\bar{h}^\circ + \bar{h}_T - \bar{h}_{537})_{H_2O} \\
& + 94(\bar{h}^\circ + \bar{h}_T - \bar{h}_{537})_{N_2} \\
& + 12.5(\bar{h}_T - \bar{h}_{537})_{O_2} \\
= {}& 8(-169\ 290 + \bar{h}_{T_{CO_2}} - 4030) \\
& + 9(-104\ 040 + \bar{h}_{T_{H_2O}} - 4268) \\
& + 94(\bar{h}_{T_{N_2}} - 3730) + 12.5(\bar{h}_{T_{O_2}} - 3745)
\end{aligned}$$

We now substitute H_R and H_P in Eq. (*a*) and assume values of T until the equation is satisfied. The iterations are:

$$\begin{aligned}
T = 3000°R \qquad H_P = {}& 8(-173\ 320 + 34\ 806) \\
& + 9(-108\ 308 + 28\ 386) \\
& + 94(22\ 761 - 3730) \\
& + 12.5(23\ 817 - 3725) \\
= {}& +310\ 000
\end{aligned}$$

This value is too high since $H_R = -107\ 530$. We, therefore, try a lower temperature for the products:

$$\begin{aligned}
T = 2000°R \qquad H_P = {}& 8(-173\ 320 + 21\ 018) \\
& + 9(-108\ 308 + 17\ 439) \\
& + 94(14\ 534 - 3730) \\
& + 12.5(15\ 164 - 3725) \\
= {}& -877\ 673
\end{aligned}$$

This value is too low, so we try an intermediate value determined by interpolation:

$$T = 2600°R \quad H_P = 8(-173\ 320 + 29\ 187)$$
$$+ 9(-108\ 308 + 23\ 869)$$
$$+ 94(19\ 415 - 3730)$$
$$+ 12.5(20\ 311 - 3725)$$
$$= -231\ 000$$

This value is still too low, but a quick interpolation yields

$$T = 2700°R \quad (1500\ K)$$

An additional iteration would show this value to be in good agreement with $H_R = -107\ 515$ Btu $(-113\ 434$ kJ$)$.

It is well to remark that this temperature would not be achieved in practice because the products would dissociate. Section 11-9 discusses the chemical equilibrium conditions which must be applied in such circumstances.

EXAMPLE 11-7M Calculation of adiabatic flame temperature

Calculate the adiabatic flame temperature for liquid octane burning with 200 percent theoretical air at 298 K.

SOLUTION The combustion equation for this reaction is

$$C_8H_{18}(l) + (2)(12.50)O_2 + (2)(12.5)(3.76)N_2 \rightarrow 8CO_2 + 9H_2O$$
$$+ (2)(12.5)(3.76)N_2 + (12.5)O_2$$

Since the process is adiabatic, we must have

$$H_R = H_P \tag{a}$$

The final temperature of the products is obtained by an iterative technique outlined in the following. The total enthalpy of the reactants must be that of liquid octane at 298 K. For octane in the gaseous state the enthalpy of formation is given in Table A-6 as $-208\ 450$ kJ/kg mol. The enthalpy of vaporization is given in Table 11-1 as 41 460 kJ/kg mol, and so the enthalpy in the liquid state would be

$$H_R = \bar{h}°_{C_8H_{18}} = -208\ 450 - 41\ 460 = -249\ 910 \text{ kJ/kg mol fuel}$$

and the relation for the enthalpy of the products is

$$H_p = 8(\bar{h}° + \bar{h}_T - \bar{h}_{298})_{CO_2}$$
$$+ 9(\bar{h}° + \bar{h}_T - \bar{h}_{298})_{H_2O}$$
$$+ 94(\bar{h}° + \bar{h}_T - \bar{h}_{298})_{N_2}$$
$$+ 12.5(\bar{h}_T - \bar{h}_{298})_{O_2}$$
$$= 8[-393\ 520 + (\bar{h}_T - \bar{h}_{298})_{CO_2}]$$
$$+ 9[-241\ 820 + (\bar{h}_T - \bar{h}_{298})_{H_2O}]$$
$$+ 94(\bar{h}_T - \bar{h}_{298})_{N_2} + 12.5(\bar{h}_T - \bar{h}_{298})_{O_2}$$

We now substitute H_R and H_P in Eq. (a) and assume values of T until the equation is satisfied. The iterations are:

$$
\begin{aligned}
T = 1700\text{ K} \quad H_P &= 8(-393\ 520 + 73\ 492) \\
&\quad + 9(-241\ 820 + 57\ 685) \\
&\quad + 94(45\ 430) + 12.5(47\ 970) \\
&= +652\ 606
\end{aligned}
$$

This value is too high since $H_R = -249\ 910$. We, therefore, try a lower temperature for the products:

$$
\begin{aligned}
T = 1100\text{ K} \quad H_P &= 8(-393\ 520 + 38\ 894) \\
&\quad + 9(-241\ 820 + 30\ 167) \\
&\quad + 94(24\ 757) + 12.5(26\ 217) \\
&= -2\ 087\ 015
\end{aligned}
$$

This value is too low, so we try an intermediate value determined by interpolation:

$$
\begin{aligned}
T = 1400\text{ K} \quad H_P &= 8(-393\ 520 + 55\ 907) \\
&\quad + 9(-241\ 820 + 43\ 447) \\
&\quad + 94(34\ 936) + 12.5(36\ 966) \\
&= -740\ 202
\end{aligned}
$$

This value is still too low, but a quick interpolation yields

$$
T = 1506\text{ K} \quad (2712°\text{R})
$$

An additional iteration would show this value to be in good agreement with

$$
H_R = 249\ 910\text{ kJ/kg mol} \quad (107\ 442\text{ Btu/lbm mol})
$$

It is well to remark that this temperature would not be achieved in prac-

TABLE 11-2
Sample adiabatic flame temperatures, reactants at 298 K, 1 atm, stoichiometric mixture

Fuel	Oxygen		Air	
	K	°R	K	°R
H_2	3079	5542	2384	4291
CH_4	3054	5497	2227	4009
C_8H_{18}	3108	5594	2277	4098

tice because the products would dissociate. The following section discusses the chemical equilibrium conditions which must be applied in such circumstances.

Values of some adiabatic flame temperatures which do take into account dissociation are given in Table 11-2.

11-8 GENERAL PRODUCT ENTHALPIES

Tabulations of product enthalpies for a general hydrocarbon $(CH_2)_n$ in combustion with 200 and 400 percent theoretical air have been assembled by Keenan and Kaye [12] and are given in Table A-18. Use of these tables can speed computations considerably. It should be noted that the molal enthalpies listed do not include enthalpies of formation for CO_2 or H_2O in the products and assume that the fuel is in the gaseous state. Although these tables have been developed for the $(CH_2)_n$ hydrocarbon, it turns out that they give a very satisfactory representation for the products of combustion with other hydrocarbons as well. Example 11-8 illustrates their application to an adiabatic flame temperature calculation for octane (C_8H_{18}).

EXAMPLE 11-8

Use the general combustion products table to calculate adiabatic flame temperature for the octane in Example 11-7.

SOLUTION We have already calculated the enthalpy of the reactants as

$$H_R = -249\ 910 \text{ kJ/kg mol} = -107\ 442 \text{ Btu/lbm mol} \qquad (a)$$

The enthalpy of the products is the sum of the formation enthalpies for CO_2 and H_2O and the product enthalpies to be taken from Table A-18.

$$H_P = H_P^\circ + n_P[\bar{h}_P(T) - \bar{h}_P(537)] \qquad (b)$$

Thus, we calculate

$$\begin{aligned} H_P^\circ &= (n\bar{h}_f^\circ)_{CO_2} + (n\bar{h}_f^\circ)_{H_2O} \\ &= (8)(-169\ 300) + (9)(-104\ 040) \\ &= -2\ 290\ 760 \text{ Btu} \end{aligned} \qquad (c)$$

The total number of moles of products is

$$n_P = 8 + 9 + 94 + 12.5 = 123.5 \text{ mol/mol fuel} \qquad (d)$$

and, from Table A-18 for 200 percent excess air,

$$\bar{h}_P(537) = 3774.9 \text{ Btu/lbm mol}$$

Inserting the numerical values in Eq. (b) gives

$$\bar{h}_P(T) = 21\ 454 \text{ Btu/lbm mol}$$

and from Table A-18 the corresponding temperature for 200 percent excess air is

$$T_P = 2718°R = 1510 \text{ K}$$

Thus, we see that there is good agreement between this calculation and that of Example 11-7. We may anticipate that the combustion products tables can be used to advantage where a more speedy calculation is desired.

EXAMPLE 11-9

Using the general combustion products table, estimate the adiabatic flame temperature for propylene (gas) C_3H_6 burning with 200 percent excess air at 1 atm. Assume reactants at 77°F (25°C).

SOLUTION　The general reaction equation for a hydrocarbon of the form $(CH_2)_n$ with 200 percent excess air is

$$(CH_2)_n + 3xO_2 + 3.76(3x)N_2 \rightarrow aCO_2 + bH_2O$$
$$+ 2xO_2 + 3.76(3x)N_2 \qquad (a)$$

from the carbon balance $a = n$ and from the hydrogen balance $b = n$. Then, the oxygen balance yields $x = 3n/2$. The total number of moles of products is, therefore,

$$n_P = n + n + 2\left(\frac{3n}{2}\right) + (3.76)(3)\left(\frac{3n}{2}\right)$$
$$= 21.92n \text{ mol product/mol fuel} \qquad (b)$$

For propylene $n = 3$ and $n_P = 65.76$ mol/mol fuel. The enthalpy of the reactants is just the formation enthalpy for propylene or

$$H_R = \bar{h}_f^°(C_3H_6) = 8790 \text{ Btu/lbm mol}$$

For the products, we first calculate the component of enthalpy due to the formation enthalpies of CO_2 and $H_2O(g)$

$$H_P^° = (n\bar{h}_f^°)_{CO_2} + (n\bar{h}_f^°)_{H_2O}$$
$$= (3)(-169\ 290) + (3)(-104\ 040)$$
$$= -819\ 990 \text{ Btu/mol fuel} \qquad (c)$$

The other component of the enthalpy can be obtained from Table A-18 for 200 percent excess air. Thus, the total product enthalpy is

$$H_P = H_{P_0}^\circ + H_P(T) - H_P(537)$$
$$= H_P^\circ + n_P[\bar{h}_P(T) - \bar{h}_P(537)] \tag{d}$$

Because we seek the adiabatic flame temperature

$$H_P = H_R = 8790 \text{ Btu/mol fuel}$$

From Table A-18

$$\bar{h}_P(537) = 3744.9 \text{ Btu/lbm mol}$$

Inserting the numerical values in Eq. (d) we have

$$8790 = -819\,990 + (65.76)[\bar{h}_P(T) - 3774.9]$$

and

$$\bar{h}_P(T) = 16\,378 \text{ Btu/lbm mol}$$

Consulting Table A-18 again we find the corresponding temperature for the products to be

$$T_P = 2140°\text{R}$$

EXAMPLE 11-9M   Irreversibility in a burner

Gaseous octane (C_8H_{18}) is burned with air at a pressure of 4 atm. Both the fuel and air enter at 4 atm. Fuel entrance temperature is 298 K and the air enters at 710 K. The products leave the combustion chamber at 1100 K, and the overall process is adiabatic. Calculate the fuel/air ratio and the irreversibility for the process with $T_0 = 25°C$.

SOLUTION   This process is what is normally experienced in the burner for a gas turbine cycle (see Sec. 12-12). Because the process is adiabatic we have

$$H_R = H_P \tag{a}$$

The combustion reaction for 1 mol of fuel is written as

$$C_8H_{18} + xO_2 + 3.76xN_2 \rightarrow aCO_2 + 6H_2O + cO_2 + 3.76xN_2 \tag{b}$$

Now, x is unknown because we do not know how much excess air is supplied. The molal balances yield:

C: $8 = a$
H: $18 = 2b$ $b = 9$
O: $2x = 2a + b + 2c$ $x = 12.5 + c$
       or $c = x - 12.5$

To determine the AF ratio we combine Eq. (a), the energy balance, with Eq. (b) to solve for x and thence the AF ratio. For the reactants

$$H_R = (1)\overline{h^\circ_f}_{\text{C}_8\text{H}_{18}} + (4.76x)(28.97)(h_{710} - h_{298})_{\text{air}}$$

where we are referencing to 298 K. From Tables A-6 and A-17M this gives

$$H_R = (1)(-208\ 450) + (4.76x)(28.97)(713.27 - 298.19)$$
$$= -208\ 450 + 57\ 239x \quad \text{kJ/kg mol fuel} \tag{c}$$

For the products, referenced to 298 K

$$H_P = (8)(\overline{h^\circ_f} + \overline{h}_{1100} - \overline{h}_{298})_{\text{CO}_2} + (9)(\overline{h^\circ_f} + \overline{h}_{1100} - \overline{h}_{298})_{\text{H}_2\text{O}}$$
$$+ (x - 12.5)(\overline{h}_{1100} - \overline{h}_{298})_{\text{O}_2} + (3.76x)(\overline{h}_{1100} - \overline{h}_{298})_{\text{N}_2}$$

Consulting Tables A-6 and A-18M we obtain

$$H_P = (8)(-393\ 520 + 38\ 894) + (9)(-241\ 820 + 30\ 167)$$
$$+ (x - 12.5)(26\ 217) + (3.76x)(24\ 757)$$
$$= -5\ 069\ 598 + 119\ 303x \quad \text{kJ} \tag{d}$$

Equating (c) and (d) gives

$$-208\ 450 + 57\ 239x = -5\ 069\ 598 + 119\ 303x$$

and

$$x = 78.32$$
$$c = x - 12.5 = 65.82$$

For stoichiometric combustion $x = 12.5$, so the percent theoretical air is $78.32/12.5 = 626$ percent. The air/fuel ratio is thus

$$AF = \frac{(4.76x)(28.97)}{114} = 94.74 \text{ kg air/kg fuel}$$

The change in availability is

$$\Delta B = B_P - B_R = H_P - H_R - T_0(S_P - S_R) \tag{e}$$

But, $H_P = H_R$ so that

$$\Delta B = -T_0(S_P - S_R) \tag{e}$$

The reaction occurs at 4 atm, so we must change the tabular values of entropies accordingly. We shall assume ideal-gas behavior. For the reactants

C_8H_{18}: $\quad \bar{s} = \bar{s}_f^\circ - \mathscr{R} \ln\left(\dfrac{4.0}{1.0}\right)$

$\qquad\qquad = 466.73 - (8.314) \ln 4 = 455.20 \text{ kJ/kg·mol·K}$

Air: $\quad \bar{s} = (28.97)[\phi_{710} - (0.287) \ln (4) + 4.1869]$

$\qquad\qquad = (28.97)[3.4014 - (0.287) \ln 4 + 4.1869]$

$\qquad\qquad = 208.31 \text{ kJ/kg·mol·K}$

$S_R = (1)(455.20) + (4.76)(78.32)(208.31)$

$\qquad = 78\ 113 \text{ kJ/K} \tag{f}$

To determine the entropy for the products we must first determine the mole fractions and partial pressures of the constituents. The total number of moles of products is

$$n_p = a + b + c + 3.76x = 377.30 \tag{g}$$

The mole fractions are then

$x_{CO_2} = \dfrac{8}{377.30} = 0.0212 \qquad\qquad x_{O_2} = \dfrac{65.82}{377.30} = 0.174\ 45$

$x_{H_2O} = \dfrac{9}{377.30} = 0.023\ 85 \qquad\quad x_{N_2} = \dfrac{294.48}{377.30} = 0.780\ 49$

The partial pressures are then these values multiplied by the total pressure of 4 atm. In using Table A-18M we note that the reference pressure is 0.1 MPa and *not* 1 atm (0.101 32 MPa). The entropies of formation are already included in these tables, and we need only correct for the actual partial pressures. We therefore use the relation

$$\bar{s} = \bar{s} \text{ (table)} - \mathscr{R} \ln \dfrac{p}{p_{\text{ref}}}$$

So, for the products at 1100 K,

CO_2: $\quad \bar{s} = 274.55 - 8.314 \ln\left[\dfrac{(4)(0.101\ 32)(0.0212)}{0.1}\right]$

$\qquad\qquad = 294.96 \text{ kJ/kg·mol·K}$

H_2O: $\quad \bar{s} = 236.694 - 8.314 \ln\left[\dfrac{(4)(0.101\ 32)(0.023\ 85)}{0.1}\right]$

$\qquad\qquad = 256.12 \text{ kJ/kg·mol·K}$

O_2: $\quad \bar{s} = 246.922 - 8.314 \ln\left[\dfrac{(4)(0.101\ 32)(0.174\ 45)}{0.1}\right]$

$\qquad\qquad = 249.80 \text{ kJ/kg·mol·K}$

$$N_2: \quad \bar{s} = 231.302 - 8.314 \ln \left[\frac{(4)(0.101\ 32)(0.780\ 49)}{0.1} \right]$$
$$= 221.73 \text{ kJ/kg·mol·K}$$

The total entropy of these products is thus

$$S_P = S_{CO_2} + S_{H_2O} + S_{O_2} + S_{N_2}$$
$$= (8)(294.96) + (9)(256.12) + (65.82)(249.80) + (294.48)(221.73)$$
$$= 864\ 02 \text{ kJ/K}$$

Then from Eq. (e),

$$\Delta B = -(298)(86\ 402 - 78\ 113) = -2\ 470\ 122 \text{ kJ}$$

The maximum work *output* is the negative of this quantity, or

$$\dot{W}_{max} = +2\ 470\ 122 \text{ kJ}$$

The *actual* work output is zero, so the irreversibility is

$$\dot{I} = \dot{W}_{max} - \dot{W}_{act} = +2\ 470\ 122 \text{ kJ/kg mol fuel}$$

For 1 mol of fuel the mass of reactants is

$$m_R = (1)(114) + (4.76)(78.32)(28.97) = 10\ 914 \text{ kg reactants/kg mol fuel}$$

so, an alternate answer for the irreversibility would be

$$\dot{I} = \frac{2\ 470\ 122}{10\ 914} = 226.31 \text{ kJ/kg reactants} = 97.26 \text{ Btu/lbm reactants}$$

EXAMPLE 11-10   Irreversibility in combustion reaction
Gaseous methane at 77°F is burned with the theoretical air at 77°F and used to heat another fluid. The discharge temperature for the combustion products is 340°F. Calculate the irreversibility of this process. Assume the process occurs at 1 atm and the lowest available temperature is 77°F.

SOLUTION   The theoretical combustion equation for methane is

$$CH_4 + 2O_2 + (2)(3.76)N_2 \rightarrow CO_2 + 2H_2O + (2)(3.76)N_2$$

The enthalpy of the reactants for 1 mol of CH_4 is

$$H_R = \bar{h}^\circ_{CH_4} = -32\ 210 \text{ Btu} \qquad (-33\ 981 \text{ kJ}) \qquad (a)$$

For the products,

$$
\begin{aligned}
H_P &= H_{CO_2} + H_{H_2O} + H_{N_2} \\
&= (1)[-169\ 290 + 6552.9 - 4030.2]_{CO_2} \\
&\quad + (2)[-104\ 040 + 6396.9 - 4258.3]_{H_2O} \\
&\quad + (7.52)[0 + 5564.4 - 3729.5]_{N_2} \\
&= -166\ 767 - 203\ 803 + 13\ 798 \\
&= -356\ 771 \text{ Btu} \qquad (-376\ 390 \text{ kJ}) \qquad (b)
\end{aligned}
$$

The heat *released* is thus

$$
\begin{aligned}
Q = H_R - H_P &= 324\ 562 \text{ Btu/lbm mol fuel} \\
&= 20\ 285 \text{ Btu/lbm fuel} \qquad (c)
\end{aligned}
$$

To calculate the irreversibility we must evaluate the availability function for the inlet and exit streams; i.e., we must evaluate

$$b = h - T_0 s \qquad (d)$$

where the enthalpy includes the energy of formation. Entropies must be calculated from

$$\bar{s} = \bar{s}^\circ - \mathcal{R} \ln \left(\frac{p}{p_{ref}} \right) \qquad (e)$$

The reference pressure for tabular values of $\bar{s}^\circ$ is 1 atm. We assume that both the methane and air enter at 1 atm. The entropy of methane is thus

$$\bar{s}_{CH_4} = \bar{s}^\circ \text{ at } 77°F = 44.50 \text{ Btu/lbm·mol·°R}$$

We may compute the entropy for the air from the mole fractions of oxygen and nitrogen and their respective reference entropies. For 1 mol of air we have

$$x_{O_2} = \frac{1}{4.76} = 0.21 \qquad x_{N_2} = \frac{3.76}{4.76} = 0.79$$

and the entropy at 77°F is calculated using Eq. (e)

$$
\begin{aligned}
\bar{s} &= 0.21\bar{s}_{O_2} + 0.79\bar{s}_{N_2} \\
&= (0.21)[48.982 - 1.986 \ln (0.21)] \\
&\quad + (0.79)[45.743 - 1.986 \ln (0.79)] \\
&= 47.444 \text{ Btu/lbm·mol·°R}
\end{aligned}
$$

We assume that the *total* products are at 1 atm and that each constituent therein exists at its individual partial pressure. The partial pressure is calculated in terms of mole fractions. The total number of moles of products is

$$n_{\text{Tot}} = 1 + 2 + 7.52 = 10.52$$

so that the mole fractions are

$$x_{\text{CO}_2} = \frac{1}{10.52} = 0.0951$$

$$x_{\text{H}_2\text{O}} = \frac{2}{10.52} = 0.1901$$

$$x_{\text{N}_2} = \frac{7.52}{10.52} = 0.7148$$

The molal entropies of the product constituents at 340°F (800°R) are now calculated with the use of Table A-18 and Eq. (*e*)

CO$_2$: $\bar{s} = 54.839 - 1.986 \ln (0.0951) = 59.512$ Btu/lbm·mol·°R
H$_2$O: $\bar{s} = 48.316 - 1.986 \ln (0.1901) = 51.613$
N$_2$: $\bar{s} = 48.552 - 1.986 \ln (0.7148) = 49.219$

We may now evaluate the availability functions for the various streams in reactants and products.

Reactants

CH$_4$: $(1)(\bar{h} - T_0\bar{s})$ $= (1)[-32\ 210 - (537)(44.50)]$
                     $= -56\ 107$
Air: $(2)(4.76)(\bar{h} - T_0\bar{s}) = (9.52)[0 - (537)(47.444)]$
                     $= -242\ 545$
Total reactants $B_R = -298\ 652$ Btu/lbm mol fuel
                   $= -694\ 660$ kJ/kg mol fuel

Products

CO$_2$: $(1)(\bar{h} - T_0\bar{s}) = (1)[-166\ 767 - (537)(59.512)]$
                     $= -198\ 725$
H$_2$O: $(2)(\bar{h} - T_0\bar{s}) = (2)[-101\ 901 - (537)(51.613)]$
                     $= -259\ 234$
N$_2$: $(7.52)(\bar{h} - T_0\bar{s}) = (7.52)[1834.9 - (537)(49.219)]$
                     $= -184\ 960$
Total products $B_P = -642\ 919$ Btu/lbm mol fuel

The overall *decrease* in availability is, therefore,

$$B_R - B_P = 344\ 267 \text{ Btu/lbm mol fuel}$$
$$= 800\ 765 \text{ kJ/kg mol fuel}$$

Because there is no actual work in the process this decrease is equal to the irreversibility

$$I = +344\ 267 \text{ Btu/lbm mol fuel}$$
$$= 21\ 516 \text{ Btu/lbm fuel}$$
$$= 50\ 046 \text{ kJ/kg fuel}$$

11-9 EQUILIBRIUM

The second law of thermodynamics and the principle of increase of entropy furnish the basic notions for analysis of problems in equilibrium. In an isolated system the entropy always tends toward its maximum value; for this reason we shall say that in an isolated system the most stable state is one which maximizes the entropy. For processes which are nonadiabatic the considerations are not so simple. Let us first consider a simple compressible substance undergoing a general process in which both heat and work may be exchanged with the surroundings. In this process the total entropy change of the universe is expressed as a sum of the entropy change of the system and its surroundings. Thus, from the principle of increase of entropy [Eq. (6-22)],

$$dS_0 + dS_{\text{sys}} > 0 \qquad\qquad (11\text{-}18)$$

In this process we assume that the system is in heat communication with the surroundings at a constant temperature of T_0, as indicated in Fig. 11-4. The entropy change of the surroundings is

$$dS_0 = -\frac{d'Q}{T_0} \qquad\qquad (11\text{-}19)$$

where $d'Q$ is the heat transferred to the system from the surroundings. The

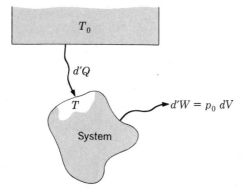

FIG. 11-4 *Equilibrium of simple compressible substance.*

work which the system does against the surroundings is $p_0\,dV$ where p_0 is the pressure of the surroundings and dV represents the change in volume of the system. The heat transfer to the system is then written as

$$d'Q = dU + p_0\,dV \tag{11-20}$$

Introducing Eqs. (11-19) and (11-20) into Eq. (11-18) gives for the inequality,

$$dU_{sys} + p_0\,dV_{sys} - T_0\,dS_{sys} < 0 \tag{11-21}$$

The inequality expressed by Eq. (11-21) presents a fairly general criterion for determining the direction that an irreversible process will take to seek the most probable equilibrium state. For an isolated system there is no work done ($dV_{sys} = 0$) and no internal energy change ($dU_{sys} = 0$), and the relation reduces to the simple increase of entropy principle.

Let us consider a somewhat simpler case in which the system remains in temperature and pressure equilibrium with its surroundings, i.e., $T = T_0$, $p = p_0$. Then we have

$$dU + p\,dV - T\,dS < 0 \tag{11-22}$$

where, now, all properties refer to the system. In effect, we have removed the irreversibilities resulting from interactions with the surroundings by writing the inequality in this manner. Equation (11-22) is just a refined way of restating the simple relation

$$dS \geqslant \frac{d'Q}{T}$$

for an irreversible process. Equation (11-22) may be more conveniently written in terms of the Helmholtz and Gibbs functions. Recalling that

$$A = U - TS \tag{11-23}$$
$$G = H - TS = U + pV - TS \tag{11-24}$$

and

$$dA = dU - T\,dS - S\,dT \tag{11-25}$$
$$dG = dU + p\,dV + V\,dp - T\,dS - S\,dT \tag{11-26}$$

then

$$\begin{aligned}
dU - T\,dS &= dA + S\,dT \\
&= dG - p\,dV - V\,dp + S\,dT
\end{aligned}$$

and Eq. (11-22) may be written as

$$dA + S\, dT + p\, dV < 0 \tag{11-27}$$

$$dG + S\, dT - V\, dp < 0 \tag{11-28}$$

For a process at constant temperature and constant volume it is clear that the Helmholtz function must approach its minimum value. Similarly, for a process at constant pressure and temperature, the Gibbs function must approach its minimum value. Thus our criterion for approach to equilibrium may be expressed as

$$(dA)_{T,V} < 0 \tag{11-29}$$

$$(dG)_{T,p} < 0 \tag{11-30}$$

At this point the reader is cautioned to remember that the only work modes we have considered are those involving $p\, dV$ type of work, i.e., work done by a simple compressible substance.

The equilibrium criterion expressed by Eq. (11-30) is an important one. It states, simply, that in a process at constant temperature and pressure the Gibbs function will decrease and seek its minimum value. Once this minimum is attained, the process will cease and equilibrium will be attained with no further change in the Gibbs function. Thus at equilibrium,

$$(dG)_{T,p} = 0 \tag{11-31}$$

The net result of Eq. (11-31) is that, if a system is to change from one equilibrium state to another equilibrium state *at the same pressure and temperature*, the Gibbs function must remain constant. A classic example of this principle is the equilibrium of phases of a pure substance. If the system is to change from a saturated liquid to a saturated vapor, for example, we would obtain

$$g_f = g_g \tag{11-32}$$

Expanding the Gibbs functions,

$$h_f - Ts_f = h_g - Ts_g$$

or

$$h_{fg} = Ts_{fg}$$

a relation we have utilized previously. We may also employ the equilibrium principle to rederive the Clausius-Clapeyron relation. As long as a system of vapor and liquid remains in equilibrium, we may write

$$dg_f = dg_g \tag{11-33}$$

when the system changes from one equilibrium saturation state to another. From Eq. (11-26),

$$dg = v \, dp - s \, dT$$

and

$$dg_f = v_f \, dp - s_f \, dT = dg_g = v_g \, dp - s_g \, dT \tag{11-34}$$

Collecting terms, we obtain

$$\frac{dp}{dT} = \frac{s_g - s_f}{v_g - v_f} = \frac{s_{fg}}{v_{fg}} = \frac{h_{fg}}{Tv_{fg}} \tag{11-35}$$

The application of Eq. (11-35) to single-component systems is fairly straightforward, as illustrated in previous chapters.

To obtain a criterion for equilibrium for a multicomponent mixture we simply rewrite Eq. (11-31) as

$$d\left(\sum_i n_i \bar{g}_i\right)_{T,p} = 0 \tag{11-36}$$

where, now, n_i is the number of moles of the ith constituent and $\bar{g}_i$ is the molal Gibbs function of that constituent. Equation (11-36) will serve as our basis for determining equilibrium concentrations in chemical reactions.

To calculate the Gibbs function of a substance we must choose a standard reference state for the entropy as well as the enthalpy. The usual practice is to take the same reference we used for enthalpy of formation, i.e., 1-atm pressure and 25°C (77°F). Then

$$\bar{g}_f^\circ = \bar{h}_f^\circ - T_0 \bar{s}_f^\circ \tag{11-37}$$

where T_0 is taken as 25°C. Values of both $\bar{g}_f^\circ$ and $\bar{s}_f^\circ$ are given in Table A-6 of the Appendix for several substances. One also can use the symbol $\bar{g}^\circ$ to designate the Gibbs function at some reference state to be defined.

11-10 CHEMICAL EQUILIBRIUM OF IDEAL GASES

Consider a simple reaction involving four ideal gases:

$$\nu_1 A_1 + \nu_2 A_2 \rightleftharpoons \nu_3 A_3 + \nu_4 A_4 \tag{11-38}$$

where A_1 and A_2 represent the reactants and A_3 and A_4 represent the products. The values of the ν's are to be determined in accordance with the requirements for equilibrium at the temperature and pressure for the reaction. To evaluate these molal quantities we shall employ the equilibrium condition given by Eq. (11-36). Let us first show how the Gibbs function for an ideal gas is calculated. We have

$$dg = v\, dp - s\, dT \tag{11-39}$$

If we assume that pressure is expressed in atmospheres and that values of the Gibbs function at the standard state $\bar{g}°$ are known, Eq. (11-38) may be integrated to give

$$\bar{g} - \bar{g}° = \int_{p=1}^{p} \bar{v}\, dp - \int_{T_{ref}}^{T} \bar{s}\, dT \tag{11-40}$$

The Gibbs function *at the reference temperature* is then written as

$$\bar{g} - \bar{g}° = \int_{p=1}^{p} \bar{v}\, dp = \int_{p=1}^{p} \mathscr{R}T\frac{dp}{p} = \mathscr{R}T \ln p \tag{11-41}$$

In both Eq. (11-40) and (11-41) we are choosing the reference state at 1-atm pressure.

Now let us consider the physical circumstance represented by Eq. (11-38). The reaction proceeds toward the right, and constituents 1 and 2 are steadily depleted while the products 3 and 4 build up. During this process the Gibbs function of the mixture is steadily decreasing. Eventually, the Gibbs function attains its minimum value and equilibrium is established. If the reaction proceeds at constant temperature and pressure, Eq. (11-36) describes the equilibrium condition, and this relation is equivalent to stating that, *at equilibrium,*

$$-dG_{\text{reactants}} = dG_{\text{products}}$$

or

$$dG_{\text{total}} = 0$$

When a reaction involving stoichiometric proportions goes to completion, only A_3 and A_4 will appear and A_1 and A_2 are completely consumed. If stoichiometric proportions are not present, some of the initial reactants may remain even though the reaction goes to completion. The term *degree of reaction* ϵ indicates how far the reaction has gone to completion, and is usually defined in terms of one of the initial constituents. Based on constituent 1,

$$\epsilon = \frac{n_{1,\text{max}} - n_1}{n_{1,\text{max}} - n_{1,\text{min}}} \tag{11-42}$$

where n_1 is the number of moles of constituent 1 present in the mixture, $n_{1,\text{max}}$ is usually the number of moles at the start of the reaction, and $n_{1,\text{min}}$ is the number at completion. If only one initial constituent is involved, ϵ is called the *degree of dissociation*. If a single initial constituent is ionized, ϵ is the *degree of ionization*.

Again, it is important to realize that at completion the reactants are not necessarily completely consumed. Consider the stoichiometric reaction for combustion of carbon:

$$C + O_2 \rightarrow CO_2 \qquad\qquad (11\text{-}43)$$

It is clear that all the O_2 will not be consumed if more than 1 mol is supplied for each mole of carbon. We could have, for example,

$$C + 2O_2 \rightarrow CO_2 + O_2 \qquad\qquad (11\text{-}44)$$

In this case the maximum and minimum number of moles of O_2 would be 2 and 1, respectively, and these values would be employed in conjunction with Eq. (11-42) to calculate the degree of reaction.

Returning now to the matter of equilibrium, we can envision the reaction proceeding, at constant temperature, as shown in Fig. 11-5, until

$$\left(\frac{\partial G_{\text{total}}}{\partial n_1}\right)_T = 0 \qquad\qquad (11\text{-}45)$$

If Eq. (11-38) is a stoichiometric mixture, then the ν's are so arranged that $n_{1,\text{max}}$ is the initial number of moles and $n_{1,\text{min}}$ is zero, since the original constituents are completely consumed. Some algebraic manipulation will show that the change in the number of moles of each constituent is a linear

FIG. 11-5   *Behavior of Gibbs function at equilibrium.*

function of the initial number of moles, the ν's, and the change in the degree of reaction. Thus

$$dn_1 = -\nu_1 \, d\epsilon$$
$$dn_2 = -\nu_2 \, d\epsilon$$
$$dn_3 = \nu_3 \, d\epsilon$$
$$dn_4 = \nu_4 \, d\epsilon$$

Accordingly, Eq. (11-36) may be written as

$$d \left(\sum_i n_i \bar{g}_i \right)_{T,p} = \left(\sum_i \bar{g}_i \, dn_i \right)_{T,p} = 0$$

or

$$(-\bar{g}_1 \nu_1 - \bar{g}_2 \nu_2 + \bar{g}_3 \nu_3 + \bar{g}_4 \nu_4) d\epsilon = 0 \qquad (11\text{-}46)$$

Making use of Eq. (11-41) for the Gibbs function at the reference temperature, this relation becomes

$$-\mathscr{R}T(\nu_3 \ln p_3 + \nu_4 \ln p_4 - \nu_1 \ln p_1 - \nu_2 \ln p_2)$$
$$= \nu_3 \bar{g}_3^\circ + \nu_4 \bar{g}_4^\circ - \nu_1 \bar{g}_1^\circ - \nu_2 \bar{g}_2^\circ \qquad (11\text{-}47)$$

The grouping of terms on the right side of this equation is clearly the difference in standard Gibbs functions of the products and reactants, and is sometimes called the *free-energy change* for the reaction designated by the symbol ΔG°. We should note once again that $\bar{g}^\circ$ is defined for some specified reference state which in this case we shall choose as 1-atm pressure and the temperature of the mixture. In reality then

$$\Delta G^\circ = f(T, p_0)$$

where p_0 is the reference pressure taken as 1 atm.

Equation (11-47) is now written more compactly as

$$-\mathscr{R}T \ln \frac{p_3^{\nu_3} p_4^{\nu_4}}{p_1^{\nu_1} p_2^{\nu_2}} = \Delta G^\circ \qquad (11\text{-}48)$$

The *equilibrium constant* K_p is defined as

$$K_p = \frac{p_3^{\nu_3} p_4^{\nu_4}}{p_1^{\nu_1} p_2^{\nu_2}} \qquad (11\text{-}49)$$

so that

TABLE 11-3

Logarithms to the base 10 of the equilibrium constant K_p according to Refs. [1, 2].

$K_p = p_3^{\nu_3} p_4^{\nu_4} / p_1^{\nu_1} p_2^{\nu_2}$ for reaction $\nu_1 A_1 + \nu_2 A_2 \leftrightarrows \nu_3 A_3 + \nu_4 A_4$

T, K	$H_2 \leftrightarrows 2H$	$O_2 \leftrightarrows 2O$	$N_2 \leftrightarrows 2N$	$H_2O(g) \leftrightarrows H_2 + \frac{1}{2}O_2$	$H_2O(g) \leftrightarrows OH + \frac{1}{2}H_2$	$CO_2 \leftrightarrows CO + \frac{1}{2}O_2$	$\frac{1}{2}O_2 + \frac{1}{2}N_2 \leftrightarrows NO$	$CO_2 + H_2 \leftrightarrows CO + H_2O(g)$
298	-71.228	-81.208	-159.600	-40.048	-46.054	-45.066	-15.171	-5.018
500	-40.318	-45.880	-92.672	-22.886	-26.130	-25.025	-8.783	-2.139
1000	-17.292	-19.614	-43.056	-10.062	-11.280	-10.221	-4.062	-0.159
1500	-9.514	-10.790	-26.434	-5.725	-6.284	-5.316	-2.487	+0.409
1800	-6.896	-7.836	-20.874	-4.270	-4.613	-3.693	-1.962	+0.577
2000	-5.582	-6.356	-18.092	-3.540	-3.776	-2.884	-1.699	+0.656
2200	-4.502	-5.142	-15.810	-2.942	-3.091	-2.226	-1.484	+0.716
2400	-3.600	-4.130	-13.908	-2.443	-2.520	-1.679	-1.305	+0.764
2500	-3.202	-3.684	-13.070	-2.224	-2.270	-1.440	-1.227	+0.784
2600	-2.836	-3.272	-12.298	-2.021	-2.038	-1.219	-1.154	+0.802
2800	-2.178	-2.536	-10.914	-1.658	-1.624	-0.825	-1.025	+0.833
3000	-1.606	-1.898	-9.716	-1.343	-1.265	-0.485	-0.913	+0.858
3200	-1.106	-1.340	-8.664	-1.067	-0.951	-0.189	-0.815	+0.878
3500	-0.462	-0.620	-7.312	-0.712	-0.547	+0.190	-0.690	+0.902
4000	+0.400	+0.340	-5.504	-0.238	-0.011	+0.692	-0.524	+0.930
4500	+1.074	+1.086	-4.094	+0.133	+0.408	+1.079	-0.397	+0.946
5000	+1.612	+1.686	-2.962	+0.430	+0.741	+1.386	-0.296	+0.956

$$\Delta G° = -\mathscr{R}T \ln K_p \qquad (11\text{-}50)$$

Equation (11-50) is called the *law of mass action*. Tables 11-3 and 11-4 list values of the equilibrium constant as a function of temperature for several simple reactions, and Fig. 11-6 gives a graphical presentation of the equilibrium constant for some of these reactions.

For a constant pressure circumstance $\Delta G°$ is a function of only the properties of the constituents at temperature T, so it follows that the equilibrium constant is also a function of temperature alone. We reemphasize that these equations are for perfect gases. In this development the reader should note that

1   Pressure is expressed in *atmospheres*.
2   The values of the v's employed for the calculation of the equilibrium constant are the stoichiometric values for the particular reaction.

At equilibrium the partial pressures of the constituents may be expressed in terms of the mole fractions of these constituents and the total pressure. Thus

TABLE 11-4
Logarithms to the base 10 of the equilibrium constants for ionization of Na and Cs, as calculated from Saha equation

T, K	$Na \rightleftharpoons Na^+ + e^-$	$Cs \rightleftharpoons Cs^+ + e^-$
298	−32.3	−25.1
400	−24.3	−17.5
600	−14.6	−10.0
800	−9.58	−6.15
1000	−6.54	−3.79
1200	−4.47	−2.18
1400	−2.97	−1.010
1600	−1.819	−0.108
1800	−0.913	+0.609
2000	−0.175	+1.194
2200	+0.438	+1.682
2400	+0.956	+2.098
2600	+1.404	+2.46
2800	+1.792	+2.77
3000	+2.13	+3.05
3200	+2.44	+3.29
3500	+2.84	+3.62
4000	+3.38	+4.07
4500	+3.82	+4.43
5000	+4.18	+4.73

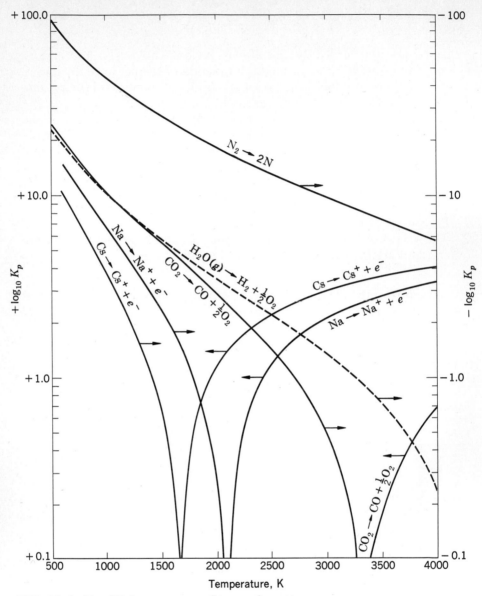

FIG. 11-6 *Equilibrium constants for several reactions.*

$$p_i = x_i p_{\text{total}}$$

and the relation for the equilibrium constant becomes

$$K_p = \frac{x_3^{\nu_3} x_4^{\nu_4}}{x_1^{\nu_1} x_2^{\nu_2}} \, p_{\text{total}}^{\nu_3 + \nu_4 - \nu_1 - \nu_2} \tag{11-57}$$

A quick inspection of Table 11-3 shows that for the simple reactions indicated, the equilibrium constant increases with temperature. In terms of Eq. (11-38) this means that these reactions proceed further toward completion as the temperature is elevated.

It is clear that the value of $\Delta G°$ depends on temperature and, therefore, the equilibrium constant pertains to equilibrium of the reaction for the temperature at which $\Delta G°$ is evaluated and on the pressure of the mixture.

EXAMPLE 11-11

Determine the equilibrium compositions for dissociation of CO_2 into CO and O_2 at 3200 K and 1 atm. Also determine the degree of reaction (or degree of dissociation).

SOLUTION

The chemical equation for dissociation is

$$CO_2 \rightleftharpoons CO + \tfrac{1}{2}O_2 \tag{a}$$

It is to be noted that this expression indicates the stoichiometric coefficients employed for calculation of the equilibrium constant. Thus, if we rewrite Eq. (a) as

$$\nu_1 CO_2 \rightleftharpoons \nu_3 CO + \nu_4 O_2 \tag{b}$$

we may express the equilibrium constant as

$$K_p = \frac{x_{CO}^{\nu_3} \times x_{O_2}^{\nu_4}}{x_{CO_2}^{\nu_1}} \, p_{total}^{\nu_3+\nu_4-\nu_1}$$

where $\nu_1 = 1$, $\nu_3 = 1$, $\nu_4 = \tfrac{1}{2}$, and $p_{total} = 1$ atm. Then

$$K_p = \frac{x_{CO} x_{O_2}^{1/2}}{x_{CO_2}} (1)^{1/2} \tag{c}$$

From Table 11-2 we have log $K_p = -0.189$ and $K_p = 0.647$. As the reaction proceeds we do *not* have stoichiometric proportions present in the mixture; i.e., $\nu_1 \neq 1$, $\nu_3 \neq 1$, etc. Instead, we have the CO_2 dissociated to some degree ϵ, so that

$$CO_2 \rightarrow (1 - \epsilon)CO_2 + \epsilon CO + \frac{\epsilon}{2} O_2$$

The total number of moles in the equilibrium mixture is therefore

$$n_{total} = (1 - \epsilon) + \epsilon + \frac{\epsilon}{2} = 1 + \frac{\epsilon}{2} \qquad (d)$$

so that the equilibrium mole fractions are

$$x_{CO_2} = \frac{1 - \epsilon}{1 + \epsilon/2} \cdot$$

$$x_{CO} = \frac{\epsilon}{1 + \epsilon/2}$$

$$x_{O_2} = \frac{\epsilon/2}{1 + \epsilon/2}$$

Using Eq. (c) we obtain

$$K_p = \frac{\dfrac{\epsilon}{1 + \epsilon/2} \left(\dfrac{\epsilon/2}{1 + \epsilon/2}\right)^{1/2}}{\dfrac{1 - \epsilon}{1 + \epsilon/2}}$$

or

$$K_p^2 = \frac{\epsilon^3}{(1 - \epsilon)^2(2 + \epsilon)} = (0.647)^2 \qquad (e)$$

Equation (e) may be solved by iterative methods to yield

$$\epsilon = 0.578$$

The equilibrium concentrations are:

For CO_2,

$$\frac{1 - \epsilon}{1 + \epsilon/2} = 0.327$$

For CO,

$$\frac{\epsilon}{1 + \epsilon/2} = 0.448$$

For O_2,

$$\frac{\epsilon/2}{1 + \epsilon/2} = 0.225$$

EXAMPLE 11-12

CO_2 is heated at constant pressure of 1 atm from 77°F to 3200 K. Calculate the heat required per mole and compare with the heating energy which would be required if there were no dissociation.

SOLUTION We have already determined the degree of dissociation for CO_2 at this temperature in Example 11-11. The basic energy balance is

$$H_R + Q = H_P \qquad (a)$$

where the reactant is 1 mol of CO_2 and the products are the concentrations of CO_2, CO, and O_2 which may be determined from Example 11-11. The total moles of products are $1 + \epsilon/2 = 1 + 0.578/2 = 1.289$ and the number of moles of each constituent of the products is

$$n_{CO_2} = 1 - \epsilon = 0.422$$
$$n_{CO} = \epsilon = 0.578$$

$$n_{O_2} = \frac{\epsilon}{2} = 0.289$$

Because the reactants are at 77°F

$$H_R = \bar{h}^\circ_{CO_2} = -169\ 290 \text{ Btu/lbm mol}$$

The temperature of the products is 3200 K = 5760°R, and the highest temperature entry in the gas tables A-18 is 5300°R; however, the specific heats may be assumed very nearly constant between these two temperatures so that the tables may be extrapolated to give at 5760°R

$$\bar{h}_{CO_2} = 75\ 357 \text{ Btu/lbm mol} \qquad (175\ 273 \text{ kJ/kg mol})$$
$$\bar{h}_{CO} = 47\ 147 \text{ Btu/lbm mol} \qquad (109\ 659 \text{ kJ/kg mol})$$
$$\bar{h}_{O_2} = 49\ 283 \text{ Btu/lbm mol} \qquad (114\ 626 \text{ kJ/kg mol})$$

The total enthalpy of the products is

$$
\begin{aligned}
H_p &= 0.422(\bar{h}^\circ + \bar{h}_T - \bar{h}_{537})_{CO_2} \\
&\quad + 0.578(\bar{h}^\circ + \bar{h}_T - \bar{h}_{537})_{CO} \\
&\quad + 0.289(\bar{h}_T - \bar{h}_{537})_{O_2} \\
&= (0.422)(-169\ 290 + 75\ 357 - 4030) \\
&\quad + (0.578)(-47\ 540 + 47\ 147 - 3730) \\
&\quad + (0.289)(49\ 283 - 3725) \\
&= -30\ 557 \text{ Btu/mol reactant} \qquad (-71\ 072 \text{ kJ/mol})
\end{aligned}
$$

From Eq. (*a*) the heat transfer is

$$Q = H_p - H_R = -30\ 557 - (-169\ 290) = 138\ 733 \text{ Btu/lbm mol}$$

If there were no dissociation, the heating of CO_2 could be calculated from the entries in the gas tables as

$$Q = \bar{h}_T - \bar{h}_{537} = \bar{h}_{5760} - \bar{h}_{537}$$
$$= 75\ 357 - 4030$$
$$= 71\ 327\ \text{Btu/lbm mol} \qquad (165\ 899\ \text{kJ/kg mol})$$

The extra energy of $138\ 733 - 71\ 327 = 67\ 406$ Btu/lbm mol is required to dissociate the CO_2.

From this example we can see that the effects of dissociation in a combustion process would be to produce a lower combustion temperature than the adiabatic flame temperature calculated with no dissociation. When dissociation is involved part of the heat of combustion is absorbed in the dissociation reaction(s), and thus less energy is available to heat the products to higher temperatures.

EXAMPLE 11-13

Determine the equilibrium compositions for ionization of cesium at 2000 K and 1-atm pressure.

SOLUTION The chemical equation for ionization is

$$Cs \rightleftharpoons Cs^+ + e^- \qquad (a)$$

In terms of stoichiometric coefficients,

$$\nu_1 Cs \rightleftharpoons \nu_3 Cs^+ + \nu_4 e^- \qquad (b)$$

and the equilibrium constant is expressed as

$$K_p = \frac{x_{Cs^+}^{\nu_3} x_{e^-}^{\nu_4}}{x_{Cs}^{\nu_1}} p_{\text{total}}^{\nu_3 + \nu_4 - \nu_1}$$

We have $\nu_1 = \nu_3 = \nu_4 = 1$, so that

$$K_p = \frac{x_{Cs} x_e}{x_{Cs}} \qquad (c)$$

From Table 11-3 we have $\log K_p = 1.194$ and $K_p = 15.63$. For the incomplete reaction we have

$$Cs \rightarrow (1 - \epsilon)Cs + \epsilon Cs^+ + \epsilon e^- \qquad (d)$$

where ϵ is the degree of reaction.
The total number of moles in the equilibrium mixture is thus

$$n_{\text{total}} = (1 - \epsilon) + \epsilon + \epsilon = 1 + \epsilon$$

and the equilibrium mole fractions are

$$x_{Cs} = \frac{1 - \epsilon}{1 + \epsilon}$$

$$x_{Cs^+} = \frac{\epsilon}{1 + \epsilon}$$

$$x_{e^-} = \frac{\epsilon}{1 + \epsilon}$$

Inserting these relations into Eq. (c) results in

$$K_p = \frac{\dfrac{\epsilon}{1 + \epsilon}\dfrac{\epsilon}{1 + \epsilon}}{(1 - \epsilon)/(1 + \epsilon)}$$

$$= \frac{\epsilon^2}{1 - \epsilon^2}$$

Inserting $K_p = 15.63$ results in

$$\epsilon = 0.97$$

The equilibrium concentrations are:

For Cs,

$$\frac{1 - \epsilon}{1 + \epsilon} = 0.0152$$

For Cs$^+$,

$$\frac{\epsilon}{1 + \epsilon} = 0.492$$

For e^-,

$$\frac{\epsilon}{1 + \epsilon} = 0.492$$

From these results it is evident that the ionization of cesium is almost complete at 2000 K.

11-11 EFFECTS OF NONREACTING GASES

In a mixture of gases it is quite common for only one constituent to be involved in a dissociation process because of the temperature level of the mixture. For such circumstances the other components of the mixture carry through the reaction unaltered; however, their presence affects the calculation of mole fractions and partial pressures for the reacting constituent and thus influences the final equilibrium concentrations. For example, one might have a mixture of 1 mol of cesium and 1 mol of nitrogen at 2000 K. Cesium is almost completely ionized at this temperature but there is hardly any dissociation of nitrogen ($\log K_p = -18.092$). We would therefore have

$$Cs + N_2 \rightleftharpoons Cs^+ + e^- + N_2$$

Using the degree of reaction ϵ based on consumption of Cs

$$Cs + N_2 \rightarrow (1 - \epsilon)Cs + \epsilon Cs^+ + \epsilon e^- + N_2$$

The total number of moles in the mixture is therefore

$$n_{total} = (1 - \epsilon) + \epsilon + \epsilon + 1 = 2 + \epsilon$$

For a mixture with 1 mol of Cs and 2 mol of N_2 the value of n_{total} would be $3 + \epsilon$, and so on. The mole fractions of the quantities used in the equilibrium constant are also changed so that

$$x_{Cs} = \frac{1 - \epsilon}{2 + \epsilon} \qquad x_{Cs^+} = \frac{\epsilon}{2 + \epsilon}$$

$$x_{e^-} = \frac{\epsilon}{2 + \epsilon}$$

We can insert these values in Eq. (c) of Example 11-13 to obtain

$$K_p = \frac{\dfrac{\epsilon}{1 + \epsilon}\dfrac{\epsilon}{2 + \epsilon}}{\dfrac{1 - \epsilon}{2 + \epsilon}} = \frac{\epsilon^2}{2 - \epsilon - \epsilon^2}$$

Carrying through with a numerical calculation at 2000 K we have $K_p = 15.63$ from Example 11-13, and so a solution for ϵ in the presence of an inert gas gives

$$\epsilon = 0.98$$

There is not much difference between this result and the value of $\epsilon = 0.97$ for pure ionization of cesium because the ionization is so nearly complete in both cases. It is interesting then to make the calculation at a lower temperature, say 1400 K, to make a further comparison.

At 1400 K, log K_p = -1.010 and K_p = 0.0977 for Cs. For pure ionization we have from Example 11-13

$$K_p = \frac{\epsilon^2}{1 - \epsilon^2}$$

and with the numerical value of K_p inserted

$$\epsilon = 0.298$$

With 1 mol of nitrogen in the mixture the value of ϵ would be ϵ = 0.378.
So, at this lower temperature there is a marked difference in the degree of
ionization of the cesium depending on whether or not a nonreacting gas is
present.

11-12 EQUILIBRIUM IN MULTIPLE REACTIONS

The previous section has shown how we may use equilibrium principles to
establish the degree of reaction of ideal gases and ionization processes. Now
let us complicate the situation by placing additional gases in the mixture so
that more than one reaction may take place. One example of such multiple
reactions would be the dissociation of H_2O by the two reactions

$$H_2O \leftrightharpoons H_2 + \tfrac{1}{2}O_2$$
$$H_2O \leftrightharpoons \tfrac{1}{2}H_2 + OH$$

Another example might be the combined dissociation of CO_2 and N_2

$$CO \leftrightharpoons CO + \tfrac{1}{2}O_2$$
$$\tfrac{1}{2}O_2 + \tfrac{1}{2}N_2 \leftrightharpoons NO$$

We may write for the two general reactions

$$\nu_1 A_1 + \nu_2 A_2 \leftrightharpoons \nu_3 A_3 + \nu_4 A_4 \tag{11-52}$$

$$\nu_5 A_1 + \nu_6 A_6 \leftrightharpoons \nu_7 A_7 + \nu_8 A_8 \tag{11-53}$$

where we are taking one reactant A_1 as common to both equations but with a
possibly different stoichiometric proportion, that is, ν_1 does not necessarily
equal ν_5. We shall designate Eq. (11-52) as reaction 1 and Eq. (11-53) as reaction 2 with their corresponding degrees of reaction ϵ_1 and ϵ_2. Both degrees of
reaction are based on consumption of reactant A_1 in each equation. For
reaction 1 we have

$$dn_1 = -\nu_1 \, d\epsilon_1$$
$$dn_2 = -\nu_2 \, d\epsilon_1$$
$$dn_3 = +\nu_3 \, d\epsilon_1$$
$$dn_4 = +\nu_4 \, d\epsilon_1$$

while for reaction 2

$$dn_1 = -\nu_5 \, d\epsilon_2$$
$$dn_6 = -\nu_6 \, d\epsilon_2$$
$$dn_7 = +\nu_7 \, d\epsilon_2$$
$$dn_8 = +\nu_8 \, d\epsilon_2$$

For the combined reactions the total consumption of reactant A_1 is

$$dn_1 = -\nu_1 \, d\epsilon_1 - \nu_5 \, d\epsilon_2$$

while the other constituents remain independent. We may now apply Eq. (11-36) as the condition for equilibrium at constant temperature and pressure and write

$$\left(\sum \bar{g}_i \, dn_i \right)_{T,p} = 0$$

or

$$(-\bar{g}_i \nu_1 - \bar{g}_2 \nu_2 + \bar{g}_3 \nu_3 + \bar{g}_4 \nu_4) \, d\epsilon_1$$
$$+ (-\bar{g}_1 \nu_5 - \bar{g}_6 \nu_6 + \bar{g}_7 \nu_7 + \bar{g}_8 \nu_8) \, d\epsilon_2 = 0 \qquad (11\text{-}54)$$

Because we are assuming an ideal-gas mixture we can use Eq. (11-41) to express the Gibbs function at the reference temperature so that we obtain two relations analogous to that for the single reaction:

$$-\mathscr{R}T \ln \left(\frac{p_3^{\nu_3} p_4^{\nu_4}}{p_1^{\nu_1} p_2^{\nu_2}} \right) = \Delta G_1^\circ \qquad (11\text{-}55)$$

$$-\mathscr{R}T \ln \left(\frac{p_7^{\nu_7} p_8^{\nu_8}}{p_1^{\nu_5} p_6^{\nu_6}} \right) = \Delta G_2^\circ \qquad (11\text{-}56)$$

where the free energies are

$$\Delta G_1^\circ = \nu_3 \bar{g}_3^\circ + \nu_4 \bar{g}_3^\circ - \nu_1 \bar{g}_1^\circ - \nu_2 \bar{g}_2^\circ \qquad (11\text{-}57)$$
$$\Delta G_2^\circ = \nu_7 \bar{g}_7^\circ + \nu_8 \bar{g}_8^\circ - \nu_5 \bar{g}_1^\circ - \nu_6 \bar{g}_6^\circ \qquad (11\text{-}58)$$

Equations (11-54) and (11-55) may be arranged in the equilibrium constant form as

$$\ln K_{p_1} = \frac{-\Delta G_1^\circ}{\mathscr{R}T} \tag{11-59}$$

$$\ln K_{p_2} = \frac{-\Delta G_2^\circ}{\mathscr{R}T} \tag{11-60}$$

where we recognize that both equilibrium constants may be expressed in terms of mole fractions of the constituents in accordance with Eq. (11-51). To obtain the equilibrium concentrations the two nonlinear equations must be solved simultaneously. For systems involving many simultaneous reactions this becomes a rather formidable task and is best accomplished with a computer.

EXAMPLE 11-14

Calculate the equilibrium concentrations for the following two reactions at 1 atm and 3000 K starting with 1 mol of CO_2 and $\frac{1}{2}$ mol each of O_2 and N_2:

$$CO_2 \rightleftharpoons CO + \tfrac{1}{2}O_2 \tag{1}$$

$$\tfrac{1}{2}O_2 + \tfrac{1}{2}N_2 \rightleftharpoons NO \tag{2}$$

SOLUTION   The two dissociation reactions are

$$CO_2 \rightarrow (1 - \epsilon_1)CO_2 + \epsilon_1 CO + \frac{\epsilon_1}{2}O_2$$

$$\tfrac{1}{2}O_2 + \tfrac{1}{2}N_2 \rightarrow (1 - \epsilon_2)\tfrac{1}{2}O_2 + (1 - \epsilon_2)\tfrac{1}{2}N_2 + \epsilon_2 NO$$

where ϵ_1 and ϵ_2 are the two degrees of reaction. The equilibrium mixture is the sum of the right sides of these equations or

$$(1 - \epsilon_1)CO_2 + \epsilon_1 CO + (\tfrac{1}{2} - \tfrac{1}{2}\epsilon_2 + \tfrac{1}{2}\epsilon_1)O_2 + (\tfrac{1}{2} - \tfrac{1}{2}\epsilon_2)N_2 + \epsilon_2 NO$$

The total number of moles in the equilibrium mixture is therefore

$$n_{total} = 1 - \epsilon_1 + \epsilon_1 + \tfrac{1}{2} - \tfrac{1}{2}\epsilon_2 + \tfrac{1}{2}\epsilon_1 + \tfrac{1}{2} - \tfrac{1}{2}\epsilon_2 + \epsilon_2$$
$$= 2 + \tfrac{1}{2}\epsilon_1$$

and the mole fractions at equilibrium are

$$x_{CO_2} = \frac{1 - \epsilon_1}{2 + \epsilon_1/2} \qquad x_{CO} = \frac{\epsilon_1}{2 + \epsilon_1/2}$$

$$x_{O_2} = \frac{1 - \epsilon_2 + \epsilon_1}{4 + \epsilon_1} \qquad x_{N_2} = \frac{1 - \epsilon_2}{4 + \epsilon_1}$$

$$x_{NO} = \frac{\epsilon_2}{2 + \epsilon_1/2}$$

At 3000 K we have

$$\begin{aligned}\log K_{p_1} &= -0.485 & K_{p_1} &= 0.3272 & (CO_2) \\ \log K_{p_2} &= -0.913 & K_{p_2} &= 0.1222 & (O_2, N_2)\end{aligned}$$

Because the reaction occurs at 1 atm we may apply Eq. (11-51) to obtain

$$K_{p_1} = 0.3273 = \frac{\left(\dfrac{\epsilon_1}{2 + \epsilon_1/2}\right)^1 \left(\dfrac{1 - \epsilon_2 + \epsilon_1}{4 + \epsilon_1}\right)^{1/2}}{\left(\dfrac{1 - \epsilon_1}{2 + \epsilon_1/2}\right)^1} \tag{a}$$

and

$$K_{p_2} = 0.1222 = \frac{\left(\dfrac{\epsilon_2}{2 + \epsilon_1/2}\right)^1}{\left(\dfrac{1 - \epsilon_2 + \epsilon_1}{4 + \epsilon_1}\right)^{1/2} \left(\dfrac{1 - \epsilon_2}{4 + \epsilon_1}\right)^{1/2}} \tag{b}$$

Equations (a) and (b) must now be solved simultaneously to obtain the values of ϵ_1 and ϵ_2. The results are

$$\epsilon_1 = 0.3736 \qquad \epsilon_2 = 0.05644$$

and the resulting mole fractions at equilibrium are obtained as

$$\begin{aligned}x_{CO_2} &= 0.2864 & x_{CO} &= 0.1708 \\ x_{O_2} &= 0.3012 & x_{N_2} &= 0.2157 \\ x_{NO} &= 0.0258\end{aligned}$$

The final molal reaction is

$$CO_2 + \tfrac{1}{2}O_2 + \tfrac{1}{2}N_2 \rightarrow$$
$$0.6264CO_2 + 0.3736CO + 0.6587O_2 + 0.4717N_2 + 0.0564NO$$

11-13   THE VAN'T HOFF EQUATION

We can extend our discussion of equilibrium constants to derive a very useful relation between the enthalpy of reaction and the equilibrium constant for a constant-pressure reaction. Let us now use the symbols K_p to designate the equilibrium constant at total pressure p and $\Delta G°$ (T, p_0) to designate the corresponding Gibbs free energy. Then we may write Eq. (11-50) as

$$\Delta G^\circ (T, p_0) = -\mathscr{R}T \ln K_p \tag{11-61}$$

Differentiating with respect to T, at constant p, gives

$$\frac{d(\Delta G^\circ)}{dT} = -\mathscr{R}T \frac{1}{K_p} \frac{dK_p}{dT} - \mathscr{R} \ln K_p$$

which may be rearranged and combined with Eq. (11-61) to give

$$\frac{dK_p}{dT} = \frac{K_p}{\mathscr{R}T^2} \left[\Delta G^\circ(T, p_0) - T \frac{d(\Delta G^\circ)}{dT} \right] \tag{11-62}$$

We can express the Gibbs free energy as

$$\Delta G^\circ = \Delta H^\circ - T \, \Delta S^\circ \tag{11-63}$$

From Eq. (7-15b) we know that

$$dg = -s \, dT + v \, dp$$

so for a reaction at constant pressure

$$\frac{dg}{dT} = -s$$

and

$$\frac{d(\Delta G^\circ)}{dT} = -\Delta S^\circ \tag{11-64}$$

Now, substituting Eqs. (11-63) and (11-64) into (11-62) gives

$$\frac{dK_p}{dT} = \frac{K_p}{\mathscr{R}T^2} \Delta H^\circ$$

or written in logarithmic form

$$\frac{d(\ln K_p)}{dT} = \frac{\Delta H^\circ(T, p_0)}{\mathscr{R}T^2} \tag{11-65}$$

where we have now inserted p_0 again to emphasize the reference pressure basis.

Equation (11-65) is called the van't Hoff isobar equation and is quite useful in that it enables us to calculate enthalpy change in a reaction from the equilibrium compositions as determined by K_p. In a reciprocal way, energy balances on reactions can yield important information on the variation of K_p with temperature. Let us rearrange Eq. (11-65) as a function of $1/T$. Then

$$\frac{d(\ln K_p)}{d(1/T)} = \frac{-\Delta H^\circ(T, p_0)}{\mathcal{R}} \tag{11-66}$$

Equation (11-66) indicates to us that the slope of the curve in a plot of $\ln K_p$ versus $1/T$ will yield the value of $\Delta H^\circ/\mathcal{R}$.

For many reactions the value of ΔH° is very nearly constant over a rather wide range of temperatures. In such cases Eq. (11-65) can be integrated between two temperatures T_1 and T_2 to yield

$$\ln \left(\frac{K_{p_2}}{K_{p_1}} \right) = \frac{\Delta H^\circ}{\mathcal{R}} \left(\frac{1}{T_1} - \frac{1}{T_2} \right) \tag{11-67}$$

and if one knows the value of K_{p_1} at T_1 it is an easy task to calculate the value K_{p_2} at temperature T_2.

The van't Hoff equation can also be helpful in evaluating the endothermic or exothermic nature of a reaction. If the equilibrium constant increases with temperature, that is, $K_{p_2} > K_{p_1}$ for $T_2 > T_1$, Eq. (11-65) indicates that ΔH° must be positive and the reaction is endothermic. If K_p decreases with an increase in temperature, ΔH° will be negative and the reaction will be exothermic.

EXAMPLE 11-15

Estimate the enthalpy of reaction of CO_2 to form CO and O_2 at 2000 K.

SOLUTION We shall use the van't Hoff equation to make this estimate and assume that the enthalpy of reaction is constant over the range 1800 to 2200 K. From Table 11-3

$$T_1 = 1800 \text{ K} = 3240°\text{R} \qquad \log K_{p_1} = -3.693 \qquad K_{p_1} = 2.028 \times 10^{-4}$$
$$T_2 = 2200 \text{ K} = 3960°\text{R} \qquad \log K_{p_2} = -2.226 \qquad K_{p_2} = 1.306 \times 10^{-3}$$

Therefore

$$\log \left(\frac{K_{p_2}}{K_{p_1}} \right) = -2.226 - (-3.693) = 1.467$$

and

$$\frac{K_{p_2}}{K_{p_1}} = 29.309$$

We evaluate ΔH° at 2000 K, the mean between 1800 and 2200 K, and rearrange Eq. (11-67) to give

$$\Delta H^\circ = \frac{\mathcal{R} \ln (K_{p2}/K_{p1})}{1/T_1 - 1/T_2}$$

$$= \frac{(1545) \ln (29.309)}{(\frac{1}{3240} - \frac{1}{3960})(778)}$$

$$= 119\ 500\ \text{Btu/lbm mol} \qquad \cdot \quad (277\ 943\ \text{kJ/kg mol})$$

11-14 THE CHEMICAL POTENTIAL AND PHASE EQUILIBRIUM

Consider a homogeneous mixture of different constituents the composition of which may change as a result of a change in temperature, pressure, or other intensive state variables. If there are i constituents, we may express the internal energy in functional form as

$$U = U(S, V, n_1, n_2, \ldots, n_i) \tag{11-68}$$

and a differential change in the internal energy is given by

$$dU = \left(\frac{\partial U}{\partial S}\right)_{V, n_i} dS + \left(\frac{\partial U}{\partial V}\right)_{S, n_i} dV + \sum_i \left(\frac{\partial U}{\partial n_i}\right)_{V, S, n_j} dn_i \tag{11-69}$$

For a single-component mixture or one for which the composition does not change it is clear that $dn_i = 0$ and Eq. (11-69) reduces to the familiar relation

$$dU = T\ dS - p\ dV$$

whereupon we have

$$T = \left(\frac{\partial U}{\partial S}\right)_{V, n_i} \qquad -p = \left(\frac{\partial U}{\partial V}\right)_{S, n_i} \tag{11-70}$$

The partial derivatives in the summation of Eq. (11-69) are defined as the *chemical potential* μ_i:

$$\mu_i = \left(\frac{\partial U}{\partial n_i}\right)_{V, S, n_j} \qquad n_j \neq n_i \tag{11-71}$$

Now we may rewrite Eq. (11-69) as

$$dU = T\ dS - p\ dV + \sum_i \mu_i\ dn_i \tag{11-72}$$

Relations similar to Eq. (11-72) can be derived for other thermodynamic functions. Consider the Gibbs function for the mixture which we assume to take the functional form

$$G = G(p, T, n_1, n_2, \ldots, n_i)$$

Then

$$dG = \left(\frac{\partial G}{\partial p}\right)_{T,n_i} dp + \left(\frac{\partial G}{\partial T}\right)_{p,n_i} dT + \sum_i \left(\frac{\partial G}{\partial n_i}\right)_{p,T,n_j} dn_i \qquad (11\text{-}73)$$

For a mixture of fixed composition we have, from Chap. 7 [Eq. (7-16b)],

$$V = \left(\frac{\partial G}{\partial p}\right)_{T,n_i} \qquad -S = \left(\frac{\partial G}{\partial T}\right)_{p,n_i} \qquad (11\text{-}74)$$

so that Eq. (11-73) becomes

$$dG = V\,dp - S\,dT + \sum_i \left(\frac{\partial G}{\partial n_i}\right)_{T,p,n_j} dn_i \qquad (11\text{-}75)$$

Now, for a single component,

$$\begin{aligned} dG &= dH - T\,dS - S\,dT \\ &= dU + p\,dV + V\,dp - T\,dS - S\,dT \end{aligned} \qquad (11\text{-}76)$$

Equating dG as obtained from Eqs. (11-75) and (11-76) and comparing with Eq. (11-72) indicates that

$$\mu_i = \left(\frac{\partial G}{\partial n_i}\right)_{T,p,n_j} \qquad n_j \neq n_i \qquad (11\text{-}77)$$

whereupon we have

$$dG = V\,dp - S\,dT + \sum_i \mu_i\,dn_i \qquad (11\text{-}78)$$

In a similar manner the enthalpy and Helmholtz functions may be expressed by

$$dH = T\,dS + V\,dp + \sum_i \mu_i\,dn_i \qquad (11\text{-}79)$$

$$dA = -S\,dT - p\,dV + \sum_i \mu_i\,dn_i \qquad (11\text{-}80)$$

We therefore have the following equivalent relations for the chemical potential:

$$\mu_i = \left(\frac{\partial U}{\partial n_i}\right)_{V,S,n_j} = \left(\frac{\partial G}{\partial n_i}\right)_{T,p,n_j} = \left(\frac{\partial A}{\partial n_i}\right)_{T,V,n_j} \tag{11-81}$$

Now let us examine the relationships between the chemical potentials when we have an equilibrium mixture of several components which may exist in more than one phase. In general, a mixture could have many phases but in most practical circumstances only two or three might be present. For equilibrium at constant temperature and pressure we have

$$dG_{T,p} = 0 \tag{11-82}$$

The total Gibbs function is formed as the sum of the Gibbs functions for each phase which we assume are homogeneous in order to apply our reasoning used to derive the chemical potential. For simplicity, consider a mixture of two constituents a and b in two phases 1 and 2. For a closed *total* system but one in which mass may be transferred between phases we would have

$$n_{a_1} + n_{a_2} = n_{a,\text{total}} = \text{const.}$$
$$n_{b_1} + n_{b_2} = n_{b,\text{total}} = \text{const.}$$

or, in differential form,

$$dn_{a_1} + dn_{a_2} = 0 \tag{11-83}$$
$$dn_{b_1} + dn_{b_2} = 0 \tag{11-84}$$

If both phases are homogeneous, we could write the Gibbs function for each in accordance with Eq. (11-78):

$$dG_1 = V\,dp - S\,dT + \mu_{a_1}\,dn_{a_1} + \mu_{b_1}\,dn_{b_1} \tag{11-85}$$
$$dG_2 = V\,dp - S\,dT + \mu_{a_2}\,dn_{a_2} + \mu_{b_2}\,dn_{b_2} \tag{11-86}$$

Now, for equilibrium at constant T and p

$$dG|_{T,p} = dG_1|_{T,p} + dG_2|_{T,p} = 0 \tag{11-87}$$

Combining Eqs. (11-83) and (11-87) gives

$$(\mu_{a_1} - \mu_{a_2})\,dn_{a_1} + (\mu_{b_1} - \mu_{b_2})\,dn_{b_1} = 0$$

which indicates that, for equilibrium at constant T and p

$$\mu_{a_1} = \mu_{a_2} \qquad \mu_{b_1} = \mu_{b_2} \tag{11-88}$$

We could extend the analysis to the general case of i constituents and k phases and would find that

$$\mu_{i_1} = \mu_{i_2} = \cdots = \mu_{i_k} \tag{11-89}$$

for equilibrium with all phases at the same pressure and temperature. *Simply stated, the chemical potential must have the same value in all phases for equilibrium at constant temperature and pressure.*

Now let us consider a *pure phase,* i.e., one which consists of only one constituent. At constant T and p the change in the Gibbs function from Eq. (11-78) reduces to

$$dG_{T,p} = \mu \, dn \tag{11-90}$$

which indicates that the change in the Gibbs function must come from an increase in mass of the phase (increase in dn). Because the total Gibbs function G is directly proportional to the total number of moles in the phase we have

$$\mu = \frac{G}{n} = \bar{g} \tag{11-91}$$

or very simply, the chemical potential for a pure phase is equal to the molal Gibbs function for the phase. As a simple example of this point we can see that for equilibrium of a pure substance (like water) at constant T and p the Gibbs functions must be equal in each phase, as we have already noted in Eq. (11-32).

11-15 THE GIBBS PHASE RULE

From the foregoing discussion it should be obvious that the determination of properties of mixtures which may exist in multiple phases is a complicated affair, and even the number of properties necessary to fix the state is not easily apparent. One of the most famous relations of thermodynamics is the phase rule developed by Josiah Willard Gibbs in 1875, which we shall now discuss. We shall first consider an equilibrium mixture without the possibility of chemical reaction. Let the mixture consist of C constituents and P phases. For equilibrium at constant temperature and pressure we would write for each phase, corresponding to Eq. (11-78),

$$dG \mid_{T,p} = \sum_i \mu_i \, dn_i = 0 \tag{11-92}$$

We would thus have C values of μ_i but only $C - 1$ of these are independent because the equation is subject to the restriction that the total number of moles is fixed for a closed system. Additionally, we have a choice of T and p at our disposal to fix the state of the system for one phase or

$$(C - 1) + 2 = C + 1$$

independent properties. Now, for an equation written for each phase the pressure and temperature are the same so they do not add new independent variables, and there are additional restrictions added because of the conditions expressed in Eq. (11-89). Thus, for two phases we add the restriction that

$$\mu_{i_1} = \mu_{i_2}$$

For three phases we have the *two* restrictions that

$$\mu_{i_1} = \mu_{i_2} = \mu_{i_3}$$

and the general notion is that there are $P - 1$ such restrictions for the P phases. Taking all the factors into account the total number of *degrees of freedom* F or *number of variants* of the system is

$$F = C + 1 - (P - 1) = C - P + 2 \qquad (11\text{-}93)$$

This is the *phase rule of Gibbs*, and a quick application to the simple water system can illustrate its validity. In the vapor-liquid saturation region there are two phases $(P = 2)$, one component $(C = 1)$ so

$$F = 1 - 2 + 2 = 1$$

This means that only *one* intensive property can be independently varied to maintain the equilibrium region; saturation temperature or pressure is normally selected. For the case where there are three phases in equilibrium

$$F = 1 - 3 + 2 = 0$$

which indicates to us that there are no choices open to vary this equilibrium state. Of course, that is exactly the case at the triple point. In a single-phase region, as with a gas or vapor, a single-component substance has

$$F = 1 - 1 + 2 = 2$$

or, there are two properties which can be varied independently. This, of course, is what we know to be the case for an ideal gas or superheated vapor, and temperature and pressure are usually selected as the independent variables for defining the state of the system.

PHASE RULE FOR CHEMICAL REACTIONS

If it is possible for the different constituents to enter into chemical reactions with one another the matter of phase equilibrium becomes even more complicated because the various constituents may be consumed or generated depending on the extent of reaction. Let N_c be the number of chemical substances which are present in the overall system. For each reaction equilibrium equation the number of independent intensive properties will be reduced by one, and if there are r reaction equations the number of components C to be used in the phase rule becomes

$$C = N_c - r \qquad (11\text{-}94)$$

So, in the general context, C is *the minimum number of chemical substances which can produce the overall system*. The phase rule then becomes

$$F = (N_c - r) - P + 2 \qquad (11\text{-}95)$$

EXAMPLE 11-16

As an example consider an equilibrium mixture of solid carbon $C(s)$, O_2, CO, and CO_2 at elevated temperatures where chemical reactions may occur. There are two phases, gas and solid, so $P = 2$. The reaction equations for formation of the chemical substances are

$$C(s) + O_2(g) \rightarrow CO_2(g) \qquad (a)$$
$$C(s) + \tfrac{1}{2}O_2(g) \rightarrow CO(g) \qquad (b)$$

Because we have stipulated that both carbon monoxide and oxygen are present in the system neither can be eliminated from the reaction equations as, for example, in writing

$$CO + \tfrac{1}{2}O_2 \rightarrow CO_2 \qquad (c)$$

Therefore, we must have at least two of these reaction equations, so $r = 2$. The value of N_c is 4, so

$$C = N_c - r = 4 - 2 = 2$$

The phase rule then yields

$$F = C - P + 2 = 2 - 2 + 2 = 2 \qquad (d)$$

11-16 THE THIRD LAW OF THERMODYNAMICS

In the analysis of many chemical reactions it is necessary to fix a reference state for the entropy. We may always choose some arbitrary reference level when only one component is involved; 32°F is chosen for the conventional steam tables. On the basis of observations by Nernst, and others, Planck stated the third law of thermodynamics in 1912 as:

> *The entropy of all perfect crystalline solids is zero at absolute zero of temperature.*

A "perfect" crystal is one which is in thermodynamic equilibrium. Consequently, the third law is usually stated in a more general form as:

> *The entropy of any pure substance in thermodynamic equilibrium approaches zero as the absolute temperature approaches zero.*

The importance of the third law is clear. It furnishes a basis for calculation of the absolute entropies of substances which may then be utilized in the appropriate equations to determine the direction of chemical reactions.

A statistical thermodynamic interpretation of the third law is rather simple. The entropy was defined in Chap. 5 as

$$S = k \ln \Omega$$

where k is Boltzmann's constant and Ω is the thermodynamic probability. In view of the foregoing discussion, the third law is equivalent to stating that

$$\Omega \rightarrow 1 \qquad \text{as } T \rightarrow 0$$

This means that there is only *one* way the lowest energy state can occur for a substance obeying the third law.

There are several instances reported in the literature where calculations based on the third law are in disagreement with experiment. However, in all cases it is possible to explain the disagreement on the basis that the substance is not "pure," i.e., there may be two or more isotopes, or the presence of different molecules, or else a nonequilibrium distribution of molecules. In these cases there is more than one quantum state at absolute zero, and the entropy does not approach zero.

11-17 THE CONCEPT OF MAXIMUM WORK

As an additional topic for this chapter we consider a calculation of the maximum work which a system can perform as it changes from one equilibrium state to another. This concept is related to the notions of available energy and availability discussed in Chap. 6 but includes work interactions the system may have with its surroundings. Because we shall be interested in the maximum work *output* of a process we shall take the work as positive when it is done *by* the system. Consider a closed system which receives heat from its surroundings at the constant temperature T_0. The system does work in the amount $d'W$, part of which may be $p\,dV$ work. From the first law

$$d'W = d'Q - dU \tag{11-96}$$

From the principle of increase of entropy we have

$$dS_0 + dS_{\text{sys}} \geq 0 \tag{11-97}$$

and the change in entropy for the surroundings is given by

$$dS_0 = -\frac{d'Q}{T_0} \tag{11-98}$$

Inserting Eq. (11-98) in (11-97) and thence in (11-96) gives

$$d'W = d'Q - dU \leq T_0\,dS - dU \tag{11-99}$$

where dS pertains to the system. The inequality holds for irreversible processes and the equality for reversible processes. If the overall process involving the system and its surroundings is to be reversible there must be no heat transfer across a finite temperature difference; i.e., the temperature of the system must equal the temperature of the surroundings. Thus, Eq. (11-98) becomes, for $T = T_0$,

$$d'W \leq T\,dS - dU \tag{11-100}$$

where the inequality is retained to account for possible internal irreversibilities in the system. Clearly $d'W$ is maximized when the equality sign holds, so it is the reversible process which produces the maximum work. Then

$$d'W_{\text{max}} = T\,dS - dU \tag{11-101}$$

In terms of the Helmholtz and Gibbs functions

$$d'W_{\text{max}} = -dA - S\,dT \tag{11-102}$$

$$d'W_{\text{max}} = -dG + V\,dp + p\,dV - S\,dT \tag{11-103}$$

For a process at constant temperature

$$(d'W_{max})_T = (-dA)_T \tag{11-104}$$

Eq. (11-103) may be rewritten

$$d'W_{max} - p\,dV = -dG + V\,dp - S\,dT$$

Then, for a process at constant pressure and temperature

$$(d'W_{max} - p\,dV)_{T,p} = (-dG)_{T,p} \tag{11-105}$$

Equations (11-104) and (11-105) now furnish us with the results:

1 The maximum work *output* in a constant-temperature process is equal to the decrease in the Helmholtz function for the system.
2 The maximum work *output, exclusive of p dV work*, in a process at constant temperature and pressure is equal to the decrease in the Gibbs function for the system.

Equation (11-105) will be useful in analyzing the performance of fuel cells in Chap. 15 since it may be used to set an upper limit on the electrical work output of the device.

REVIEW QUESTIONS

1 What is meant by the term *stoichiometric proportions?*

2 Define the meaning of excess air.

3 For what purpose is an Orsat apparatus used?

4 What is the base, or zero level, chosen for measuring the energy level of chemical substances?

5 Define enthalpy of formation.

6 Describe endothermic and exothermic reactions.

7 How does heat of reaction differ from enthalpy of formation?

8 What is the distinguishing difference between higher heating value and lower heating value?

9 Define adiabatic flame temperature. What influence does excess air have on its value? What effect does dissociation have on its value?

10 What reference state is used to determine the standard Gibbs functions?

11 What is the criterion for equilibrium for a reaction occurring at constant temperature and pressure? At constant temperature and volume?

12 Define degree of reaction.

13 What is meant by the term *free-energy change?*

14 What is the law of mass action?

15 How is the equilibrium constant defined?

16 How is it possible for the presence of nonreacting gases to affect the equilibrium concentrations of a reacting component in a mixture?

17 How is the van't Hoff isobar used to advantage?

18 State the third law of thermodynamics. How is it interpreted in a microscopic thermodynamic sense?

19 Discuss the concept of reference levels for thermodynamic properties.

20 What is the phase rule of Gibbs?

21 Discuss the factors which influence the generation of irreversibility in processes involving chemical reactions.

PROBLEMS (English units)

11-1 Determine the dew-point temperature for the water vapor in the products of combustion for the following reactions at atmospheric pressure:
(a) Benzene with 50 percent excess air
(b) Pentane with stoichiometric proportions
(c) A fuel with a volumetric analysis of 70 percent CH_4 and 30 percent C_4H_{10} along with 100 percent excess air
(d) Ethyl alcohol with 20 percent excess air

11-2 A certain hydrocarbon fuel is burned with excess air, and the analysis of the dry products of combustion yields the following volumetric proportions:

$$CO_2 = 10.4 \text{ percent}$$
$$CO = 1 \text{ percent}$$
$$O_2 = 5 \text{ percent}$$
$$N_2 = 83.6 \text{ percent}$$

Estimate the approximate composition of the fuel and the percent of theoretical air.

11-3 An analysis of the dry products of combustion for CH_4 yields:

$$CO_2 = 10 \text{ percent}$$
$$CO = 0.7 \text{ percent}$$
$$O_2 = 2.0 \text{ percent}$$
$$N_2 = 87.3 \text{ percent}$$

Calculate the air/fuel ratio.

11-4 A mixture of gaseous fuels is used in a certain combustion process. The volumetric analysis of the fuel is:

$$CH_4 = 35 \text{ percent}$$
$$C_3H_8 = 25 \text{ percent}$$
$$C_5H_{12} = 40 \text{ percent}$$

A volumetric analysis of the dry products of combustion yields:

CO$_2$ = 12 percent
CO = 1 percent
O$_2$ = 4 percent
N$_2$ = 83 percent

Calculate the percent theoretical air and the dew point of the products assuming combustion at atmospheric pressure.

11-5. Calculate the heat liberated per mole of fuel when gaseous propane is burned with the stoichiometric proportions of air at 77°F and 1 atm.

11-6 Using the data of Table 11-1 calculate the lower heating value for *liquid* propane at 77°F.

11-7 Calculate the enthalpy of combustion for gaseous octane at 540°F.

11-8 Calculate the energy liberated when C$_{12}$H$_{26}$ is burned at 77°F with theoretical oxygen and the water in the products is in the liquid state.

11-9 Propane gas is burned with 50 percent excess air. The reactants enter the combustion chamber at 100°F and leave at 1200°F. Calculate the heat liberated per mole of fuel.

11-10 Liquid octane is burned with a certain quantity of excess air such that the adiabatic flame temperature is 1500°F. Calculate the percent excess air if the reactants enter the combustion chamber at 77°F.

11-11 Calculate the adiabatic flame temperature for gaseous methane burning with the theoretical air. Assume complete combustion.

11-12 Liquid octane fuel at 77°F is mixed in a combustion chamber with 400 percent theoretical air at 300°F. The products of combustion leave the chamber at 1200°F. Calculate the heat loss per mole of fuel.

11-13 Determine the adiabatic flame temperature for solid carbon burning with 200 percent theoretical air. What percent excess air would be required to produce an adiabatic flame temperature of 2000°R? Assume complete combustion.

11-14 Determine the enthalpy of combustion of gaseous methane and liquid octane at 77°F when liquid hydrogen peroxide is used as an oxidizer for the combustion reaction.

11-15 Determine the adiabatic flame temperature for hydrogen burning with 300 percent theoretical air. Assume complete combustion.

11-16 1 lbm mol/min of butane gas is burned with 100 percent excess air in a steady-flow process at 1-atm pressure. The reactants enter the combustion chamber at 77°F. Calculate the maximum temperature for which the ducts carrying the products of combustion should be designed if the heat loss in the combustion is 30 000 Btu/min.

11-17 Consider the combustion of liquid fuels—octane, methyl alcohol, propane, and ethyl alcohol—with the theoretical quantities of air at 77°F. Compare the reactions on the basis of:

(a) Volume of products per unit energy liberated

      (b) Volume of air per unit energy liberated
      (c) Energy liberated per unit mass of fuel
Discuss these comparisons.

11-18   Consider the dissociation of diatomic oxygen. Calculate the equilibrium concentrations of O_2 and O at 6000°R for system pressures of 1, 5, and 10 atm.

11-19 CO_2 is heated at a constant pressure of 5 atm from 77°F to 2800 K. Calculate the heat transfer for the process and determine the final concentrations at the end of heating.

11-20   Assume that the human metabolic system can be approximated by a simple combustion reaction occurring at 100°F. For a total energy consumption of 3000 kcal/day what quantity of methane and air at 20°C would be required to carry on the combustion process? How much carbon dioxide would be produced?

11-21   Some authorities claim that hydrogen produced by electrolysis of water will become a widely used fuel as more and more of our hydrocarbon fuel reserves are depleted. Calculate the higher and lower heating value for gaseous hydrogen at 77°F and 1 atm. What size tank for storage at 2000 psia and 70°F would be required to provide the same energy as a 25-gal tank of gasoline (octane)?

11-22   What adiabatic flame temperature would be obtained by burning hydrogen with 400 percent excess air?

11-23   Nitrogen oxides (that is, NO, etc.) are present in the stack discharge gas from power plants because they are "frozen" in the flow when the hot combustion gases come in rapid contact with the relatively cool surface of the boiler tubes and sufficient time is not available to perform the reverse reaction back to N_2 and O_2. Suppose the combustion gases are at 3140°F and the boiler tubes are at 1340°F. Estimate the percentage of NO present at the higher temperature that will be retained after sudden cooling to the lower temperature.

11-24   In a home heating furnace methane is burned with 25 percent excess air, and the manufacturer claims that 70 percent of the higher heating value of the fuel is recovered as useful heat. What is the outlet temperature of the products of combustion in this circumstance if the air and fuel enter at 77°F?

11-25   In a certain power-plant combustion, air enters the furnace at 250°F, and methane fuel enters at 77°F. Twenty percent excess air is used, and the products of combustion leave the furnace at 2140°F. How much energy is obtained per pound mass of fuel?

11-26   The burner for a gas-turbine plant is designed to burn liquid propane at 77°F with air at 350°F. The outlet temperature of the products of combustion is to be 2040°F. Calculate the air/fuel ratio required and percent excess air for steady flow and no heat loss from the burner.

11-27   A radiant heater for use in large warehouse areas burns gaseous butane with 25 percent excess air. Both fuel and air enter the heater at 77°F. The products of combustion leave the device at 1240°F. How much heat is lost by the heater per pound mass of fuel? What percent of the higher heating value does this represent?

11-28 Calculate the maximum work output as saturated liquid water changes to saturated vapor at 100 psia in (a) a process at constant temperature and (b) exclusive of $p\,dV$ work in a process at constant temperature and pressure. How do these values compare with the $p\,dV$ work done in the process?

11-29 What is the maximum work output exclusive of $p\,dV$ work when CO is burned with O_2 to form CO_2 at constant pressure of 1 atm and constant temperature of 77°F?

11-30 Methane (CH_4) is burned with the stoichiometric proportion of oxygen at 1 atm and 77°F. What is the maximum work output in this process? What would be the actual work output if the process occurs under steady-flow conditions?

11-31 A mixture of air and water vapor can be cooled so that liquid is present along with the vapor, or at low enough temperatures a solid phase (ice) will be present along with the vapor. Apply the Gibbs phase rule to this system to determine the number of degrees of freedom for the three regimes: (a) air and vapor, (b) air, vapor, and liquid, and (c) air, vapor, and ice.

11-32 How many degrees of freedom are there for a mixture of N_2 and O_2 at room temperature?

11-33 How many degrees of freedom are there for an equilibrium mixture of N_2, O_2, and NO at elevated temperatures?

11-34 What is the number of degrees of freedom for an equilibrium mixture of CO, CO_2, H_2, and H_2O at elevated temperatures?

11-35 What is the maximum number of homogeneous phases which can exist for (a) a one-component, (b) a two-component, and (c) a three-component system?

PROBLEMS (Metric units)

11-1M Calculate the heat liberated per mole of fuel when gaseous propane is burned with the stoichiometric proportions of air at 25°C and 1 atm.

11-2M Using the data of Table 11-1, calculate the lower heating value for *liquid* propane at 25°C.

11-3M Calculate the enthalpy of combustion for gaseous octane at 280°C.

11-4M Calculate the energy liberated when $C_{12}H_{26}$ is burned at 25°C with theoretical oxygen and the water in the products is in the liquid state.

11-5M Propane gas is burned with 50 percent excess air. The reactants enter the combustion chamber at 40°C and leave at 650°C. Calculate the heat liberated per mole of fuel.

11-6M Liquid octane is burned with a certain quantity of excess air such that the adiabatic flame temperature is 1100 K. Calculate the percent excess air if the reactants enter the combustion chamber at 298 K.

11-7M   Calculate the adiabatic flame temperature for gaseous methane burning with the theoretical air. Assume complete combustion.

11-8M   Liquid octane fuel at 298 K is mixed in a combustion chamber with 400 percent theoretical air at 425 K. The products of combustion leave the chamber at 900 K. Calculate the heat loss per mole of fuel.

11-9M   Determine the adiabatic flame temperature for solid carbon burning with 200 percent theoretical air. What percent excess air would be required to produce an adiabatic flame temperature of 1100 K. Assume complete combustion.

11-10M   Determine the enthalpy of combustion of gaseous methane and liquid octane at 25°C when liquid hydrogen peroxide is used as an oxidizer for the combustion reaction.

11-11M   0.5 kg/min of butane gas is burned with 100 percent excess air in a steady-flow process at 1-atm pressure. The reactants enter the combustion chamber at 25°C. Calculate the maximum temperature for which the ducts carrying the products of combustion should be designed if the heat loss in the combustion is 530 kW.

11-12M   Consider the combustion of liquid fuels—octane, methyl alcohol, propane, and ethyl alcohol—with the theoretical quantities of air at 25°C. Compare the reactions on the basis of
　　　　　(a) Volume of products per unit energy liberated
　　　　　(b) Volume of air per unit energy liberated
　　　　　(c) Energy liberated per unit mass of fuel
　　　　Discuss these comparisons.

11-13M   Consider the dissociation of diatomic oxygen. Calculate the equilibrium concentrations of O_2 and O at 3300 K for system pressures of 1, 5, and 10 atm.

11-14M CO_2 is heated at a constant pressure of 500 kPa from 298 to 2800 K. Calculate the heat transfer for the process and determine the final concentrations at the end of heating.

11-15M   1 mol of CO is mixed with 1 mol of water vapor at 298 K and 1-atm pressure. The mixture is then heated to 1800 K. Calculate the heating required and the composition of the final mixture.

11-16M   Consider the dissociation of water vapor into H_2 and O_2. How much heat must be added to 1 mol of water vapor at 200 K to raise the temperature in a steady-flow process to 2500 K?

11-17M   What temperature is necessary to dissociate diatomic hydrogen to a point that the monatomic hydrogen comprises 15 percent of the total number of moles of the mixture?

11-18M   1 mol of CO is mixed with 1 mol of water vapor at 25°C and 1-atm pressure.

The mixture is then heated to 1800 K. Calculate the heating required and the composition of the final mixture.

11-19M   Consider the ionization reaction for cesium as discussed in Example 11-13. Calculate the number density of electrons (no./m³) for conditions of the example. Repeat for a total pressure of 0.1 atm and 1000 K.

11-20M   In magnetohydrodynamic power-generation systems (see Chap. 15) it is desirable to employ gases which will ionize at modest temperatures. One technique for producing such gases is to "seed" a gas with a substance like cesium which ionizes at a low temperature. Consider argon as the carrier gas at a total pressure of 1 atm and 1000 K. Calculate the electron number density (no./m³) for cesium mass concentrations of 0.1, 0.5, and 1.0 percent. The molecular weight for cesium is 132.91. Repeat the calculation for a total pressure of 10 atm. Assume that the electrons behave like an ideal monatomic gas.

11-21M   Repeat Prob. 11-20M using sodium as the seeding gas.

11-22M   Using the concepts discussed in Sec. 9-13, calculate the electric conductivity for the systems in Probs. 11-20M and 11-21M, assuming $E = 10$ V/m.

11-23M   Calculate the equilibrium constant for the reaction $CO_2 \rightarrow CO + \frac{1}{2}O_2$ at 1000 K using heat of reaction data calculated from Table A-6 and the equilibrium constant at 298 K from Table 11-3.

11-24M   How much energy is required to ionize cesium at 2400 K, at 2800 K, and at 4000 K?

11-25M   Calculate the degree of ionization of cesium in a mixture of 1 mol cesium and 2 mol nitrogen at 2000 K and 10 atm.

11-26M   In deriving Eq. (11-67) the heat of reaction was assumed constant over a range of temperatures. How accurate is this assumption for dissociation of CO_2 in the temperature range from 1500 to 3000 K?

11-27M   Estimate the energy required to heat an equimolal mixture of nitrogen and oxygen from 298 to 3000 K.

11-28M   Assume that the human metabolic system can be approximated by a simple combustion reaction occurring at 37°C. For a total energy consumption of 13 MJ/day what quantity of methane and air at 20°C would be required to carry on the combustion process? How much carbon dioxide would be produced?

11-29M   Some authorities claim that hydrogen produced by electrolysis of water will become a widely used fuel as more and more of our hydrocarbon fuel reserves are depleted. Calculate the higher and lower heating value for gaseous hydrogen at 298 K and 100 kPa. What size tank for storage at 14 MPa and 295 K would be required to provide the same energy as a 95-liter tank of gasoline (octane)?

11-30M   Nitrogen oxides (that is, NO, etc.) are present in the stack discharge gas from power plants because they are "frozen" in the flow when the hot combustion gases come in rapid contact with the relatively cool surface of the boiler tubes and sufficient time is not available to perform the reverse reaction back to N_2 and O_2. Suppose the combustion gases are at 2000 K and the boiler tubes are at 1000 K. Estimate the percentage of NO present at the higher temperature that will be retained after sudden cooling to the lower temperature.

11-31M   In a home heating furnace methane is burned with 25 percent excess air, and the manufacturer claims that 70 percent of the higher heating value of the fuel is recovered as useful heat. What is the outlet temperature of the products of combustion in this circumstance if the air and fuel enter at 25°C?

11-32M   In a certain power plant combustion air enters the furnace at 400 K, and methane fuel enters at 298 K. Twenty percent excess air is used and the products of combustion leave the furnace at 1450 K. How much energy is obtained per kilogram of fuel?

11-33M   The burner for a gas-turbine plant is designed to burn liquid propane at 298 K with air at 450 K. The outlet temperature of the products of combustion is to be 1400 K. Calculate the air/fuel ratio required and percent excess air for steady flow and no heat loss from the burner.

11-34M   A radiant heater for use in large warehouse areas burns gaseous butane with 25 percent excess air. Both fuel and air enter the heater at 298 K. The products of combustion leave the device at 950 K. How much heat is lost by the heater per kilogram of fuel? What percent of the higher heating value does this represent?

11-35M   Calculate the maximum work output as saturated liquid water changes to saturated vapor at 700 kPa in (a) a proces at constant temperature and (b) exclusive of $p\,dV$ work in a process at constant temperature and pressure. How do these values compare with the $p\,dV$ work done in the process?

11-36M   What is the maximum work output exclusive of $p\,dV$ work when CO is burned with O_2 to form CO_2 at constant pressure of 1 atm and constant temperature of 298 K?

11-37M   Methane (CH_4) is burned with the stoichiometric proportion of oxygen at 1 atm and 298 K. What is the maximum work output in this process? What would be the actual work output if the process occurs under steady-flow conditions?

11-38M   Calculate the energy released per kilogram of coal for the conditions in Prob. 11-43M.

11-39M   A gaseous fuel contains 76 percent CH_4 and 24 percent C_2H_6 on a molal basis. Determine the air/fuel ratio for stoichiometric combustion. Also determine the higher- and lower-heating values at 1 atm, 25°C, per kilogram of fuel and per kilogram of air/fuel mixture.

11-40M   Gaseous ethane, C_2H_6, is burned with 200 percent theoretical air at 1 atm.

The reactants enter at 25°C and leave at 1000 K in a steady-flow process. Calculate the total energy released for a fuel flow of 15 kg/s. Make the calculation using individual component enthalpies, and also using Table A-18.

11-41M Liquid octane, C_8H_{18}, is burned adiabatically with 400 percent theoretical air at 1 atm and 25°C. Calculate the adiabatic flame temperature using (a) enthalpies of each compound and (b) Table A-18.

11-42M CO is burned with the theoretical oxygen at 4 atm and 25°C in a steady-flow process. Calculate the adiabatic flame temperature taking into account dissociation.

11-43M An analysis is performed on a coal-fired boiler to estimate air flow requirements for combustion and the amount of waste products which must be moved for disposal. The analysis of the coal on a mass basis is:
    75% C
    7% H_2
    7% O_2
    2% N_2
    4% H_2O
    5% Ash
An Orsat analysis of the dry products of combustion gives a volumetric analysis as 13% CO_2, 1.5 percent CO, 6.5 percent O_2, and 79 percent N_2. A mass analysis of the waste material indicates 21 percent C and 79 percent ash. For 1.0 kg of incoming coal, determine the volume of air required at 1 atm, 30°C, in liters/s. Also calculate the mass of refuse per kilogram of coal. If the products are discharged at 150°C, calculate their volume flow per kilogram of coal.

11-44M 1.0 kg mol of CH_4 and 2.0 kg mol of O_2 at 1 atm and 25°C react in a steady-flow process to produce products of 1 mol of CO_2 and 2 mol of H_2O (vapor) at 1 atm and 25°C. Calculate the change in entropy for the reaction, the energy released, and the decrease in availability for $T_0 = 25$°C.

11-45M A mixture of H_2 and O_2 is allowed to react to H_2O in a rigid tank. The mass of H_2 is 1.0 kg and the mass of O_2 is 29 kg, and the initial condition of the reactants is 1 atm and 35°C. Calculate the volume of the tank. Calculate the amount of condensation if the products are cooled to 25°C.

11-46M 4 kg mol of CH_4 are burned with 9 kg mol of O_2. Calculate the percent excess of O_2 and the dew point of the products for a total pressure of 1 atm.

11-47M For the reactants in Prob. 11-46M, calculate the energy released if the reactants enter at 1 atm, 25°C, and products leave at 1 atm, 200°C, in a steady-flow device. Also calculate the irreversibility for the reaction.

11-48M 3 kg mol/h of propane, C_3H_8, are burned with the theoretical amount of air at 1 atm and 25°C. The products leave at 150°C and 1 atm in a steady-flow process. Calculate the energy released in this process, the volume flow rate of reactants and products in liters per hour, and the irreversibility for the reaction.

REFERENCES

1 JANAF Thermochemical Tables, *Document PB 168–370*, Clearinghouse for Federal Scientific and Technical Information, August 1965.

2 Selected Values of Chemical Thermodynamic Properties, *NBS Technical Notes*, 270-1 and 270-2, 1955.

3 Selected Values of Physical and Thermodynamic Properties of Hydrocarbons and Related Compounds, *API Res. Project 44*, Carnegie Press, Carnegie Institute of Technology, Pittsburgh, Pa.

4 Callen, H. B.: "Thermodynamics," John Wiley & Sons, Inc., New York, 1960.

5 Kirkwood, J. G., and I. Oppenheim: "Chemical Thermodynamics," McGraw-Hill Book Company, New York, 1961.

6 Lewis, G. N., and M. Randall: "Thermodynamics," 2d ed. (revised by K. S. Pitzer and L. Brewer), McGraw-Hill Book Company, New York, 1961.

7 Reynolds, W. C.: "Thermodynamics," 2d ed., McGraw-Hill Book Company, New York, 1968.

8 Wark, Kenneth: "Thermodynamics," 2d ed., McGraw-Hill Book Company, New York, 1971.

9 Van Wylen, G. T., and R. E. Sonntag: "Fundamentals of Classical Thermodynamics," John Wiley & Sons, Inc., New York, 1965.

10 Sears, F. W.: "Thermodynamics," 2d ed., Addison-Wesley Publishing Company, Inc., Reading, Mass., 1953.

11 Hatsopoulos, G. N., and J. H. Keenan: "Principles of General Thermodynamics," John Wiley & Sons, Inc., New York, 1965.

12 Keenan, J. H., and J. Kaye: "Gas Tables," John Wiley & Sons, Inc., New York, 1948.

chapter twelve
Power and refrigeration cycles

12-1 INTRODUCTION

One of the important objectives of engineering practice is to convert energy from one form to another—chemical energy of a fuel to heat through a combustion process, heat energy to electric energy in a steam power plant, gas turbine, internal-combustion engine, or other devices. We have seen that the second law of thermodynamics serves as a guide in predicting the permissible direction of an energy-conversion process. Our objective in this chapter is to describe some of the practical arrangements employed for large-scale energy conversion and to illustrate the methods of analysis applicable to these devices. It is clear that a key item of interest is the effectiveness or efficiency of a particular energy-conversion process. We shall, therefore, want to establish a consistent basis for calculation of such quantities so that various processes may be compared in a meaningful manner.

The term *power cycle* in this chapter title describes a process in which devices are used that produce power continuously, i.e., a repetitive cyclic process which converts energy into mechanical or electrical work. Mechanical and electrical work are, of course, the most practicable forms of energy in that they can be adapted to a variety of applications.

The applicability of any power cycle is eventually related to cost and efficiency. Environmental effects associated with power production must also be taken into account, and the cost of pollution-control equipment is becoming a major factor in automotive, aircraft, and electric power applications. In some cases environmental considerations will influence the choice of fuel alternatives: fuel oil or natural gas in favor of coal in a power plant, natural gas in favor of fuel oil in homes and apartments, and the long-term advantages of nuclear fuel with its minimum air pollution.

A fundamental point often overlooked in environmental considerations is that the amount of pollution in most power-production schemes is directly proportional to the amount of fuel consumed which, in turn, is related to the efficiency of energy conversion. The higher the efficiency is, the lower the fuel consumption, and thus the lower the total amount of pollution. What this means is that a thermodynamic analysis or design which is capable of improving efficiency will also help to minimize negative environmental side effects.

In this chapter we are primarily concerned with technical analysis; economic and environmental factors are beyond the scope of our discussion. It is necessary to bear in mind, nonetheless, that these variables must be kept in sharp focus in any practical power plant design.

12-2 GENERAL CONSIDERATIONS

The basis for a great number of power cycles is the Carnot cycle discussed in Chap. 6, because, for given temperature limits, the thermal efficiency of such a cycle is the maximum obtainable; that is,

$$\eta_t = 1 - \frac{T_L}{T_H} \tag{12-1}$$

From Eq. (12-1) we sense two basic notions:

1 The higher the temperature at which heat is added, the higher the efficiency.
2 The lower the temperature at which heat is rejected, the higher the efficiency.

These two notions carry over into practical power cycles. The basic idea of efficiency, it will be remembered, is

$$\eta = \frac{\text{useful effect}}{\text{energy that must be purchased}} \tag{12-2}$$

We shall be considering both reversible and actual processes in the following sections and will show the effects of irreversibility on efficiency.

A refrigeration cycle is just the opposite of a power cycle; there is a net work input and a net heat output. Compare the Carnot cycles in Fig. 12-1. The efficiency of the power cycle is given by Eq. (12-1), and the characteristic parameter for a refrigeration cycle is the coefficient of performance (COP) defined by:

$$\text{COP} = \frac{\text{refrigeration effect}}{\text{work input}} \tag{12-3}$$

For the reversible Carnot refrigeration cycle this becomes

$$\text{COP} = \frac{Q_L}{W} = \frac{Q_L}{Q_H - Q_L} = \frac{1}{T_H/T_L - 1} \tag{12-4}$$

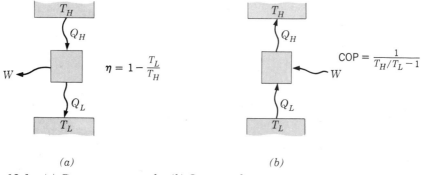

(a)   (b)

FIG. 12-1   (a) Carnot power cycle; (b) Carnot refrigerator.

It is not uncommon for the COP to be greater than unity. This fact is the reason why it is not called an efficiency even though it is a "useful effect" divided by "energy that must be purchased."

12-3 VAPOR POWER CYCLES

To initiate our discussion of power cycles consider the schematic arrangement shown in Fig. 12-2a. Heat is supplied to the boiler where liquid is converted to vapor. The vapor is then expanded adiabatically in the turbine to produce a work output. Vapor leaving the turbine then enters the condenser where heat is removed until the vapor is condensed into the liquid state. The condensation process is the heat-rejection mechanism for the cycle. Saturated liquid is delivered to a pump where its pressure is raised to the saturation pressure corresponding to the boiler temperature, and the high-pressure liquid is delivered to the boiler where the cycle repeats itself. For a reversible cycle the turbine and pump processes are isentropic as indicated in Fig. 12-2b.

The vapor power cycle is perhaps the most widely used thermal cycle and accounts for most of the electric energy generation in the world. The basic *Rankine* cycle is shown in Fig. 12-2. We shall assume that the working fluid is steam although, in some specialized applications, other fluids might be employed. The basic Rankine cycle shown differs from a Carnot cycle in that the heat-addition process does not occur at constant temperature. If it were possible to operate the cycle either as 1-2-3'-4" or 1-2-3-4', a Carnot cycle would be obtained exactly. Practical difficulties prohibit these two arrangements. The process 3'-4" is impractical because it would involve the compression of a wet mixture into a saturated liquid. It simply does not seem practical to design a turbine pump to accomplish this compression. The process 3-4' would involve liquid compressions to enormous pressures, and the pumping machinery to accomplish this feat would also be highly impractical. The average temperature at which heat is added to the Rankine cycle is less than T_1 because part of the process involves heating the compressed liquid up to the saturation temperature at the high pressure. Thus the efficiency is less than would be attained by a Carnot cycle operating between T_1 and T_2.

From the discussion so far we can see that the shape of the T-s diagram for the particular fluid governs the adaptability of that fluid to use with the Rankine cycle. It would be most desirable to have a fluid with a high critical temperature and with very steep saturated-liquid and -vapor lines so that the Rankine cycle would more closely approximate the Carnot performance. At the same time, the pressure limits required should be reasonable, and availability and cost of the fluid are also important factors to consider. The total quantity of fluid which must be circulated will depend on the enthalpy of vaporization because this governs the energy which may be delivered to the

(a)

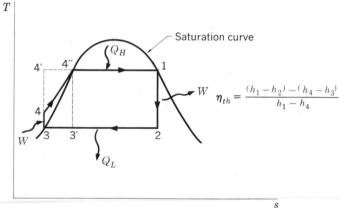

$$\eta_{th} = \frac{(h_1 - h_2) - (h_4 - h_3)}{h_1 - h_4}$$

(b)

FIG. 12-2 *Basic Rankine vapor power cycle. (a) Schematic; (b) temperature-entropy diagram.*

fluid at the high temperature. Taking all these factors into account, water is a good choice of working fluid for the Rankine power cycle. It is readily available at low cost, it has a very high enthalpy of vaporization (~900 Btu/lbm), it has a reasonably high critical temperature (705°F), and it is non-toxic. Corrosion problems with water are significant but can be controlled with careful engineering design.

One way of improving the efficiency of the Rankine cycle is to continue the constant-pressure heating process past point 1 and into the superheat region, as shown in Fig. 12-3. Such a heating process has two advantages:

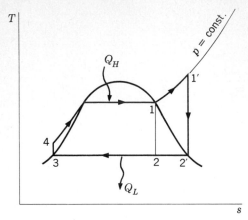

FIG. 12-3 *Temperature-entropy diagram for Rankine cycle with superheat.*

1   It increases the average temperature at which heat is added, thereby increasing the cycle efficiency.
2   It results in an operating condition which involves less moisture in the turbine expansion process. A major source of wear on turbine blades is the erosion caused by liquid droplets. The superheat process produces less moisture and, thus, serves to alleviate this difficulty.

Another way to improve the efficiency of the Rankine cycle is to engage in a *reheat* operation, as shown in Fig. 12-4. In this situation the steam is expanded in the first turbine stage until it approximately reaches the saturation line at p_2. Then it is returned to the boiler and reheated at constant pressure to T_3. The steam is then expanded in the second-stage turbine to p_4. The reheat arrangement has the same two advantages as the basic superheat cycle and is even more effective in preventing moisture from entering the turbine.

HEAT RATE

A term in common use among power plant engineers is the *heat rate* defined as

$$\text{Heat rate} = \text{heat input to boiler (Btu) per electric energy output (kWh)}$$

or in terms of the thermal efficiency

$$\text{Heat rate} = \frac{3413}{\eta} \text{ Btu/kWh} \qquad (12\text{-}5)$$

EXAMPLE 12-1   Rankine cycle
A Rankine cycle operates between pressure limits of 500 and 2 psia. The temperature of the steam entering the turbine is 700°F. Calcu-

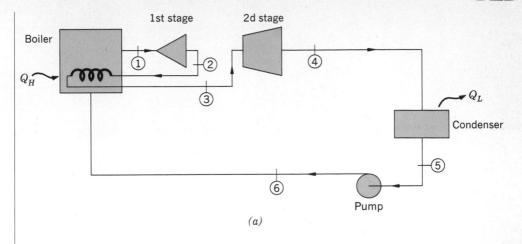

1st stage   2d stage

Boiler

Q_H

①   ②

③

Condenser Q_L

④

⑤

⑥

Pump

(a)

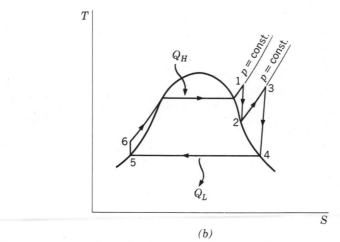

(b)

FIG. 12-4 *Rankine cycle with reheat.* (a) *Schematic;* (b) *temperature-entropy diagram.*

late the thermal efficiency of the cycle assuming isentropic expansion in the turbine.

SOLUTION   Figures 12-2a and 12-3 apply to this problem. Using either the steam tables or a Mollier diagram, we have, for the steam properties,

$h_1 = 1357$ (500 psia, 700°F)
$h_2 = 935$ ($s_1 = s_2$, $p_2 = 2$ psia)
$h_3 = 94.02$ (saturated liquid at 2 psia)

The pump work, $h_4 - h_3$, is computed from (see Example 4-2)

$$h_4 - h_3 = v_f(p_4 - p_3)$$
$$= \frac{(0.016\ 23)(500 - 2)(144)}{778}$$
$$= 1.496\ \text{Btu/lbm}$$

Thus

$$h_4 = 95.52\ \text{Btu/lbm}$$

The net work output of the cycle is the work output of the turbine less the work input to the pump, or $(h_1 - h_2) - (h_4 - h_3)$. The thermal efficiency of the cycle is based on the net work output and is calculated as

$$\eta_{th} = \frac{W}{Q_H} = \frac{(h_1 - h_2) - (h_4 - h_3)}{h_1 - h_4}$$
$$= \frac{(1357 - 935) - 1.496}{1357 - 95.52}$$
$$= 33.3\ \text{percent}$$

EXAMPLE 12-2   Rankine cycle with reheat

A steam cycle operates between the same upper and lower pressure and temperature limits as in Example 12-1. The steam is extracted at 100 psia and reheated to 700°F. Calculate the thermal efficiency under this new arrangement.

SOLUTION   The reheat cycle is shown schematically in Fig. 12-4. Using the nomenclature of this figure, the various enthalpies may be found from a Mollier diagram as:

$h_1 = 1357$ (500 psia, 700°F)
$h_2 = 1194$ ($s_1 = s_2$, $p_2 = 100$ psia)
$h_3 = 1379$ (100 psia, 700°F)
$h_4 = 1047$ ($s_3 = s_4$, $p_4 = 2$ psia)
$h_5 = 94.02$ (saturated liquid at 2 psia)
$h_6 = 95.52$ (see Example 12-1)

The thermal efficiency is now calculated as

$$\eta_{th} = \frac{W}{Q} = \frac{(h_1 - h_2) + (h_3 - h_4) - (h_6 - h_5)}{(h_1 - h_6) + (h_3 - h_2)}$$
$$= \frac{(1357 - 1194) + (1379 - 1047) - 1.496}{(1357 - 95) + (1379 - 1194)}$$
$$= 34.2\ \text{percent}$$

Neither the superheat nor reheat arrangement alleviates the efficiency "loss" which results from the low-temperature heat-addition process encountered as the liquid is heated from the exit temperature at the pump up to the saturation temperature at the upper pressure limit. This problem may be solved by making use of *regeneration*. An ideal regenerative cycle is shown in Fig. 12-5. The basic idea is that the high-temperature steam expanding in the turbine is employed to preheat the liquid before it enters the boiler. In this way, the heat transferred to the overall cycle occurs at constant temperature, and the arrangement will have an efficiency equal to the Carnot value. In terms of the *T-s* diagram in Fig. 12-6, the actual cycle 1-2-3-4-4' is equivalent to the Carnot cycle 1-2'-3'-4'. We might think of the ideal regenerative cycle as a Carnot cycle which is distorted sideways. The term *regeneration* means that heat is generated internally to the cycle by the heat-exchange process.

There are practical difficulties with the ideal regenerative cycle. It is not possible to design a turbine which will serve as a power-production device and a heat exchanger as well. Accordingly, practical heat exchangers must be introduced which cause the actual regeneration cycle to have an efficiency somewhat less than the ideal value.

A practical regenerative cycle arrangement with one feedwater heater is shown in Fig. 12-7. We have shown the high-pressure steam expanded through one turbine stage. Then, a quantity m is extracted and fed to a feedwater heater. For unit mass leaving the boiler we will have $1 - m$ mass to be expanded in the second turbine stage. In this instance the heater is an *open* type; this means that the steam extraction from the turbine is mixed directly with the incoming water from the low-pressure pump.

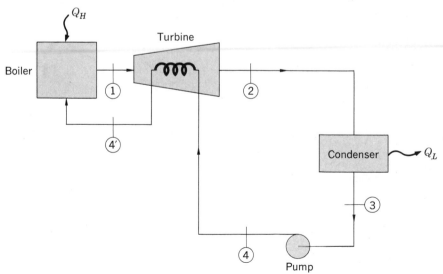

FIG. 12-5 *Ideal-vapor regenerative cycle.*

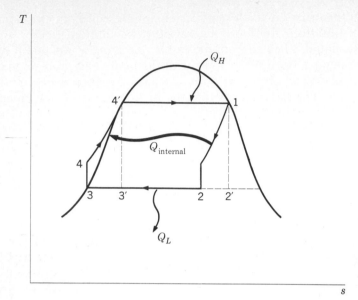

FIG. 12-6 *Temperature-entropy diagram for ideal regenerative cycle.*

For an ideal open feedwater heater we assume that the mixing of steam and cool water occurs at constant pressure. In terms of the cycle shown in Fig. 12-7 this means that $p_5 = p_6 = p_2$. The best we can hope to accomplish in this heater is to produce saturated liquid at point 6. We could not produce a higher energy level because a wet mixture of steam would be encountered and the high-pressure pump, designed for liquids, would be ren-

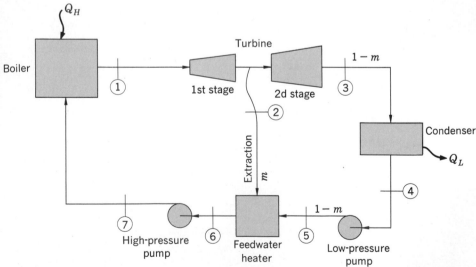

FIG. 12-7 *Practical regenerative cycle with one open feedwater heater.*

FIG. 12-8 *Closed feedwater heater with condensate pump.*

dered inoperative. In an actual steam cycle there could be some pressure
drop through the heater, and the outlet temperature would be controlled at
a value somewhat below the saturation temperature. However, for the cycle
analysis here, we shall assume no pressure drop and saturation temperature
at exit from the heater.

A *closed* feedwater heater could also be employed, and a schematic of
such a device is shown in Fig. 12-8. The term *closed* means that the incoming feedwater is not mixed with the extraction steam. A separate smaller
pump is used to pump the condensate up to the boiler pressure. There is
only one main feedwater pump in this instance. An alternative scheme for
using a closed feedwater heater is shown in Fig. 12-9. In this example the
condensate is allowed to pass through a steam trap back to the condenser

FIG. 12-9 *Closed feedwater heater with condensate trapped to condenser hot well.*

hot well. A steam trap is a device with a float-operated valve which permits the passage of liquid but does not allow passage of vapor.

The closed feedwater heater is different in operation from its open counterpart. As we have said, the main pump raises the liquid water to the boiler pressure so that $p_5 = p_7$ = boiler pressure in Fig. 12-8. The extraction steam at point 2 is at a lower pressure because of passage through the first turbine stage. It may be superheated or in the wet-mixture region, depending on the particular cycle conditions. In the closed heater the extraction steam is exposed to tubes through which flows the cool, high-pressure water. As a result, the steam condenses on the tubes, and the energy released in the condensation process raises the temperature of the water to be delivered to the boiler. The best we can hope to accomplish in this process is to heat the high-pressure water up to the *saturation temperature corresponding to the extraction pressure*; this is the temperature at which condensation takes place. Thus, at point 7 a compressed liquid will be experienced.

We can summarize our discussion of the exit temperatures from feedwater heaters with the following table

Type of heater	Exit feedwater pressure	Exit feedwater temperature
Open	Extraction pressure	Saturation temperature corresponding to extraction pressure (saturated liquid)
Closed	Boiler pressure	Saturation temperature corresponding to extraction pressure (compressed liquid)

In practice, several regeneration stages may be employed for closer approximation of the ideal regenerative cycle. A schematic for a cycle using two closed feedwater heaters is shown in Fig. 12-10. For unit mass leaving the boiler $1 - m_1$ will be left for expansion in the second stage after the m_1 is extracted for the first heater. Then, when m_2 is extracted for the second heater there will remain $1 - m_1 - m_2$ for the third turbine stage.

The number of feedwater heaters employed in an actual power plant is determined by cost considerations. In general, the increased capital expenditures for additional heaters must be more than compensated by reduced operating costs resulting from higher cycle efficiencies, if the additional expense is to be justified.

To compute the work output and efficiencies of regenerative cycles we must make energy balances on the cycle components, as in any other cycle. To perform such energy balances on the turbines and pumps requires that we determine the values of the extraction rate(s). These quantities may be obtained from energy balances on the feedwater heaters in accordance with the exit temperatures we discussed above. For example, we could write energy balances for the two feedwater heaters in Fig. 12-10 as

FIG. 12-10 *Schematic of vapor power cycle with two closed feedwater heaters.*

First heater: $(1)h_7 + m_1h_2 = (1)h_8 + m_1h_9$
Second heater: $(1)h_6 + m_2h_3 = (1)h_7 + m_2h_{11}$

Once the enthalpies are determined, m_1 and m_2 are easily calculated.
     We can now see the general pattern. For each feedwater heater there is
an extraction point and an unknown extraction flow rate. For each heater
we can write an energy balance so that we will obtain an equation for each
extraction flow. The solution of this set of equations will establish the particular extraction flows required.

EXAMPLE 12-3   Regenerative steam cycle

A regenerative steam cycle operates between the same upper and lower
pressure and temperature limits as in Example 12-1. Steam extraction
occurs at 100 psia and an open feedwater heater is employed. Calculate
the thermal efficiency of the cycle assuming isentropic expansion in the
turbine.

SOLUTION   This cycle is shown schematically in Fig. 12-7. For isentropic expansion in the turbine,

$$s_1 = s_2 = s_3$$

From a Mollier diagram we obtain

$$h_1 = 1357 \qquad \text{(500 psia, 700°F)}$$
$$h_2 = 1194 \qquad \text{(100 psia)}$$
$$h_3 = 935 \qquad \text{(2 psia)}$$
$$h_4 = 94.02 \qquad \text{(saturated liquid at 2 psia)}$$

The low-pressure pump raises the pressure of the condensate to 100 psia, so that

$$h_5 - h_4 = v_f(p_5 - p_4)$$
$$= \frac{(0.016\ 23)(100 - 2)(144)}{778}$$
$$= 0.294 \text{ Btu/lbm}$$

Thus

$$h_5 = 0.294 + 94.02$$
$$= 94.31 \text{ Btu/lbm}$$

For an ideal open feedwater heater, $p_2 = p_5 = p_6$ and saturated-liquid conditions are assumed to exist at point 6. Thus

$$h_6 = 298.6 \qquad \text{(saturated liquid at 100 psia)}$$

For the high-pressure pump we have

$$h_7 - h_6 = v_f(p_7 - p_6)$$
$$= \frac{(0.017\ 74)(500 - 100)(144)}{778}$$
$$= 1.313 \text{ Btu/lbm} \qquad \text{(3.0589 kJ/kg)}$$

and

$$h_7 = 1.313 + 298.6$$
$$= 299.9 \text{ Btu/lbm} \qquad \text{(697.07 kJ/kg)}$$

We are now in a position to calculate the extraction rate m with an energy balance on the feedwater heater. The energy balance on the feedwater heater is

$$m h_2 + (1 - m)(h_5) = (1)(h_6) \qquad (a)$$

Inserting the proper numerical values,

$$m(1194) + (1 - m)(94.3) = 298.6$$

and

$$m = 0.186 \text{ lbm extracted/lbm total flow}$$

The thermal efficiency of the cycle is now calculated with

$$\eta_{th} = \frac{W}{Q_H}$$

$$= \frac{(1)(h_1 - h_2) + (1 - m)(h_2 - h_3) - (1 - m)(h_5 - h_4) - (1)(h_7 - h_6)}{h_1 - h_7}$$

$$= \frac{1}{1357 - 299.9} [(1)(1357 - 1194) + (1 - 0.186)(1194 - 935)$$

$$- (1 - 0.186)(0.3) - (1)(1.313)]$$

$$= 35.2 \text{ percent}$$

12-4 DEVIATIONS FROM THEORETICAL VAPOR CYCLES

All of the vapor cycles discussed in the foregoing are idealized in that the expansion and compression processes are assumed to be reversible. It has been tacitly assumed that no pressure losses occur between cycle elements as a result of fluid friction. Perhaps the most significant deviation from ideal behavior occurs in the turbine expansion process. In Fig. 12-11 this deviation is illustrated on an h-s or Mollier diagram. Ideally, the process would occur reversibly and adiabatically as in 1-2. Actually, various influences of fluid friction would cause the process to behave as in 1-2'. The process would still be

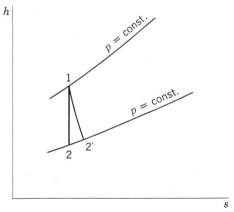

FIG. 12-11 *Comparison of actual and ideal turbine expansion processes.*

adiabatic, but, as a result of irreversibilities, an increase in entropy would be experienced.

The turbine efficiency is defined by

$$\eta_{turb} = \frac{\text{actual work}}{\text{isentropic work}} = \frac{h_1 - h_{2'}}{h_1 - h_2} \tag{12-6}$$

A similar definition is employed to take account of irreversibilities in the pumps.

$$\eta_{pump} = \frac{\text{isentropic work}}{\text{actual work}} \tag{12-7}$$

In Eq. (12-7) it is to be noted that the isentropic work for a pump is less than the actual work because the device absorbs power. In general, turbine and pump efficiencies may only be determined from experiment. The effects of pump and turbine efficiencies on the T-s diagram for the basic Rankine cycle are shown in Fig. 12-12.

Modern steam power plants are much more complicated than might be gathered from the above discussions. Each component of the cycle is an intricate piece of machinery which has variable performance depending on load, speed, and different pressure and temperature conditions. Elaborate control systems must be provided to ensure reliable and stable operation as well as to protect the machines from too rapid changes in loads, speeds, or other variables.

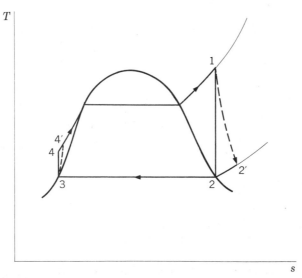

FIG. 12-12   *Influence of actual turbine and pump processes on T-s diagram of basic Rankine cycle.*

As we have seen in Chap. 11, combustion processes can be involved affairs, and this is especially true for a large steam boiler. The fuel and air flow rates must be carefully controlled to avoid excessive temperatures which can destroy the tubes in the boiler. At the same time, environmental constraints may be placed on allowable discharges from the combustion process which can restrict the modes of operation. If coal is the fuel, the process of getting it to the boiler in satisfactory form can be quite a demanding engineering problem. It must be mined, shipped, and then stored at the plant site. It must then be pulverized to a proper size and transported to the boiler by a conveyor system. Not the least of the problems is the removal and disposal of the huge quantities of ash which accumulate after coal is burned; and, if high sulfur coal is burned, one must use elaborate methods to remove SO_2 from the stack discharge gases. Despite all the above complications modern steam power plants operate throughout the world to produce dependable electric power at all times.

At this point the reader may have temporarily lost sight of the purpose of reheat, feedwater heaters, etc. The objective is to increase efficiency by increasing the average temperature at which heat is added to the cycle. One might ask: Why not build a hotter fire in the boiler? The answer is that metallurgical considerations limit the temperatures and/or pressures which the boiler tubes and turbine blades can tolerate. As better materials are developed, we may indeed raise the boiler and turbine temperatures and achieve a higher efficiency; still, regeneration and reheat will be employed for their contribution to efficiency gain.

We have noted that efficiency can also be increased by lowering the temperature of heat rejection; however, this temperature is essentially limited by local atmospheric conditions. Because of the large amounts of energy which must be removed from the condenser it is usually desirable to locate a power plant adjacent to a lake or river for cooling purposes. In some cases power companies will construct a lake for this purpose when the power plant is built. In some locales lakes or rivers are not available, and the heat-rejection process is accomplished with large cooling towers as described in Sec. 10-8. In some parts of the world, water is simply unavailable for use as a coolant, and the heat from the condenser must be dissipated directly to the surrounding air. Enormous fans and heat exchangers are required in these instances.

EXAMPLE 12-4   Effects of turbine efficiencies

Calculate the thermal efficiency for the reheat cycle of Example 12-2, assuming that the turbines have efficiencies of 80 percent and the expansion processes are no longer isentropic.

SOLUTION   The expansion processes for the turbines are shown in schematic form on the accompanying figure. The isentropic processes are shown as dashed lines, and the actual expansion processes are indicated with the solid lines. The overall cycle schematic is indicated in Fig. 12-4. Using the definition of turbine efficiency from Eq. (12-6), we have for the first turbine stage,

$$\eta = \frac{h_1 - h_{2'}}{h_1 - h_2}$$

Also, for the second turbine stage,

$$\eta = \frac{h_3 - h_{4'}}{h_3 - h_4}$$

From Example 12-2,

$$h_1 = 1357 \qquad h_4 = 1047$$
$$h_2 = 1194 \qquad h_5 = 94.02$$
$$h_3 = 1379 \qquad h_6 = 95.52$$

Then

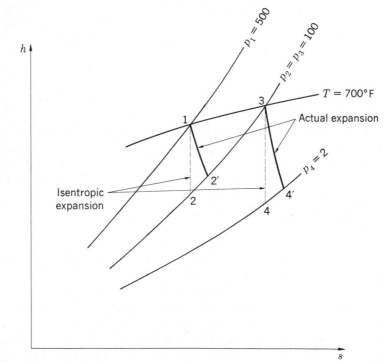

FIG. EXAMPLE 12-4

$$0.8 = \frac{1357 - h_{2'}}{1357 - 1194}$$

and

$$0.8 = \frac{1379 - h_{4'}}{1379 - 1047}$$

so that

$$h_{2'} = 1227$$
$$h_{4'} = 1113$$

The thermal efficiency is now calculated as

$$\eta_{th} = \frac{W}{Q_H} = \frac{(h_1 - h_{2'}) + (h_3 - h_{4'}) - (h_6 - h_5)}{(h_1 - h_6) + (h_3 - h_{2'})}$$
$$= \frac{(1357 - 1227) + (1379 - 1113) - 1.496}{(1357 - 95.5) + (1379 - 1227)}$$
$$= 27.9 \text{ percent}$$

It should be noted that the total heat added is reduced as a result of the non-isentropic expansion in the first turbine stage. This reduction, however, is not enough to offset the reduced work output of the turbines.

NUCLEAR POWER CYCLES

Nuclear power plants which are in operation today, and those projected for the future, all use the Rankine steam cycle for conversion of heat to useful work. Instead of the combustion of fossil fuels, a nuclear reactor serves as the heat source. Because of radioactive hazards, elaborate schemes are used to prevent contamination of the main steam power cycle. Liquid sodium is frequently employed as an intermediate fluid to remove heat from the reactor and then vaporize the high-pressure water. Special piping and pumps are required for these applications to prevent leaks and any direct contact between sodium and water.

Nuclear power plants have two obvious advantages. First, they do not produce air pollution because there are no products of combustion to be dispersed in the atmosphere, and second, they alleviate demands on petroleum and natural gas fuels which may be required in home, small industrial, and automotive applications. But the nuclear plants are not without their disadvantages too. With an increase in the amount of power produced by nuclear means there is the growing problem of disposal of radioactive waste products. Burial in deep underground caverns or at sea is a possibility but needs careful evaluation to determine the eventual environmental impact.

Because of the fluids which must be circulated through the reactor and the materials which must be used to contain the fissionable substances, the practical upper temperature limit of a nuclear power plant is not as high as its fossil fuel counterpart. As a consequence its efficiency will be lower and for a given power output more heat must be rejected to the surrounding atmosphere. Thus, the heating of streams and lakes, a serious environmental problem with any power plant, is even more critical with a nuclear plant. The upper limit of efficiency for a fossil fuel plant is about 40 percent, and only about 30 percent for a nuclear plant. We can see how this reduction in efficiency affects waste heat rejection by making a simple calculation for two plants, each with 500-MW output. For the fossil fuel plant

$$Q_H = \frac{500 \text{ MW}}{0.4} = 1250 \text{ MW}$$

$$Q_L = Q_H - W = 1250 - 500 = 750 \text{ MW}$$

And for the nuclear plant

$$Q_H = \frac{500 \text{ MW}}{0.3} = 1667 \text{ MW}$$

$$Q_L = Q_H - W = 1667 - 500 = 1167 \text{ MW}$$

Thus, in this case a reduction of efficiency from 40 to 30 percent increases the heat rejection from 750 to 1167 MW or *56 percent*. So, we see that there must be trade-offs made between the advantages and disadvantages of nuclear power production. However, the important point to remember is that a thermodynamic analysis of the Rankine cycle will proceed in the same fashion, whatever the application.

12-5 POWER–CYCLE ANALYSIS

From the foregoing discussion we can see that vapor power cycles can have many components, and one must be careful to make the proper energy balances. Too often, one is tempted to think of each cycle as a special case with its own peculiar analysis technique. A far better procedure, and one which will be adaptable to other types of power cycles in this chapter, is to view cycle analysis in a general way. This procedure will almost always yield correct results.

1 Draw a schematic block diagram for the cycle. Label all points in the cycle. The type of nomenclature does not matter, e.g., 1, 2, 3; *a*, *b*, *c*; *x*, *y*, *z*.
2 Draw arrows indicating heat or work flows delivered to or rejected from the cycle components.

3 Designate mass flows in, out, or through the various components.
4 Write down the known properties, states, and work or heat flows in terms of the nomenclature devised above.
5 Write down the desired results of the analysis, again, in terms of the above nomenclature.
6 Write out energy balances for works, heat flows, etc.
7 Write any energy balances which may be necessary to establish flow rates, as in a feedwater heater.
8 Determine needed properties to execute the energy balance calculations using process information (isentropic process, constant-pressure process, etc.).
9 Assemble the information and calculations to obtain the desired results.

EXAMPLE 12-5   Regeneration with two feedwater heaters

A regenerative steam cycle employs two closed feedwater heaters and three turbine stages. The condensate from the first heater flows through a steam trap into the second heater, and the condensate from the sec-

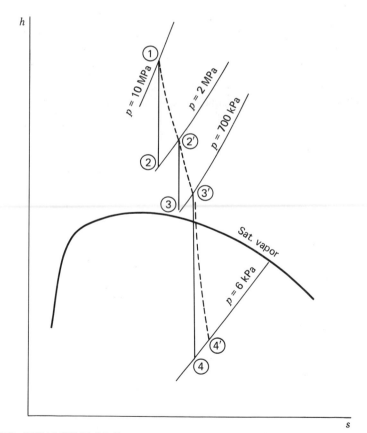

FIG. EXAMPLE 12-5a

ond heater is pumped up to the boiler pressure. The steam enters the turbine at 10 MPa, 540°C, extraction occurs at 2 MPa and 700 kPa, and discharge from the third turbine stage at 6 kPa. The turbine efficiencies are 85 percent, and the pump efficiencies are 100 percent. Calculate the thermal efficiency and steam flow required for a total electrical output from the turbine of 20 000 kW. The schematic diagram is shown in the accompanying figure, along with a Mollier diagram for the turbine processes.

SOLUTION From the problem statement

$$p_1 = p_{13} = p_{12} = p_8 = p_7 = p_6 = 10 \text{ MPa}$$
$$T_1 = 540°C \qquad p_2 = p_{11} = 2 \text{ MPa}$$
$$p_3 = p_9 = 700 \text{ kPa} \qquad p_4 = p_5 = 6 \text{ kPa}$$

From the Mollier diagram

$$h_1 = 3475 \qquad h_2 = 3005 \text{ kJ/kg}$$

Using the turbine efficiency,

$$0.85 = \frac{h_1 - h_{2'}}{h_1 - h_2} \qquad h_{2'} = 3076 \text{ kJ/kg} \qquad (a)$$

Then,

$$h_3 = 2828$$

and

$$0.85 = \frac{h_{2'} - h_{3'}}{h_{2'} - h_3} \qquad h_{3'} = 2865 \text{ kJ/kg} \qquad (b)$$

Now,

$$h_4 = 2136$$

and

$$0.85 = \frac{h_{3'} - h_{4'}}{h_{3'} - h_4} \qquad h_{4'} = 2245 \text{ kJ/kg} \qquad (c)$$

The liquid enthalpies are now calculated

$$h_5 = 151.53 \qquad (h_f \text{ at } 6 \text{ kPa})$$

$$h_6 - h_5 = v_5(p_6 - p_5) = (1.0064 \times 10^{-3})(10\ 000 - 6)$$
$$= 10.06 \text{ kJ/kg} \qquad (d)$$
$$h_6 = 161.59 \text{ kJ/kg}$$
$$h_9 = 697.22 \qquad (h_f \text{ at } 700 \text{ kPa}, T = 165.0°C)$$
$$h_{11} = h_{10} = 908.79 \qquad (h_f \text{ at } 2 \text{ MPa}, T = 212.4°C)$$

$$h_{12} - h_9 = v_9(p_{12} - p_9) = (1.1080 \times 10^{-3})(10\ 000 - 700)$$
$$= 10.3 \text{ kJ/kg} \qquad (e)$$
$$h_{12} = 707.5 \text{ kJ/kg}$$
$$h_7 = 703.18 \qquad (\text{Table A-10M, 165°C, 10 MPa})$$
$$h_8 = 912.07 \qquad (\text{Table A-10M, 212.4°C, 10 MPa})$$

The energy balances on the heaters are

$$m_1 h_{2'} + (1 - m_1 - m_2)h_7 = (1 - m_1 - m_2)h_8 + m_1 h_{11} \qquad (f)$$

$$m_2 h_{3'} + (1 - m_1 - m_2)h_6 + m_1 h_{10} = (1 - m_1 - m_2)h_7$$
$$+ (m_1 + m_2)h_9 \ (g)$$

The appropriate enthalpy values may be inserted, and the equations solved simultaneously to give:

$$m_1 = 0.0721 \text{ kg} \qquad m_2 = 0.1799 \text{ kg} \qquad (h)$$

We are now in a position to calculate the thermal efficiency of the power cycle.

FIG. EXAMPLE 12-5b

$$\eta = \frac{(h_1 - h_{2'}) + (1 - m_1)(h_{2'} - h_{3'}) + (1 - m_1 - m_2)(h_{3'} - h_{4'})}{h_1 - h_{13}}$$

$$\frac{-(1 - m_1 - m_2)(h_6 - h_5) - (m_1 + m_2)(h_{12} - h_9)}{h_1 - h_{13}} \qquad (i)$$

The enthalpy at point 13 is obtained from the energy balance

$$(1 - m_1 - m_2)h_8 + (m_1 + m_2)h_{12} = (1)h_{13} \qquad (j)$$
$$h_{13} = 860.5 \text{ kJ/kg}$$

Inserting the enthalpy values in Eq. (i) we obtain the efficiency as

$$\eta = \frac{1048.4}{2614.5} = 40.1\% \qquad (k)$$

The net work output is 1048.4 kJ/kg so the steam flow required for 20 000 kW is

$$\dot{m} = \frac{20\ 000}{1048.4} = 19.08 \text{ kg/s} = 6.87 \times 10^4 \text{ kg/h} \qquad (l)$$

12-6 GAS COMPRESSORS

A gas compressor may be considered as an energy-conversion device since it converts mechanical work into stored potential energy resulting from high pressure. We initiate our discussion with the reciprocating type of machine illustrated in Fig. 12-13. The ideal pressure-volume diagram for such a machine is shown in Fig. 12-14. At point 4 the piston has just completed the compression and delivery of the gas in the cylinder and the exhaust value E has just closed. The volume of gas remaining in the cylinder at this time V_4 is called the *clearance volume*. As the piston moves back from the top-dead-center position (tdc) this volume expands until the pressure p_1 is reached. At this point the intake valve I is opened, and the gas is pulled into

FIG. 12-13   *Reciprocating air compressor.*

FIG. 12-14 *p-v diagram for ideal reciprocating air compressor.*

the cylinder by the continuing backward motion of the piston. When the piston reaches bottom-dead-center (bdc) at point 2 the intake valve is closed. The gas is then compressed from 2 to 3 and the exhaust valve is opened at point 3, thereby allowing the piston to force the gas out of the cylinder in process 3-4.

The actual compression and expansion processes in the cylinder do not follow the nice theoretical processes shown in Fig. 12-14. Intake and exhaust valves do not open instantaneously, and various pressure-loss and frictional effects may cause substantial deviations from the ideal processes. Figure 12-15 indicates the shape an actual *p-v* diagram might take.

Figure 12-16 illustrates the basic operation of a centrifugal compressor. Air enters the casing and is turned by the rotating-vane impeller. Because of

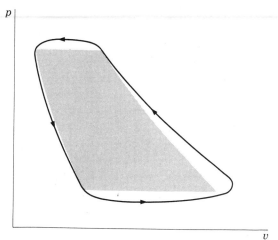

FIG. 12-15 *Actual p-v diagram for reciprocating air compressor.*

FIG. 12-16 *Contrifugal compressor. (a) Radial impeller vanes; (b) backward curved vanes; (c) forward-curved vanes. Diagram according to Wood [11].*

the centrifugal force action the pressure is increased at the outer periphery. In (*a*) the rotating vanes are constructed in a radial fashion while the stationary diffuser section in the casing has vanes which are curved forward in the direction of rotation. Centrifugal compressors may also be constructed with backward- or forward-curved moving vanes as illustrated in (*b*) and (*c*), depending on the pressure-speed characteristics desired. Of course, a centrifugal fan or squirrel cage blower is just a compressor to handle large flow volumes with a small pressure rise.

The operation of an axial flow compressor is illustrated in Fig. 12-17. The machine consists of a series of rotating blades (rotor) alternating with stationary blade sections (stator). A stage is comprised of one rotor and one stator. The rotors are all connected to a common shaft while the stators are fixed to the enclosing casing. The blades are special airfoil sections, and the pressure rise results from the lifting action on the blades. For this reason, the pressure rise per stage is rather small, and axial compressors always consist of many stages in series in order to accomplish any substantial compression ratio. The design of the airfoil sections in the rotor and stator is a crucial matter, of course, and will vary significantly depending on the application for the compressor; e.g., an axial compressor for use in the inlet of a high-speed jet engine would be substantially different from a compressor for low-speed air in a stationary power plant.

Gas compressors are basically steady-flow devices. In reciprocating ma-

FIG. 12-17 *Axial flow compressor. Diagram according to Wood* [11].

chines a pulsating steady flow is experienced, but the work input is calculated from the steady-flow energy equation. Neglecting any cooling which may be performed on the cylinder walls, the machine is adiabatic, and the work is usually computed as the change in enthalpy for the air. For large volumes of air and modest pressure ratios, centrifugal or turbine-type compressors are normally employed. The turbojet is one application for turbine compressors. Centrifugal compressors are widely used in large air conditioning systems.

If very large pressures are to be attained by the compression process the task is very difficult to accomplish with a single piston-cylinder arrangement. First, compression to high pressures will result in prohibitively high temperatures in the cylinder; e.g., isentropic compression from 1 atm and 70°F to 1500 psia produces a temperature of 1530°F. Second, if the cylinder is large enough to handle the volume of air at the low pressure the proportionate clearance volume at tdc will be very small with serious leakage problems encountered around the piston rings. These difficulties can be alleviated by performing the compression in stages as indicated in Fig. 12-18. The high-temperature problem is solved by cooling the gas between stages, and the difficulty with leakage at high pressure is minimized because the second-stage piston cylinder can be smaller since the input volume of gas to it is reduced. Of course, many stages of compression can be employed if very high outlet pressures are to be attained.

The interstage cooler is appropriately called an *intercooler*, and for the ideal arrangement of Fig. 12-18 we assume that the gas is cooled to the initial

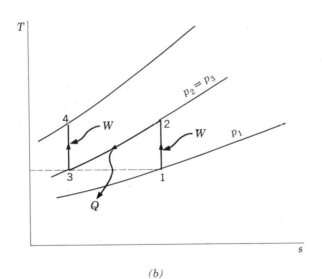

FIG. 12-18 *Intercooler arrangement for air compressor. (a) Schematic; (b) temperature-entropy diagram.*

inlet temperature before admittance to the next stage. In Fig. 12-18 this
means that $T_1 = T_3$ for an ideal intercooler.

In a multistage compressor it is desirable to select the interstage pressure so that the total work input is minimized. Using the nomenclature of
Fig. 12-18, the stage works for isentropic compression are

$$W_{12} = h_2 - h_1 = c_p(T_2 - T_1) = c_pT_1\left[\left(\frac{p_2}{p_1}\right)^{(\gamma-1)/\gamma} - 1\right]$$

$$W_{34} = h_4 - h_3 = c_p(T_4 - T_3) = c_pT_3\left[\left(\frac{p_4}{p_3}\right)^{(\gamma-1)/\gamma} - 1\right]$$

The total work is the sum of these stage works. We wish to find the interstage
pressure p_2 which minimizes this total work. Thus, we form the derivative

$$\frac{\partial W_{total}}{\partial p_2} = 0 = \frac{\partial}{\partial p_2} \left\{ c_p T_1 \left[\left(\frac{p_2}{p_1}\right)^{(\gamma-1)/\gamma} + \left(\frac{p_4}{p_2}\right)^{(\gamma-1)/\gamma} - 2 \right] \right\}$$

where $p_2 = p_3$ and $T_1 = T_3$ for ideal intercooling. Performing the indicated differentiation we obtain

$$\frac{p_2}{p_1} = \frac{p_4}{p_3} \tag{12-8}$$

or, the pressure ratio is the same across each stage. We have developed this relation for a two-stage compressor but a further analysis would show that the result applies to multiple stages in general; i.e., minimum work is obtained when the pressure ratio is the same across each stage, provided isentropic compression and ideal intercooling are experienced.

EXAMPLE 12-6

A reciprocating compressor receives atmospheric air at 14.7 psia and 70°F and discharges at 50 psia. The clearance volume is 5 percent of the piston displacement, and the compression and expansion processes are assumed to occur isentropically. Calculate:

(a) The piston displacement in cubic feet per minute necessary to compress 100 ft³/min of air at inlet conditions
(b) The *volumetric efficiency* defined as

$$\eta_{vol} = \frac{\text{actual mass of gas compressed}}{\begin{array}{c}\text{mass of gas occupying piston displacement}\\ \text{at inlet pressure and temperature}\end{array}}$$

(c) The horsepower required

SOLUTION   Figure 12-14 applies to this problem. The *piston displacement* is the volume swept out by the piston as it moves from tdc to bdc. Thus

$$\text{P.D.} = V_2 - V_4 \tag{a}$$

V_4 is the clearance volume, so that

$$V_4 = 0.05(V_2 - V_4) \tag{b}$$

The volume of free air which is pulled into the cylinder is $V_2 - V_1$, so that

$$V_2 - V_1 = 100 \text{ ft}^3/\text{min} \qquad (2832 \text{ liters/min}) \tag{c}$$

Since the compression and expansion processes are isentropic, we have

$$\frac{V_1}{V_4} = \left(\frac{p_4}{p_1}\right)^{1/\gamma} = \left(\frac{50}{14.7}\right)^{1/1.4} = 2.395 \tag{d}$$

$$\frac{V_2}{V_3} = \left(\frac{p_3}{p_2}\right)^{1/\gamma} = \left(\frac{50}{14.7}\right)^{1/1.4} = 2.395 \tag{e}$$

Combining Eqs. (b), (c), and (d),

$$
\begin{aligned}
V_4 &= 5.38 \text{ ft}^3/\text{min} & (152.3 \text{ liters/min}) \\
V_1 &= 12.9 \text{ ft}^3/\text{min} & (365.3 \text{ liters/min}) \\
V_2 &= 112.9 \text{ ft}^3/\text{min} & (3196.8 \text{ liters/min})
\end{aligned}
$$

The piston displacement is, thus,

$$\text{P.D.} = V_2 - V_4 = 107.5 \text{ ft}^3/\text{min} \qquad (3044 \text{ liters/min})$$

The volumetric efficiency may be calculated as

$$\eta_{vol} = \frac{p_1(V_2 - V_1)/RT_1}{p_1(V_2 - V_4)/RT_1} = \frac{V_2 - V_1}{V_2 - V_4} = \frac{100}{107.5}$$

$$= 93 \text{ percent}$$

Of course, if there were no clearance volume the volumetric efficiency would be 100 percent, but from a practical standpoint it is not possible to design a compressor with zero clearance. We may calculate the work by integrating the area on the p-V diagram. For a closed cycle we may either compute

$$\oint p\, dV$$

or, as is a bit easier in this case,

$$W = \oint V\, dp$$

For the isentropic processes we have

$$
\begin{aligned}
\int_2^3 V\, dp &= \frac{\gamma}{1 - \gamma}(p_3 V_3 - p_2 V_2) \\
&= \frac{\gamma p_2 V_2}{1 - \gamma}\left[\left(\frac{p_3}{p_2}\right)^{(\gamma-1)/\gamma} - 1\right] \\
&= \frac{(1.4)(14.7)(144)(112.9)}{1 - 1.4}\left[\left(\frac{50}{14.7}\right)^{0.4/1.4} - 1\right] \\
&= -3.51 \times 10^5 \text{ ft·lbf/min} \qquad (-7.93 \text{ kW}) \tag{f}
\end{aligned}
$$

$$\int_4^1 V\,dp = \frac{\gamma p_4 V_4}{1-\gamma}\left[\left(\frac{p_1}{p_4}\right)^{(\gamma-1)/\gamma}-1\right]$$

$$= \frac{(1.4)(50)(144)(5.38)}{1-1.4}\left[\left(\frac{14.7}{50}\right)^{0.4/1.4}-1\right]$$

$$= 4.0\times 10^4 \text{ ft·lbf/min}\qquad(0.904\text{ kW})\tag{g}$$

For the constant-pressure processes

$$\int_3^4 V\,dp = \int_1^2 V\,dp = 0\tag{h}$$

Thus

$$W_{\text{total}} = -3.51\times 10^5 + 4.0\times 10^4$$
$$= -3.11\times 10^5 \text{ ft·lbf/min}$$

or

$$W_{\text{total}} = -9.42\text{ hp}\qquad(-7.024\text{ kW})$$

This is the work done on the face of the piston by the gas. The work *input* to the compressor would be the negative of this, or +9.42 hp. We could also recognize that the compressor is a pulsating steady-flow device, and for an adiabatic process the rate of doing work could be calculated from

$$W = \dot m(h_2 - h_1)\tag{i}$$

where now subscript 1 refers to inlet and subscript 2 to outlet conditions. For ideal-gas behavior with constant specific heats

$$W = \dot m c_p(T_2 - T_1)\tag{j}$$

We have, for isentropic compression,

$$T_2 = T_1\left(\frac{p_2}{p_1}\right)^{(\gamma-1)/\gamma} = (530)\left(\frac{50}{14.7}\right)^{0.286} = 752°\text{R}\qquad(418\text{ K})$$

The mass flow rate is calculated from the volumetric flow at inlet

$$\dot m = \frac{p_1\dot V_1}{RT_1} = \frac{(14.7)(144)(100)}{(53.35)(530)} = 7.486\text{ lbm/min}\qquad(3.396\text{ kg/min})$$

The work input is then obtained from Eq. (j) as

$$W = (7.486)(0.24)(752 - 530) = 399 \text{ Btu/min}$$
$$= 9.42 \text{ hp} \qquad (7.024 \text{ kW})$$

EXAMPLE 12-7 Two-stage compressor

A two-stage steady-flow compressor compresses air from 14.7 psia and 70°F to 100 psia. An ideal intercooler is used along with optimum staging. Calculate the horsepower required if the mass flow of air is 10 lbm/min. Also calculate the heat removed in the intercooler. Assume constant specific heats. Compare with the work required for a single-stage compressor.

SOLUTION   For this problem the schematic diagrams of Fig. 12-18 apply. We have

$$T_1 = T_3 = 70°F = 530°R$$
$$p_2 = p_3 \qquad p_1 = 14.7 \text{ psia} \qquad p_4 = 100 \text{ psia}$$

For ideal staging

$$\frac{p_2}{p_1} = \frac{p_4}{p_3}$$

or

$$p_2^2 = p_1 p_4$$
$$p_2 = 38.3 \text{ psia}$$

For isentropic compressions

$$\frac{T_2}{T_1} = \left(\frac{p_2}{p_1}\right)^{(\gamma-1)/\gamma} \qquad \frac{T_4}{T_3} = \left(\frac{p_4}{p_3}\right)^{(\gamma-1)/\gamma}$$

Thus

$$T_2 = (530)\left(\frac{38.3}{14.7}\right)^{0.286}$$
$$= 697°R \qquad (387.2 \text{ K})$$

Since the pressure ratios are equal in each stage, we also have

$$T_4 = 697°R$$

The work input per pound mass of air for the two stages is

$$W = h_2 - h_1 + h_4 - h_3$$
$$= (0.24)(697 - 530) + (0.24)(697 - 530)$$
$$= 80.2 \text{ Btu/lbm} \qquad (186.5 \text{ kJ/kg})$$

The total work is, therefore, for a flow of 10 lbm/min,

$$W_{\text{total}} = 802 \text{ Btu/min} \qquad (14.1 \text{ kW})$$

Since 1 hp = 2545 Btu/h,

$$\text{hp} = \frac{(802)(60)}{2545}$$
$$= 18.9$$

The heat removed in the intercooler is

$$Q = \dot{m}_a c_p (T_2 - T_3)$$
$$= (10)(0.24)(697 - 530)$$
$$= 401 \text{ Btu/min} \qquad (7.05 \text{ kW})$$

For a single-stage isentropic compression the exit temperature from the compressor would be

$$T_{4'} = T_1 \left(\frac{p_4}{p_1}\right)^{(\gamma-1)/\gamma}$$
$$= 530 \left(\frac{100}{14.7}\right)^{0.286}$$
$$= 917°R \qquad (509 \text{ K})$$

In this circumstance the work would be

$$W = \dot{m}_a (h_{4'} - h_1)$$
$$= (10)(0.24)(917 - 530)$$
$$= 928 \text{ Btu/min} \qquad (16.31 \text{ kW})$$

The advantage of the intercooler and multistaging is obvious; it results in a reduction in work of 13.5 percent (928 to 802 Btu/min). In a practical situation the reduced power requirements must be balanced against the increased cost of an intercooler-multistage arrangement.

12-7 INTERNAL COMBUSTION ENGINE CYCLES

In this section we shall discuss several power cycles which form a theoretical basis for spark-ignition and compression-ignition (diesel) engines. These cycles are employed in *reciprocating* engines. Cycles using a steady flow of fluid are discussed in connection with gas turbines in Sec. 12-12. From a practical standpoint, it is worthwhile to discuss the meaning of four-stroke and two-stroke cycles. Consider the two typical reciprocating engine configurations shown in Fig. 12-19. Let us first examine the four-cycle engine. When the piston is at tdc, the intake valve *I* opens and a charge of fuel and air is drawn into the cylinder as the piston moves downward (intake stroke). At bdc the intake valve closes and the piston moves upward to compress the mixture (compression stroke). At tdc the ignition device fires, thereby burning the fuel and forcing the piston downward again (power stroke). At bdc, the exhaust valve *E* opens and the piston moves upward to dispel the products of combustion (exhaust stroke). Finally, when the piston reaches tdc the exhaust valve closes and the intake valve opens, thereby starting the cycle once again. It is to be noted that *two* complete revolutions of the crankshaft are involved for each power stroke and firing of the ignition device.

The two-cycle engine combines functions so that a power stroke is obtained for each revolution of the crankshaft. Let us start the cycle at tdc just as the ignition device is fired. The piston moves downward in the power stroke. In the two-cycle engine the crankcase is sealed and serves as a storage reservoir for fuel-air mixture. Thus, as the piston moves downward in the power stroke, the fuel-air mixture is compressed in the crankcase. As the piston passes the exhaust and intake ports, the exhaust gases move outward and the compressed fuel-air mixture shoots into the cylinder. Thus this stroke is a combined power-exhaust-intake stroke. Upon completion of

(a)   (b)

FIG. 12-19 *Basic reciprocating internal-combustion-engine arrangements.* (a) *Four-cycle;* (b) *two-cycle.*

these processes, the piston moves upward again in a compression stroke until at tdc the spark plug fires to start the cycle over again.

In the two-cycle engine the unavoidable mixing of inlet fuel-air mixture with exhaust gases and even some loss of fuel out the exhaust port makes the engine generally less efficient than its four-cycle counterpart. But the fact that a power stroke is obtained with each revolution of the crankshaft means that two-cycle engines are usually smaller and lighter in weight for a given power output. For this reason we normally see two-cycle engines employed in applications where weight and size are important factors, viz., outboard motors, chain saws, and motorcycles.

There are several factors which cause problems in practical internal-combustion engines: incomplete exhaust, incomplete combustion, mixing of exhaust and intake gases, improper valve timing, too rapid combustion, too slow combustion, etc. However, a consideration of these factors is beyond the scope of our discussion.

In the following we shall investigate several theoretical cycles which may be used to describe internal-combustion-engine behavior. For purposes of analysis we shall assume that the working fluid is air, in which case the cycles are called *air-standard cycles*.

12-8 THE OTTO CYCLE

Consider first the air-standard Otto cycle shown in Fig. 12-20. The compression process is represented by 1-2, followed by a constant-volume heat addition 2-3, which is an approximation to the rapid combustion occurring in a spark-ignition engine. The power stroke is then accomplished in 3-4, fol-

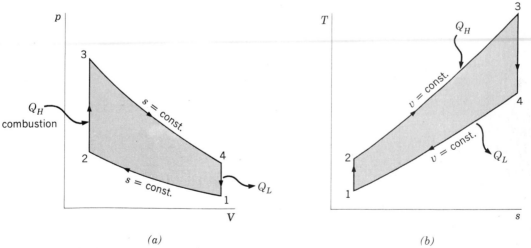

FIG. 12-20 *Cycle diagrams for air-standard Otto cycle.* (a) *p-V diagram;* (b) *T-s diagram.*

lowed by a constant-volume heat rejection in 4-1, which corresponds approximately to the dispelling of combustion products through the exhaust valve. The thermal efficiency of the cycle is calculated from

$$\eta = \frac{Q_H - Q_L}{Q_H} = 1 - \frac{Q_L}{Q_H} = 1 - \frac{mc_v(T_4 - T_1)}{mc_v(T_3 - T_2)} \qquad (12\text{-}9)$$

In the isentropic processes

$$\frac{T_2}{T_1} = \left(\frac{v_1}{v_2}\right)^{\gamma-1} = \left(\frac{v_4}{v_3}\right)^{\gamma-1} = \frac{T_3}{T_4} \qquad (12\text{-}10)$$

Making appropriate substitutions for temperatures in Eq. (12-9), the efficiency becomes

$$\eta = 1 - \frac{T_1}{T_2} = 1 - \left(\frac{v_1}{v_2}\right)^{1-\gamma} \qquad (12\text{-}11)$$

The ratio v_1/v_2 is called the *compression ratio r*, so that the efficiency may be expressed as

$$\eta = 1 - r^{1-\gamma} \qquad (12\text{-}12)$$

It is clear that the thermal efficiency of the Otto cycle increases with increasing compression ratio.

A higher compression ratio means a higher temperature at which heat addition is started, and we once again find that an increase in the temperature for heat addition produces an increase in efficiency. For many years automotive engine manufacturers used this principle to increase the efficiencies of their engines through successive increases in compression ratio without large increases in physical size of the engine. High compression ratios are not without disadvantage, however. The higher pressures and temperatures encountered at tdc cause very rapid burning of gasoline fuels, so that engine knock results from detonation waves in the cylinder. To cure these ills, more expensive processing of the fuel must be performed to increase the burning times and antiknock characteristics.

In particular, tetraethyl lead was used as a fuel additive for this purpose. To alleviate air pollution caused by automobiles, catalytic converters are now required on the exhaust system of most passenger vehicles. The use of lead fuel additives renders the converters inoperative so manufacturers must use unleaded gasolines, and correspondingly produce engines with lower compression ratios. This is an example of the balance which must be drawn between environmental protection and efficiency in arriving at a proper engineering design. Internal combustion engines are major sources of air pollution, and stringent efforts are made to minimize the discharge of pollutants from them. Cost of pollution-control equipment and increased fuel consumption caused by its installation are important factors which must be considered in final designs.

EXAMPLE 12-8 Efficiency of Otto cycle

An air-standard Otto cycle operates between temperature limits of 70 and 1500°F. Calculate the thermal efficiency of the cycle for compression ratios of (a) 10:1, (b) 12:1, and (c) 15:1. Compare with the appropriate Carnot efficiency.

SOLUTION   We immediately calculate the efficiencies with Eq. (12-12), using $\gamma = 1.4$.

$$\eta = 1 - r^{1-\gamma}$$

At $r = 10$,

$$\eta = 1 - (10)^{-0.4} = 60.2 \text{ percent}$$

At $r = 12$,

$$\eta = 1 - (12)^{-0.4} = 63 \text{ percent}$$

At $r = 15$

$$\eta = 1 - (15)^{-0.4} = 66.2 \text{ percent}$$

These efficiencies could have been read directly from Fig. 12-22. A Carnot cycle operating between the given temperature limits would have an efficiency of

$$\eta_{\text{Carnot}} = 1 - \frac{T_L}{T_H}$$
$$= 1 - \frac{530}{1960}$$
$$= 73 \text{ percent}$$

12-9   THE DIESEL CYCLE

Now consider the air-standard diesel cycle shown in Fig. 12-21. The diesel cycle is similar to the Otto cycle in that isentropic compression is assumed as well as constant-volume heat rejection of the exhaust gases from 4-1. It differs from the Otto cycle in that heat addition occurs in a constant-pressure process 2-3. This means that ignition and combustion form part of

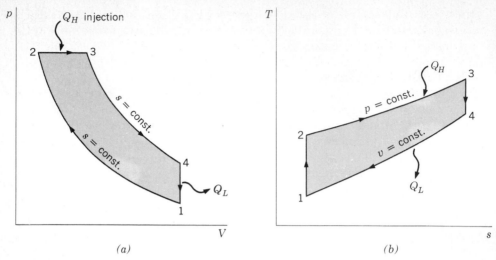

FIG. 12-21 *Cycle diagrams for air-standard diesel cycle. (a) p-V diagram; (b) T-s diagram.*

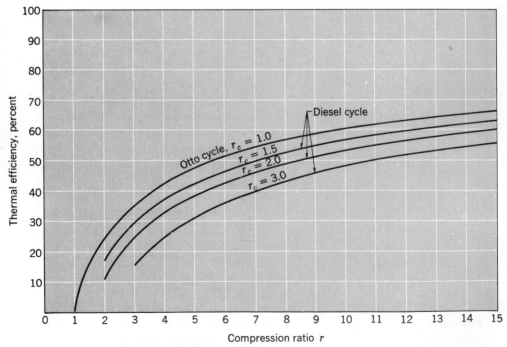

FIG. 12-22 *Comparison of efficiencies of air-standard Otto and diesel cycles for $\gamma = 1.4$.*

the total power stroke 2-3-4. The diesel cycle is the basis of practical compression-ignition engines. In these engines air is compressed to high pressures (point 2) and correspondingly high temperatures. A fuel with a low ignition point is then injected into the cylinder at tdc. The high temperature of compression is sufficient to ignite the fuel, and the burning continues from 2 to 3 as more fuel is injected into the cylinder. The hot products of combustion then expand in the cylinder in process 3-4. In practical diesel engines the injection process is a crucial matter, and considerable engineering effort is devoted to the design of injector mechanisms.

The efficiency of the diesel cycle is given by

$$\eta = 1 - \frac{Q_L}{Q_H} = 1 - \frac{c_v(T_4 - T_1)}{c_p(T_3 - T_2)} \tag{12-13}$$

Using the isentropic relations

$$\frac{T_2}{T_1} = \left(\frac{v_1}{v_2}\right)^{\gamma-1} \qquad \frac{T_3}{T_4} = \left(\frac{v_4}{v_3}\right)^{\gamma-1}$$

the efficiency expression becomes

$$\eta = 1 - \frac{r^{1-\gamma}(r_c^{\gamma} - 1)}{\gamma(r_c - 1)} \tag{12-14}$$

where $r = v_1/v_2$ is the compression ratio and $r_c = v_3/v_2$ is called the *cut-off ratio*.

Efficiencies of the Otto and diesel cycles are plotted in Fig. 12-22. It is clear that, for a given compression ratio, the Otto cycle is more efficient than the diesel cycle. The physical reason for this behavior is that the constant-volume heat-addition process results in heat being added at a higher temperature than in a constant-pressure process for the same compression ratio. This does not mean, however, that an Otto-cycle or spark-ignition engine is always to be preferred over a diesel. In practice, the diesel can operate at much higher compression ratios than a spark-ignition engine because of detonation and preignition problems encountered with compression of fuel-air mixtures to high pressures. If a fuel-air mixture is compressed to pressures which are too high, it may ignite before the piston reaches tdc and, consequently, reduce the power output of the engine. The diesel does not encounter such problems because only air is compressed and fuel does not enter the cylinder until the appropriate time for combustion.

The diesel has a further advantage over its spark-ignition counterpart in that diesel fuels are usually much less expensive than fuels for spark-ignition engines. But, one does not select an Otto (spark-ignition) or diesel cycle strictly on the basis of the thermodynamic information given above. Consideration must be given to the specific design of each type of engine. Most diesel engines employ a carefully metered injection system which injects just the right amount of fuel into the cylinder. Of course, the time available for injection, and thus the injection pressure, is strongly dependent on the engine speed. It turns out that if optimum performance is to be achieved the diesel engine should operate over a fairly narrow speed range. In contrast, a spark-ignition engine can operate efficiently over a very wide speed range with a properly designed carburetion system. For these reasons we usually find diesel engines used in applications where fairly constant speeds are required, as in locomotives, large trucks, stationary electric generators, and the like, while spark-ignition engines are used for wide-speed ranges of personal automobiles.

We may expect that electronic microprocessor control of the injection process will be employed extensively with diesels of the future, thereby enabling an incumbent improvement in variable speed performance.

EXAMPLE 12-9   Diesel cycle

Calculate the work output per pound mass for an air-standard diesel cycle with $r = 15$ and $r_c = 2.0$. The inlet conditions are 14.7 psia and 70°F. Also calculate the efficiency.

SOLUTION   Figure 12-21 gives the process and cycle schematic for the diesel cycle. We have

$$T_1 = 70°F = 530°R$$
$$p_1 = 14.7 \text{ psia}$$

From the isentropic relations we have

$$\frac{T_2}{T_1} = \left(\frac{v_1}{v_2}\right)^{\gamma-1}$$

$$T_2 = (530)(15)^{0.4} = 1565°R \qquad (869 \text{ K})$$

For the constant-pressure heat-addition process

$$\frac{v_3}{v_2} = \frac{T_3}{T_2}$$

so that

$$T_3 = (1565)(2.0) = 3130°R \qquad (1739 \text{ K})$$

The temperature at the end of the power stroke is determined from

$$\frac{T_4}{T_3} = \left(\frac{v_3}{v_4}\right)^{\gamma-1}$$

But $v_3 = 2v_2$, so that

$$\frac{v_3}{v_4} = \frac{2v_2}{v_4} = \frac{2v_2}{v_1} = \frac{2}{15}$$

Thus

$$T_4 = (3130)(\tfrac{2}{15})^{0.4}$$
$$= 1395°R \qquad (775 \text{ K})$$

The heat added and rejected may now be calculated:

$$Q_H = c_p(T_3 - T_2)$$
$$= (0.24)(3130 - 1565) = 375 \text{ Btu/lbm} \qquad (872.2 \text{ kJ/kg})$$
$$Q_L = c_v(T_4 - T_1)$$
$$= (0.1715)(1395 - 530) = 148 \text{ Btu/lbm} \qquad (344.2 \text{ kJ/kg})$$

The work output per pound mass is, therefore,

$$W = Q_H - Q_L = 375 - 148$$
$$= 227 \text{ Btu/lbm} \qquad (528 \text{ kJ/kg})$$

The efficiency is

$$\eta = \frac{W}{Q_H} = \frac{227}{375} = 60.5 \text{ percent}$$

This value may be checked by referring to Fig. 12-22.

12-10 THE WANKEL ENGINE

A rotary-type internal combustion engine with increasing numbers of applications is the Wankel engine shown in Fig. 12-23. In this arrangement a three-lobed rotor is mounted eccentrically on the shaft which in turn is connected to a planetary gear train so that the rotor turns at one-third the crankshaft speed. The internal surface of the housing is a special trochoid shape generated by the radius R. As the rotor moves clockwise, simultaneous processes occur in the three chambers as follows: As the rotor moves by the intake port a fresh charge of fuel-air mixture is drawn in and with further movement the intake port is closed and the mixture is compressed.

FIG. 12-23 *Schematic of a Wankel rotary combustion engine. Diagram according to Wood* [11].

When the tip of the rotor arrives at the approximate location of the spark plug, combustion and power expansion occur in the lower right chamber. Finally, the products of combustion are forced out the discharge port in the lower left chamber. The device is very smooth in operation with lower vibration and noise levels than its Otto cycle counterpart. It has the additional advantage of very lightweight construction for a given power output. The main practical design problems are those associated with rubbing and end seals for the rotor.

12-11   THE STIRLING CYCLE

We shall now consider a cycle which offers opportunity for high efficiency as well as reduced emissions from combustion products. It is called the Stirling cycle and is depicted on the *p-V* and *T-s* diagrams of Fig. 12-24.

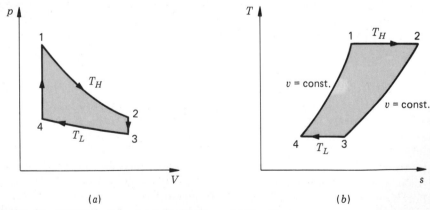

FIG. 12-24 *Stirling cycle.* (a) *p-V diagram;* (b) *T-s diagram.*

FIG. 12-25 *Stirling cycle with regenerator.*

The working fluid is a gas, and we assume reversible processes for an ideal cycle. In process 1-2 heat is added at the constant temperature T_H. Process 2-3 is a constant-volume heat rejection; process 3-4 is a constant-temperature heat rejection at T_L, and finally, 4-1 is a constant-volume heat addition. We assume a closed system and ideal-gas behavior. Because processes 2-3 and 4-1 occur between the same temperature limits the heat rejected in 2-3 must be numerically equal to the heat added in 4-1. This fact immediately suggests the use of regeneration.

If we can transfer the heat from 2-3 over to process 4-1, internal to the cycle, then we will have a reversible cycle with heat addition and rejection at constant temperatures. Such a cycle will have an efficiency equal to that of a Carnot cycle, or

$$\eta = 1 - \frac{T_L}{T_H} \tag{12-15}$$

The Stirling cycle with regeneration is illustrated in Fig. 12-25. The cycle can operate as an *external* combustion device and has been investigated as a means to reduce exhaust emissions in comparison to the internal combustion engine. There are problems with the cycle. To achieve acceptable power to weight ratios, very high pressures must be developed and the regenerator involves stringent heat-transfer design requirements. Some devices have been developed using air, helium, and hydrogen as working fluids.

12-12 GAS–TURBINE CYCLES

The thermodynamic cycle employed with gas turbines is called the *Brayton cycle* and is illustrated in Fig. 12-26. In the *open* cycle the compressor raises the pressure of the air, and heat is added at the high pressure by burning a fuel with the air. The high-temperature products of combustion are then expanded in the turbine to produce a work output. Part of the turbine work is

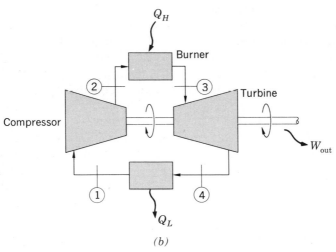

FIG. 12-26 *Basic Brayton cycle for gas turbines.* (a) *Open-cycle;* (b) *closed-cycle.*

used to drive the compressor, and the remainder is available for driving external mechanisms. The *closed* gas-turbine cycle recirculates the working fluid while heat is added at the high pressure and rejected at the low pressure. Temperature-entropy diagrams for the ideal cycles are shown in Fig. 12-27.

In the *air-standard* Brayton cycle the analysis is performed by assuming that the working fluid is air in both the compressor and turbine. In an actual cycle the mass flow in the turbine would be somewhat greater than the mass flow in the compressor because of the added mass of the fuel. In addition, the properties of the products of combustion may be different from those of air; however, most gas turbines use a substantial amount of excess air, so that an analysis of the air-standard cycle offers reasonable estimates of the performance of actual cycles.

The gas-turbine cycle is a steady-flow device and, as such, the energy

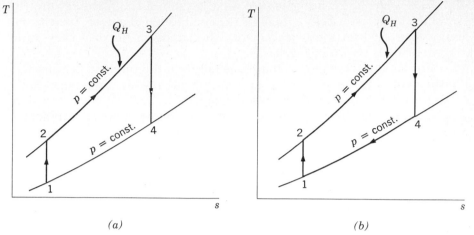

FIG. 12-27 *Temperature-entropy diagram for Brayton cycles. (a) Open-cycle; (b) closed-cycle.*

quantities may be determined with enthalpy differences. Both the compressor and turbine work processes are assumed adiabatic so the net work output of the cycle is the work output of the turbine less the work input to the compressor, or $(h_3 - h_4) - (h_2 - h_1)$. This is the shaft work which the cycle can deliver to other things. The burner is assumed to operate as a constant-pressure heating device so the net heat input is $h_3 - h_2$. Accordingly, the thermal efficiency for the ideal air-standard cycle may be calculated as:

$$\eta = \frac{W_{net}}{Q_{in}} = \frac{(h_3 - h_4) - (h_2 - h_1)}{h_3 - h_2} \tag{12-16}$$

where the subscript nomenclature refers to Fig. 12-26. Expressing the enthalpy differences in terms of temperature differences and using the isentropic relations.

$$\frac{p_2}{p_1} = \left(\frac{T_2}{T_1}\right)^{\gamma/(\gamma-1)} = \frac{p_3}{p_4} = \left(\frac{T_3}{T_4}\right)^{\gamma/(\gamma-1)} \tag{12-17}$$

gives for the efficiency

$$\eta = 1 - \left(\frac{p_2}{p_1}\right)^{(1-\gamma)/\gamma} \tag{12-18}$$

Most gas turbines operate on the open cycle, and the exhaust gases from the turbine are dissipated to the surroundings. In the simple cycle shown in the foregoing this represents a substantial loss of high-energy gas.

In a *regenerator* these hot gases are used to preheat the high-pressure air from the compressor before it enters the combustion chamber, thereby requiring less fuel for the overall cycle. The arrangement for a cycle with regeneration is shown in Fig. 12-28.

For an ideal regenerator the temperature of the products leaving the regenerator is just equal to the incoming high-pressure air temperature. This means that in an ideal case the air would be heated to point 5' (equal to T_4) and the products would be cooled to point 6' (equal to T_2). An actual regenerator does not accomplish such an ideal heat exchange, and the efficiency of the regenerator is normally expressed by

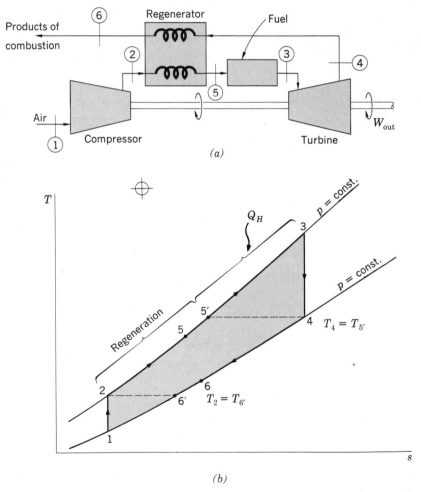

(a)

(b)

FIG. 12-28 *Gas-turbine cycle with ideal regeneration.* (a) *Schematic;* (b) *T-s diagram.*

$$\eta_{\text{reg}} = \frac{\text{actual heat transfer}}{\text{maximum heat transfer}} = \frac{h_5 - h_2}{h_{5'} - h_2} \tag{12-19}$$

For constant specific heats this becomes

$$\eta_{\text{reg}} = \frac{T_5 - T_2}{T_{5'} - T_2} \tag{12-20}$$

Actual gas-turbine cycles differ from theoretical cycles in that compressor and turbine efficiencies must be taken into account. The T-s diagram for an actual cycle would resemble the one shown in Fig. 12-29.

It is possible to improve the efficiency of the gas-turbine cycle by using multiple-stage compression with intercooling as well as multiple-stage turbines with reheat. A schematic for an ideal cycle employing two stages of compression and expansion is shown in Fig. 12-30. The efficiency for this cycle would be expressed by

$$\eta = \frac{\text{net work output}}{\text{net heat input}}$$
$$= \frac{(h_5 - h_6) + (h_7 - h_8) - (h_4 - h_3) - (h_2 - h_1)}{(h_5 - h_4) + (h_7 - h_6)} \tag{12-21}$$

Obviously, it is possible to incorporate a regenerator along with the intercooler and reheat arrangement.

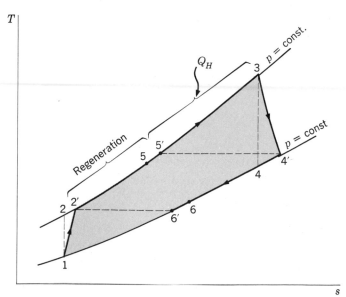

FIG. 12-29 *Effect of turbine and compressor efficiencies on T-s diagram of gas-turbine cycle with regeneration.*

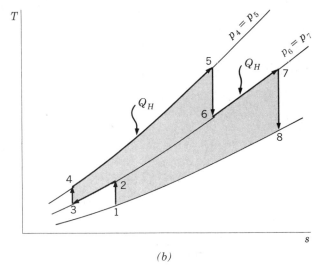

FIG. 12-30   *Two-stage gas-turbine cycle with intercooling and reheat. (a) Schematic; (b) T-s diagram.*

At this point it is well to remind the reader that our discussion has been limited to air-standard cycles and some of the ideal-efficiency relations have involved the further assumption of constant specific heats. The examples are based on these assumptions. A first refinement in the analysis may be made by employing the air tables to compute the various enthalpies. Still further refinement may be made by considering actual fuel-air mixtures and the corresponding products of combustion. In this case we would employ the appropriate relations of Chap. 11 to compute flame temperature, etc. Of course, we are no longer dealing with an air-standard cycle when actual combustion processes are considered, but actual gas-cycle processes are beyond the scope of our discussion. Despite its simplicity, air-standard cycle

analysis does give a very good approximation to the behavior and performance of real cycles.

EXAMPLE 12-10 Brayton cycle efficiency
An air-standard Brayton cycle receives air at 14.7 psia and 70°F. The
upper pressure and temperature limits of the cycle are 60 psia and
1500°F, respectively. The compressor efficiency is 85 percent, and the
turbine efficiency is 90 percent. Calculate the thermal efficiency of the
cycle, assuming constant specific heats.

SOLUTION We shall employ the diagram in Fig. 12-27*a* for this problem,
except that the actual compression and expansion processes are nonisentropic. From the problem statement,

$$p_1 = p_4 = 14.7 \text{ psia} \qquad T_1 = 70°F = 530°R$$
$$p_2 = p_3 = 60 \text{ psia} \qquad T_3 = 1500°F = 1960°R$$

For isentropic processes we have

$$T_2 = T_1 \left(\frac{p_2}{p_1}\right)^{(\gamma-1)/\gamma}$$

$$= (530) \left(\frac{60}{14.7}\right)^{0.286}$$

$$= 792°R \qquad (440 \text{ K})$$

$$T_4 = T_3 \left(\frac{p_4}{p_3}\right)^{(\gamma-1)/\gamma}$$

$$= (1960) \left(\frac{14.7}{60}\right)^{0.286}$$

$$= 1310°R \qquad (728 \text{ K})$$

Using the given efficiencies, with constant specific heats,

$$\eta_{comp} = \frac{W_{theo}}{W_{act}} = \frac{T_2 - T_1}{T_{2'} - T_1} = 0.85$$

where $T_{2'}$ is the actual exit temperature from the compressor. Inserting the
proper numerical values,

$$\frac{792 - 530}{T_{2'} - 530} = 0.85 \qquad T_{2'} = 838°R$$

For the turbine,

$$\eta_{turb} = 0.90 = \frac{W_{act}}{W_{theo}} = \frac{T_3 - T_{4'}}{T_3 - T_4}$$

$$0.90 = \frac{1960 - T_{4'}}{1960 - 1310} \qquad T_{4'} = 1375°R \qquad (764 \text{ K})$$

The thermal efficiency of the cycle is now calculated as

$$\eta_{th} = \frac{W_{act}}{Q_H}$$

$$= \frac{(h_3 - h_{4'}) - (h_{2'} - h_1)}{h_3 - h_{2'}}$$

$$= \frac{(0.24)(1960 - 1375) - (0.24)(838 - 530)}{(0.24)(1960 - 838)}$$

$$= 24.7 \text{ percent}$$

EXAMPLE 12-11   Gas turbine with regenerator

A regenerator having an efficiency of 83 percent is installed on the gas-turbine cycle of Example 12-10. Calculate the thermal efficiency of the cycle under these conditions. Make the calculation using constant specific heats and also using the air tables.

SOLUTION   We use the schematic diagram of Fig. 12-29 for this problem. From Example 12-10,

$$T_1 = 530°R$$
$$T_{2'} = 838°R = T_{6'}$$
$$T_3 = 1960°R$$
$$T_{4'} = 1375°R = T_{5'}$$

Now, from Eq. (12-20) for the regenerator efficiency,

$$\eta_{reg} = \frac{T_5 - T_{2'}}{T_{5'} - T_{2'}}$$

$$0.83 = \frac{T_5 - 838}{1375 - 838}$$

$$T_5 = 1284°R \qquad (713 \text{ K})$$

The thermal efficiency of the cycle is now written as

$$\eta_{th} = \frac{(h_3 - h_{4'}) - (h_{2'} - h_1)}{h_3 - h_5}$$

$$= \frac{(0.24)(1960 - 1375) - (0.24)(838 - 530)}{(0.24)(1960 - 1284)}$$

$$= 41 \text{ percent}$$

We now must repeat the calculations using the air tables which take into account variable specific heats. We still have

$$T_1 = 530°R \qquad T_3 = 1960°R$$

Using the air tables for the ideal isentropic processes

$$\frac{p_2}{p_1} = \frac{P_{r_2}}{P_{r_1}} \qquad P_{r_1} = 1.3004 \text{ at } 530°R$$

$$P_{r_2} = \left(\frac{60}{14.7}\right)(1.3004) = 5.308$$

Therefore $T_2 = 790.8°R$. Also,

$$\frac{p_4}{p_3} = \frac{P_{r_4}}{P_{r_3}} \qquad P_{r_3} = 160.37 \text{ at } 1960°R$$

$$P_{r_4} = \left(\frac{14.7}{60}\right)(160.37) = 39.291$$

and $T_4 = 1388.1°R$. We now can find the corresponding enthalpies from the air tables (Table A-17).

$$h_1 = 126.67 \text{ Btu/lbm} \qquad h_2 = 189.57 \text{ Btu/lbm}$$
$$h_3 = 493.64 \text{ Btu/lbm} \qquad h_4 = 339.79 \text{ Btu/lbm}$$

Because the specific heats are no longer constant the turbine and compressor efficiencies must be expressed in terms of enthalpies. Thus,

$$\eta_{\text{comp}} = 0.85 = \frac{h_2 - h_1}{h_{2'} - h_1}$$

Solving for $h_{2'}$ we obtain

$$h_{2'} = 200.67 \text{ Btu/lbm}$$

and for the turbine

$$\eta_{\text{turb}} = 0.90 = \frac{h_3 - h_{4'}}{h_3 - h_4}$$

so that

$$h_{4'} = 355.18 \text{ Btu/lbm}$$

The regenerator effectiveness is also written in terms of enthalpies:

$$\eta_{\text{neg}} = 0.83 = \frac{h_5 - h_{2'}}{h_{5'} - h_{2'}} \qquad \text{with } h_{5'} = h_{4'}$$

Then,

$$h_5 = (0.83)(355.18 - 200.67) + 200.67$$
$$= 328.31 \text{ Btu/lbm}$$

The relation for the thermal efficiency is the same as before.

$$
\begin{aligned}
\eta_{\text{th}} &= \frac{(h_3 - h_{4'}) - (h_{2'} - h_1)}{h_3 - h_5} \\
&= \frac{(493.64 - 355.18) - (200.67 - 126.67)}{493.64 - 328.31} \\
&= 39 \text{ percent}
\end{aligned}
$$

So, there is a small variation in the calculation depending on whether constant specific heats are assumed. It should be noted that the use of a regenerator can produce very striking increases in thermal efficiency for the gas-turbine cycle.

EXAMPLE 12-12   Closed-cycle gas turbine

Almost all current applications of gas-turbine cycles employ the open cycle with combustion of a hydrocarbon fuel providing the heat source. Should it become possible to harness the thermonuclear reaction (nuclear fusion), new techniques may be desired to extract the energy from the reaction and convert it to useful work. It is expected that the thermonuclear reaction will produce a hot plasma gas at billions of degrees temperature, and one proposed arrangement calls for initial removal of energy from the reaction with a liquid lithium coolant, which could then be used to supply heat to a closed helium gas-turbine cycle as shown in Fig. 12-31. Helium is chosen for its high specific heat and inert nature, and many of the design factors for such an installation are discussed in Ref. [9]. The upper temperature and pressure limits for our example cycle are 1700°F and 500 psia and the lower limits are 100°F and 200 psia. Ideal intercooling and interstaging are used in the two stages of compression along with a regenerator having an effectiveness of 91 percent. The turbine efficiency is 93 percent, and the compressor efficiencies are 88 percent. Calculate the thermal efficiency of the cycle and the mass flow of helium required per megawatt of power output. Also calculate the cooling water–flow requirements assuming a 10°F water temperature rise in both the intercooler and main cooler. Assume constant specific heats.

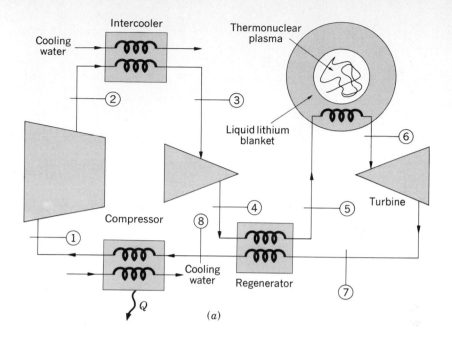

Cooling
water

Intercooler

Thermonuclear
plasma

② ③

Liquid lithium
blanket

⑥

Compressor

⑧

④ ⑤

Turbine

①

Cooling
water

Regenerator

⑦

Q

(a)

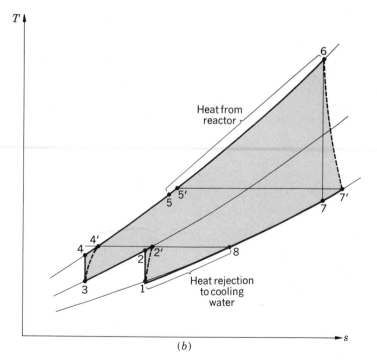

T

6

Heat from
reactor

5 5'

7'
7

4 4'

2 2'

8

3 1

Heat rejection
to cooling
water

s

(b)

FIG. 12-31 *A closed helium gas-turbine cycle for power production with a thermonuclear reactor. (a) Schematic; (b) T-s diagram.*

SOLUTION Using the nomenclature of Fig. 12-31*a*

$$T_1 = T_3 = 100°F = 560°R \qquad T_6 = 1700°F = 2160°R$$
$$p_1 = 200 \text{ psia} = p_8 = p_7 \qquad p_4 = p_5 = p_6 = 500 \text{ psia}$$

For ideal interstaging, with $p_2 = p_3$

$$\frac{p_2}{p_1} = \frac{p_4}{p_3}$$

$$\frac{p_2}{p_1} = (\tfrac{500}{200})^{1/2} = 1.581$$

For helium we obtain from Table 2-2

$$c_p = 1.25 \text{ Btu/lbm·°F} \qquad \gamma = 1.66$$

For ideal interstaging the isentropic temperatures at the discharge are the same because the pressure ratios and inlet temperatures are the same. Thus,

$$\frac{T_2}{T_1} = \left(\frac{p_2}{p_1}\right)^{(\gamma-1)/\gamma} = (1.581)^{0.398} = 1.120$$
$$T_2 = (560)(1.120) = 672°R = T_3 \qquad (373 \text{ K})$$

The *T-s* diagram for this cycle is shown in Fig. 12-31*b*. The actual discharge temperatures are determined by using the compressor efficiencies. For constant specific heats,

$$\eta_{comp} = \frac{T_2 - T_1}{T_{2'} - T_1} = 0.88$$

so that

$$T_{2'} = 687°R = T_{4'} \qquad (382 \text{ K})$$

For the turbine the isentropic exit temperature is

$$T_7 = T_6 \left(\frac{p_7}{p_6}\right)^{\gamma-1/\gamma} = (2160)(\tfrac{200}{500})^{0.398} = 1500°R \qquad (833 \text{ K})$$

and using the turbine efficiency, the actual exit temperature is obtained from:

$$\eta_{turb} = \frac{T_6 - T_{7'}}{T_6 - T_7} = 0.93$$
$$T_{7'} = 1546°R \quad (859\ K)$$

For constant specific heats the regenerator effectiveness is

$$\eta_{reg} = 0.91 = \frac{T_5 - T_{4'}}{T_{5'} - T_{4'}} \text{ with } T_{5'} = T_{7'}$$

so that

$$0.91 = \frac{T_5 - 687}{1546 - 687} \quad T_5 = 1469°R \quad (816\ K)$$

The overall thermal efficiency is therefore

$$\eta_{th} = \frac{\text{net work output}}{\text{heat added}} = \frac{(h_6 - h_{7'}) - (h_{2'} - h_1) - (h_{4'} - h_3)}{h_6 - h_5}$$

which becomes, for constant specific heats,

$$\eta_{th} = \frac{(1.25)[(2160 - 1546) - (687 - 560) - (687 - 560)]}{(1.25)(2160 - 1469)}$$
$$= \frac{450\ Btu/lbm}{863.8\ Btu/lbm} = 52.1 \text{ percent}$$

The net work output per pound mass is 450 Btu so the required mass flow for 1 MW is

$$\dot{m}_{HE}(450) = 10^6\ W = 3.413 \times 10^6\ Btu/h$$
$$\dot{m}_{He} = 7584\ lbm/h \cdot MW$$

This same mass flow moves through the intercooler and main cooler. The heat given up in the intercooler is

$$Q_{int} = \dot{m}_{He}c_p(T_{2'} - T_3)$$
$$= (7584)(1.25)(672 - 560) = 1.062 \times 10^6\ Btu/h \quad (311.1\ kW)$$

To calculate the heat rejected in the main cooler we must first determine T_8, the helium temperature at inlet to the cooler. From the energy balance on the regenerator, assuming constant specific heats,

$$T_{7'} - T_8 = T_5 - T_{4'}$$

so that

$$T_8 = 1546 - 1469 + 687 = 764°R \qquad (424 \text{ K})$$

The heat lost by the helium in the main cooler is then

$$\begin{aligned} Q_{\text{cooler}} &= \dot{m}_{\text{He}} c_p (T_8 - T_1) \\ &= (7584)(1.25)(764 - 560) = 1.934 \times 10^6 \text{ Btu/h} \qquad (566.6 \text{ kW}) \end{aligned}$$

The total heat rejection is therefore

$$Q_{\text{tot}} = (1.062 + 1.934) \times 10^6 = 2.996 \times 10^6 \text{ Btu/h} \qquad (877.8 \text{ kW})$$

which must be picked up by the cooling water. Recalling that $c_p \approx 1.0$ Btu/lbm·°F for liquid water, we have

$$Q_{\text{tot}} = \dot{m}_w c_w \, \Delta T_w = m_w(1.0)(10) = 2.996 \times 10^6$$

and

$$\dot{m}_w = 2.996 \times 10^5 \text{ lbm/h·MW} \qquad (1.359 \times 10^5 \text{ kg/h·MW})$$

This is a flow of about 600 gal/min for each megawatt of power output. For these working temperatures a moderate-size power plant of 500 MW would require a cooling water flow of 300 000 gal/min. Even with the very favorable efficiency figure of 52 percent, this example illustrates the enormous cooling load placed on the environment by large power plants. In many cases the load is dissipated to the atmosphere by evaporation in a cooling tower. At normal atmospheric temperature the enthalpy of vaporization is about 1000 Btu/lbm, so the 500-MW plant would require evaporation of about

$$\frac{(2.996 \times 10^6 \text{ Btu/h·MW})(500 \text{ MW})}{1000 \text{ Btu/lbm}} = 1.498 \times 10^6 \text{ lbm/h}$$

$$(6.8 \times 10^5 \text{ liters/h})$$

This is an evaporation rate of about 3000 gal/min, quite a large amount of water. In extremely calm weather such evaporation rates may cause ground fog in the surroundings adjacent to the power plant.

From our discussions and the examples, we see that gas-turbine cycles offer the advantage of high efficiency when regeneration is employed. In many cases they also are more economical on the basis of initial capital expenditure for production of power in the range of 25 to 100 MW than competing equipment. Because they operate with high excess air they produce fewer air pollutants in the form of unburned hydrocarbons than other internal combustion cycles, and the high-temperature flow processes in the turbine help to reduce NO_x emissions.

The high efficiency of gas-turbine cycles result, of course, from the high upper-temperature limits which may be accommodated in the burner and turbine processes. As better high-temperature materials are developed for turbine blades, one may expect further increases in efficiency. Much of the development of high-temperature materials will come as a result of research on advanced turbojet engine designs for aircraft.

12-13 SECOND-LAW ANALYSIS OF POWER CYCLES

In Chap. 6 we discussed the concepts of availability and irreversibility as applied to steady-flow problems. We made further examination of these concepts as they might apply to combustion processes in Chap. 11. Let us now consider how they may apply to the exposition of this chapter, specifically to power cycles.

As we have stated several times, real processes are not reversible; we cannot have frictionless pumps, compressors, or turbines, and heat-exchange devices always involve heat transfer across a finite temperature difference with some resultant irreversibility. Ultimately, we might choose as our thermodynamic objective a minimization of the overall irreversibility generation by a particular power cycle. Generally speaking, we may expect this objective to also coincide with a condition of minimum fuel consumption and energy costs. For example, a steam-power cycle with feedwater heaters will have less total irreversibility than one without; a gas-turbine cycle with regeneration will have less irreversibility than one without; and so on. However, these facts do not tell us anything about the economic cost saving which may result. The advantage of second-law analysis of power cycles is that it may point the way to *possible* alternatives which can result in savings in fuel or equipment costs.

A second-law analysis of a power cycle is normally performed with the following sequence of operations:

1 The availability function $b = h - T_0 s$ is computed for each point in the cycle.
2 The maximum work output for each *component* is calculated as the decrease in the availability across the component.
3 The actual work output is compared with the maximum output, and the irreversibility of each component is computed from Eq. (6-49):

$$\dot{I} = \dot{W}_{max} - \dot{W}_{act}$$

4 The overall cycle irreversibility is calculated by summing the irreversibilities for all the components.
5 Comparisons can then be drawn between different cycles, different operating conditions within a cycle, and alteration of component performance within a cycle.

EXAMPLE 12-13 Irreversibility in a boiler

The combustion system in Example 11-10 is used to heat water from 500 psia, 200°F liquid, to superheated conditions of 500 psia, 1000°F. What is the overall irreversibility for this process. (*Note:* The high steam temperature at exit is possible because the combustion temperature can be quite high inside the boiler).

SOLUTION The overall energy release per pound of fuel was calculated as 20 285 Btu. In accordance with the schematic diagram, this energy must go to raise the enthalpy of the water. Thus,

$$q = 20\ 285\ \text{Btu} = \dot{m}_w(h_2 - h_1) \tag{a}$$

From the steam tables

$$h_1 = 167.65\ \text{Btu/lbm} \quad \text{(compressed liquid)}$$
$$s_1 = 0.293\ 41\ \text{Btu/lbm·°R}$$
$$h_2 = 1520.7\ \text{Btu/lbm}$$
$$s_2 = 1.7471\ \text{Btu/lbm·°R}$$

Thus

$$\dot{m}_w = 14.99\ \text{lbm/lbm fuel}$$

The decrease in the availability of the water (steam) is

$$\begin{aligned}
B_1 - B_2 &= \dot{m}_w[(h_1 - T_0 s_1) - (h_2 - T_0 s_2)] \\
&= (14.99)[(167.65 - 1520.7) - (537)(0.292\ 41 - 1.7471)] \\
&= -8573\ \text{Btu/lbm fuel} \quad (-19\ 940\ \text{kJ/kg fuel}) \tag{b}
\end{aligned}$$

The decrease in the availability for the combustion process was 21 516 Btu/lbm fuel, so the overall decrease for the boiler is

FIG. EXAMPLE 12-13

$$B_R - B_P + B_1 - B_2 = 21\ 516 - 8573 = 12\ 943 \text{ Btu/lbm fuel}$$
$$= 30\ 105 \text{ kJ/kg fuel}$$

This is equal to the irreversibility because there is no actual work done.

$$\dot{I} = +12\ 943 \text{ Btu/lbm fuel}$$
$$= +30\ 105 \text{ kJ/kg fuel}$$

EXAMPLE 12-14 Availability analysis of gas-turbine cycle

Let us examine the cycle of Example 12-11 from the standpoint of the concepts of availability and irreversibility. We shall perform the analysis only for constant specific heats and $T_0 = 70°F$, but for two variations:

(a) Using the turbine, compressor, and regenerator efficiencies of Example 12-11

(b) Assuming 100 percent efficiencies for the cycle components

SOLUTION The availability function for steady flow was defined in Chap. 6 as

$$b = h - T_0 s \tag{a}$$

and the irreversibility for a process with unit mass was

$$\dot{I} = \dot{W}_{\substack{\text{max} \\ \text{output}}} - \dot{W}_{\substack{\text{actual} \\ \text{output}}} \tag{b}$$
$$= -\Delta b - \dot{W}_{\text{act}}$$
$$= -\Delta h + T_0 \Delta S - \dot{W}_{\text{act}} \tag{c}$$

For ideal gases the entropy changes may be computed with

$$\Delta s = c_p \ln \left(\frac{T_2}{T_1}\right) - R \ln \left(\frac{p_2}{p_1}\right) \tag{d}$$

(a) For this case, corresponding to Example 12-11, we go around the cycle and compute the needed quantities. For the compressor

$$h_{2'} - h_1 = c_p(T_{2'} - T_1) = (0.24)(838 - 530)$$
$$= 73.92 \text{ Btu/lbm} \qquad (165.3 \text{ kJ/kg})$$

$$s_{2'} - s_1 = (0.24) \ln \left(\frac{838}{530}\right) - \frac{53.35}{778} \ln \left(\frac{60}{14.7}\right)$$
$$= +0.0135 \text{ Btu/lbm·°R} \qquad (0.0565 \text{ kJ/kg·K})$$
$$b_{2'} - b_1 = 73.92 - (530)(0.0135) = 66.76 \text{ Btu/lbm} \qquad (155.3 \text{ kJ/kg})$$

This is the negative of the maximum work output.

The *actual* work *output* is the negative of the enthalpy change or

$$\dot{W}_{\substack{act \\ output}} = -(h_{2'} - h_1) = -73.92 \text{ Btu/lbm}$$

so that the irreversibility is

$$\dot{I}_{1,2'} = -\Delta b_{1,2'} - W_{act\ 1,2'}$$
$$= -66.76 - (-73.92) = +7.16 \text{ Btu/lbm} \qquad (16.65 \text{ kJ/kg})$$

Regenerator

As the air is heated in the regenerator there is zero actual work for the constant-pressure process. We therefore calculate

$$h_5 - h_{2'} = (0.24)(1284 - 838) = 107.04 \text{ Btu/lbm} \qquad (248.98 \text{ kJ/kg})$$
$$s_5 - s_{2'} = (0.24) \ln \left(\frac{1284}{838}\right) - \frac{53.35}{778} \ln \left(\frac{60}{60}\right)$$
$$= +0.1024 \text{ Btu/lbm·°R}$$
$$b_5 - b_{2'} = 107.04 - (530)(0.1024) = 52.76 \text{ Btu/lbm} \qquad (122.7 \text{ kJ/kg})$$

On the other side of the regenerator the hot exhaust gases are cooled from point 4′ to point 6. The temperature at point 6 is obtained from the energy balance on the regenerator

$$h_5 - h_{2'} = h_{4'} - h_6$$

so that

$$T_6 = T_{4'} - T_5 + T_{2'} = 1375 - 1284 + 838 = 929°R$$

or, the temperature rise of the air from the compressor is equal to the temperature drop of the hot exhaust gases. We may now calculate the properties of interest:

$$h_6 - h_{4'} = (0.24)(929 - 1375) = -107.04 \text{ Btu/lbm}$$
$$s_6 - s_{4'} = (0.24) \ln \left(\frac{929}{1375}\right) - \frac{53.35}{778} \ln \left(\frac{14.7}{14.7}\right)$$
$$= -0.0941 \text{ Btu/lbm·°R}$$
$$b_6 - b_{4'} = -107.04 - (530)(-0.0941)$$
$$= -57.16 \text{ Btu/lbm}$$

Now, the overall irreversibility for the regenerator is the negative of the overall increase in availability less the actual work output (zero) or

$$\dot{I}_{reg} = -(b_5 - b_{2'}) - (b_6 - b_{4'}) - \dot{W}_{act}$$
$$= -(+52.76) - (-57.16) - 0$$
$$= +4.40 \ \text{Btu/lbm} \qquad (10.23 \ \text{kJ/kg})$$

Burner

In the burner the air is heated from point 5 to point 3 (for an air-standard cycle). From a practical standpoint this would be accomplished by burning a fuel, and some irreversibility would result as described in Chap. 11. In this problem we do not have enough information to determine the irreversibility of the combustion process, so we shall just calculate the change in availability of the air.

$$h_3 - h_5 = (0.24)(1960 - 1284) = 162.24 \ \text{Btu/lbm}$$
$$s_3 - s_5 = (0.24) \ln \left(\frac{1960}{1284}\right) - \frac{53.35}{778} \ln \left(\frac{60}{60}\right)$$
$$= +0.1015 \ \text{Btu/lbm·°R}$$
$$b_3 - b_5 = 162.24 - (530)(+0.1015)$$
$$= 108.44 \ \text{Btu/lbm} \qquad (252.23 \ \text{kJ/kg})$$

Turbine

The calculations for the turbine are

$$h_{4'} - h_3 = (0.24)(1375 - 1960) = -140.4 \ \text{Btu/lbm}$$
$$s_{4'} - s_3 = (0.24) \ln \left(\frac{1375}{1960}\right) - \frac{53.35}{778} \ln \left(\frac{14.7}{60}\right)$$
$$= +0.011 \ 37 \ \text{Btu/lbm·°R} \qquad (0.0476 \ \text{kJ/kg·K})$$
$$b_{4'} - b_3 = -140.4 - (530)(0.011 \ 37)$$
$$= -146.43 \ \text{Btu/lbm} \qquad (-340.6 \ \text{kJ/kg})$$

The actual work *output* is $h_3 - h_4' = 140.4$ Btu/lbm, so the irreversibility is

$$\dot{I}_{3,4'} = -\Delta b_{3,4'} - W_{act \ 3,4'}$$
$$= -(-146.43) - 140.4$$
$$= +6.03 \ \text{Btu/lbm} \qquad (14.03 \ \text{kJ/kg})$$

The various quantities are shown in the table on the next page to illustrate the overall behavior of the cycle.

As mentioned above, the combustion process in the burner has not been considered in this analysis. To obtain an estimate of the effect of combustion we might consult the results of Example 11-9M. In that example the entrance and exit temperatures and pressures are nearly the same as those here. Summarizing that example

Availabilities for actual gas-turbine cycle

Component	Process(es)	Δh, Btu/lbm	Δb, Btu/lbm	$\dot{W}_{max}$, Btu/lbm	$\dot{W}_{act}$, Btu/lbm	$\dot{I}$, Btu/lbm
Compressor	1-2'	73.92	66.76	−66.76	−73.92	7.16
Regenerator	2-5	107.04	52.76			
	4'-6	−107.04	−57.16			
	Total	0	−4.40	4.40	0	4.40
Burner	5-3	162.24	108.44	—	0	—
Turbine	3-4'	−140.4	−146.43	146.43	140.4	6.03
Overall cycle excluding combustion process		95.76	24.37	84.07	66.48	17.59

Octane fuel in at 4 atm, 25°C (58.8 psia, 77°F)
Air in at 4 atm, 710 K (58.8 psia, 1278°R)
Products out at 4 atm, 1100 K (58.8 psia, 1980°R)
Irreversibility of combustion reaction = 226.31 kJ/kg reactants
= 97.29 Btu/lbm reactants

This result is very close to the change in availability calculated in the present example. Of course, the air fuel ratio in Example 11-9M is so high (94.74) that the reaction is closely represented by a calculation on the basis of air alone.

(b) For the ideal cycle, assuming 100 percent efficiencies in the components, we will notice a marked change in performance. Both the turbine and compressor will have zero irreversibility (because the processes are reversible), and the actual work output will be equal to the maximum possible. Using the nomenclature of Fig. 12-28, the regenerator temperature for η_{reg} = 100 percent will be

$$T_4 = T_5 = 1310°R \quad \text{and} \quad T_6 = T_2 = 792°R$$

We then have the following calculations:

$$h_2 - h_1 = b_2 - b_1 = -\dot{W}_{max} = \dot{W}_{act} \quad \text{(because } s_1 = s_2\text{)}$$
$$= (0.24)(792 - 530) = 62.88 \text{ Btu/lbm}$$
$$h_4 - h_3 = b_4 - b_3 = -\dot{W}_{max} = \dot{W}_{act} \quad (s_3 = s_4)$$
$$= (0.24)(1310 - 1960) = -156 \text{ Btu/lbm}$$

For the regenerator

$$h_5 - h_2 = (0.24)(1310 - 792) = 124.32 \text{ Btu/lbm}$$
$$s_5 - s_2 = (0.24) \ln \left(\frac{1310}{792}\right) - \frac{53.35}{778} \ln \left(\frac{60}{60}\right)$$
$$= 0.1208 \text{ Btu/lbm·°R} \quad (0.5058 \text{ kJ/kg·K})$$
$$b_5 - b_2 = 60.31 \text{ Btu/lbm} \quad (140.3 \text{ kJ/kg})$$
$$h_6 - h_4 = (0.24)(792 - 1310) = -124.32 \text{ Btu/lbm}$$

$$s_6 - s_4 = (0.24) \ln \left(\frac{792}{1310}\right) - \frac{53.35}{778} \ln \left(\frac{14.7}{14.7}\right)$$

$$= -0.1208 \text{ Btu/lbm·°R}$$

$$b_6 - b_4 = -60.31 \text{ Btu/lbm}$$

so that the overall $\Delta b = 0$ along with $\dot{W}_{\text{act}} = 0$ and

$$\dot{I}_{\text{reg}} = \dot{W}_{\text{max}} - \dot{W}_{\text{act}} = 0$$

In the burner we have

$$h_3 - h_5 = (0.24)(1960 - 1310) = 156 \text{ Btu/lbm}$$

$$s_3 - s_5 = (0.24) \ln \left(\frac{1960}{1310}\right) - \frac{53.35}{778} \ln \left(\frac{60}{60}\right)$$

$$= 0.0967 \text{ Btu/lbm} \qquad (0.4049 \text{ kJ/kg·K})$$

$$b_3 - b_5 = 104.75 \text{ Btu/lbm} \qquad (243.6 \text{ kJ/kg})$$

We may now tabulate the properties for the ideal cycle in the following table:

Availability for ideal gas-turbine cycle

Component	Process(es)	Δh, Btu/lbm	Δb, Btu/lbm	$\dot{W}_{\text{max}}$, Btu/lbm	$\dot{W}_{\text{act}}$, Btu/lbm	$\dot{I}$, Btu/lbm
Compressor	1-2	62.88	62.88	−62.88	−62.88	0
Regenerator	2-5	124.32	60.31			
	4-6	−124.32	−60.31			
	Total	0	0	0	0	0
Burner	5-3	156	104.75	—	0	—
Turbine	3-4	−156	−156	156	156	0
Overall cycle excluding combustion process		62.88	11.63	93.12	93.12	0

The thermal efficiency for this ideal cycle is

$$\eta = \frac{\dot{W}_{\text{act}}}{Q_{\text{burner}}} = \frac{93.12}{156} = 59.7 \text{ percent}$$

12-14 HYBRID CYCLES

Depending on the application, it may be advantageous to combine the gas-turbine and vapor-power cycle in a hybrid arrangement as shown in Fig. 12-32. Instead of using a regenerator, the hot exhaust gases are used to supply heat to the boiler for driving the vapor cycle. If the exhaust gases are the only source of energy for the boiler, then we would have for the energy balance on the boiler

FIG. 12-32 *Hybrid cycle schematic for gas-turbine–vapor-power cycle.*

$$\dot{m}_a(h_4 - h_5) = \dot{m}_s(h_1 - h_9) \tag{12-22}$$

If supplemental heat Q is supplied to the boiler, then that value must be added to the left side of the equation.

The advantage of the hybrid or combined cycle is that we take advantage of high-temperature heat addition for the gas turbine as well as the low-temperature heat rejection for the vapor cycle. We have the added advantage that the products of combustion from the gas turbine are "cleaner" owing to the high excess air rates which can be used. For the cycle in Fig. 12-32 the total work output of the cycle would be

$$W_{\text{total}} = W(\text{gas turbine}) - W(\text{air compressor}) + W(\text{steam turbine})$$
$$- W(\text{pump})$$
$$= \dot{m}_a[(h_3 - h_4) - (h_2 - h_1)] + \dot{m}_s[(h_6 - h_7) - (h_9 - h_8)] \tag{12-23}$$

The relation between $\dot{m}_a$ and $\dot{m}_s$ is obtained from the boiler energy balance, Eq. (12-22). For no supplemental heat addition to the boiler the only heat added to the combined cycle *from external sources* occurs in the burner from point 2 to point 3. Thus,

$$Q_H = \dot{m}_a(h_3 - h_2) \tag{12-24}$$

and the thermal efficiency is the ratio of Eq. (12-23) to Eq. (12-24).

The only heat rejection *for the combined cycle* occurs in the vapor-condenser process 7-8 and can be calculated as

$$Q_L = \dot{m}_s(h_7 - h_8) \tag{12-25}$$

From an overall energy-balance standpoint the total work output could also be calculated from

$$W = Q_H - Q_L \tag{12-26}$$

Of course, a hybrid cycle can also include regeneration for the vapor-cycle portion and multistage compression with intercooling on the gas-cycle portion.

12-15 THE GAS TURBINE FOR JET PROPULSION

The turbojet engine which is extensively used for aircraft propulsion is a simple modification of the open gas-turbine cycle and is shown schematically in Fig. 12-33. The hot combustion gases are expanded in the turbine only far enough to generate work to drive the compressor. The remaining energy is then converted to high-velocity kinetic energy by expanding it in a nozzle downstream from the turbine. The jet thrust results from the difference in momentum of the entering air flow to the compressor and the high-velocity exhaust gases. This thrust is given by

$$F = \frac{\dot{m}}{g_c} (V_1 - V_5) \tag{12-27}$$

where $\dot{m}$ is the mass flow in an air-standard cycle and V_1 and V_5 are the flow velocities corresponding to the points in Fig. 12-33. We shall discuss the principles of high-speed compressible flow systems in Chap. 13 with explanations of the design factors which must be considered in establishing the configuration of the nozzle and flow exit geometries. For now we can be content with an example which illustrates the thermodynamic principles involved in the turbojet cycle.

EXAMPLE 12-15 Turbojet engine
A turbojet engine operates between pressure limits of 5 and 50 psia. The inlet air temperature to the compressor is $-40°F$, and the upper temperature limit for the engine is 2000°F. Calculate the thrust for 1 lbm/s of air flow, assuming isentropic compression and expansion and an inlet velocity of 300 ft/s. Assume constant specific heats. Also calculate the heat input per pound mass of air.

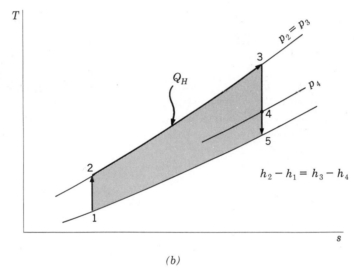

(b)

FIG. 12-33 *Gas-turbine cycle for jet propulsion. (a) Schematic; (b) T-s diagram.*

SOLUTION   The schematic diagram of Fig. 12-33 is used for the solution of this problem. We have

$$T_1 = -40°F = 420°R \quad (233 \text{ K})$$
$$T_3 = 2000°F = 2460°R \quad (1367 \text{ K})$$
$$V_1 = 300 \text{ ft/s} \quad (91.44 \text{ m/s})$$
$$p_1 = p_5 = 5 \text{ psia} \quad (34.5 \text{ kPa})$$
$$p_2 = p_3 = 50 \text{ psia} \quad (345 \text{ kPa})$$

For the isentropic compression process,

$$T_2 = T_1 \left(\frac{p_2}{p_1}\right)^{(\gamma-1)/\gamma}$$

$$= (420) \left(\frac{50}{5}\right)^{0.286}$$

$$= 812°R \quad (451 \text{ K})$$

Since the compressor and turbine works are equal

$$h_2 - h_1 = h_3 - h_4$$

or, for constant specific heats,

$$T_2 - T_1 = T_3 - T_4$$

Thus

$$T_4 = 2460 - 812 + 420 = 2068°R \quad (1149 \text{ K})$$

For the isentropic turbine-exhaust nozzle process

$$T_5 = T_3 \left(\frac{p_5}{p_3}\right)^{\gamma-1/\gamma}$$

$$= (2460) \left(\frac{5}{50}\right)^{0.286}$$

$$= 1272°R \quad (707 \text{ K})$$

For the nozzle,

$$h_4 + \frac{V_4^2}{2g_c} = h_5 + \frac{V_5^2}{2g_c}$$

We shall assume that V_4 is small, so that

$$V_5 = [2g_c(h_4 - h_5)]^{1/2}$$
$$= [2g_c c_p(T_4 - T_5)]^{1/2}$$
$$= [(2)(32.2)(0.24)(2068 - 1272)(778)]^{1/2}$$
$$V_5 = 3093 \text{ ft/s} \quad (943 \text{ m/s})$$

The thrust is now calculated from Eq. (12-27) as

$$F = T = \dot{m}(V_1 - V_5)$$

$$= \frac{(1)(300 - 3093)}{32.2}$$

$$T = -96.8 \text{ lbf}$$

The negative sign indicates that the thrust is in the opposite direction to the velocity.

The heat added per pound mass is

$$Q_H = \dot{m}(h_3 - h_2)$$
$$= (1)(0.24)(2460 - 812)$$
$$= 395 \ \text{Btu/s}$$

If we assume that the turbojet moves with a velocity equal to the inlet velocity, the power developed by the thrust force is

$$P = TV_1$$
$$= (86.8)(300) = 26 \ 000 \ \text{ft·lbf/s}$$

or

$$P = 33.46 \ \text{Btu/s}$$
$$= 47.3 \ \text{hp} \qquad (35.3 \ \text{kW})$$

These calculations are for the 1.0 lbm/s flow rate and would be proportionately higher for higher flows.

12-16 THE RAMJET

The ramjet is a device employed for aircraft propulsion at very high speeds. It operates on the same principle as the turbojet, except that a flow passage is used to convert the kinetic energy of the incoming flow into pressure. The device is shown schematically in Fig. 12-34. The incoming flow velocity is reduced in the inlet nozzle in process 1-2. Fuel is burned at constant pressure in process 2-3, thereby increasing the temperature. Finally, the flow is expanded to a high velocity in the outlet nozzle. Because of the energy which has been added to the flow, the velocity at exit V_4 is greater than that at inlet. The net thrust is given as the change in momentum flux,

$$T = F = \frac{\dot{m}}{g_c} (V_4 - V_1) \qquad\qquad (12\text{-}28)$$

where $\dot{m}$ is the mass rate of flow. The ramjet is only feasible for high inlet-flow velocities; otherwise it would not be possible to achieve a high enough compression ratio in the inlet to counteract the losses incurred due to friction and irreversibilities in the engine.

(a)

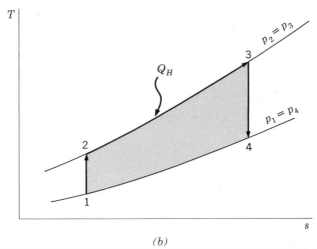

(b)

FIG. 12-34 *The ramjet.* (a) *Schematic;* (b) *T-s diagram.*

12-17 THE ROCKET

A rocket is a simple device whereby a fuel and oxidizer are supplied to a combustion chamber (burner) at high pressure. The high-pressure products of combustion are then expanded through a nozzle to high velocities, thereby producing thrust. The rocket engine is shown schematically in Fig. 12-35. Both liquid and solid propellants (fuel) may be used. The thrust is again given as the change in momentum flux, or

$$T = F = \frac{\dot{m}}{g_c} V_3 \qquad (12\text{-}29)$$

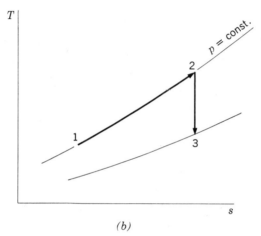

(b)

FIG. 12-35   *The rocket.* (a) *Schematic;* (b) *T-s diagram.*

Of course, the total mass of the rocket engine steadily decreases as more and more combustion products leave the exhaust nozzle.

12-18   SPECIFIC IMPULSE

A parameter frequently used to compare propulsion systems is the specific impulse defined as

$$I_{sp} = \text{specific impulse} = \frac{\text{thrust}}{\text{mass flow rate}} \tag{12-30}$$

Table 12-1 gives a comparison of approximate specific impulses for the three propulsion devices described in the preceding sections. The reader is cautioned against comparing the systems on the basis of I_{sp} alone. A ramjet, for

TABLE 12-1
Approximate specific impulses for some propulsion devices

Turbojet	1500–2000
Chemical rocket	160–400
Ramjet	Very large when used at extreme altitudes

example, is completely inoperable for takeoff of an aircraft. Both the turbojet and ramjet are completely inoperable in space flight, and the rocket consumes huge amounts of expensive fuel. Each of these systems has its own place of application, but a discussion of the details involved is beyond the scope of this exposition.

12-19 VAPOR REFRIGERATION CYCLES

The vapor refrigeration cycle is a practical device which attempts to duplicate the Carnot refrigeration cycle. A schematic of the basic practical arrangement is shown in Fig. 12-36. In the ideal case compression is isentropic. Heat is added to the cycle (Q_L) by evaporating fluid at the lower temperature limit while heat rejection (Q_H) takes place by condensing the vapor at the higher temperature limit. In the evaporator section heat is added until a saturated or slightly superheated vapor is obtained at point 1.

Condensation at the high temperature produces a saturated liquid at point 3 which must then be expanded to low pressure in order to accommodate the low-temperature evaporation process. An isentropic expansion is practically unfeasible so the simple alternative of throttling the fluid from point 3 to point 4 is selected. For small kinetic energies an adiabatic throttling process is one of constant enthalpy, so $h_3 = h_4$. The process is shown as a dashed line on the T-s diagram because it is irreversible. In an actual cycle the compression process would also be irreversible, and the outlet from the compressor might appear as point 2'.

To select a working fluid for refrigeration cycles the behavior of the fluid at low temperatures is an important factor. Water, for example, could not be used below 32°F, and even at moderately low temperatures a vacuum would need to be maintained in the evaporator section. Ammonia (NH_3) is widely used as a refrigerant at very low temperatures but is not normally used in air conditioning applications because of its toxic nature should leaks occur in the piping or evaporator systems. The fluorinated hydrocarbons (Freons) are most often used in air conditioning work because of their low cost, inert behavior, and nontoxic effect on humans. They have an additional advantage in that they have a strong affinity for oil and, thus, can serve as self-lubricating agents in the compressor. Both ammonia and some of the Freons have saturation pressures above 1 atm even at low temperatures, thereby alleviating any necessity to maintain vacuums. Properties of ammo-

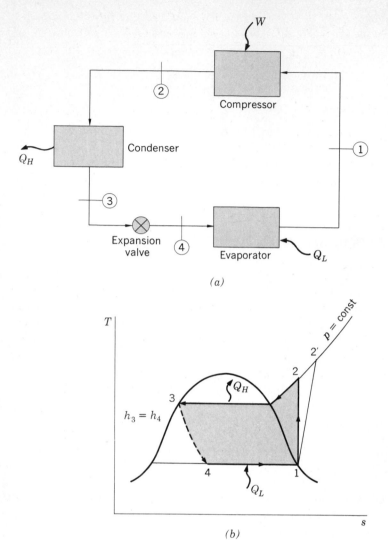

FIG. 12-36 *Vapor refrigeration cycle.* (*a*) *Schematic;* (*b*) *T-s diagram.*

nia and dichlorodifluoromethane (Freon 12) are given in the Appendix for calculation purposes.

The coefficient of performance is

$$\text{COP} = \frac{\text{refrigeration effect}}{\text{work supplied}} = \frac{Q_L}{W} \qquad (12\text{-}31)$$

$$\text{COP} = \frac{h_1 - h_4}{h_2 - h_1}$$

where the subscripts refer to the points in Fig. 12-36. A common unit used in practice to describe the refrigeration effect is the *ton*, defined as the heat rate to melt a ton of ice in a 24-h period. This definition reduces to

$$1 \text{ ton} = 12\ 000 \text{ Btu/h} = 3.516 \text{ kW} \tag{12-32}$$

The following example illustrates the overall analysis of a refrigeration cycle.

EXAMPLE 12-16   Freon refrigerator

An ideal-vapor refrigeration cycle uses Freon 12 as the working fluid. The saturation temperature in the condenser is 120°F, and the evaporator temperature is 30°F. Calculate the coefficient of performance and the horsepower input necessary to produce a cooling effect of 5 tons.

SOLUTION   The schematic diagram of Fig. 12-36 applies to this problem. We may consult the Freon 12 property tables in the Appendix to obtain

$h_1 = 80.419$ Btu/lbm   (saturated vapor at 30°F)
$s_1 = 0.16648$ Btu/lbm·°R
$h_3 = h_4 = 36.013$ Btu/lbm   (saturated liquid at 120°F)
$p_3 = p_2 = 172.35$ psia   (saturation pressure at 120°F)

Since $s_2 = s_1$ we may consult the superheat tables to obtain

$h_2 = 91.5$ Btu/lbm   (212.8 kJ/kg)

We may now calculate the COP from Eq. (12-31) as

$$\text{COP} = \frac{h_1 - h_4}{h_2 - h_2}$$
$$= \frac{80.4 - 36.0}{91.5 - 80.4}$$
$$\text{COP} = 4.0$$

The work input required per pound mass of fluid is

$$W = h_2 - h_1 = 11.1 \text{ Btu/lbm} \quad (25.82 \text{ kJ/kg})$$

The total flow rate of fluid required may be obtained from

$$Q_L = \dot{m}(h_1 - h_4)$$
$$(5)(12\,000) = \dot{m}(80.4 - 36.0)$$
$$\dot{m} = 1350 \text{ lbm/h} \quad (612.4 \text{ kg/h})$$

The total work required is therefore

$$W_{total} = \dot{m}(h_2 - h_1)$$
$$= (1350)(11.1)$$
$$= 15\,000 \text{ Btu/h}$$
$$W = 5.9 \text{ hp} \quad (4.39 \text{ kW})$$

12-20 THE HEAT PUMP

In viewing the vapor refrigeration cycle we could have thought of the device as a mechanism to "pump" heat uphill on the temperature scale and given it the name "heat pump." We used the term *refrigeration* cycle because the cooling was the useful effect we were trying to achieve. The same device may be used to accomplish another objective, that of *delivering heat* Q_H at the condenser. Thus, one might employ the same machine for both summer cooling and winter heating in residences and smaller industrial or commercial buildings. In the heating mode it is normal practice to define the coefficient of performance as

$$COP_H = \frac{\text{useful effect}}{\text{energy that must be purchased}}$$
$$= \frac{\text{heating effect}}{\text{work input}} = \frac{Q_H}{W} = \frac{1}{1 - Q_L/Q_H} \tag{12-33}$$

For a Carnot heat pump the corresponding COP_H would be

$$COP_H(\text{Carnot}) = \frac{1}{1 - T_L/T_H} \tag{12-34}$$

The heat pump has an obvious advantage over electric resistance heating in that one obtains the "multiplier effect" through the COP; viz., a unit with a cooling COP of 3.0 would have

$$3.0 = \frac{Q_L}{W} \qquad Q_L = 3W$$

so that

$$Q_H = Q_L + W = 4.0 \text{ W}$$

and

$$COP_H = 4.0$$

Thus, we would only have to *purchase* one unit of energy for each *four* units of heat delivered. This is not getting something for nothing! One must make a capital investment in the machine in order to have a future savings in electric costs.

The heat pump may or may not offer economic advantages over natural gas- or oil-fired heating systems. Final cost comparisons are strongly dependent on climate. In moderate climates the heat pump is usually competitive with these other modes of heating, while it is not in very cold climates. Capital costs for heat pump installations are about 20 to 25 percent greater than for a comparable air conditioner and natural gas furnace.

We may expect that future heat pump designs will make extensive use of solar energy as a heat source on the evaporator. By significantly raising the evaporator temperature above ambient outside temperature, it is possible to make dramatic improvements in the overall coefficient of performance.

EXAMPLE 12-17 Carnot heat pump

A Carnot heat pump operates between an outside temperature of $-15°C$ (5°F) and inside room heating unit temperature of 38°C (100.4°F). Calculate the percent of improvement in performance if a solar energy collector is installed so that the low-temperature limit is raised to 10°C (50°F).

SOLUTION   At the first condition we have

$$T_L = -15°C = 258 \text{ K}$$
$$T_H = 38°C = 311 \text{ K}$$

so that the heating coefficient of performance is

$$COP_H = \frac{1}{1 - T_L/T_H} = \frac{1}{1 - 258/311}$$
$$= 5.868$$

When the low temperature is raised to 10°C we have

$$T_L = 10°C = 283 \text{ K}$$
$$T_H = 38°C = 311 \text{ K}$$

And

$$COP_H = \frac{1}{1 - 283/311} = 11.107$$

or an increase of 89 percent. Thought of another way, by raising the low side temperature from -15 to $10°C$, we can produce the same heating output with a reduction in power input of 47 percent. Naturally, real heat pumps will not give the high values of COP illustrated in this example; however, one may expect about the same *percentage* improvements when the evaporator (low side) temperature is raised.

EXAMPLE 12-18 Cost comparisons: Heat pump vs. fossil fuel heater
Freon 12 is employed as the working fluid in a heat pump with an evaporator temperature (outdoor temperature) of $40°F$ and a condenser temperature (indoor temperature) of $120°F$. The compressor effiency is 85 percent. The heating requirements are $100\ 000$ Btu/h. If electricity costs $0.053/kWh, compare the cost of operating the heat pump with a fossil fuel heater in which fuel costs $3.25 per 10^6 Btu. Assume the efficiency of the fossil fuel heater is 70 percent; i.e., only 70 percent of the fuel energy is delivered to the indoor space to be heated.

SOLUTION   The heat pump is a vapor refrigeration device and the schematic is shown in Fig. 12-36. The enthalpies are determined as

$h_1 = 81.436$ Btu/lbm   (saturated vapor at $40°F$, Table A-15)
$s_1 = 0.16586$ Btu/lbm·°R $= s_2$
$h_3 = h_4 = 36.013$ Btu/lbm   (saturated liquid at $120°F$)
$p_3 = p_2 = 172.35$ psia   (saturation pressure at $120°F$)

Using $s_1 = s_2$, we consult the superheat tables (Table A-16) at 172.35 psia and obtain by interpolation

$h_2 = 90.64$ Btu/lbm

Because of the compressor efficiency we must write

$$0.85 = \frac{h_2 - h_1}{h_{2'} - h_1} = \frac{90.64 - 81.436}{h_{2'} - 81.436}$$
$$h_{2'} = 92.264 \text{ Btu/lbm} \tag{a}$$

The heating capacity is $Q_H = 100\ 000$ Btu/h, so

$$Q_H = \dot{m}(h_{2'} - h_3)$$

and the flow rate of Freon is calculated as

$$\dot{m} = \frac{100\ 000}{92.264 - 36.013} = 1778 \text{ lbm/h} \tag{b}$$

The corresponding work input is

$$W = \dot{m}(h_{2'} - h_1) = (1778)(92.264 - 81.436)$$
$$= 19\ 252\ \text{Btu/h} = 5.64\ \text{kW} \tag{c}$$

At $0.053/\text{kWh}$ ($15.53/10^6\ \text{Btu}$) the corresponding cost is

$$C_{HP} = (0.053)(5.64) = \$0.2989/\text{h}$$

The unit heating cost is

$$C_{HP}/10^6\ \text{Btu} = \frac{\$0.2989/\text{h}}{0.1 \times 10^6\ \text{Btu/h}} = \$2.989/10^6\ \text{Btu} \tag{d}$$

For the fossil fuel heater

$$E_{\text{fuel}}(0.7) = 100\ 000\ \text{Btu/h}$$
$$E_{\text{fuel}} = 142\ 860\ \text{Btu/h}$$

and, at $3.25/10^6\ \text{Btu}$ the cost is

$$C_F = (\$3.25/10^6\ \text{Btu})(0.142\ 86 \times 10^6\ \text{Btu/h})$$
$$= \$0.4643/\text{h}$$

The cost per 10^6 Btu *delivered to the inside* is

$$C_F/10^6\ \text{Btu} = \frac{0.4643}{0.1} = \$4.643/10^6\ \text{Btu} \tag{e}$$

Thus, we find that the unit heating costs for the heat pump are about 35 percent cheaper than for the fossil fuel heater even though the unit cost of energy *purchased* is much higher for the heat pump ($15.53/10^6$ Btu for the heat pump versus $3.25/10^6$ Btu for fuel). This results of course from the coefficient of performance effect. As a matter of interest we can calculate the heating coefficient of performance as

$$\text{COP}_H = \frac{Q_H}{W} = \frac{100\ 000}{19\ 252} = 5.194 \tag{f}$$

12-21 ABSORPTION REFRIGERATION CYCLES

In many plants and buildings a ready supply of steam or hot flue gases are available as a by-product of some process or as unused capacity of a heating system which is inoperative in off seasons. In such cases there is a strong economic motivation to make use of the heat supply for other purposes. The absorption refrigeration system provides a vehicle for making use of waste heat, either for air conditioning in the summer or for an appropriate refrigeration storage system.

It is clear that a pressure differential must be involved in a refrigeration system to accomplish a temperature difference for heat addition and rejection. In the vapor cycle discussed in the preceding section the pressure differential is accomplished with the vapor compressor. Such a device requires a relatively large power input per unit mass of fluid circulated. A liquid pump would not require such a large amount of power. The absorption system represents a scheme whereby the vapor compressor is replaced by a liquid pump and a suitable heater. The basic absorption system is shown in Fig. 12-37. For the sake of discussion we shall assume that the refrigerant is ammonia, NH_3; this is the most widely used refrigerant in absorption systems. The condenser, evaporator, and expansion valve are the same as in the vapor compression cycle, so we are primarily concerned with the elements replacing the compressor—the absorber, pump, and generator.

The basic principle involved is that water absorbs ammonia and liberates heat in the process, provided the temperature is held constant. Conversely, when heat is added to a strong solution of NH_4OH, it will break down into NH_3 vapor and H_2O. In the absorber, NH_3 vapor is bubbled through H_2O and heat Q_R is removed to maintain the temperature constant. The strong liquid solution is then taken off at point 5 and pumped to a higher pressure at point 6. The strong solution is then placed in the generator where heat Q_A is added to boil off the NH_3 at the elevated pressure. The remaining weak solution of liquid is then returned to the absorber to acquire a fresh charge of NH_3.

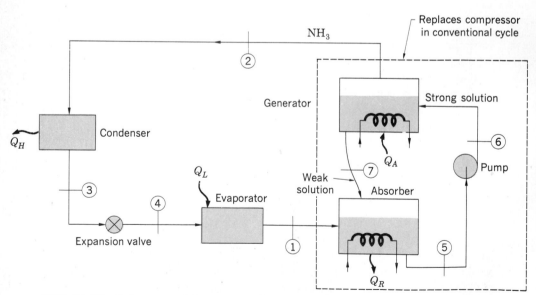

FIG. 12-37 *Schematic of absorption refrigeration system.*

In the overall absorption cycle the work input to the pump is very small, and the major source of energy which must be supplied from the outside is Q_A for the generator. It is to be noted, however, that the absorption cycle requires substantially more heat-rejection facilities than the vapor compression cycle because of the necessity to dissipate the heat of absorption Q_R in addition to the quantity Q_H. Consideration must be given to such matters in an overall economic analysis of a system to be used in practice.

Another pair of fluids which may be used in absorption systems for air conditioning applications where evaporator temperatures are not too low is the combination of water as the refrigerant and lithium bromide as an absorber. With water as the refrigerant, a vacuum must be maintained in the evaporator to achieve low temperatures; an absolute pressure of 0.12 psia would have to be maintained to achieve an evaporator temperature of 40°F. Such vacuums are normally maintained by steam ejectors.

The schematic for a practical LiBr–H_2O system is shown in Fig. 12-38. Both the evaporator and absorber operate on the low-pressure side of the system and thus are enclosed in the common shell with a connection for purging to maintain vacuum conditions. The condenser and generator operate on the high-pressure side and so are similarly enclosed in a common shell. This system is normally used in large cooling systems which require cooling rates above 10^6 Btu/h, so a substantial amount of cooling water must be supplied to remove the heat from the absorber and condenser. Steam is commonly employed to supply energy to the generator, and the evaporator is used to produce chilled water.

FIG. 12-38 *Schematic of a practical lithium bromide–water absorption refrigeration system.* (*Carrier Corporation.*)

Now let us trace through the processes involved. At point 6 water vapor in the lower shell is being absorbed in the weak LiBr solution arriving at point 5. After the absorption, and accompanying heat release to the cooling water, the strong LiBr–H$_2$O solution at point 1 is pumped through pump 1 to the generator. The three-way valve and heat exchanger will be discussed in a moment. The strong solution enters the bottom of the upper shell where it is heated with steam coils, thereby causing water vapor to come out of solution at point 3. This vapor is subsequently condensed in the pan at the top of the upper shell. The weak LiBr solution is then returned to the lower shell at points 4 and 5. The pressure difference between the two shells is not large because the temperature for condensation is modest, say 85°F, where the saturation pressure is 0.6 psia. For an evaporator temperature of 40°F the pressure difference would be $0.6 - 0.12 = 0.48$ psia. Cooling is accomplished when the liquid water flashes to wet vapor in the lower shell, and the U arrangement in the liquid water line can maintain the pressure differential between the two shells.

Low pressure is maintained in the lower shell by the rapid absorption of water vapor in the LiBr solution and by the purge valve, when necessary. Pump 2 is provided to keep a fresh spray of cool water over the chilled water coils to enhance the heat-transfer rates. A similar spray system is provided in the bypass of pump 1 to enhance the absorption process at point 6.

Steam heating is normally supplied at temperatures of 210 to 220°F, so the solution at point 4 is quite warm. The solution at point 1 is much cooler, and the heat exchanger shown improves overall performance in a manner similar to the regenerators used in power cycles. The high-temperature liquid at 4 is used to preheat the cool fluid from 1 to 2 so that less heat must be added in the generator section. There is also a saving in cooling water requirements for the absorber because the solution entering at 5 is cooler than it would have been without the heat exchanger. The three-way valve serves as a load-control device to increase or decrease liquid flow between the two shells in accordance with the chilled water cooling requirements.

REVIEW QUESTIONS

1   What are the primary factors which influence thermal efficiency of a power cycle or coefficient of performance of a refrigeration cycle?

2   How is the thermal efficiency of a power cycle related to pollution generated by the cycle?

3   Why is it advantageous to employ superheat in the basic Rankine cycle?

4   Describe some important factors in the selection of a working fluid for a vapor power cycle.

5   How does a regenerator improve the performance of a vapor power cycle?

6 Distinguish between closed and open feedwater heaters.

7 Define turbine efficiency.

8 Why does a lower thermal efficiency result in more "thermal pollution"?

9 What advantages are offered by intercoolers in air-compression machines?

10 What is the general principle which is applied to selection of interstage pressure(s) in compressors with ideal intercooling?

11 Define volumetric efficiency.

12 Why is it possible to analyze an air compressor as a steady-flow device even though the compression occurs in the closed confines of a piston-cylinder arrangement?

13 Distinguish between the Otto and diesel cycles.

14 What is an air-standard cycle?

15 Compare the spark-ignition and compression-ignition cycles on the basis of their respective advantages and disadvantages.

16 How does compression ratio affect thermal efficiency of an internal combustion engine?

17 Describe some advantages of gas-turbine cycles for power production.

18 How does a turbojet work?

19 Define regenerator efficiency.

20 How does a ramjet work?

21 What is coefficient of performance?

22 What is a "ton"?

23 How does an absorption refrigeration system work?

24 What are the main factors affecting the coefficient of performance of a refrigeration machine?

25 What is a "heat pump"?

26 What is a Stirling cycle? How is it possible for it to have a thermal efficiency approaching the Carnot value?

27 What are some advantages of hybrid cycles combining the gas-turbine with the vapor-power cycle?

PROBLEMS (English units)

12-1 In a Rankine cycle, steam enters the single-stage turbine at 1000 psia and 800°F. Heat is rejected in the condenser at 80°F. Calculate the thermal efficiency of the cycle. Repeat the calculation for reheat of the steam at 500 psia back to 800°F.

12-2 A certain Rankine cycle operates with two reheat stages. Steam enters the turbine at 1500 psia and 1200°F. Reheat to 1200°F occurs at 1000 and 500 psia. The lower-pressure limit of the cycle is 1.0 psia. Each turbine stage has an efficiency of 90 percent. Calculate the thermal efficiency of the cycle and compare with the Carnot efficiency between the same temperature limits.

12-3 A small steam turbine is designed to operate with 200 psia of steam. The turbine is to be employed in a simple Rankine cycle which rejects heat at 90°F. To what temperature must the high-pressure steam be heated to ensure that the moisture content does not exceed 5 percent in the turbine? Calculate the thermal efficiency of the cycle under these conditions. Assume isentropic expansion in the turbine.

12-4 Repeat Prob. 12-3 for a turbine efficiency of 90 percent.

12-5 A regenerative steam cycle is to operate between 1000 psia, 800°F, and 2 psia. One open feedwater heater is to be employed. Isentropic expansion occurs in the turbine, and the extraction pressure is chosen such that the enthalpy drop is the same in each of the two stages. Calculate the thermal efficiency of the cycle.

12-6 Repeat Prob. 12-5, using a closed feedwater heater with the condensate trapped back to the condenser hot well.

12-7 A regenerative steam cycle operates between 1500 psia, 1200°F, and 2 psia. Two closed feedwater heaters are used with extraction at 800 and 200 psia. The condensate is trapped back to the condenser hot well. Calculate the thermal efficiency of the cycle, assuming (a) isentropic expansion in the turbine and (b) a turbine efficiency of 85 percent for each of the three stages.

12-8 A combined reheat-regenerative steam cycle has the following operating conditions:

Steam inlet to first turbine stage: 1000 psia, 1000°F
Condenser pressure: 2 psia
Extraction pressure: 100 psia

After a portion of the steam is extracted for use in an open feedwater heater, the remainder is reheated to 1000°F and then expanded in the second turbine stage. The turbine efficiency for both stages is 85 percent, and the high-pressure pump efficiency is 60 percent. Calculate the thermal efficiency of the cycle.

12-9 A reciprocating air compressor receives air at 14 psia and 100°F and discharges at 45 psia. The clearance volume is 5 percent of the piston displacement. The compressor has a 4-in diameter bore and a 5-in stroke and rotates at 500 rpm. Calculate the volume of air delivered at inlet conditions, the volumetric efficiency, and the horsepower required for (a) isentropic compression and (b) polytropic compression with $n = 1.30$.

12-10 A two-stage machine is used to compress air from 14.7 psia, 70°F, to 500 psia. The compressor handles 110 ft³/min at inlet conditions, and ideal interstage pressure is used. Each stage is adiabatic but has an efficiency of 70 percent.

An ideal intercooler is employed between the stages. Calculate the horse-power required if the machine operates as a steady-flow device. What horse-power would be required if a single-stage machine were used?

12-11   A three-stage compressor is used to compress air from 14.7 psia, 70°F, to 1000 psia. Calculate the work required per pound mass of air compressed, assuming ideal interstaging and intercooling. Assume that the compression processes occur isentropically.

12-12   An air-standard Otto cycle has a compression ratio of 9:1. At the start of the compression stroke the conditions are 14.0 psia and 90°F. The maximum pressure in the cycle is 560 psia. Determine the maximum temperature in the cycle and the thermal efficiency. Compare this efficiency with that of a Carnot cycle operating between the same temperature limits.

12-13   An air-standard Otto cycle operates such that the compression and expansion processes follow polytropic relations with $n = 1.30$. Calculate the thermal efficiency for a cycle with a compression ratio of 8:1, inlet conditions of 14 psia, 80°F, and a maximum temperature limit of 2000°F.

12-14   An air-standard Otto cycle operates with a compression ratio of 10:1 and conditions at the start of the compression stroke of 14.5 psia, 100°F. During the combustion process the heat addition rate is 300 Btu/lbm of air. Calculate the temperature and pressure at all points in the cycle and the thermal efficiency, assuming (a) constant specific heats and (b) variable specific heats based on the air tables. Compare with the efficiency of a Carnot cycle operating between the same temperature limits.

12-15   The compression ratio for an air-standard diesel cycle is 19:1, and the heat input per pound mass of air is 400 Btu. The conditions at the start of compression are 80°F and 14.5 psia. Calculate the temperature and pressure at all points in the cycle, the cutoff ratio, and the thermal efficiency. Compare with the efficiency of a Carnot cycle operating between the same temperature limits.

12-16   An air-standard diesel cycle has a compression ratio of 17:1 with conditions at the beginning of the compression stroke of 14.0 psia and 50°F. The maximum temperature in the cycle is 3500°R. Calculate the temperature and pressure at each point in the cycle, the cutoff ratio, and the thermal efficiency. Compare with the efficiency of a Carnot cycle operating between the same temperature limits.

12-17   Calculate the thermal efficiency of an air-standard Otto cycle having the same upper pressure limit as the diesel cycle of Prob. 12-15.

12-18   Calculate the thermal efficiency of an Otto cycle having the same upper temperature limit as the diesel cycle of Prob. 12-16.

12-19   What conclusions can be drawn from the results of Probs. 12-17 and 12-18?

12-20   An air-standard Brayton cycle receives air at 14 psia and 60°F. The pressure ratio is 4:1, and the upper temperature limit of the cycle is 1800°F. The compressor and turbine efficiencies are 85 percent. Calculate the thermal efficiency of the cycle and the net work output per pound mass of air.

12-21 Repeat Prob. 12-20, using air tables which take into account variable specific heats.

12-22 An ideal regenerator is employed with the Brayton cycle of Prob. 12-20. Calculate the thermal efficiency under these conditions.

12-23 A gas-turbine cycle uses two stages of compression and expansion along with an ideal intercooler for the compression stages. Inlet conditions are 14.7 psia and 70°F, and the maximum pressure is 150 psia. The temperature at the inlet to each turbine stage is 1700°F. Calculate the thermal efficiency for the cycle. Assume isentropic processes in the compression and turbine stages.

12-24 A regenerator having an efficiency of 80 percent is installed on the two-stage cycle of Prob. 12-23. Calculate the thermal efficiency under these conditions and compare with the efficiency of a Carnot cycle operating between the same temperature limits.

12-25 Rework Prob. 12-24, using compressor and turbine efficiencies of 80 and 85 percent, respectively, for each stage.

12-26 One way of making use of the energy contained in the hot exhaust gases of a gas-turbine cycle is to employ them for heating steam in a Rankine cycle. Consider the gas-turbine cycle of Prob. 12-23. The hot exhaust gases are used to supply heat to a Rankine cycle operating between 500 and 2 psia. The upper temperature in the steam cycle is 700°F, and isentropic expansion is assumed in the turbine. The exhaust gases leave the steam heater at 500°F. Calculate:

    (a) The circulation of steam per pound mass of air in the gas-turbine cycle

    (b) The net work output of the combination gas-turbine-steam cycle per pound mass of air

    (c) The thermal efficiency of the combined cycle

12-27 Rework Prob. 12-26, assuming steam reheat to 700°F at 100 psia.

12-28 A turbojet engine takes in air at 14.5 psia and 70°F while in a stationary position on the ground. The pressure ratio in the compressor is 5:1, and the upper temperature limit in the cycle is 1800°F. The compressor and turbine efficiencies are 80 and 85 percent, respectively, and the expansion in the exhaust nozzle may be assumed isentropic. Calculate the thrust and heat input per pound mass of air, neglecting the inlet velocity.

12-29 Rework Prob. 12-28 for flight at an altitude where the inlet conditions are 3 psia, $-60°F$, and 800 ft/s.

12-30 An ideal-vapor refrigeration cycle uses Freon 12 as the working fluid. The unit is used for an air conditioning application such that the evaporator temperature is 45°F and the condenser saturation temperature is 90°F. Calculate the coefficient of performance and the horsepower required per ton of cooling.

12-31 A low-temperature refrigeration system uses ammonia as the working fluid. The evaporator temperature is $-35°F$, and the upper pressure limit of the cycle is 180.6 psia. The compressor efficiency is 80 percent. Calculate the COP and the horsepower required per ton of cooling. Compare these figures with the values for a Carnot refrigerator operating between the same temperature limits.

12-32 Suppose chilled water is available for cooling the condenser of the cycle in Prob. 12-31 such that the upper pressure limit of the cycle is reduced to 80.00 psia. Calculate the COP and kW input per ton under these new conditions.

12-33 A certain absorption air-conditioning system uses water as the working fluid and lithium bromide as the absorber. The evaporator temperature for the cycle is 40°F, and the condenser saturation temperature is 100°F. Calculate the cooling effect per pound mass of water.

12-34 An automobile air conditioner uses a Freon 12 refrigeration system. The evaporator temperature at high-speed driving conditions may be as low as 20°F while the saturation temperature in the condenser may be as high as 140°F. Calculate the horsepower required to provide 3 tons of cooling with a compressor efficiency of 80 percent. What is the coefficient of performance under these conditions?

12-35 A regenerative steam cycle employs one closed feedwater heater and two turbine stages. Steam enters the turbine at 800 psia and 700°F, and extraction occurs at 100 psia. Discharge pressure is 1 psia. The efficiency of each turbine stage is 85 percent, and a condensate pump pumps the condensate from the heater up to the boiler pressure. The pump efficiencies are 100 percent. Calculate the thermal efficiency of the cycle and the pump horsepowers required if the total electric output from the turbine is 20 000 kW.

12-36 A gas-turbine cycle receives air at 14.7 psia and 70°F and operates with an upper pressure limit of 250 psia. Two compression states are employed with ideal intercooling and the interstage pressure selected accordingly. Two turbine stages are also used with reheat between stages. A regenerator having an efficiency of 85 percent is connected to the discharge from the second turbine and the high-pressure air entering the first burner. Compressor efficiencies are 88 percent, and turbine efficiencies are 90 percent. The upper temperature limit of the cycle is 1900°F. Calculate the thermal efficiency of the cycle and the mass flow of air required for an electrical output of 10 000 kW. Assume air as the working fluid with constant specific heats. If a fuel having a heating value of 20 000 Btu/lbm is available at a price of $0.075/lbm, calculate the cost of electric energy generated in cents per kWh.

12-37 A regenerative steam cycle employs two closed feedwater heaters and three turbine stages. The condensate from the first heater flows through a steam trap into the second heater, and the condensate from the second heater is pumped up to the boiler pressure. The steam enters the turbine at 1500 psia, 1000°F, and extraction occurs at 300 and 100 psia, and discharge from the third turbine stage is at 1 psia. The turbine efficiencies are 85 percent, and the pump efficiencies are 100 percent. Calculate the thermal efficiency and steam flow required for a total electrical output from the turbine of 20 000 kW.

12-38   In a certain refrigeration system for low-temperature applications a two-stage operation is desirable which employs a Freon 12 system in combination with an ammonia system. The ammonia operates between a low-temperature limit of $-40°F$ and an upper temperature of 30°F. The condenser for the ammonia cycle (at 30°F) is also the evaporator for the Freon 12 cycle which then rejects heat in its own condenser at 100°F. The combined ammonia condenser and Freon evaporator is insulated so that no heat is gained from the surroundings, and all the heat rejected by the ammonia goes to evaporate the Freon. The overall system is to produce a refrigeration effect of 5 tons at the $-40°F$ condition. Assuming isentropic compression processes, calculate the mass flows of each fluid required, the power input to each compressor, and the coefficient of performance for the overall system.

12-39   In some parts of the western United States geothermal energy, i.e., underground hot water, may be an economical source of power. Suppose that the hot water is available at 200°F and the surroundings temperature for dissipation of heat is 85°F. A special Rankine vapor cycle is devised to operate between these limits using Freon 12 as the working fluid and a maximum pressure limit of 250 psia. The turbine efficiency is 85 percent. Assuming that the hot water temperature drops to 140°F as it is giving up heat to the Freon in the boiler, calculate the water flow rate required for a power output of 25 MW.

12-40   A regenerative steam cycle operates with steam inlet to the turbine at 1000 psia and 1200°F and a condenser temperature of 100°F. One open feedwater heater is to be employed with the steam extraction pressure selected to provide the maximum increase in cycle efficiency. Select several extraction pressures and examine their effect on thermal efficiency. Make the calculations for (a) isentropic expansion and (b) turbine stage efficiencies of 87 percent.

12-41   A nuclear power plant has an upper temperature limit of 850°F because of metallurgical considerations in the reactor vessel. In one particular installation the following design information is supplied:

Upper pressure limit:	1500 psia
Feedwater heaters:	Two, closed, condensate from first trapped to second, condensate from second pumped to boiler
Extraction pressures:	500 and 150 psia
Reheat:	Reheat to 850°F after each extraction
Condenser temperature:	100°F
Turbine efficiencies:	90 percent
Power output:	500,000 kW

Calculate the mass flow of steam required in the boiler, the thermal efficiency of the cycle, and the waste heat which must be dissipated in the condenser.

12-42   A small gas-turbine cycle is to be designed for automotive applications with the following data:

Maximum temperature: 1800°F
Turbine efficiency: 90 percent
Compressor efficiency: 85 percent
Regenerator efficiency: 70 percent
Inlet air temperature: 120°F, 14.7 psia
Power output: 300 hp
Pressure ratio: 3.5:1

Assuming an air-standard cycle, calculate the thermal efficiency of the cycle and mass flow of air required. If a fuel is used with a heating value of 140 000 Btu/gal, calculate the fuel consumption rate at the maximum power output. Make the calculations assuming constant specific heats and also using the air tables.

12-43 A closed gas-turbine cycle using helium as the working fluid is to be designed to accomplish the same power output as the cycle in Prob. 12-42. The same conditions may be assumed to apply except that the helium inlet conditions to the compressor are 300°F and 150 psia. Calculate the thermal efficiency of this cycle and the mass flow of helium required assuming constant specific heats. Also calculate the heat which must be dissipated to the surroundings.

12-44 To operate at high-compression ratios spark-ignition engines (Otto cycles) require special tetraethyl lead additives to the fuel which environmentalists claim are serious health hazards. To reduce these hazards some engine designs employ lower compression ratios. Estimate the percentage increase in fuel consumption which would result from lowering the compression ratio from 10.75:1 to a value of 8.5:1.

12-45 A steady-flow device is designed to compress helium from 1.5 atm, 25°C, to 100 atm in three stages. Ideal intercooling is employed with optimum staging and the stage efficiencies are 90 percent. Calculate the work input in kW to compress 10 kg/min.

12-46 A reciprocating air compressor has a volumetric efficiency of 95 percent and is designed to compress 500 ft³/min of air measured at inlet conditions of 1 atm and 70°F. The discharge pressure is 75 psia, and isentropic compression and expansion may be assumed. Calculate the clearance volume as a percentage of piston displacement and the horsepower input required.

12-47 A closed-cycle gas-turbine plant using helium as the working fluid is to be designed for a marine installation where heat may be rejected to seawater. An overall pressure ratio of 6:1 is used with two stages of compression and ideal intercooling to 70°F. Inlet conditions to the first stage are 70°F and 150 psia, and the upper temperature limit of the cycle is 2000°F. Only a single turbine stage is employed with a regenerator having an efficiency of 80 percent. Both compressor and turbine efficiencies are 90 percent. Calculate the thermal efficiency of the cycle and the mass flow of helium necessary to produce a 25 000-hp output. What is the total amount of heat which must be dissipated to the seawater, including that removed in the intercooler?

12-48 A large diesel engine is to produce a power output of 5000 hp and operates with a compression ratio of 18:1. The cutoff ratio is 2.0, and isentropic compression and expansion may be assumed. Assuming an air/fuel ratio of 16:1 and a fuel with a heating value of 20 000 Btu/lbm, calculate the fuel consumption rate necessary to produce the indicated power output. Make the calculations on the basis of an air-standard cycle.

12-49   A "heat pump" is a reversed refrigeration machine which may be used for heating in cold weather. The condenser becomes the heating unit inside the building while the evaporator receives heat from the cold air outdoors. Normally, a reciprocating compressor provides the work input to power the cycle. Consider a Freon 12 unit which has the following specifications:

      Condenser temperature: 100°F
      Evaporator temperature: 20°F
      Heating required: 125 000 Btu/h at 100°F

Assuming isentropic compression calculate the work input to produce the indicated heating and the mass flow rate of Freon 12. If electric energy costs $0.042/kWh, what is the cost of operation of the device? How does this compare with the cost of performing the same heating with fuel oil costing $4.00 per million Btu's?

12-50   The heat pump of Prob. 12-49 is to be operated as an air conditioning unit in the summer with the compressor providing the same mass flow rate capabilities, but the condenser rejects heat to the outdoors at 120°F (an air-cooled unit) while the evaporator functions at 55°F indoors. Calculate the cooling rate in tons for these conditions and the required horsepower input. Also calculate the electric operating cost of the unit for these conditions.

PROBLEMS (Metric units)

12-1M   In a Rankine cycle, steam enters the single-stage turbine at 8.0 MPa and 440°C. Heat is rejected in the condenser at 25°C. Calculate the thermal efficiency of the cycle. Repeat the calculation for reheat of steam at 4.0 MPa back to 440°C.

12-2M   A certain Rankine cycle operates with two reheat stages. Steam enters the turbine at 10 MPa and 640°C. Reheat to 640°C occurs at 8.0 MPa and 4.0 MPa. The lower pressure limit of the cycle is 6.0 kPa. Each turbine stage has an efficiency of 90 percent. Calculate the thermal efficiency of the cycle and compare with the Carnot efficiency between the same temperature limits.

12-3M   A small steam turbine is designed to operate with 1.5 MPa of steam. The turbine is to be employed in a simple Rankine cycle which rejects heat at 30°C. To what temperature must the high-pressure steam be heated to ensure that the moisture content does not exceed 5 percent in the turbine? Calculate the thermal efficiency of the cycle under these conditions. Assume isentropic expansion in the turbine.

12-4M   A regenerative steam cycle is to operate between 8 MPa, 440°C, and 10 kPa. One open feedwater heater is to be employed. Isentropic expansion occurs in the turbine, and the extraction pressure is chosen such that the enthalpy drop is the same in each of the two stages. Calculate the thermal efficiency of the cycle.

12-5M A regenerative steam cycle operates between 10 MPa, 640°C, and 10 kPa. Two closed feedwater heaters are used with extraction at 500 kPa and 1.5 MPa. The condensate is trapped back to the condenser hot well. Calculate the thermal efficiency of the cycle, assuming (a) isentropic expansion in the turbine and (b) a turbine efficiency of 85 percent for each of the three stages.

12-6M A combined reheat-regenerative cycle has the following operating conditions:

Steam inlet to first turbine stage: 8 MPa, 560°C
Condenser pressure: 10 kPa
Extraction pressure: 700 kPa

After a portion of the steam is extracted for use in an open feedwater heater, the remainder is reheated to 560°C and then expanded in the second turbine stage. The turbine efficiency for both stages is 85 percent, and the high-pressure pump efficiency is 60 percent. Calculate the thermal efficiency of the cycle.

12-7M A reciprocating air compressor receives air at 95 kPa and 38°C and discharges at 300 kPa. The clearance volume is 5 percent of the piston displacement. The compressor has a 10-cm diameter bore, a 12.5-cm stroke, and rotates at 500 rpm. Calculate the volume of air delivered at inlet conditions, the volumetric efficiency, and the power required for (a) isentropic compression and (b) polytropic compression with $n = 1.30$.

12-8M A two-stage machine is used to compress air from 1 atm, 20°C, to 3.5 MPa. The compressor handles 50 liters/s at inlet conditions, and ideal interstage pressure is used. Each stage is adiabatic but has an efficiency of 70 percent. An ideal intercooler is employed between the stages. Calculate the power required if the machine operates as a steady-flow device. What power would be required if a single-stage machine were used?

12-9M A three-stage compressor is used to compress air from 1 atm, 20°C, to 8.0 MPa. Calculate the work required per kilogram of air compressed, assuming ideal interstaging and intercooling. Assume that the compression processes occur isentropically.

12-10M An air-standard Otto cycle has a compression ratio of 9:1. At the start of the compression stroke the conditions are 95 kPa and 30°C. The maximum pressure in the cycle is 3.8 MPa. Determine the maximum temperature in the cycle and the thermal efficiency. Compare this efficiency with that of a Carnot cycle operating between the same temperature limits.

12-11M An air-standard Otto cycle operates such that the compression and expansion processes follow polytropic relations with $n = 1.30$. Calculate the thermal efficiency for a cycle with a compression ratio of 8:1, inlet conditions of 95 kPa, 25°C, and a maximum temperature limit of 1100°C.

12-12M   An air-standard Otto cycle operates with a compression ratio of 10:1 and conditions at the start of the compression stroke of 100 kPa, 35°C. During the combustion process the heat-addition rate is 650 kJ/kg of air. Calculate the temperature and pressure at all points in the cycle and the thermal efficiency, assuming (a) constant specific heats and (b) variable specific heats based on the air tables. Compare with the efficiency of a Carnot cycle operating between the same temperature limits.

12-13M   The compression ratio for an air-standard diesel cycle is 19:1, and the heat input per pound mass of air is 800 kJ/kg. The conditions at the start of compression are 25°C and 100 kPa. Calculate the temperature and pressure at all points in the cycle, the cutoff ratio, and the thermal efficiency. Compare with the efficiency of a Carnot cycle operating between the same temperature limits.

12-14M   An air-standard diesel cycle has a compression ratio of 17:1 with conditions at the beginning of the compression stroke of 95 kPa and 10°C. The maximum temperature in the cycle is 1900 K. Calculate the temperature and pressure at each point in the cycle, the cutoff ratio, and the thermal efficiency. Compare with the efficiency of a Carnot cycle operating between the same temperature limits.

12-15M   Calculate the thermal efficiency of an air-standard Otto cycle having the same upper-pressure limit as the diesel cycle of Prob. 12-13M.

12-16M   An automobile manufacturer produces two engines, each having a displacement of 5.7 liters. One engine operates as a diesel with a compression ratio of 22 and a cutoff ratio of 1.5. The other is a spark-ignition engine with a compression ratio of 8.0. The claim is made that the diesel will consume about 20 percent less fuel per kilometer traveled than the spark-ignition engine. Is this a reasonable claim?

12-17M   In some locales it is possible to use hot pressurized water from geothermal sources for the production of power. In one application water is available from the ground at 20 MPa and 300°C. This water is throttled to a pressure of 1.5 MPa to produce a wet mixture of steam, and this mixture is subsequently passed through a device to remove the liquid droplets (separator) so that saturated vapor at 1.5 MPa is available to enter a steam turbine. Discharge from the turbine is at 1 atm. Calculate the flow of groundwater needed to produce a 10 000-kW power output from the turbine. Assume isentropic flow in the turbine.

12-18M   Calculate the thermal efficiency of an Otto cycle having the same upper temperature limit as the diesel cycle of Prob. 12-14M.

12-19M   What conclusions can be drawn from the results of Prob. 12-15M and 12-18M?

12-20M   An air-standard Brayton cycle receives air at 95 kPa and 15°C. The pressure ratio is 4:1, and the upper temperature limit of the cycle is 980°C. The compressor and turbine efficiencies are 85 percent. Calculate the thermal efficiency of the cycle and the net work output per kilogram of air.

12-21M   An ideal regenerator is employed with the Brayton cycle of Prob. 12-20M. Calculate the thermal efficiency under these conditions.

12-22M   A gas-turbine cycle uses two stages of compression and expansion along with an ideal intercooler for the compression stages. Inlet conditions are 1 atm and 20°C, and the maximum pressure is 1.0 MPa. The temperature at the inlet to each turbine stage is 925°C. Calculate the thermal efficiency for the cycle. Assume isentropic processes in the compression and turbine stages.

12-23M   A regenerator having an efficiency of 80 percent is installed on the two-stage cycle of Prob. 12-22M. Calculate the thermal efficiency under these conditions and compare with the efficiency of a Carnot cycle operating between the same temperature limits.

12-24M   Rework Prob. 12-23M using compressor and turbine efficiencies of 80 and 85 percent, respectively, for each stage.

12-25M   One way of making use of the energy contained in the hot exhaust gases of a gas-turbine cycle is to employ them for heating steam in a Rankine cycle. Consider the gas-turbine cycle of Prob. 12-22M. The hot exhaust gases are used to supply heat to a Rankine cycle operating between 4.0 MPa and 10 kPa. The upper temperature in the steam cycle is 370°C, and isentropic expansion is assumed in the turbine. The exhaust gases leave the steam heater at 260°C. Calculate:
   (a) The circulation of steam per kilogram mass of air in the gas-turbine cycle
   (b) The net work output of the combination gas-turbine-steam cycle per kilogram mass of air
   (c) The thermal efficiency of the combined cycle

12-26M   Rework Prob. 12-25M, assuming steam reheat to 370°C at 700 kPa.

12-27M   A turbojet engine takes in air at 100 kPa and 20°C while in a stationary position on the ground. The pressure ratio in the compressor is 5:1, and the upper temperature limit in the cycle is 980°C. The compressor and turbine efficiencies are 80 and 85 percent, respectively, and the expansion in the exhaust nozzle may be assumed isentropic. Calculate the thrust and heat input per kilogram mass of air, neglecting the inlet velocity.

12-28M   Rework Prob. 12-27M for flight at an altitude where the inlet conditions are 20 kPa, −50°C, and 275 m/s.

12-29M   A regenerative steam cycle employs one closed feedwater heater and two turbine stages. Steam enters the turbine at 6.0 MPa and 370°C, and extraction occurs at 700 kPa. Discharge pressure is 6 kPa. The efficiency of each turbine stage is 85 percent, and a condensate pump pumps the condensate from the heater up to boiler pressure. The pump efficiencies are 100 percent. Calculate the thermal efficiency of the cycle and the pump horsepowers required if the total electrical output from the turbine is 20 000 kW.

12-30M   A gas-turbine cycle receives air at 1 atm and 20°C and operates with an upper pressure limit of 1.7 MPa. Two compression states are employed with ideal intercooling, and the interstage pressure is selected accordingly. Two turbine stages are also used with reheat between stages. A regenerator having an efficiency of 85 percent is connected to the discharge from the second turbine and the high-pressure air entering the first burner. Compressor efficiencies are 88 percent and turbine efficiencies are 90 percent. The upper temperature limit of the cycle is 1050°C. Calculate the thermal efficiency of the cycle and the mass flow of air required for an electrical output of 10 000 kW. Assume air as the working fluid with constant specific heats. If a fuel having a heating value of 44 000 kJ/kg is available at a price of $0.165/kg, calculate the cost of electric energy generated in cents per kilowatt hour.

12-31M   A regenerative steam cycle employs two closed feedwater heaters and three turbine stages. The condensate from the first heater flows through a steam trap into the second heater, and the condensate from the second heater is pumped up to boiler pressure. The steam enters the turbine at 10 MPa, 560°C, extraction occurs at 2 MPa and 700 kPa, and discharge from the third turbine stage is at 6 kPa. The turbine efficiencies are 85 percent, and the pump efficiencies are 100 percent. Calculate the thermal efficiency and steam flow required for a total electrical output from the turbine of 20 000 kW.

12-32M   A regenerative steam cycle operates with steam inlet to the turbine at 8.0 MPa and 640°C and a condenser temperature of 35°C. One open feedwater heater is to be employed with the steam extraction pressure selected to provide the maximum increase in cycle efficiency. Select several extraction pressures and examine their effect on thermal efficiency. Make the calculations for (a) isentropic expansion and (b) turbine stage efficiencies of 87 percent.

12-33M   A nuclear power plant has an upper temperature limit of 450°C because of metallurgical considerations in the reactor vessel. In one particular installation the following design information is supplied:

Upper pressure limit:	10.0 MPa
Feedwater heaters:	Two closed, condensate from first trapped to second; condensate from second pumped to boiler
Extraction pressures:	3.5 MPa and 1.0 MPa
Reheat:	Reheat to 450°C after each extraction
Condenser Temperature:	35°C
Turbine efficiencies:	90 percent
Power output:	500 000 kW

Calculate the mass flow of steam required in the boiler, the thermal efficiency of the cycle, and the waste heat which must be dissipated in the condenser.

12-34M   A small gas-turbine cycle is to be designed for automotive applications with the following data:

Maximum temperature:	1000°C
Turbine efficiency:	90 percent
Compressor efficiency:	85 percent
Regenerator efficiency:	70 percent
Inlet air temperature:	50°C, 1 atm
Power output:	225 kW
Pressure ratio:	3.5:1

Assuming an air-standard cycle, calculate the thermal efficiency of the cycle and mass flow of air required. If a fuel is used with a heating value of 36 000 kJ/liter, calculate the fuel consumption rate at the maximum power output. Make the calculations assuming constant specific heats and also using the air tables.

12-35M A closed gas-turbine cycle using helium as the working fluid is to be designed to accomplish the same power output as the cycle in Prob. 12-34M. The same conditions may be assumed to apply except that the helium inlet conditions to the compressor are 150°C and 1.0 MPa. Calculate the thermal efficiency of this cycle and the mass flow of helium required assuming constant specific heats. Also calculate the heat which must be dissipated to the surroundings.

12-36M A reciprocating air compressor has a volumetric efficiency of 95 percent and is designed to compress 235 liters/s of air measured at inlet conditions of 1 atm and 20°C. The discharge pressure is 500 kPa, and isentropic compression and expansion may be assumed. Calculate the clearance volume as a percentage of piston displacement and the horsepower input required.

12-37M A closed-cycle gas-turbine plant using helium as the working fluid is to be designed for a marine installation where heat may be rejected to seawater. An overall pressure ratio of 6:1 is used with two stages of compression and ideal intercooling to 20°C. Inlet conditions to the first stage are 20°C and 1.0 MPa, and the upper temperature limit of the cycle is 1100°C. Only a single turbine stage is employed with a regenerator having an efficiency of 80 percent. Both compressor and turbine efficiencies are 90 percent. Calculate the thermal efficiency of the cycle and the mass flow of helium necessary to produce a 18 700-kW output. What is the total amount of heat which must be dissipated to the seawater, including that removed in the intercooler?

12-38M A large diesel engine is to produce a power output of 3700 kW and operates with a compression ratio of 18:1. The cutoff ratio is 2.0, and isentropic compression and expansion may be assumed. Assuming an air/fuel ratio of 16:1 and a fuel with a heating value of 44 000 kJ/kg, calculate the fuel-consumption rate necessary to produce the indicated power output. Make the calculations on the basis of an air-standard cycle.

12-39M An ammonia refrigeration machine is designed to produce 25 tons of cooling at −20°F. Outlet from the compressor is at 220 psia and 360°F, and the throttling (expansion) valve receives the liquid at 100°F. Calculate the compressor efficiency, flow rate of ammonia, and coefficient of performance.

12-40M   A certain Freon 12 heat pump is designed to heat 460 liters/s of air at 1 atm from 5°C to 30°C. The evaporator temperature (outside) is 20°F, and the condenser pressure is 140 psia. For isentropic compression calculate the coefficient of performance, work input, and cost per hour for heating with an electric power cost of $0.07/kWh.

12-41M   A Carnot refrigerator has a coefficient of performance of 5.4 when the low temperature is $-20°C$. The change in entropy for heat addition and rejection is 0.40 kJ/K. Determine the refrigeration effect and work input.

12-42M   A certain refrigeration device is designed to produce cooling of 1 ton with a coefficient of performance of 4.0. Determine the power input in kilowatts. Suppose the same device is used for heating (heat pump). What would be its *heating* COP?

12-43M   A reversed Carnot cycle is to be employed for heating (heat pump) in which the heating requirements are 6000 kJ/min. The heat added to the cycle is 4500 kJ/min. What power input in kilowatts is required?

12-44M   A jet aircraft travels at a speed of 270 m/s and the velocity of the exhaust gases is 600 m/s. A net thrust of 18 kN is produced. Calculate the mass flow of air required.

12-45M   Inlet conditions for Prob. 12-44M are 30 kPa, 0°C, and the exhaust conditions are 30 kPa and 670°C. Compute the overall irreversibility for the aircraft assuming $T_0 = 20°C$.

12-46M   The work input to a reversed Carnot cycle is 5.0 kW. What is the refrigeration effect if the coefficient of performance is 4.0? What would be the coefficient of performance if the unit were used for heating? What is the upper temperature if the lower temperature is 0°C?

12-47M   Air having an initial velocity of 20 m/s is compressed from 1 atm, 25°C, to 300 kPa. The final velocity is 80 m/sec. Calculate the inlet volume flow if 1900-kW power is required and the process is reversible.

12-48M   The air flow in Prob. 12-47M is compressed from the same inlet conditions to 300 kPa in an irreversible manner, and the discharge temperature is 160°C. Calculate the power input.

12-49M   An air-standard gas turbine is designed to produce a power output of 3.7 MW with air inlet at 1 atm, 30°C, and a pressure ratio of 10.0. Inlet to the turbine is at 840°C, and the compressor and turbine efficiencies are both 85 percent. Calculate the air flow required, compressor and turbine works, and cycle thermal efficiency.

12-50M   Repeat Prob. 12-49M for a regenerator installed with an efficiency of 90 percent. Compare overall cycle irreversibilities for the two cycles. For Prob. 12-49M assume that the turbine discharge is cooled at constant pressure to 30°C.

12-51M   A gas turbine employs one compression and two turbine stages. The high-pressure turbine produces just enough work to drive the compressor, while the low-pressure turbine produces the net cycle output. Inlet air is at

1 atm, 25°C, and the overall pressure ratio is 6.5. Inlet temperature to each turbine is 800°C. Isentropic compression and expansion may be assumed, and a regenerator with an efficiency of 90 percent is employed. If a fuel with heating value of 40 MJ/kg is used, calculate the thermal efficiency and fuel flow for a power output of 10 MW.

12-52M A small gas turbine is designed to operate on a closed cycle using argon as the working fluid. The compressor and turbine efficiencies are 88 percent and 82 percent, respectively. The gas enters the compressor at 400 kPa, 40°C, and the compression ratio is 10.0. The temperature at entry to the turbine is 1100°C. Calculate the argon flow rate for a total power output of 6.0 MW. Also calculate the irreversibility for the cycle with $T_0 = 20$°C.

12-53M A closed-cycle gas turbine uses argon as the working fluid with the following conditions:

Entry to compressor:	60°C, 475 kPa
Exit from compressor:	375°C, 1.9 MPa
Entry to turbine:	1100°C
Exit from turbine:	620°C

Calculate the work output per unit mass flow and the compressor and turbine efficiencies. Also calculate the irreversibility for the cycle with $T_0 = 20$°C.

12-54M Steam enters the turbine of a Rankine cycle at a pressure of 3.0 MPa and discharges at a pressure of 30 kPa. Calculate the thermal efficiency of the cycle for three inlet temperatures: (a) saturated vapor, (b) 320°C, and (c) 700°C.

12-55M Measurements are performed on an actual regenerative-reheat steam power cycle. One closed feedwater heater is employed with condensate pumped to the boiler. The following data are obtained:

Rated power output:	50 000 kW
Steam entrance to turbine:	$p = 9.0$ MPa, $T = 490$°C
Extraction properties:	$p = 900$ kPa, $T = 210$°C
Fraction extracted:	24 percent
Steam entrance to second turbine stage after reheat:	$p = 850$ kPa, $T = 490$°C
Condenser temperature:	30°C
Discharge temperature from feedwater heater:	175°C
Turbine efficiency for second stage:	78 percent

For these data calculate the steam flow required and the thermal efficiency of the cycle.

12-56M In modern steam power plants it is common practice to use supercritical pressures in the boiler. Consider an ideal cycle with steam leaving the boiler at 40 MPa and 800°C. Five closed feedwater heaters are employed with extractions at 8.0 MPa, 5.0 MPa, 1.5 MPa, 800 kPa, and 100 kPa. Discharge from the turbine is at 8 kPa. Assume an isentropic turbine and the condensate from each heater trapped to the next heater, and that the condensate from the last heater is pumped to the boiler. Calculate the thermal efficiency of the cycle. What power output may be obtained with a flow rate of 500 000 kg/h.

12-57M   An ideal reheat cycle employs two stages of reheat. Initially, steam enters the turbine at 20.0 MPa and 540°C and is reheated back to 540°C at pressures of 4.0 and 1.0 MPa. Discharge from the turbine is at 20 kPa. Calculate the thermal efficiency of the cycle and the power output for a steam flow of 10 kg/s.

12-58M   Proposals have been made to use a Rankine cycle for automotive applications with Freon 12 as the working fluid. Suppose a power output of 38 kW is desired and that the Freon 12 enters an isentropic turbine at 500 psia and discharges as a saturated vapor at 100°F. Calculate the thermal efficiency of the cycle and the flow rate of Freon 12 required for the rated output.

12-59M   One proposal for power production would obtain methane gas from offshore geothermal sources to fuel a hybrid gas-turbine steam power cycle. The plant is to be installed on an offshore platform so that heat rejection can be to the ocean. The following conditions are assumed to apply:

Inlet to compressor:	1 atm, 20°C
Compression ratio:	10:1
Temperature at inlet to gas turbine:	1400 K
Inlet to steam turbine:	6.0 MPa, 320°C
Condensing temperature for steam:	15°C
Compressor efficiency:	87 percent
Gas-turbine efficiency:	90 percent
Steam-turbine efficiency:	92 percent
Gas-exit temperature from boiler and water preheater:	100°C

Calculate the thermal efficiency for the hybrid cycle, the flow rate of air and steam required for a power output of 500 MW, and the heat rejected to the ocean for this power level. If the fuel supplies 38 kJ/liter at 1 atm, 20°C, calculate the flow rate required at 10 atm, 50°C, to produce the required power output.

12-60M   Rework Prob. 12-59M assuming two stages of compression and expansion with ideal interstaging and intercooling. Assume the temperature at inlet to each turbine stage is 1400 K.

REFERENCES

1   Gaffert, G. A.: "Steam Power Stations," 4th ed., McGraw-Hill Book Company, New York, 1952.

2   Hill, P. G., and C. R. Peterson: "Mechanics and Thermodynamics of Propulsion," Addison-Wesley Publishing Company, Inc., Reading, Mass., 1965.

3 Jones, J. B., and G. A. Hawkins: "Engineering Thermodynamics," John Wiley & Sons, Inc., New York, 1960.

4 Obert, E. F., and R. A. Gaggioli: "Thermodynamics," McGraw-Hill Book Company, New York, 1963.

5 Doolittle, J. S.: "Thermodynamics for Engineers," 2d ed., International Textbook Company, Scranton, Pa., 1964.

6 Stoecker, W. F.: "Refrigeration and Air Conditioning," McGraw-Hill Book Company, New York, 1958.

7 Barron, Randall: "Cryogenic Systems," McGraw-Hill Book Company, New York, 1966.

8 Culp, A. W.: "Principles of Energy Conversion," McGraw-Hill Book Company, New York, 1979.

9 Fraas, A. P.: Problems in Coupling a Gas Turbine to a Thermonuclear Reactor, *ASME*, 72-GT-98, May, 1972.

10 Cole, D. E.: The Wankel Engine, *Scientific American*, vol. 227, pp. 14–23, August 1972.

11 Wood, B. D.: "Applications of Thermodynamics," Addison-Wesley, 1969.

chapter thirteen
Thermodynamics of compressible flow

13-1 INTRODUCTION

We have already applied thermodynamic principles to a number of flow processes and analyzed the behavior of certain open systems with the control-volume concept and appropriate mass and energy balances. In this chapter we wish to extend the analysis to situations which involve the very high-speed flow of a compressible fluid: in most cases, air. In such systems the concepts of the Mach number and shock waves become important. Our discussion will be confined to one-dimensional compressible flows, i.e., those flows which involve only a single velocity direction.

13-2 THE STAGNATION STATE

When a fluid is brought to rest from some velocity we say that it has attained a stagnation state. The manner in which the slowing-down process is accomplished obviously influences the final stagnation state which will be experienced. The reversible adiabatic, or isentropic, stagnation state will be of great interest later. For such a stagnation process the steady-flow energy equation is

$$h + \frac{V^2}{2g_c} = h_0 \tag{13-1}$$

where h_0 designates the enthalpy of the fluid after it is brought to rest and h is the fluid enthalpy when it has the velocity V. If the fluid is brought to rest adiabatically but *irreversibly*, the stagnation enthalpy would still be the same because Eq. (13-1) is an energy balance and does not depend on the reversibility of the process. However, the final stagnation *state* would not be the same because an increase in entropy would occur in the irreversible process. The two processes are illustrated in Fig. 13-1 on an h-s diagram. The main difference between the two is that the isentropic stagnation pressure p_0 is higher than p', the value which might be attained in some actual stagnation process. Although both stagnation states involve zero velocity we shall reserve the zero subscript to designate the isentropic stagnation state.

The pressure in the flow which would be measured by an observer moving along with the fluid is usually termed the *static pressure* and is the pressure which determines the thermodynamic state of the moving fluid. The choice of the word *static* is somewhat unfortunate because it implies a pressure of the fluid after it is brought to rest, but, as we have already discussed, it is the *stagnation* properties that are measured at zero velocity. To keep the two types of pressure straight the reader is urged to recall the first sentence of this paragraph. Anytime we use the term static applied to a thermodynamic property it will mean a value measured by an observer as he

FIG. 13-1   *Stagnation states.*

moves at the local flow velocity. Extended discussions of this matter may be found in the references at the end of the chapter.

13-3   SONIC VELOCITY AND THE MACH NUMBER

The velocity at which a small pressure disturbance will propagate in a fluid is called the velocity of sound and designated by the symbol c. We shall now derive an expression for calculating this velocity for an ideal gas in terms of the thermodynamic state of the gas. Consider the tube shown in Fig. 13-2. In (a) we are viewing the wave front as a stationary observer, watching the infinitesimal wave move by at a velocity c. Across the front there are small changes in the thermodynamic properties, as illustrated. In Fig. 13-2b, an alternate viewpoint is taken of moving along with the wave front and observing the flow in and out of the control surface enclosing the front. We assume that the flow through the front is adiabatic so a small change in enthalpy dh will cause a change in velocity dV because of energy conservation.

Now let us make a mass and energy balance on the control volume surrounding the wave front in Fig. 13-2b. From the mass-continuity relation of Chap. 4 we have

$$\rho A c = (\rho + d\rho)A(c - dV) \tag{13-2}$$

or

FIG. 13-2   *Sonic wave.* (*a*) *Stationary observer and* (*b*) *observer moving with wave front.*

$$c \, d\rho - \rho \, dV = 0 \qquad (13\text{-}3)$$

where A is the flow cross-sectional area. We know that

$$T \, ds = dh - v \, dp = dh - \frac{dp}{\rho} \qquad (13\text{-}4)$$

If the process is reversible as well as adiabatic Eq. (13-4) becomes

$$0 = dh - \frac{dp}{\rho} \qquad (13\text{-}5)$$

The energy balance on the wave front is

$$h + \frac{c^2}{2g_c} = (h + dh) + \frac{(c - dV)^2}{2g_c}$$

or, clearing terms,

$$dh - \frac{c \, dV}{g_c} = 0 \qquad (13\text{-}6)$$

Combining Eqs. (13-5) and (13-6) gives

$$\frac{dp}{\rho} - \frac{c \, dV}{g_c} = 0 \qquad (13\text{-}7)$$

Now, combining Eqs. (13-3) and (13-7) gives

$$\frac{c^2}{g_c} = \frac{dp}{d\rho}$$

Because we have assumed an isentropic process this should be written

$$\frac{c^2}{g_c} = \left(\frac{\partial p}{\partial \rho}\right)_s \qquad\qquad (13\text{-}8)$$

Equation (13-8) now provides a mechanism for calculating the speed of sound in a fluid medium. For an ideal gas further simplification is possible. We have shown that the isentropic process for an ideal gas can be represented by

$$pv^\gamma = \text{const.} \qquad \text{or} \qquad p\left(\frac{1}{\rho}\right)^\gamma = \text{const.} \qquad\qquad (13\text{-}9)$$

We can therefore perform a differentiation of Eq. (13-9) to obtain

$$\frac{dp}{d\rho} = \left(\frac{\partial p}{\partial \rho}\right)_s = \frac{\gamma p}{\rho} \qquad\qquad (13\text{-}10)$$

Now, combining Eqs. (13-8) and (13-10) we obtain, for an ideal gas,

$$c^2 = \frac{\gamma p g_c}{\rho}$$

But $p/\rho = RT$ for the ideal gas so that

$$c^2 = \gamma g_c RT \qquad\qquad (13\text{-}11)$$
$$c = \sqrt{\gamma g_c RT} \qquad\qquad (13\text{-}12)$$

Thus the speed of sound in an ideal gas is a function of temperature alone. Recalling our discussions of Chaps. 8 and 9 wherein various mean molecular velocities were proportional to $T^{1/2}$, it is all the more interesting to note that the velocity of sound is also proportional to $T^{1/2}$. We showed that various transport processes were related to the mean molecular speed available to convey energy, momentum, and mass. It is not surprising, therefore, to find that the rate at which we can propagate a small pressure disturbance, i.e., the velocity of sound, follows the same temperature dependence. From a molecular viewpoint, of course, it is molecular collisions that must propagate the pressure disturbance, and the collision rates are proportional to mean molecular speed.

The Mach number M is a dimensionless ratio which we shall find useful in later analytical developments and is defined as

$$M = \frac{V}{c} \tag{13-13}$$

where V is the flow velocity. Different flow regions are designated in terms of values of the Mach number as supersonic for $M > 1$, subsonic for $M < 1$, and sonic flow for $M = 1$.

EXAMPLE 13-1   Isentropic stagnation

What temperature and pressure will air attain if it is stagnated isentropically from $M = 4$, $T = -60°F$, and $p = 0.1$ atm?

SOLUTION   The speed of sound at the given conditions is calculated from Eq. (13-12):

$$c = (\gamma g_c RT)^{1/2} = [(1.4)(32.2)(53.35)(460 - 60)]^{1/2}$$
$$= 980.8 \text{ ft/s} = 299 \text{ m/s}$$

For $M = 4$ the corresponding flow velocity is

$$V = (4)(980.8) = 3923 \text{ ft/s} = 1196 \text{ m/s}$$

Using the energy balance for the stagnation process

$$h + \frac{V^2}{2g_c} = h_0$$

or

$$c_p(T_0 - T) = \frac{V^2}{2g_c}$$

$$T_0 = -60 + \frac{(3923)^2}{(2)(0.24)(32.2)(778)}$$

$$= -60 + 1279 = 1219°F = 660°C$$

The final stagnation pressure is obtained from the isentropic relation

$$\frac{p_0}{p} = \left(\frac{T_0}{T}\right)^{\gamma/(\gamma-1)}$$

$$p_0 = (0.1) \left(\frac{1679}{400}\right)^{3.5} = 15.15 \text{ atm} = 2227 \text{ psia}$$

EXAMPLE 13-2   Speed of sound

Calculate the speed of sound in nitrogen at 200°C.

SOLUTION   We shall use SI units for this calculation. The gas constant is

$$R = \frac{\mathscr{R}}{M} = \frac{8315}{28} = 297 \text{ J/kg·K}$$

From Table 2-2, $\gamma = 1.40$ for nitrogen, so

$$
\begin{aligned}
c &= (\gamma g_c RT)^{1/2} \\
&= [(1.4)(1.0 \text{ kg·m/N·s}^2)(297 \text{ J/kg·K})(200 + 273)]^{1/2} \\
&= 443.5 \text{ m/s} = 1455 \text{ ft/s}
\end{aligned}
$$

13-4   STEADY, ONE-DIMENSIONAL ISENTROPIC FLOW

Let us now derive some useful relations for a general one-dimensional isentropic flow in steady state. Such a flow system may be approximated in a number of practical cases, most often when an ideal gas flows in a smooth channel, nozzle, or diffuser. At high velocities the sonic velocity and Mach number become important parameters which determine how the flow will react to changes in pressure, flow cross-sectional area, etc. To analyze the problem we have four general notions at our disposal: the energy relation for the flow, the mass-continuity equation, an equation representing the thermodynamic equation of state, and finally a process representation which, in this case, is obtained from our previous development for the sonic velocity.

For isentropic steady flow the energy equation is

$$h + \frac{V^2}{2g_c} = \text{const.}$$

or, in differential form

$$dh + \frac{V \, dV}{g_c} = 0 \tag{13-14}$$

The mass-continuity relation is

$$\rho AV = \dot{m} = \text{const.}$$

or, again in differential form,

$$AV \, d\rho + \rho A \, dV + \rho V \, dA = 0$$

or

$$\frac{d\rho}{\rho} + \frac{dV}{V} + \frac{dA}{A} = 0 \tag{13-15}$$

Our thermodynamic property relation for the isentropic case is

$$T \, ds = dh - v \, dp = dh - \frac{dp}{\rho} = 0 \tag{13-16}$$

If we combine the energy and property relations [Eqs. (13-14) and (13-16)] we obtain

$$dh = \frac{dp}{\rho} = \frac{-V \, dV}{g_c} \tag{13-17}$$

or

$$\frac{V \, dV}{g_c} + \frac{1}{\rho} \, dp = 0 \tag{13-18}$$

Equation (13-18) is the differential form of *Bernoulli's equation* and indicates to us that as the velocity increases ($dV > 0$) the pressure must decrease ($dp < 0$). In a physical sense this means that the pressure force serves to accelerate the flow.

Because the flow is isentropic we can obtain from our equation for the speed of sound

$$\left(\frac{\partial p}{\partial \rho}\right)_s = \frac{dp}{d\rho} = \frac{c^2}{g_c}$$

or

$$dp = \frac{c^2}{g_c} \, d\rho \tag{13-19}$$

We may now combine the differential continuity relation, Eq. (13-15), the property relation, Eq. (13-16), and Eq. (13-19) to obtain

$$\frac{dA}{A} = \frac{dp}{\rho V^2 / g_c} \left(1 - \frac{V^2}{c^2}\right) = \frac{dp}{\rho V^2 / g_c} (1 - M^2) \tag{13-20}$$

Equation (13-20) provides us with the basic information to determine the shape which a channel must assume for subsonic or supersonic isentropic flow. These shapes are illustrated in Fig. 13-3. Consider a subsonic flow case where we wish to *increase* the velocity (a nozzle). From Eq. (13-18) $dp < 0$ for $dV > 0$. For $M < 1$ and $dV > 0$ the right side of Eq. (13-20) is therefore

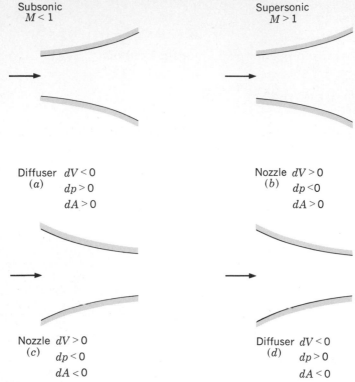

FIG. 13-3   *Subsonic and supersonic nozzles and diffusers.*

negative and dA must be negative, so we must decrease the flow area to cause an increase in velocity as illustrated in Fig. 13-3c. For $M > 1$ just the opposite is the case because the right side of Eq. (13-20) becomes positive for $dV > 0$, and we must therefore *increase* the flow area to increase the velocity in supersonic flow, as illustrated in Fig. 13-3b. To effect a decrease in velocity (a diffuser) requires the opposite changes in flow area as shown in Fig. 13-3a and d for $M < 1$ and $M > 1$, respectively.

When $M = 1$ the flow area attains its minimum value and we say that a *throat* condition is attained. A nozzle which is to accelerate the fluid from subsonic to supersonic conditions therefore must have first a converging and then a diverging section as shown in Fig. 13-4. Such a flow channel is appropriately called a converging-diverging nozzle, and $M = 1$ is experienced at the minimum flow area, or throat.

We can now express the temperature variation in the above isentropic flow situations through the use of the energy equation

$$h_0 - h = c_p(T_0 - T) = \frac{V^2}{2g_c} \tag{13-21}$$

Noting that

Flow

$M < 1$ $M = 1$ $M > 1$

FIG. 13-4 *Converging-diverging supersonic nozzle.*

$$c_p = \frac{\gamma R}{\gamma - 1} \quad \text{and} \quad c^2 = \gamma g_c RT$$

Eq. (13-21) can be written

$$\frac{V^2}{c^2} = M^2 = 2g_c \frac{\gamma RT}{(\gamma - 1)(\gamma g_c RT)} \left(\frac{T_0}{T} - 1\right)$$

Clearing terms,

$$\frac{T_0}{T} = 1 + \frac{\gamma - 1}{2} M^2 \tag{13-22}$$

For isentropic flow the pressure and density are obtained from

$$\frac{p_0}{p} = \left(\frac{T_0}{T}\right)^{\gamma/(\gamma-1)} = \left(1 + \frac{\gamma - 1}{2} M^2\right)^{\gamma/(\gamma-1)} \tag{13-23}$$

$$\frac{\rho_0}{\rho} = \left(\frac{T_0}{T}\right)^{1/(\gamma-1)} = \left(1 + \frac{\gamma - 1}{2} M^2\right)^{1/(\gamma-1)} \tag{13-24}$$

The flow properties at the throat are designated with a superscript asterisk (*) and correspond to $M = 1$. Therefore,

$$\frac{T^*}{T_0} = \frac{2}{\gamma + 1} \tag{13-25}$$

$$\frac{p^*}{p_0} = \left(\frac{2}{\gamma + 1}\right)^{\gamma/(\gamma-1)} \tag{13-26}$$

$$\frac{\rho^*}{\rho_0} = \left(\frac{2}{\gamma + 1}\right)^{1/(\gamma-1)} \tag{13-27}$$

It is common practice to refer to the throat properties as *critical* properties (distinguished from the critical properties of Chap. 7) and the above ratios are *critical ratios*. Their values for several values of γ are indicated in Table 13-1. Values of the various temperature, pressure and density ratios are tabulated in Table A-19 of the Appendix.

TABLE 13-1
Critical pressure, temperature, and density ratio for
isentropic flow of an ideal gas

	$\gamma = 1.2$	$\gamma = 1.3$	$\gamma = 1.4$	$\gamma = 1.67$
$\dfrac{p^*}{p_0}$	0.5644	0.5457	0.5283	0.4867
$\dfrac{T^*}{T_0}$	0.9091	0.8696	0.8333	0.7491
$\dfrac{\rho^*}{\rho_0}$	0.6209	0.6276	0.6340	0.6497

For steady flow the mass flow rate is constant and may be calculated from

$$\dot{m} = \rho A V = \frac{p}{RT} VA$$

$$= \frac{p}{RT} M\sqrt{\gamma g_c RT}\, A \qquad (13\text{-}28)$$

Using the relation for T_0/T [Eq. (13-22)] this becomes

$$\frac{\dot{m}}{A} = \frac{pM}{\sqrt{T_0}} \sqrt{\frac{\gamma g_c}{R}} \left(1 + \frac{\gamma - 1}{2} M^2\right)^{1/2} \qquad (13\text{-}29)$$

Substituting for p in terms of p_0 from Eq. (13-23) gives

$$\frac{\dot{m}}{A} = \frac{p_0 M}{\sqrt{T_0}} \sqrt{\frac{\gamma g_c}{R}} \frac{1}{\left(1 + \dfrac{\gamma - 1}{2} M^2\right)^{(\gamma+1)/2(\gamma-1)}} \qquad (13\text{-}30)$$

At the throat $M = 1$ and $A = A^*$ so that

$$\frac{\dot{m}}{A^*} = p_0 \sqrt{\frac{\gamma g_c}{RT_0}} \left(\frac{2}{\gamma + 1}\right)^{(\gamma+1)/2(\gamma-1)} \qquad (13\text{-}31)$$

This yields the interesting result that the total mass flow in the nozzle is a function of only the stagnation properties and the throat area, which leads us to the concept of *choked flows*.

For a given set of stagnation properties there is a maximum flow which can be forced through the nozzle as governed by the throat area. No matter what one does downstream from the throat in the way of decreasing the pressure, or increasing the flow area, the flow rate will remain the same if sonic conditions are established at the throat. If the flow has not been accelerated enough to establish $M = 1$ at the minimum flow area then choked flow conditions are never established and the above reasoning does not apply.

The overall area variation through the nozzle may be determined by combining Eqs. (13-30) and (13-31) to give

$$\frac{A}{A^*} = \frac{1}{M} \left[\left(\frac{2}{\gamma + 1} \right) \left(1 + \frac{\gamma - 1}{2} M^2 \right) \right]^{(\gamma+1)/2(\gamma-1)} \tag{13-32}$$

Values of A/A^* are also given in Table A-19 of the Appendix.

EXAMPLE 13-3   Converging nozzle for flow measurement

A sonic nozzle (converging section only) is sometimes used as a flow-measurement device because for $M = 1$ at the exit the flow rate is dependent only on the upstream stagnation pressure and temperature. What is the maximum air flow rate for such a device with exit area of 0.25 in in diameter and stagnation properties of 100 psia and 400°F?

SOLUTION   For this calculation we can make use of Eq. (13-31) to calculate the flow rate with

$$A^* = \frac{\pi(0.25)^2}{4} = 0.0491 \text{ in}^2 = 3.41 \times 10^{-4} \text{ ft}^2$$

$$p_0 = 100 \text{ psia} \qquad T_0 = 400°F = 860°R$$

Then, substituting in Eq. (13-31)

$$\dot{m} = (3.41 \times 10^{-4})(100)(144) \left[\frac{(1.4)(32.2)}{(53.35)(860)} \right]^{1/2} \left(\frac{2}{1.4 + 1} \right)^{(1.4+1)/(2)(1.4-1)}$$

$$= 0.0891 \text{ lbm/s}$$

EXAMPLE 13-4   Converging-diverging nozzle

A small converging-diverging nozzle having an exit area of 4.0 cm² is to be designed to produce an exit flow of air at atmospheric pressure and $M = 2.5$. The upstream stagnation temperature is 125°C. Calculate (a) the stagnation pressure, (b) the exit temperature, (c) the throat area, and (d) the mass flow rate.

SOLUTION   For this calculation we could make use of the appropriate equations which have been derived but it is much simpler to employ the tables in the Appendix for $\gamma = 1.4$. Designating the exit with the subscript e we have the given data:

$$A_e = 4.0 \text{ cm}^2 \qquad M_e = 2.5 \qquad T_0 = 125°C = 398 \text{ K} \qquad p_e = 1 \text{ atm}$$

Entering Table A-19 at $M = 2.5$ we find

$$\frac{A_e}{A^*} = 2.6367 \qquad \frac{p_e}{p_0} = 0.05853 \qquad \frac{T_e}{T_0} = 0.44444$$

The stagnation pressure is therefore

$$p_0 = \frac{1}{0.058\ 53} = 17.09 \text{ atm} = 1.73 \text{ MPa}$$

and the exit temperature is

$$T_e = (398)(0.444\ 44) = 176 \text{ K} = -96°C(-141°F)$$

The throat area is

$$A^* = \frac{4.0}{2.6367} = 1.517 \text{ cm}^2$$

There are several formulas available to us for calculating the mass flow rate. We shall use Eq. (13-29) applied at exit conditions. Because we are working in SI units we recall that $g_c = 1.0 \text{ kg·m/N·s}^2$.
    For air

$$R = \frac{\mathcal{R}}{28.97} = \frac{8315}{28.97} = 287 \text{ J/kg K}$$

Substituting in Eq. (13-29),

$$m = \frac{(4.0 \times 10^{-4} \text{ m}^2)(1.0132 \times 10^5 \text{ N/m}^2)(2.5)}{(398)^{1/2}}$$

$$\times \left[\frac{(1.4)(1.0 \text{ kg·m/N·s}^2)}{287 \text{ J/kg·K}}\right]^{1/2} \left[1 + \frac{(1.4 - 1)(2.5)^2}{2}\right]^{1/2}$$

$$= 0.532 \text{ kg/s} \qquad (1.171 \text{ lbm/s})$$

We would obtain the same answer if we had applied Eq. (13-31) to the throat conditions.

EXAMPLE 13-5

For the nozzle of Example 13-4 what would be the flow area, temperature, pressure, and velocity at the section where $M = 1.5$?

SOLUTION   This calculation is very simple because we have already determined the throat area and stagnation properties. We simply enter Table A-19 at $M = 1.5$ and read

$$\frac{A}{A^*} = 1.1762 \qquad \frac{p}{p_0} = 0.2724 \qquad \frac{T}{T_0} = 0.689\ 65$$

so that

$$A = (1.1762)(1.517) = 1.784\ \text{cm}^2$$
$$p = (0.2764)(17.09) = 4.724\ \text{atm}$$
$$T = (0.689\ 65)(398) = 274\ \text{K} = 1°\text{C}$$

To calculate the flow velocity we first compute the speed of sound at this location using Eq. (13-12):

$$c = (\gamma g_c RT)^{1/2} = [(1.4)(1.0)(287)(274)]^{1/2} = 331.8\ \text{m/s}$$

For $M = 1.5$ the flow velocity is therefore

$$V = (1.5)(331.8) = 498\ \text{m/s} = 1517\ \text{ft/s}$$

13-5   MOMENTUM EQUATION FOR A CONTROL VOLUME

Because most of our discussion of flow processes will be concerned with control volumes (open systems) it will be useful to express Newton's momentum principle in a form which is most easily adapted to the control-volume analysis technique. Momentum is a vector quantity and a properly general analysis must be done in three dimensions; however, because we are concerned with one-dimensional flow systems in this chapter we can simplify our procedure somewhat by just analyzing the x component of momentum.

Newton's second law of motion states that the force acting on a body is proportional to the rate of change of momentum. For the x direction

$$F_x \sim \frac{d(mV_x)}{d\tau}$$

or, with the proportionality constant,

$$F_x = \frac{1}{g_c} \frac{d(mV_x)}{d\tau} \tag{13-33}$$

Equation (13-33) applies to a fixed mass. Now consider the rather general control volume shown in Fig. 13-5. In (a) the volume is indicated at time τ

FIG. 13-5   *Control volume for momentum analysis (a) at time τ and (b) at time $\tau + d\tau$.*

with flow moving in and out of the control surface defining the control-volume region. Inside the control volume is a certain mass of fluid having a certain velocity and corresponding momentum. Now note the situation in (b) at time $\tau + \Delta\tau$. The control surface is indicated with the dashed line and is in the same spatial location as before, but the mass, which was totally enclosed by the control surface, has now moved out of the control volume somewhat. We shall refer to this quantity as the *control mass*. Now let us apply Eq. (13-33) to this control mass, using only forces and momenta in the x direction. Expressing the derivative in limit form,

$$\sum F_x = \lim_{\Delta\tau \to 0} \frac{1}{g_c} \frac{(m V_x)_{\tau+\Delta\tau} - (m V_x)_\tau}{\Delta\tau} \tag{13-34}$$

At time $\tau + \Delta\tau$ the momentum of the control mass can be expressed in terms of the momenta in regions A, B, and C as

$$(m V_x)_{\tau+\Delta\tau} = (m V_x)_B + (m V_x)_C \tag{13-35}$$

Designating the control volume with the subscript σ

$$(m V_x)_{A+B} = (m V_x)_\sigma = (m V_x)_A + (m V_x)_B \tag{13-36}$$

Solving for $(m V_x)_B$ from Eq. (13-36) at time $\tau + \Delta\tau$ and substituting in (13-35) gives

$$(m V_x)_{\tau+\Delta\tau} = (m V_x)_c - (m V_x)_A + (m V_x)_{\sigma,\tau+\Delta\tau} \tag{13-37}$$

At time τ the control mass is contained entirely in the control volume so that

$$(m V_x)_\tau = (m V_x)_{\sigma,\tau}$$

Substituting the various expressions back into Eq. (13-34) gives

$$\sum F_x = \frac{1}{g_c} \lim_{\Delta\tau \to 0} \left[\frac{(m V_x)_c}{\Delta\tau} - \frac{(m V_x)_A}{\Delta\tau} + \frac{(m V_x)_{\sigma,\tau+\Delta\tau} - (m V_x)_{\sigma,\tau}}{\Delta\tau} \right]$$

As $\Delta\tau \to 0$, m_A and m_C approach δm_A and δm_C, and the ratios $\delta m_A/\delta\tau$ and $\delta m_C/\Delta\tau$ become the mass flow rates entering and leaving the control volume. The third term in the limit becomes the rate of accumulation of momentum in the control volume, and the overall expression becomes

$$\sum F_x = \frac{1}{g_c}\left[(\dot{m}V_x)_e - (\dot{m}V_x)_i + \frac{d(mV_x)_\sigma}{d\tau}\right] \tag{13-38}$$

where, now, the subscripts i and e have been used to designate inlet and exit streams, respectively. The term $\dot{m}V$ is called the momentum flux, and an equation like (13-38) could be written for both the y and z directions. Equation (13-38) may be expressed in words as

Sum of forces acting on control volume
= momentum flux out − momentum flux in + rate of accumulation
of momentum in the control volume

In steady state there will be no accumulation within the control volume. For steady state and only one stream entering and leaving the control volume we obtain, for the three force components,

$$\sum F_x = \frac{\dot{m}}{g_c}(V_{ex} - V_{ix})$$

$$\sum F_y = \frac{\dot{m}}{g_c}(V_{ey} - V_{iy}) \tag{13-39}$$

$$\sum F_z = \frac{\dot{m}}{g_c}(V_{ez} - V_{iz})$$

EXAMPLE 13-6   Force and momentum flux

A large tank is mounted on rollers, as shown, and a water jet discharges water into the tank at the rate of 8 lbm/s with a velocity of 30 ft/s. Calculate the force F_x necessary (a) to hold the tank stationary and (b) to allow it to move to the left with a velocity of 1 ft/s.

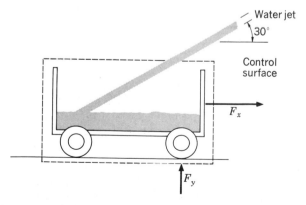

FIG. EXAMPLE 13-6

SOLUTION   This is an unsteady-state problem because mass, and therefore momentum, is accumulating in the tank with time. We select the control surface shown in the dashed line so that it always encloses the tank. The
x-direction momentum equation is

$$F_x = \frac{1}{g_c}\left[(\dot{m}V_x)_e - (\dot{m}V_x)_i + \frac{d(mV_x)_\sigma}{d\tau}\right] \qquad (a)$$

We have only an inlet stream so $(mV_x)_e = 0$, and choosing to the right as
positive gives

$$(\dot{m}V_x)_i = (8 \text{ lbm/s})(-30 \cos 30° \text{ ft/s})$$
$$= -207.8 \text{ lbm·ft/s}^2$$

For part (a) of the problem the tank is at rest and $V_{x,\sigma} = 0$, so there is no accumulation of momentum even though there is an accumulation of mass. In
this case, then, Eq. (a) becomes

$$F_x = \frac{1}{32.2 \text{ lbm·ft/lbf·s}^2}[0 - (-207.8 \text{ lbm·ft/s}^2) + 0]$$
$$= +6.45 \text{ lbf}$$

The positive sign indicates that the force is to the right because this was the
positive direction for the coordinate system. In case (b) the tank is moving to
the left, and there is a rate of increase in momentum in the control volume
amounting to

$$\frac{d(mV_x)_\sigma}{d\tau} = (8 \text{ lbm/s})(-1 \text{ ft/s}) \qquad (b)$$
$$= -8 \text{ lbm·ft/s}^2$$

The -1 ft/s is the velocity of the water inside the control volume. The value
of the force for case (b) is therefore

$$F_x = \frac{1}{32.2}[0 - (-207.8) + (-8)]$$
$$= 6.20 \text{ lbf}$$

13-6   SHOCK WAVES IN AN IDEAL GAS

We have already seen that a small pressure disturbance is propagated in a
fluid at the speed of sound. We might say that this is the speed of movement
of an infinitesimal wave in the fluid. When stronger waves are encountered

such that there are rapid and sharp changes in fluid properties in a very small region of the flow we say that a *shock wave* has developed, and we now wish to analyze the conditions which must prevail if such waves are to exist. We shall limit our discussion to *normal* shock waves, or waves that are perpendicular to the flow velocity.

Consider the one-dimensional flow channel shown in Fig. 13-6. We assume the flow is adiabatic and consider the control volume bounded by the dashed line containing the shock wave which is *stationary* in the channel. Upstream from the shock the properties are designated with the subscript x while the subscript y is used to designate the downstream properties. There is no heat or work added so the steady-flow energy equation is

$$h_{0x} = h_x + \frac{V_x^2}{2g_c} = h_y + \frac{V_y^2}{2g_c} = h_{0y} \tag{13-40}$$

The mass continuity relation is

$$\frac{\dot{m}}{A} = \rho_x V_x = \rho_y V_y \tag{13-41}$$

and the momentum equation for the control volume is obtained by setting the net pressure force equal to the increase in momentum flux through the control volume:

$$A(p_x - p_y) = \frac{\dot{m}}{g_c}(V_y - V_x) \tag{13-42}$$

We have assumed the process to be adiabatic, but there is no assurance that it is reversible so the second law of thermodynamics requires that

$$s_y - s_x \geqslant 0 \tag{13-43}$$

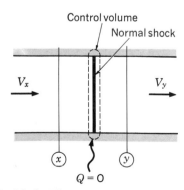

FIG. 13-6 *Flow across a normal shock wave.*

If the energy and continuity equations (13-40) and (13-41) are combined and the combination expressed in appropriate form, it is possible to plot the resulting relation on an enthalpy-entropy diagram. Such a plot is called a *Fanno* line and is illustrated in Fig. 13-7. A combination of the momentum and continuity equations (13-41) and (13-42) can also be plotted on an *h-s* diagram and is called a *Rayleigh* line. The intersections of these two lines at points x and y represent the simultaneous solution of the three equations and therefore must represent the states x and y before and after the shock wave. The second-law relation of Eq. (13-43) requires that $s_y > s_x$ so this fact establishes the specific designation of the points on the diagram.

A more detailed analysis of the Fanno and Rayleigh line relations would show that:

1 The maximum entropy on both curves (points a and b) corresponds to $M = 1$.
2 Above point a or b on a particular line the flow is subsonic, $M < 1$.
3 Below point a or b on a particular line the flow is supersonic, $M > 1$.

These considerations are illustrated in Fig. 13-7, and the net result is that if a normal shock wave is to exist the flow must change from supersonic to subsonic conditions across the shock. In the limit, if V_x is just sonic ($M = 1$), we will have an infinitesimal shock wave, or the sonic wave discussed before. The *strength* of the shock is usually measured in terms of the magnitude of the change in the Mach number across it. The adiabatic flow assumption dictates that the stagnation enthalpy remain constant across the shock, and for an ideal gas with constant specific heats this means

$$T_{0x} = T_{0y} \qquad\qquad (13\text{-}44)$$

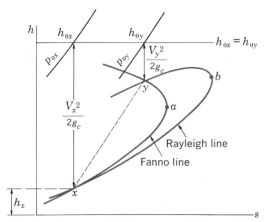

FIG. 13-7   *Fanno and Rayleigh lines for evaluation of normal shock waves.*

or the stagnation temperature remains constant regardless of the strength of the shock.

Although we have assumed an abrupt or almost step change in fluid properties across the shock for the analysis so far, we know from our experience with transport phenomena in Chap. 9 that there must always be some finite distance of the order of a mean free path required for communication of energy and momentum between gas molecules. At normal atmospheric pressure the mean free path is very small ($\sim 10^{-7}$m) and shock waves at these pressures are indeed very thin. At much lower pressures, as encountered in very high altitude flight, the mean free path is much larger and shock waves encountered in those circumstances may have substantial thicknesses. For purposes of our analysis here, we will assume that the pressures are always high enough that the shock waves can be assumed very thin. From Eq. (13-22) we have

$$\frac{T_{0x}}{T_x} = 1 + \frac{\gamma - 1}{2} M_x^2$$

$$\frac{T_{0y}}{T_y} = 1 + \frac{\gamma - 1}{2} M_y^2$$

Dividing these equations

$$\frac{T_y}{T_x} = \frac{1 + \frac{\gamma - 1}{2} M_x^2}{1 + \frac{\gamma - 1}{2} M_y^2} \tag{13-45}$$

Using the continuity relation, Eq. (13-41), along with

$$\rho_x = \frac{p_x}{RT_x} \quad \text{and} \quad \rho_y = \frac{p_y}{RT_y}$$

gives

$$\frac{T_y}{T_x} = \frac{p_y V_y}{p_x V_x} \tag{13-46}$$

But $V = cM$ and $c = \sqrt{\gamma g_c RT}$, so that Eq. (13-46) may be expressed as

$$\frac{T_y}{T_x} = \frac{p_y M_y \sqrt{T_y}}{p_x M_x \sqrt{T_x}} = \left(\frac{p_y}{p_x}\right)^2 \left(\frac{M_y}{M_x}\right)^2 \tag{13-47}$$

Equations (13-45) and (13-47) may now be combined to give the pressure ratio across the shock as

$$\frac{p_y}{p_x} = \frac{M_x \sqrt{1 + \frac{\gamma - 1}{2} M_x^2}}{M_y \sqrt{1 + \frac{\gamma - 1}{2} M_y^2}} \tag{13-48}$$

Because Eq. (13-48) results from a combination of the energy and continuity equations it is a representation of the Fanno line expressed in terms of pressure and Mach number. A similar relation for the Rayleigh line in terms of these same variables may be obtained by combining the momentum and continuity equations. We have

$$p_x - p_y = \frac{\dot{m}}{Ag_c}(V_y - V_x)$$

But $\dot{m} = \rho_x A V_x = \rho_y A V_y$ so that

$$p_x - p_y = \frac{\rho_y V_y^2 - \rho_x V_x^2}{g_c}$$

Introducing $V = cM$, $\rho = p/RT$, and $c = \sqrt{\gamma g_c RT}$ gives

$$p_x g_c + \frac{p_x M_x^2}{RT_x}(\gamma g_c RT_x) = p_y g_c + \frac{p_y M_y^2}{RT_y}(\gamma g_c RT_y)$$

Collecting terms,

$$\frac{p_y}{p_x} = \frac{1 + \gamma M_x^2}{1 + \gamma M_y^2} \tag{13-49}$$

The combination of Eqs. (13-48) and (13-49) represents the intersections of the Fanno and Rayleigh lines and gives

$$M_y^2 = \frac{M_x^2 + \frac{2}{\gamma - 1}}{\frac{2\gamma}{\gamma - 1} M_x^2 - 1} \tag{13-50}$$

Given the upstream Mach number M_x, the downstream Mach number M_y is uniquely defined. Table A-20 gives a convenient tabulation of the properties across a normal shock wave using the upstream Mach number as the primary variable. We have already mentioned that the stagnation temperature remains constant across the shock wave because of the adiabatic flow condition. The stagnation *pressure*, however, always *decreases* across the shock because of the requirement that the entropy must increase. We can see this rather easily by writing the relation for entropy change between the isentropic stagnation states before and after the shock.

$$s_{0y} - s_{0x} = c_p \ln \frac{T_{0y}}{T_{0x}} - R \ln \frac{p_{0y}}{p_{0x}} \geq 0$$

With the requirement that $T_{0y} = T_{0x}$ it is clear that p_{0y} must always be less than p_{0x} in order to produce an increase in entropy. We note in Table A-20 that the values of p_{0y}/p_{0x} are indeed always less than unity.

13-7 SHOCK WAVES IN CONVERGENT-DIVERGENT PASSAGES

We have already discussed isentropic flow in a converging-diverging passage and pointed out that a variety of flow conditions can be established depending on the particular flow-area configuration and stagnation properties available to drive the flow. But, the previous discussion considered only the ideal isentropic flow. Under suitable combinations of back pressure and nozzle design it is possible to produce shock-wave phenomena in the passage, and it is these phenomena we now wish to discuss.

Consider the converging-diverging nozzle shown in Fig. 13-8 connected to the stagnation-chamber supply pressure p_0 and discharging into another chamber where we designate the pressure as p_b and call it the back pressure. We assume that the stagnation pressure is sufficiently high to pro-

FIG. 13-8 *Influence of back pressure on flow in a supersonic nozzle.*

duce sonic flow conditions at the throat and supersonic flow in the downstream diverging section. Now let us examine the effect of changes in the back pressure on the flow in the nozzle.

As long as the back pressure remains equal to or less than the exit pressure corresponding to the exit area there is no effect on the flow inside the nozzle. There may, however, be complicated fan-type expansion waves downstream from the exit for $p_b < p_e$, as shown at point 1. As the back pressure is increased above p_e, progressively stronger "oblique" shock waves are formed downstream from the exit illustrated as points 3 and 4 in Fig. 13-8. Still, the flow inside the nozzle is unaffected. Point 5 illustrates the condition where the back pressure is just high enough to create a normal shock wave at the exit, that is, p_e/p_b at that point corresponds to the value for a normal shock with an upstream Mach number of M_e. Further increase in the back pressure causes the normal shock to move back into the nozzle. Downstream from the shock the flow is subsonic so the diverging section will cause a *decrease* in flow velocity and an increase in pressure as illustrated in point 6. Eventually, an increase in the back pressure can cause the shock to move all the way back to the throat at which point it will disappear and the flow in the diverging section becomes totally subsonic. Still further increase in the back pressure causes subsonic flow in the entire nozzle and a progressive reduction in the mass flow rate. The examples illustrate the calculation procedures for nozzle-shock phenomena.

EXAMPLE 13-7   Shock at nozzle exit

To what value must the back pressure be raised to just produce a normal shock wave at the exit of the nozzle of Example 13-4?

SOLUTION   This calculation is most easily performed by using the normal shock relations presented in Table A-20 of the Appendix. For this case $M_x = 2.5$ just before entering the shock wave at the exit. The pressure at that point is 1 atm as given in Example 13-4. Entering Table A-20 at $M_x = 2.5$ we find

$$M_y = 0.512\ 99 \qquad \frac{p_y}{p_x} = 7.125 \qquad \frac{p_{0y}}{p_x} = 8.5262$$

The pressure in the flow just downstream from the shock is therefore

$$p_y = (7.125)(1.0) = 7.125 \text{ atm}$$

The back pressure in a discharge chamber like that of Fig. 13-8 is the stagnation pressure downstream from the shock which is

$$p_{0y} = (8.5262)(1.0) = 8.5262 \text{ atm}$$

EXAMPLE 13-8   Shock in diffuser section

Suppose the back pressure were increased on the nozzle of Example 13-4 until a normal shock wave was formed in the diffuser section where $M_x = 1.5$. What back pressure would be necessary to accomplish this, and what would be the resulting exit temperature and velocity?

SOLUTION   This problem is complicated further because the flow will be reduced to subsonic downstream from the shock and will then slow down further because of the area increase. The schematic indicates the nomenclature to be employed in the calculations.

Shock wave

Flow

FIG. EXAMPLE 13-8

We can use the results of Examples 13-4 and 13-5 to obtain

$$A^* = 1.517 \text{ cm}^2 \qquad A_x = A_y = 1.784 \text{ cm}^2 \qquad p_x = 4.724 \text{ atm}$$
$$T_x = 274 \text{ K} \qquad M_x = 1.5 \qquad A_e = 4.0 \text{ cm}^2 \qquad T_{0x} = T_{0y} = 398 \text{ K}$$

Entering Table A-20 of the Appendix at $M_x = 1.5$ we find

$$M_y = 0.701 \ 99 \qquad \frac{p_y}{p_x} = 2.4583 \qquad \frac{T_y}{T_x} = 1.3202 \qquad \frac{p_{0y}}{p_x} = 3.4133$$

The back pressure in the discharge chamber is p_{0y} so

$$p_{0y} = (3.4133)(4.724) = 16.12 \text{ atm}$$

To calculate the exit temperature and flow velocity we note that the flow is isentropic as it moves from y to e, so we can use Table A-19 for our calculations. However, the effective throat area for the downstream-shock flow is *not* the same as that calculated previously; i.e., it is not 1.517 cm² because the downstream-shock flow is not the same isentropic flow as the upstream-shock flow. This results, of course, from the fact that the entropy increases across the shock.

We shall designate the effective throat in the downstream-shock region as $A^{*\prime}$ and determine it by entering Table A-19 at $M = 0.701 \ 99$. Then

$$\frac{A_y}{A^{*\prime}} = 1.093\ 25$$

and

$$A^{*\prime} = \frac{1.784}{1.093\ 25} = 1.632\ \text{cm}^2$$

Our actual exit area is 4.0 cm² so that

$$\frac{A_e}{A^{*\prime}} = \frac{4.0}{1.632} = 2.451$$

We can now determine the exit-flow conditions by entering Table A-19 at a value of A/A^* of 2.451 for *subsonic* flow. By interpolation the results are

$$\frac{T_e}{T_0} = 0.986\ 70 \qquad M_e = 0.255$$

The exit temperature is therefore

$$T_c = (0.986\ 70)(398) = 393\ \text{K} = 120°\text{C}$$

The speed of sound at exit is calculated from Eq. (13-12):

$$c_e = [(1.4)(1.0)(287)(393)]^{1/2} = 397\ \text{m/s}$$

so the actual flow velocity is

$$V_e = c_e M_e = (397)(0.255) = 101.2\ \text{m/s} = 308.6\ \text{ft/s}$$

We should remark that the mass flow rate in the nozzle is not altered by the presence of the shock because sonic conditions are still present at the throat and we still have so-called choked-flow conditions.

EXAMPLE 13-9 Nuclear ramjet

A nuclear ramjet can be idealized as shown in the accompanying sketch. The inlet flow is at $M = 4.0$, $T = -50°\text{C}$, and $p = 0.05$ atm. A normal shock "stands" at the inlet, and the flow is accelerated to $M = 0.7$ at section 2 whereupon heat is added from the nuclear reactor at constant pressure and velocity until the temperature is raised to 1200°C at point 3. The flow is subsequently expanded to $p = 0.05$ atm at the exit. Calculate the reactor heat required and thrust produced for an inlet area of 1 m².

FIG. EXAMPLE 13-9

SOLUTION To evaluate the thrust force on the ramjet we must be able to evaluate the momentum fluxes at inlet and exit as well as the pressure forces at those points. As we shall see, the momentum forces will make a major contribution to the overall thrust. Using the x and y subscripts to designate upstream and downstream conditions on the shock we have, using Table A-20,

$$M_x = 4.0 \qquad M_y = 0.434\ 96 \qquad \frac{p_{0y}}{p_x} = 21.068$$

$$\frac{T_y}{T_x} = 4.0469 \qquad T_x = -50°C = 223\ \text{K} \qquad p_x = 0.05\ \text{atm}$$

Using Table A-19 at $M = 4.0$,

$$\frac{T_x}{T_0} = 0.238\ 10 \tag{a}$$

and

$$T_{0x} = T_{0y} = \frac{223}{0.238\ 10} = 937\ \text{K} = 664°C = T_{02}$$

Also, $p_{0y} = p_{02} = p_{03} = p_{04} = (21.068)(0.05) = 1.0534$ atm. Entering Table A-19 again with $M_2 = 0.7$ gives

$$\frac{A_2}{A_2^*} = 1.094\ 37 \qquad \frac{p_2}{p_{02}} = 0.720\ 92 \qquad \frac{T_2}{T_{02}} = 0.910\ 75$$

whereupon

$$T_2 = (0.910\ 75)(937) = 853\ \text{K} = 580°C \tag{b}$$

We are given that

$$p_3 = p_2 \qquad T_3 = 1200°C = 1473 \text{ K} \qquad V_2 = V_3$$

Because the velocities are equal

$$T_3 - T_2 = T_{03} - T_{02} \tag{c}$$

or

$$T_{03} = 937 + (1200 - 580) = 1557 \text{ K} = 1284°C = T_{04}$$

and

$$\frac{T_3}{T_{03}} = \frac{1473}{1557} = 0.946\ 05 \tag{d}$$

We now have a new isentropic flow from point 3 to point 4 which we can calculate using relation (d) and Table A-19. Entering the table at the value given in (d) gives

$$M_3 = 0.5324 \qquad \frac{p_3}{p_{03}} = 0.823\ 90$$

We have

$$p_2 = p_3 = (1.0534)(0.720\ 92) = 0.759\ 42 \text{ atm} \tag{e}$$

and

$$p_{03} = \frac{0.759\ 42}{0.823\ 90} = 0.921\ 73 \text{ atm} = p_{04}$$

Now, at section 4, $p = 0.05$ atm and

$$\frac{p_4}{p_{04}} = \frac{0.05}{0.921\ 73} = 0.054\ 25 \tag{f}$$

We may now once again enter Table A-19 using the value from (f) with the result

$$M_4 = 2.551 \qquad \frac{T_4}{T_{04}} = 0.434\ 63$$

$$\frac{A_4}{A_4^*} = 2.780$$

so that

$$T_4 = (0.434\ 63)(1557) = 676 \text{ K} = 403°C$$
$$c_4 = [(1.4)(1.0)(287)(676)]^{1/2} = 521 \text{ m/s}$$

and

$$V_4 = c_4 M_4 = (521)(2.551) = 1330 \text{ m/s}$$

At the inlet, $T_{1,x} = 223$ K so that

$$c_{1,x} = [(1.4)(1.0)(287)(223)]^{1/2} = 229 \text{ m/s}$$

and

$$V_{1x} = c_{1x} M_1 = (299)(4.0) = 1197 \text{ m/s}$$

The mass flow may be calculated at entrance conditions. First, the density is evaluated as

$$\rho_1 = \frac{p_1}{RT_1} = \frac{(0.05)(1.0132 \times 10^5)}{(287)(223)} = 0.0792 \text{ kg/m}^3$$

Then

$$\dot{m} = \rho_1 A_1 V_1 = (0.0792 \text{ kg/m}^3)(1 \text{ m}^2)(1197 \text{ m/s}) = 94.80 \text{ kg/s}$$

At exit the density is

$$\rho_4 = \frac{p_4}{RT_4} = \frac{(0.05)(1.0132 \times 10^5)}{(287)(676)} = 0.0261 \text{ kg/m}^3$$

and the exit area can be determined from the mass-continuity relation

$$\dot{m} = \rho_4 A_4 V_4$$
$$A_4 = \frac{94.80}{(0.0261)(1330)} = 2.731 \text{ m}^2$$

The force balance is now made on the control volume bounded by the dashed line with the thrust force T on the fluid assumed to the right and the positive force direction taken to the right. The thrust on the ramjet structure would be to the left.

$$\sum F_x = \frac{\dot{m}}{g_c}(V_e - V_i)$$

$$T + p_1 A_1 - p_4 A_4 = \frac{\dot{m}}{g_c}(V_4 - V_1) \tag{g}$$

We now have all the numerical values for insertion in Eq. (g) to solve for the thrust

$$T = -(0.05)(1.0132 \times 10^5)(1 - 2.731) + \frac{(94.80)}{(1.0)}(1330 - 1197)$$

$$= 8769 + 12\,608 = 21\,377\ \text{N} = 4805\ \text{lbf}$$

The heat which must be supplied by the reactor is

$$Q = \dot{m}(h_3 - h_2) = \dot{m}c_p(T_3 - T_2) \tag{h}$$

The specific heat for air is

$$c_p = 1004.8\ \text{J/kg·°C}$$

so that

$$Q = (94.80)(1004.8)(1200 - 580)$$
$$= 59.06\ \text{MW}$$
$$= 2.016 \times 10^8\ \text{Btu/h}$$

As a matter of interest we might calculate the rate of doing work by the thrust force moving at the velocity V_1.

$$W = TV_1 = (21\,377\ \text{N})(1197\ \text{m/s})$$
$$= 2.559 \times 10^7\ \text{W}$$

The energy-conversion efficiency for the ramjet is therefore

$$\text{Efficiency} = \frac{W}{Q} = \frac{2.559 \times 10^7}{5.906 \times 10^7} = 43.3\ \text{percent}$$

This is a rather high value and results from the fact that the heat is added at a very high temperature (1200°C).

We have already seen from the many flow regimes which may develop in the downstream section of a nozzle that shock waves are not always the simple "normal" type. Shock waves can form at an oblique orientation to the flow velocity and may assume curved shapes under certain circumstances. In Fig. 13-9 we have a photograph of the "bow shock" produced by a high-speed bullet, and a similar wave would be formed around the nose of a high-speed aircraft. The flow just behind the most forward shock position can be calculated as that downstream from a normal shock wave. As the Mach number is increased the trailing shock wave is swept backward at pro-

FIG. 13-9 *Photograph of flight of a 0.22-caliber Remington "swift" shell. Velocity = 4100 ft/s. (Courtesy ARO, Inc., Arnold Air Force Base, Tenn.)*

gressively steeper angles. The sweep angle is approximated by the so-called Mach cone shown in Fig. 13-10.

In front of the Mach cone is a zone of silence because disturbances can only propagate at the speed of sound. For the source movement of distance $V\tau$ in time τ the sound only moves $c\tau$, or to the edge of the cone. Behind the cone, disturbances are heard. Evidently, the cone angle is given by

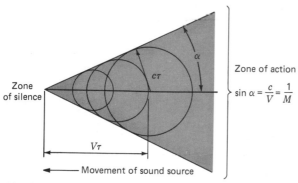

FIG. 13-10 *Mach cone and zone of silence.*

$$\sin \alpha = \frac{c}{V} = \frac{1}{M}$$

For $M < 1$ disturbances can propagate ahead of the source as well as be-hind. This means that a supersonic aircraft may have passed well overhead before an observer on the ground hears the trailing bow shock moving along with it.

As a simple example consider an aircraft at an altitude of 50 000 ft trav-eling at $M = 2$. The Mach cone angle is

$$a = \sin^{-1}\frac{1}{M} = \sin^{-1}\tfrac{1}{2} = 30°$$

When the edge of the cone arrives at the ground the aircraft has moved past the ground observer a distance x where

$$\frac{50\ 000}{x} = \tan \alpha$$

or

$$x = 86\ 602\ \text{ft} = 16.4\ \text{mi} \qquad (26.4\ \text{km})$$

13-8   NOZZLE FLOW OF A VAPOR

We have already learned that a vapor does not necessarily behave as an ideal gas, particularly when the thermodynamic state approaches the saturation region. The calculation of high-speed vapor flows in nozzles therefore must be performed differently than in the case of an ideal gas with constant spe-cific heats. Consider the h-s diagram shown in Fig. 13-11 where steam is to

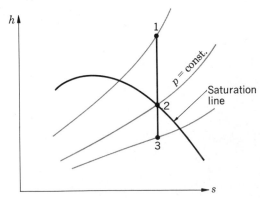

FIG. 13-11   *Enthalpy-entropy diagram for high-speed flow of a vapor.*

be expanded isentropically from state 1 to state 3 in a nozzle. As the satura-tion line is encountered at point 2 one would expect condensation to begin but this is not always the case. If the flow velocity is high enough (as it would be in a supersonic nozzle), sufficient time may not be available to form nu-cleation of droplets and accomplish the heat-transfer process associated with condensation. Once the flow moves far enough into the saturation region large enough droplets will be formed to sustain the condensation process and a "condensation shock" will be experienced. The delay of condensation produces a situation which is appropriately called *supersaturation*.

The condensation phenomenon described above is usually described in thermodynamic terms as a problem in *metastable equilibrium*. In a strict sense this means a circumstance in which the thermodynamic equilibrium state remains the same for small displacements, but then moves to a new equilibrium position for large displacements. In the above example this means that large enough droplets must be formed to effect the change. From a microscopic viewpoint things are very complicated and involve the capture of vapor molecules by the liquid surface of a droplet, droplet growth rates, and their general transport in the flow stream.

As a result of the phenomenon of supersaturation, vapor flow in a nozzle may be calculated with more precision than might otherwise be pos-sible by assuming a constant value of γ over a rather wide range of tempera-tures and pressures. For steam $\gamma = 1.3$ is a good approximation. The critical pressure ratio is then

$$\frac{p^*}{p_0} = \left(\frac{2}{\gamma + 1}\right)^{\gamma/(\gamma-1)} = 0.545 \tag{13-51}$$

If the steam enters the nozzle as a saturated vapor the critical ratio is, to a good approximation, 0.577, or that corresponding to a value of $\gamma = 1.14$.

EXAMPLE 13-10   Steam flow in a nozzle

Steam expands isentropically in a nozzle from stagnation conditions of 100 psia and 500°F to 14.7 psia. The mass flow rate is 1 lbm/s. Calculate the exit area required using the steam tables and the throat area using the constant property assumptions pertaining to Eq. (13-51).

SOLUTION   Designating exit properties with a subscript e, we have

$$T_0 = 500°F = 960°R \qquad p_e = 14.7 \text{ psia}$$
$$p_0 = 100 \text{ psia} \qquad \dot{m} = 1 \text{ lbm/s}$$

From the steam tables (Table A-9)

$$s_0 = s_e = 1.7085 \text{ Btu/lbm·°R}$$
$$h_0 = 1279.1 \text{ Btu/lbm} \tag{a}$$

At the exit pressure of 14.7 psia we find that the flow is in the saturation region with values from Table A-8 of

$$s_f = 0.3121 \qquad s_{fg} = 1.4446 \text{ Btu/lbm·°R}$$
$$h_f = 180.15 \qquad h_{fg} = 970.4 \text{ Btu/lbm}$$
$$v_f = 0.016\ 72 \qquad v_g = 26.80 \text{ ft}^3/\text{lbm}$$

Using Eq. (a), we calculate the exit quality from

$$s_e = 1.7085 = s_f + xs_{fg} = 0.3121 + x(1.4446)$$
$$x = 0.9666$$

The resulting exit enthalpy and specific volume are therefore

$$h_e = 180.15 + (0.9666)(970.4) = 1118.1 \text{ Btu/lbm}$$
$$v_e = 0.016\ 72 + (0.9666)(26.8 - 0.016\ 72) = 25.91 \text{ ft}^3/\text{lbm}$$

The exit velocity can be calculated from the steady-flow energy equation for this problem:

$$h_e + \frac{V_e}{2g_c} = h_0 \qquad\qquad (b)$$

With the numerical values inserted

$$V_e = [(2)(32.2)(778)(1279.1 - 1118.1)]^{1/2}$$
$$= 2840 \text{ ft/s} \qquad (931.8 \text{ m/s})$$

The mass rate of flow can be expressed as

$$\dot{m} = \rho A V = \frac{AV}{v} \qquad\qquad (c)$$

At exit conditions this gives a flow area of

$$A_e = \frac{(1 \text{ lbm/s})(25.91 \text{ ft}^3/\text{lbm})}{2840 \text{ ft/s}} = 9.123 \times 10^{-3} \text{ ft}^2 = 1.314 \text{ in}^2$$

Using the constant property assumptions of Eq. (13-51), the throat area is calculated using a value of $\gamma = 1.3$. This value may then be applied to Eq. (13-31) with the gas constant for water vapor taken as

$$R = \tfrac{1545}{18} = 85.8 \text{ ft·lbf/lbm·°R}$$

Using Eq. (13-31),

$$\frac{\dot{m}}{A^*} = (100)(144) \left[\frac{(1.3)(32.2)}{(85.8)(960)}\right]^{1/2} \left(\frac{2}{1.3+1}\right)^{(1.3+1)/2(1.3-1)}$$

$$= 190 \text{ lbm/s·ft}^2$$

With $\dot{m} = 1$ lbm/s the throat area required is

$$A^* = \tfrac{1}{190} = 5.263 \times 10^{-3} \text{ ft}^2 = 0.758 \text{ in}^2$$

13-9   NOZZLE AND DIFFUSER COEFFICIENTS

Up to this point we have considered one-dimensional adiabatic flow systems where compressibility effects are important. With the exception of situations involving shock-wave phenomena the flows were also considered reversible, or isentropic, for the ideal case. The adiabatic assumption is a reasonable one for actual nozzle and diffuser flow situations, but there are several factors which cause the actual flow to be irreversible. Fluid friction is always present at the wall and will cause the overall flow to be irreversible. The exact configuration of a nozzle strongly influences the nature of the frictional effects because of the viscous boundary layer buildup on the wall. A detailed treatment of these effects is beyond the scope of our discussion, but we can indicate their diverse nature by noting the two situations in Fig. 13-12. In (a) the nozzle contour is smooth and gradual, and if the flow is to be supersonic, smooth acceleration to $M = 1$ at the throat and then $M > 1$ in the diffusing section is experienced with a viscous boundary layer at the wall, as shown.

In Fig. 13-12b the flow is again accelerated smoothly to $M = 1$ at the throat, but the area change in the diffuser section downstream is much more

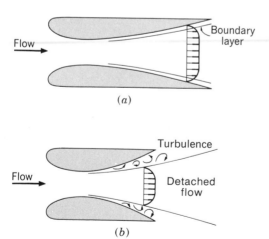

FIG. 13-12 *Effect of nozzle contour on flow.* (a) *Gradual expansion;* (b) *too rapid expansion and separated flow.*

rapid than in (*a*). As a result, the high-speed flow may not be able to follow the wall and may become "detached" or "separated" with a highly turbulent mixing region at the wall. Not surprisingly, the nature of the fluid friction effects in (*b*) can be quite different from those in (*a*). Fundamental analyses of the nozzle flow systems can sometimes predict the frictional effects, and the results are usually expressed in terms of certain performance coefficients which can be compared with experiments. Regardless of the success of the analysis, experimental data can be used to determine the values of the coefficients for design purposes, and it is the more important coefficients which we now wish to discuss.

The *nozzle efficiency* η_N is defined by

$$\eta_N = \frac{\text{actual K.E. at exit}}{\text{K.E. at exit with isentropic flow to same exit pressure}} \qquad (13\text{-}52)$$

Efficiencies range between 90 and 99 percent with larger nozzles having the higher efficiencies because the viscous boundary layer occupies a relatively smaller portion of the flow, the larger the flow area.

If the kinetic energy at inlet is small then the inlet enthalpy is very nearly h_0, and the nozzle efficiency can be expressed by

$$\eta_N = \frac{h_0 - h_e}{h_0 - h_{es}} \qquad (13\text{-}53)$$

where h_e is the actual exit enthalpy and h_{es} is the isentropic exit enthalpy for expansion to the actual exit pressure.

The *nozzle velocity coefficient C_V* is defined in a similar way:

$$C_V = \frac{\text{actual velocity at exit}}{\text{velocity at exit for isentropic flow to same exit pressure}} \qquad (13\text{-}54)$$

The *coefficient of discharge C_D* is defined as

$$C_D = \frac{\text{actual mass rate of flow}}{\text{mass rate of flow for isentropic flow}} \qquad (13\text{-}55)$$

In some cases the design objective for the flow channel is to slow the flow down and increase the pressure. For these instances it is frequently advantageous to define a *diffuser pressure recovery factor C_P* as

$$C_P = \frac{\text{actual pressure rise}}{\text{isentropic pressure rise}} \qquad (13\text{-}56)$$

An alternate method for describing diffuser performance is to define a *diffuser efficiency η_D* as

$$\eta_D = \frac{\text{isentropic enthalpy rise for discharge at}}{\text{the actual exit stagnation pressure}}{\text{inlet kinetic energy}} \qquad (13\text{-}57)$$

The reasoning behind this definition is not immediately apparent but may be clarified by referring to Fig. 13-13. In this diagram the inlet kinetic energy is

$$\frac{V_i^2}{2g_c} = h_{0i} - h_i$$

while the kinetic energy at exit is

$$\frac{V_e^2}{2g_c} = h_{0e} - h_e$$

The numerator of Eq. (13-57) represents the maximum kinetic energy which could be recovered for discharge at the actual exit stagnation pressure. So, in terms of the nomenclature of Fig. 13-13 we obtain

$$\eta_D = \frac{h_{es} - h_i}{h_{0i} - h_i} \qquad (13\text{-}58)$$

EXAMPLE 13-11   Diffuser pressure recovery

A certain diffuser having an efficiency of 90 percent is used to reduce the velocity of an air stream initially at 500 ft/s, 70°F, and 14.7 psia down to 200 ft/s. Calculate the ratio of exit area to inlet area to accomplish this. What is the value of the pressure-recovery factor for this diffuser?

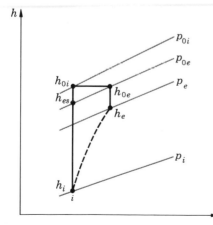

FIG. 13-13   *Parameters for evaluating diffuser efficiency.*

SOLUTION   From the steady-flow energy equation

$$h_{oi} - h_i = \frac{V_i^2}{2g_c} = \frac{(500)^2}{(2)(32.2)(778)} = 4.99 \text{ Btu/lbm} \tag{a}$$

Using the nomenclature of Fig. 13-13 and Eq. (13-58) gives

$$h_{es} - h_i = (0.9)(4.99) = 4.491 \text{ Btu/lbm} \tag{b}$$

We have $T_i = 70°F = 530°R$ so that, from (a) and (b),

$$T_{oi} = \frac{(4.99)}{0.24} + 530 = 550.8°R = 90.8°F = T_{oe}$$

$$T_{es} = \frac{4.491}{0.24} + 530 = 548.7°R = 88.7°F \tag{c}$$

The actual exit temperature can be obtained from

$$T_e = T_{oe} - \frac{1}{c_p} \left(\frac{V_e^2}{2g_c} \right)$$
$$= 550.8 - \frac{(200)^2}{(0.24)(778)(32.2)} = 547.5°R = 87.5°F$$

Now, observing Fig. 13-13 we can follow the isentropic line from point i to point es to calculate the exit stagnation pressure.

$$\frac{p_{oe}}{p_i} = \frac{p_{es}}{p_i} = \left(\frac{T_{es}}{T_i} \right)^{\gamma/(\gamma-1)}$$
$$p_{oe} = (14.7) \left(\frac{548.7}{530} \right)^{1.4/0.4} = 16.597 \text{ psia} \tag{d}$$

In a similar manner we can determine the actual exit pressure by going from point $0e$ to point e.

$$\frac{p_e}{p_{oe}} = \left(\frac{T_e}{T_{oe}} \right)^{\gamma/(\gamma-1)}$$
$$p_e = (16.597) \left(\frac{547.5}{550.8} \right)^{3.5} = 16.252 \text{ psia} \tag{e}$$

The area ratio is determined from the mass continuity relation

$$m = \rho_i A_i V_i = \rho_e A_e V_e \tag{f}$$

But $\rho = p/RT$, so that

$$\frac{A_e}{A_i} = \frac{p_i V_i T_e}{p_e V_e T_i} = \frac{(14.7)(500)(547.5)}{(16.252)(200)(530)} = 2.336$$

To calculate the pressure-recovery factor we must first determine the isentropic pressure rise. The inlet stagnation pressure is calculated from

$$\frac{p_{oi}}{p_i} = \left(\frac{T_{oi}}{T_i}\right)^{\gamma/(\gamma-1)} \tag{g}$$

$$p_{oi} = (14.7)\left(\frac{550.8}{530}\right)^{3.5} = 16.82 \text{ psia}$$

The pressure-recovery factor is calculated from Eq. (13-56):

$$C_p = \frac{p_e - p_i}{p_{oi} - p_i} = \frac{16.252 - 14.7}{16.82 - 14.7} = 0.732$$

REVIEW QUESTIONS

1   What is meant by the term *isentropic stagnation state?*

2   Why is the stagnation pressure reached in an irreversible adiabatic process lower than that reached in an isentropic process?

3   What is meant by the term *static pressure?*

4   Define the Mach number. Why is it important?

5   Give a word definition of the speed of sound.

6   Explain, in a physical sense, why the speed of sound in an ideal gas is a function of temperature alone.

7   What is Bernoulli's equation?

8   Explain the meaning of the term *choked flow.*

9   Why must the flow area be increased to cause an increase in flow velocity in the supersonic section of a nozzle?

10   Explain in a physical sense why the pressure must always decrease for an increase in velocity in an isentropic nozzle, regardless of subsonic or supersonic flow conditions.

11   What is meant by the term *momentum flux?*

12   Why cannot Newton's second law of motion be applied directly to a control volume (open system)?

13   Describe a shock wave in terms which could be understood by a layman.

14   Why are we justified in assuming that the flow across a shock wave is adiabatic?

15   What is the approximate thickness of a shock wave?

16 What are the Fanno and Rayleigh lines?

17 Why must the flow always proceed from supersonic to subsonic conditions when moving across a shock wave?

18 For an ideal gas why does the stagnation temperature remain constant across a shock wave?

19 Why does the presence of a shock wave in the diverging section of a converging-diverging nozzle not affect the mass flow rate through the nozzle?

PROBLEMS

13-1 Calculate the stagnation temperature and pressure for flow of
  (a) Helium at 2000 ft/s, 100°F, 1 psia
  (b) Air at $M = 5$, $-40°F$, 0.01 atm
  (c) Hydrogen at 1000 m/s, 100°C, 0.5 atm
  (d) Steam at 1500 ft/s, 120°F, 1 psia

13-2 Calculate the Mach number or flow velocity for each condition in Prob. 13-1, Assume $\gamma = 1.3$ for steam.

13-3 A converging-diverging nozzle is designed to produce an exit flow of air at $M = 4.0$ and 0.1 atm. The stagnation temperature is 120°F. Calculate the throat area and mass flow for an exit area of 1.0 in².

13-4 To what value must the back pressure be raised to just cause a normal shock to stand in the exit of the nozzle of Prob. 13-3?

13-5 Calculate the speed of sound for air, hydrogen, helium, carbon dioxide, and propane at 400°F.

13-6 Calculate the speed of sound in water vapor at 500 psia and 1000°F using the steam tables.

13-7 A ballistic missile may reenter the atmosphere at very high velocities of the order of $M = 10$. For a location where $T = -60°F$ and $p = 0.03$ atm estimate the stagnation temperature which will be present on the front of the missile assuming ideal-gas behavior. If a bow shock envelops the nose cone estimate the pressure which will exist at the stagnation point.

13-8 Consider a normal shock wave as a slowing-down mechanism for supersonic flow. For $M_x = 1.5$ calculate the value of a "diffuser efficiency" as defined by Eq. (13-57).

13-9 A converging-diverging nozzle is designed to operate with inlet stagnation conditions of 150 psia and 200°F with air discharging at $M = 2.5$. Assuming the stagnation conditions remain constant, to what value must the exit pressure be raised to just produce subsonic flow in the entire nozzle?

13-10 What percent reduction in flow rate would result from lowering the stagnation pressure in Prob. 13-9 to 100 psia?

13-11 A conveyor belt moves with a velocity of 10 ft/s and receives material from directly above at the rate of 900 lbm/min. What horsepower motor is required to drive the belt neglecting frictional losses in pulleys, drives, etc.?

13-12   A novel motorboat design employs a turbine pump inside the boat which sucks water through a front inlet and discharges it at the rear of the boat at a higher velocity. For a particular design it is determined that 500 pounds thrust is required to propel the boat at 30 mph. The inlet flow area is 1 ft², and the water may be assumed to enter at the boat speed. What exit flow area is required considering all the thrust to result from increased fluid momentum? What horsepower engine is required to drive the boat?

13-13   Very low temperatures can be produced by expanding high-pressure air in a supersonic nozzle. What ratio of exit to throat area is required to produce −60°F air from high-pressure air at 70°F and 20 atm?

13-14   There are design limitations placed on high-speed aircraft because of temperatures which the skin materials can withstand at high speeds. For a given speed the maximum temperature of the aircraft is the stagnation temperature. What is the maximum speed for flight at an altitude where $T = -50°F$, $p = 0.1$ atm, and there is a maximum allowable skin temperature of 700°F?

13-15   A small supersonic wind tunnel is to be designed so that it will have a test section with $M = 2.0$, $T = -50°F$, and $p = 2$ psia. The flow cross-sectional area is to be 1.0 ft², and the air flow is to be established with a high-pressure tank discharging through a converging-diverging nozzle section. What stagnation properties would be necessary in the tank, and what mass flow of air would be required? Also calculate the throat area of the nozzle.

13-16   If the local free stream temperature is 0°F for the flow around the bullet of Fig. 13-9, calculate the stagnation temperature and pressure on the nose of the bullet.

13-17   A diffuser is used to reduce the velocity of an air stream from $M = 0.8$, $p = 14.7$ psia, and $T = 70°F$ down to a value of 200 ft/s. Calculate the ratio of exit to inlet area if the diffuser efficiency is 85 percent. Also calculate the pressure-recovery factor. How much heat is added per pound mass of fluid as it flows through the diffuser?

13-18   Air enters a converging section at 200 ft/s, 50 psia, and 150°F. The exit and inlet areas are 4 and 6 in², respectively. Calculate the exit velocity, temperature, and pressure for isentropic flow.

13-19   Suppose an air flow stream at $M = 2.5$, $p = 0.1$ atm, and $T = 10°F$ is imposed on the inlet of the converging-diverging nozzle of Prob. 13-33M and that a normal shock wave develops at that point. What would be the exit flow conditions in this circumstance?

13-20   A tank filled with water and having a volume of 150 ft³ is placed on wheels whereby it can move freely on a level surface. An engine-driven pump mounted on the tank discharges water at the rate of 50 lbm/s through a nozzle having an exit area of 1.0 in². What is the net thrust force produced when the tank is moving at a speed of 20 ft/s opposite to the direction of the nozzle discharge velocity?

13-21 Steam expands isentropically in a nozzle from 500 psia, 600°F, to 200 psia. The inlet velocity is small. Calculate the exit velocity and exit area for a flow rate of 2 lbm/s.

13-22 Nitrogen expands isentropically from 150 psia and 300°F to atmospheric pressure. What throat and exit areas would be required for a mass flow of 0.1 lbm/s?

13-23 A converging-diverging nozzle having an exit to throat area ratio of 2.40 is connected to a large tank containing air at 200 psia and 70°F. What back pressure must be exerted to just produce a normal shock at the exit of the nozzle?

13-24 Suppose the steam nozzle of Prob. 13-21 has a velocity coefficient of 0.95. What would be the mass flow in this circumstance and the coefficient of discharge?

13-25 A small toy rocket is charged with air at 2000 psia and 110°F. The air is allowed to discharge through a converging-diverging nozzle having a throat area of 1.0 mm² and an exit area of 3.0 mm². Discharge takes place to atmospheric pressure. Calculate the net thrust force when the rocket is in a stationary position.

PROBLEMS (Metric units)

13-1M Calculate the stagnation temperature and pressure for flow of
(a) Helium at 610 m/s, 35°C, 7 kPa
(b) Air at $M = 5$, $-40°C$, 0.01 atm
(c) Hydrogen at 1000 m/s, 100°C, 0.5 atm
(d) Steam at 460 m/s, 49°C, 0.5 atm

13-2M A converging-diverging nozzle is designed to produce an exit flow of air at $M = 4.0$ and 0.1 atm. The stagnation temperature is 50°C. Calculate the throat area and mass flow for an exit area of 6.5 cm².

13-3M Calculate the speed of sound for air, hydrogen, helium, carbon dioxide, and propane at 200°C.

13-4M Calculate the speed of sound in water vapor at 3.5 MPa and 280°C using the steam tables.

13-5M A ballistic missile may reenter the atmosphere at very high velocities of the order of $M = 10$. For a location where $T = -50°C$ and $p = 0.03$ atm, estimate the stagnation temperature which will be present on the front of the missile assuming ideal-gas behavior. If a bow shock envelops the nose cone, estimate the pressure which will exist at the stagnation point.

13-6M A converging-diverging nozzle is designed to operate with inlet stagnation conditions of 1.0 MPa and 100°C, with air discharging at $M = 2.5$. Assuming the stagnation conditions remain constant, to what value must the exit pressure be raised to just produce subsonic flow in the entire nozzle?

13-7M Air at 900 kPa and 30°C expands isentropically in a nozzle to a pressure of

170 kPa. The inlet velocity is very low. Calculate the temperature and velocity at exit.

13-8M   Air at 1.0 kPa and 60°C is decelerated in an isentropic diffuser from $M = 3.0$ to $M = 1.0$ at exit from the diffuser. The flow rate is 12.0 kg/s. Determine the temperature and pressure at exit and the inlet and exit flow areas.

13-9M   Calculate the stagnation temperature and pressure for steam flowing at 6.0 MPa and 440°C with a velocity of 150 m/s.

13-10M In a certain jet aircraft the stagnation properties of the air are measured as $p_0 = 8.5$ MPa and $T_0 = 875$°C for a free stream pressure of 35 kPa. Calculate the speed of the aircraft and the Mach number.

13-11M A small sonic flow nozzle meters the flow of nitrogen at 5 atm and 50°C. What flow rate will the nozzle measure for an exit area of 4.0 mm² and a discharge at atmospheric pressure?

13-12M What would the flow rate be for the nozzle of Prob. 13-11M using air and helium for the same temperature and pressure?

13-13M A high-temperature gas stream of helium is to be accelerated in a converging nozzle from $T = 1000$°C, $p = 3$ atm, and $V = 100$ m/s to $M = 0.9$. Assuming isentropic flow, calculate the ratio of exit to inlet area and the exit temperature and pressure.

13-14M A cylinder of high-pressure air discharges through a small converging nozzle having an exit area of 5 mm². The conditions in the cylinder are 100 atm and 30°C and may be assumed to remain constant during the discharge process. What thrust force results, assuming isentropic flow and discharge into a large room at 1-atm pressure? What thrust force would result if the nozzle efficiency were 97 percent?

13-15M What percent reduction in flow rate would result from lowering the stagnation pressure in Prob. 13-6M to 700 kPa?

13-16M A conveyor belt moves with a velocity of 3.0 m/s and receives material from directly above at the rate of 33 kg/s. What power motor is required to drive the belt neglecting frictional losses in pulleys, drives, etc.?

13-17M A novel motor boat design employs a turbine pump inside the boat which sucks water through a front inlet and discharges it at the rear of the boat at a higher velocity. For a particular design it is determined that 2.2 kN thrust is required to propel the boat at 13 m/s. The inlet flow area is 0.1 m², and the water may be assumed to enter at the boat speed. What exit flow area is required considering all the thrust to result from increased fluid momentum? What power engine is required to drive the boat?

13-18M Very low temperatures can be produced by expanding high-pressure air in a supersonic nozzle. What ratio of exit to throat area is required to produce −50°C air from high-pressure air at 20°C and 20 atm?

13-19M There are design limitations placed on high-speed aircraft because of the temperatures which the skin materials can withstand at high speeds. For a given speed the maximum temperature of the aircraft is the stagnation temperature. What is the maximum speed for flight at an altitude where $T = -45°C$, $p = 0.1$ atm, and there is a maximum allowable skin temperature of 370°C?

13-20M   A small supersonic wind tunnel is to be designed so that it will have a test section with $M = 2.0$, $T = -45°C$, and $p = 14$ kPa. The flow cross-sectional area is to be 0.1 m², and the air flow is to be established with a high-pressure tank discharging through a converging-diverging nozzle section. What stagnation properties would be necessary in the tank, and what mass flow of air would be required? Also calculate the throat area of the nozzle.

13-21M   If the local free stream temperature is $-18°C$ for the flow around the bullet of Fig. 13-9, calculate the stagnation temperature and pressure on the nose of the bullet.

13-22M   A diffuser is used to reduce the velocity of an air stream from $M = 0.8$, $p = 1$ atm, and $T = 20°C$ down to a value of 60 m/s. Calculate the ratio of exit to inlet area if the diffuser efficiency is 85 percent. Also calculate the pressure-recovery factor. How much heat is added per kilogram of fluid as it flows through the diffuser?

13-23M   Air enters a converging section at 60 m/s, 340 kPa, and 65°C. The exit and inlet areas are 4 and 6 cm², respectively. Calculate the exit velocity, temperature, and pressure for isentropic flow.

13-24M   What minimum stagnation properties would be required to produce flow at $M = 3.0$, $p = 1$ atm, and $T = 20°C$?

13-25M   A tank filled with water and having a volume of 4200 liters is placed on wheels whereby it can move freely on a level surface. An engine-driven pump mounted on the tank discharges water at the rate of 23 kg/s through a nozzle having an exit area of 6.5 cm². What is the net thrust force produced when the tank is moving at a speed of 6 m/s opposite to the direction of the nozzle discharge velocity?

13-26M   Steam expands isentropically in a nozzle from 3.5 MPa, 320°C, to 1.5 MPa. The inlet velocity is small. Calculate the exit velocity and exit area for a flow rate of 0.9 kg/s.

13-27M   Nitrogen expands isentropically from 1 MPa and 150°C to atmospheric pressure. What throat and exit areas would be required for a mass flow of 0.05 kg/s?

13-28M   A converging-diverging nozzle having an exit-to-throat area ratio of 2.40 is connected to a large tank containing air at 1.3 MPa and 20°C. What back pressure must be exerted to just produce a normal shock at the exit of the nozzle?

13-29M   Suppose the steam nozzle of Prob. 13-26M has a velocity coefficient of 0.95. What would be the mass flow in this circumstance and the coefficient of discharge?

13-30M   A small toy rocket is charged with air at 13.5 MPa and 45°C. The air is allowed to discharge through a converging-diverging nozzle having a throat area of 1.0 mm² and an exit area of 3.0 mm². Discharge takes place to atmospheric pressure. Calculate the net thrust force when the rocket is in a stationary position.

13-31M   Calculate the maximum velocity which may theoretically be obtained by expanding air from stagnation conditions of 10 atm and 100°C.

13-32M   Suppose the converging-diverging nozzle of Example 13-4 has an inlet area of 8.0 cm² and is placed in a flow stream at $M = 3.0$, $p = 0.5$ atm, and $T = 0°C$, and that a normal shock forms at the inlet. What will be the exit temperature, pressure, and velocity under these conditions? What would be the mass flow rate under these conditions, and the net force exerted on the fluid as it moves from just upstream of the shock to the exit of the nozzle?

13-33M   A converging-diverging nozzle for use with air is constructed so that the inlet area is 8.0 cm², the minimum area is 4.0 cm², and the exit area is also 8.0 cm². What Mach number must be present at the inlet section to produce sonic conditions at the minimum area? What would be the exit Mach number under these conditions? If the inlet temperature and pressure are 25°C and 5 atm, what is the mass flow rate for sonic conditions at the minimum area?

13-34M   What are the stagnation temperature and pressure for the sonic flow of Prob. 13-33M?

13-35M   Air is contained in a 0.5-m³ tank at 5 atm and 25°C and is allowed to discharge through a small converging nozzle having an exit area of 1.0 cm². Discharge occurs into a large room at 1 atm, and the tank is perfectly insulated. How long will it take the discharge process to lower the pressure in the tank to 3 atm?

13-36M   A normal shock wave is established in helium at a point where $M = 3.0$, and the free-stream pressure and temperature are 0.2 atm and $-10°C$, respectively. Calculate the stagnation pressure downstream from the shock and the downstream-flow velocity. Compare these results with the values for air undergoing a shock at the same upstream conditions.

REFERENCES

1   Liepmann, H. W., and A. Roshko: "Elements of Gas Dynamics," John Wiley & Sons, Inc., New York, 1957.

2   Shapiro, A. H.: "The Dynamics and Thermodynamics of Compressible Flow," The Ronald Press Company, New York, 1954.

3   Thompson, P. A.: "Compressible Fluid Dynamics," McGraw-Hill Book Company, New York, 1972.

4   Chapman, A. J., and W. F. Walker: "Introductory Gas Dynamics," Holt, Rinehart and Winston, Inc., New York, 1971.

chapter fourteen
Thermodynamics of irreversible processes

14-1 INTRODUCTION

Thermodynamics is usually considered a study of systems in equilibrium and processes which may be represented by a series of equilibrium states such as the quasistatic or reversible process described in Chap. 1. It should be quite clear to the reader by now that nature does not interact in the form of reversible processes. In all cases the system under consideration must depart from equilibrium—even if only a slight departure—in order to change from one equilibrium state to another. The principles of thermodynamics, that is, energy balances, the principle of increase of entropy, etc., may be employed to predict the overall heat flow and work in various processes. But these principles will not predict the *rate* at which a change can take place. It is a fairly simple matter to predict the equilibrium temperature of a hot steel bar placed in an insulated container of water. But to predict the *time* required to attain a particular set of bar- and water-temperature combinations involves concepts which are normally treated apart from thermodynamics under the title of *transport processes*. Some of the analyses of Chap. 9 have already indicated techniques for studying transport phenomena in gases.

The purpose of this chapter is to examine some general classifications of irreversible processes and indicate common concepts involved in various rate processes. A simple application of the concepts to thermoelectric phenomena will then be given to provide a concrete example of irreversible thermodynamic analysis.

14-2 PHENOMENOLOGICAL LAWS

Irreversible processes are associated with the transport of heat, mass, momentum, and electric charge. Some examples of such transport processes have been given in Chap. 9. In each irreversible process there is a *flow* of heat, mass, electric charge, etc. Producing this flow is a *driving force* usually described by the gradient in some physical property. A temperature gradient is the driving force for a heat flow, a concentration gradient for mass transfer, and electric potential is the driving force for electric current, i.e., flow of electric charge. The flow per unit area or *flux* is then postulated to be proportional to the driving forces, and analytical expressions are written as

$$J_Q = -K_t \frac{\partial T}{\partial x} \qquad \text{Fourier's law} \qquad (14\text{-}1a)$$

$$J_m = -D \frac{\partial C}{\partial x} \qquad \text{Fick's law} \qquad (14\text{-}1b)$$

$$J_e = -K_e \frac{\partial E}{\partial x} \qquad \text{Ohm's law} \qquad (14\text{-}1c)$$

where J indicates the flux. The proportionality constants K_t, D, and K_e are called the thermal conductivity, mass diffusion coefficient, and electric conductivity, respectively. It should be fairly obvious that these equations are really not "laws" in the same sense as the first and second laws of thermodynamics. They serve as convenient analytical vehicles for describing the physical transport processes and as defining relations for the constants K_t, D, and K_e. These "constants" are frequently strong functions of material properties, as has already been noted in Chap. 9.

If only one driving force (property gradient) is present in a system, we say that there is a *single flow* associated with the property gradient, which may be described by some type of phenomenological relation.

If more than one driving force is present in the system, there will be more than one flow, but each flow in such a circumstance is not necessarily a function of a unique driving force. When two or more fluxes are present in the system, we say that we have *coupled flows*. An example of a coupled-flow system has been given in Chap. 9 in the discussion of *thermal diffusion*. In multicomponent systems a temperature gradient can give rise to a concentration gradient and, consequently, to a mass flux. The total mass flux in such a system is a function of the overall concentration gradient and the temperature gradient as well. Thus, in this instance, the mass flux is a function of *two* driving forces. The heat flux would also be a function of these same two driving forces.

The general principle governing the foregoing discussion is that each flux J_i is assumed to be related to each gradient or driving force X_i by the following linear relation:

$$J_i = L_{i_1}X_1 + L_{i_2}X_2 + \cdot \cdot \cdot = \sum_{j=1}^{n} L_{ij}X_j \tag{14-2}$$

Of course, Eqs. (14-2) must be written for each ith flux. The resulting equations are called the *linear phenomenological* equations, and the coefficients L_{ij} are called the *Onsager phenomenological coefficients*.

In Eqs. (14-2) it is to be noted that the subscript i refers to the *flux* and the subscript j refers to the driving force. The *primary* driving force is X_i and has associated with it the primary phenomenological coefficient of L_{ii}. All the other L_{ij} $(i \neq j)$ are *coupling* coefficients.

14-3 ONSAGER RECIPROCITY RELATION

A solution of Eqs. (14-2) represents a formidable problem because, in general, the coefficients L_{ij} must be determined experimentally. For some simple problems, particularly those involving gases, an analytical calculation may be possible, but even these calculations must be verified by experiment if they are to be accepted with confidence. Measurements of primary coefficients are not terribly difficult to perform, but experimental determinations of the coupling coefficients can be very tedious because of the larger number of parameters which must be kept under strict control. Further measurements are needed for a number of cases. An analytical relationship which helps to alleviate some of the difficulty is the *Onsager reciprocity relation*, which states that

$$L_{ij} = L_{ji} \tag{14-3}$$

This relation applies to systems in the absence of a magnetic field. A proof of the Onsager relation must be undertaken with statistical thermodynamics, and the reader should consult the references at the end of the chapter for a derivation. In a physical sense, the relation indicates that coupled flows are induced in a reciprocal manner between interacting driving forces.

A quick application to a two-flux system indicates the utility of the reciprocity relation. Equations (14-2) become

$$J_1 = L_{11}X_1 + L_{12}X_2 \tag{14-4}$$

$$J_2 = L_{21}X_1 + L_{22}X_2 \tag{14-5}$$

Given X_1 and X_2, a knowledge of four coefficients is necessary to determine the two fluxes. With the reciprocity relation $L_{12} = L_{21}$ only three coefficients are necessary, two of which are primary coefficients (L_{11} and L_{22}) which are usually readily available.

14-4 APPLICABILITY OF THE PHENOMENOLOGICAL RELATIONS

The reader may have tacitly assumed that the fluxes J_i and driving forces X_i may take *any* form, but such is not the case. In order for the reciprocity relation to apply, it is necessary that the total entropy production per unit time and per unit volume resulting from all the irreversible processes be represented as a linear sum of products of forces and fluxes; i.e.,

$$\left(\frac{dS}{d\tau}\right)_{\text{prod}} = J_1 \cdot X_1 + J_2 \cdot X_2 + \cdots = \sum_i J_i \cdot X_i \tag{14-6}$$

In the following sections we shall learn how to go about matching fluxes and forces so that they fulfill this requirement.

14-5 HEAT FLUX AND ENTROPY PRODUCTION

Consider a solid bar of cross-sectional area A, as shown in Fig. 14-1. The ends of the rod are exposed to the reservoirs at T_1 and T_2, and the outside of the rod is insulated. At each location in the rod the heat flux is

FIG. 14-1 *One-dimensional heat flow and entropy flux.*

$$J_Q = \frac{Q}{A} = -K_t \frac{dT}{dx}$$

and, for constant K_t,

$$J_Q = \frac{K_t(T_1 - T_2)}{L}$$

There is also an entropy flux at each location in the amount

$$J_S = \frac{J_Q}{T} = -K_t \frac{1}{T} \frac{dT}{dx} \qquad (14\text{-}7)$$

The entropy flux is not conserved, however. For the element of length dx the entropy produced is

$$\left(\frac{dS}{d\tau}\right)_{\text{prod}} = A(J_{Sx+dx} - J_{Sx})$$

$$= AJ_Q \left[\left(\frac{1}{T}\right)_{x+dx} - \left(\frac{1}{T}\right)_x\right]$$

$$= AJ_Q \frac{d(1/T)}{dx} dx \qquad (14\text{-}8)$$

and the entropy produced per unit volume is

$$\left(\frac{dS}{d\tau}\right)_{\text{prod}} = -J_Q \frac{1}{T^2} \frac{dT}{dx} \qquad (14\text{-}8a)$$

The overall entropy production is

$$\Delta S_{\text{prod}} = A(J_{S2} - J_{S1}) = AJ_Q \left(\frac{1}{T_2} - \frac{1}{T_1}\right) \qquad (14\text{-}9)$$

We may also describe the entropy production in a bar resulting from conduction of an electric current. In this instance we allow heat transfer to the surroundings in the amount of the electric energy dissipated. The system is shown in Fig. 14-2. The electric current is given by

$$\frac{I}{A} = J_e = -K_e \frac{dE}{dx}$$

At each section of the conductor the electric energy dissipated is equal to the heat transfer to the surroundings:

$$dQ = I \frac{dE}{dx} dx$$

The resultant entropy production in the volume $A\,dx$ is

$$\left(\frac{dS}{d\tau}\right)_{prod} = \frac{dQ}{T} = -\frac{J_e}{T} \frac{dE}{dx} \tag{14-10}$$

In a bar where both an electric current and a temperature gradient are present, the total entropy production per unit volume is given as the sum of Eqs. (14-8a) and (14-10):

$$\left(\frac{dS}{d\tau}\right)_{total\ prod} = -J_Q \frac{1}{T^2} \frac{dT}{dx} - J_e \frac{1}{T} \frac{dE}{dx} \tag{14-11}$$

If Eq. (14-6) is to be satisfied for a system involving electric and heat conduction, it is apparent from Eq. (14-11) that the selection of fluxes and driving forces must correspond to

$$J_Q; \quad X_Q = -\frac{1}{T^2} \frac{dT}{dx} \tag{14-12}$$

$$J_e; \quad X_e = -\frac{1}{T} \frac{dE}{dx} \tag{14-13}$$

FIG. 14-2   *One-dimensional flow of electric current.*

An alternate selection may be made by using the entropy flux $J_S = J_Q/T$. Then the appropriate forces are

$$J_S; \quad X_S = -\frac{1}{T}\frac{dT}{dx} \tag{14-14}$$

$$J_e; \quad X_e = -\frac{1}{T}\frac{dE}{dx} \tag{14-15}$$

14-6 THERMOELECTRIC PHENOMENA

We shall now employ the foregoing relations to analyze the physical phenomena associated with combined electric- and heat-conduction problems. Writing the heat flux and electric current in phenomenological form according to Eq. (14-2),

$$J_Q = -L_{qq}\frac{1}{T^2}\frac{dT}{dx} - L_{qe}\frac{1}{T}\frac{dE}{dx} \tag{14-16}$$

$$J_e = -L_{eq}\frac{1}{T^2}\frac{dT}{dx} - L_{ee}\frac{1}{T}\frac{dE}{dx} \tag{14-17}$$

For either pure thermal conduction or electric conduction these equations reduce to

$$J_Q = -L_{qq}\frac{1}{T^2}\frac{dT}{dx} = -K_t\frac{dT}{dx} \tag{14-18}$$

$$J_e = -L_{ee}\frac{1}{T}\frac{dE}{dx} = -K_e\frac{dE}{dx} \tag{14-19}$$

From these relations the conductivities may be expressed in terms of the primary coefficients by

$$K_t = \frac{L_{qq}}{T^2} \tag{14-20}$$

$$K_e = \frac{L_{ee}}{T} \tag{14-21}$$

Let us now investigate the circumstance when there is no current flow in the material; i.e., $J_e = 0$. Then, from Eq. (14-17),

$$\left(\frac{dE/dx}{dT/dx}\right)_{J_e=0} = \left(\frac{dE}{dT}\right)_{J_e=0} = \frac{-L_{eq}}{TL_{ee}} \tag{14-22}$$

Equation (14-22) expresses the fact that even with zero current flow, an electric potential will exist when a temperature gradient is present. The phenomenon is called the *Seebeck effect*, and the gradient is called the *thermoelectric power* of the material, ϵ.

$$\epsilon = - \left(\frac{dE}{dT}\right)_{J_e=0} \tag{14-23}$$

The thermoelectric power is a quantity which is reasonably easy to determine experimentally. Combining Eqs. (14-22) and (14-23) gives

$$L_{eq} = \epsilon T L_{ee} = \epsilon K_e T^2 \tag{14-24}$$

Using the reciprocity relation $L_{eq} = L_{qe}$, the heat flux becomes, for $J_e = 0$,

$$J_Q = \frac{-L_{ee}L_{qq} + L_{eq}^2}{L_{ee}T^2} \frac{dT}{dx} \tag{14-25}$$

and the thermal conductivity for this coupled system will be defined as

$$K_t = \frac{L_{ee}L_{qq} - L_{eq}^2}{L_{ee}T^2} \tag{14-26}$$

Now we may solve Eq. (14-26) for L_{qq} and write the following for $J_e \neq 0$:

$$J_Q = -(K_t + \epsilon^2 K_e T) \frac{dT}{dx} - \epsilon T K_e \frac{dE}{dx} \tag{14-27}$$

$$J_e = -\epsilon K_e \frac{dT}{dx} - K_e \frac{dE}{dx} \tag{14-28}$$

By eliminating dE/dx from these expressions we also obtain

$$J_Q = -K_t \frac{dT}{dx} + \epsilon T J_e \tag{14-29}$$

Now consider the junction of two dissimilar metals *a* and *b* shown in Fig. 14-3. A current flow J_e is forced through the junction, and we allow equilibrium to be established at isothermal conditions where the electric energy is dissipated to the surroundings.

For isothermal conditions,

$$J_e = -K_e \frac{dE}{dx} \tag{14-30}$$

In accordance with Eq. (14-29) the heat flux through each conductor is

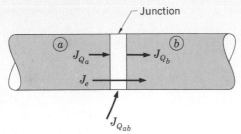

FIG. 14-3 *Heat fluxes at junction of two dissimilar metals, the Peltier effect.*

$$J_{Q_a} = T\epsilon_a J_e \tag{14-31}$$

$$J_{Q_b} = T\epsilon_b J_e \tag{14-32}$$

and the heat exchange with the surroundings is

$$J_{Q_{ab}} = J_{Q_b} - J_{Q_a} = T(\epsilon_b - \epsilon_a)J_e = \pi_{ab}J_e \tag{14-33}$$

The foregoing phenomenon is called the *Peltier effect*, and π_{ab} is called the *Peltier coefficient*. It is to be noted that the heat flux $J_{Q_{ab}}$ is due solely to the *junction* of two dissimilar metals and is *not* the result of conventional I^2R Joulean heating.

    Finally, let us consider an electric conductor as shown in Fig. 14-4. A heat flux and electric current are impressed on the conductor and heat is dissipated to the surroundings. The *total energy flux* J_E through the conductor is expressed by

$$J_E = J_Q + EJ_e$$
$$= -K_t \frac{dT}{dx} + (\epsilon T + E)J_e \tag{14-34}$$

The heat transfer to the surroundings J_{QR} results from a gradient in the total energy flux

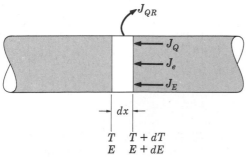

FIG. 14-4 *Combined heat flux and electric current, the Thomson effect.*

$$J_{QR} = J_{E_x} - J_{E_{x+dx}} = -\frac{dJ_E}{dx}\, dx \qquad (14\text{-}35)$$

Now combining Eqs. (14-35) and (14-34) gives

$$J_{QR} = -\frac{d}{dx}\left[-K_t\frac{dT}{dx} + (\epsilon T + E)J_e\right] dx \qquad (14\text{-}36)$$

Now let us examine this situation for the special case of a linear temperature profile. This is the case where there is no current flow or heat transfer to the surroundings, as illustrated in Fig. 14.1. Then

$$\frac{d}{dx}\left(K_t\frac{dT}{dx}\right) = 0$$

and Eq. (14-36) becomes, after a substitution for dE/dx from Eq. (14-30),

$$\frac{J_{QR}}{dx} = -TJ_e\frac{d\epsilon}{dx} + \frac{J_e^2}{K_e} \qquad (14\text{-}37)$$

From Eq. (14-37) we see that the heat transfer to the surroundings for this special case is not equal to the Joulean heat production J_e^2/K_e. The additional heat transfer over and above the Joulean heat is called the *Thomson heat*, and the Thomson coefficient τ is defined by

$$(J_Q)_{\text{Thomson}} = -TJ_e\frac{d\epsilon}{dx} = -TJ_e\frac{d\epsilon}{dT}\frac{dT}{dx} = -\tau J_e\frac{dT}{dx}$$

or

$$\tau = T\frac{d\epsilon}{dT} \qquad (14\text{-}38)$$

For a junction of dissimilar materials some algebraic transcription yields the following relation among the various thermoelectric coefficients:

$$\frac{d\pi_{ab}}{dT} = (\epsilon_b - \epsilon_a) + (\tau_b - \tau_a) \qquad (14\text{-}39)$$

14-7 THERMOELECTRIC CIRCUITS

Consider the simple thermoelectric circuit shown in Fig. 14-5. The dissimilar metals a and b are connected as shown, and the two junctions are exposed to the different temperatures T_1 and T_2. As a result of this temperature

FIG. 14-5 *Simple thermoelectric circuit.*

difference there exists an open-circuit voltage of E_{ab} resulting from the combined Seebeck effects in the two conductors. The *Seebeck coefficient* α for the two materials is defined by

$$\alpha_{ab}(T) = \lim_{\Delta T \to 0} \frac{\Delta E_{ab}}{\Delta T} = \frac{dE_{ab}}{dT} \tag{14-40}$$

The Seebeck coefficient may be expressed in terms of the thermoelectric powers of the materials in the following way. First, the overall potential difference in the circuit is written by integrating Eq. (14-23) over both the conductors. Thus

$$-E_{ab} = \int_{T_a}^{T_1} \epsilon_b(T)dT + \int_{T_1}^{T_2} \epsilon_a(T)dT + \int_{T_2}^{Tb} \epsilon_b(T)dT$$

When the output taps are at the same temperature, i.e., $T_a = T_b$, this expression reduces to

$$-E_{ab} = \int_{T_1}^{T_2} [\epsilon_a(T) - \epsilon_b(T)]dT \tag{14-41}$$

In the limit for small ΔT

$$\frac{dE_{ab}}{dT} = \epsilon_b(T) - \epsilon_a(T) = \alpha_{ab}(T) \tag{14-42}$$

From Eq. (14-33) the Peltier coefficient is

$$\pi_{ab} = T(\epsilon_b - \epsilon_a)$$

so that the Seebeck coefficient may be expressed as

$$\alpha_{ab}(T) = \frac{\pi_{ab}}{T} \tag{14-43}$$

Rewriting Eq. (14-39) in terms of α_{ab} gives

$$\frac{d(T\alpha_{ab})}{dT} = \alpha_{ab} + \tau_b - \tau_a$$

Performing the differentiation yields

$$\frac{d\alpha_{ab}}{dT} = \frac{\tau_b - \tau_a}{T} \tag{14-44}$$

The device illustrated in Fig. 14-5 is commonly called a *thermocouple*. When combined with a knowledge of values of $\alpha_{ab}(T)$ for the two materials it is extremely useful for temperature measurements. With the reference temperature T_2 known, the voltage output is a unique function of T_1 and the integrated Seebeck coefficient. Consequently, a temperature measurement may be performed through a simple voltage measurement. Thermocouple tables are available which give temperatures as a function of voltage output for various combinations of thermocouple materials. A summary of such data is given in Tables 14-1 and 14-2.

EXAMPLE 14-1   Thermocouple voltages

A chromel-constantan thermocouple employs a cold junction at 32°F. Assuming that the open-circuit voltage follows a linear variation with temperature between 32 and 200°F, calculate the thermocouple reading at 150°F, the Peltier coefficient at 150 and 200°F, and the difference in the Thomson coefficients at these temperatures.

SOLUTION   From Table 14-1 we have

$$E_{ab}(32°F) = 0$$
$$E_{ab}(200°F) = 5.87 \text{ mV}$$

A linear relation is assumed to be

$$E_{ab} = a + bT$$

Using the foregoing conditions,

TABLE 14-1
Thermal emf in absolute millivolts for commonly used thermocouple combinations according to Ref. [7] (based on reference junction at 32°F)

Tempera- ture, °F	Copper- constantan	Chromel- constantan	Iron- constantan	Chromel- alumel	Platinum- 10% rhodium
−300	−5.284	−8.30	−7.52	−5.51	
−250	−4.747		−6.71	−4.96	
−200	−4.111	−6.40	−5.76	−4.29	
−150	−3.380		−4.68	−3.52	
−100	−2.559	−3.94	−3.49	−2.65	
−50	−1.654		−2.22	−1.70	
0	−0.670	−1.02	−0.89	−0.68	
50	0.389		0.50	0.40	
100	1.517	2.27	1.94	1.52	0.221
150	2.711		3.41	2.66	0.401
200	3.967	5.87	4.91	3.82	0.595
250	5.280		6.42	4.97	0.800
300	6.647	9.71	7.94	6.09	1.017
350	8.064		9.48	7.20	1.242
400	9.525	13.75	11.03	8.31	1.474
450	11.030		12.57	9.43	1.712
500	12.575	17.95	14.12	10.57	1.956
600	15.773	22.25	17.18	12.86	2.458
700	19.100	26.65	20.26	15.18	2.977
800		31.09	23.32	17.53	3.506
1000		40.06	29.52	22.26	4.596
1200		49.04	36.01	26.98	5.726
1500		62.30		33.93	7.498
1700		70.90		38.43	8.732
2000				44.91	10.662
2500				54.92	13.991
3000					17.292

$$E_{ab} = -1.12 + 0.035T \text{ mV} \qquad T \text{ in } °F$$

At 150°F,

$$E_{ab} = -1.12 + (0.035)(150) = 4.13 \text{ mV}$$

A relation for the Peltier coefficient is now obtained from Eq. (14-43):

$$\pi_{ab} = T\alpha_{ab} = T \frac{dE_{ab}}{dT}$$

$$= 0.035T \qquad (T \text{ in } °R)$$
$$\pi_{ab}(150°F) = (0.035)(150 + 460) = 21.3 \text{ mV}$$
$$\pi_{ab}(200°F) = (0.035)(200 + 460) = 23.1 \text{ mV}$$

TABLE 14-2
Seebeck coefficient of thermocouple made of material listed against platinum at 0°C, $\mu V /°C$

Material	Coefficient
Bismuth	−72
Constantan	−35
Nickel	−15
Potassium	−9
Sodium	−2
Platinum	0
Mercury	0.6
Carbon	3
Aluminum	3.5
Lead	4
Tantalum	4.5
Rhodium	6
Silver	6.5
Copper	6.5
Gold	6.5
Tungsten	7.5
Cadmium	7.5
Iron	18.5
Nichrome	25
Antimony	47
Germanium	300
Silicon	440
Tellurium	500
Selenium	900

For this assumed linear variation of voltage output with temperature we obtain, from Eq. (14-44),

$$\tau_b - \tau_a = T \frac{d\alpha_{ab}}{dT}$$
$$= T(0) = 0$$

If a more precise relation for E_{ab} was obtained in the form of a higher-order polynomial such as

$$E_{ab} = a + bT + cT^2$$

then an appropriate relation for the difference in the Thomson coefficients could be derived.

EXAMPLE 14-2   Peltier coefficient
Calculate the Peltier coefficient of a junction of bismuth and tellurium at 0°C.

SOLUTION   From Table 14-2 the thermoelectric powers referenced to platinum are:

$$\epsilon_{Bi} - \epsilon_{Pl} = -72 \ \mu V/°C$$
$$\epsilon_{Te} - \epsilon_{Pl} = 500 \ \mu V/°C$$

so that, for the bismuth-tellurium combination,

$$\alpha_{ab} = -72 - 500 = -572 \ \mu V/°C$$

The Peltier coefficient is then calculated from Eq. (14-43) as

$$\pi_{ab} = T\alpha_{ab} = (273)(-572) = 1.56 \times 10^5 \ \mu V$$
$$= 156 \ mV$$

It is fairly easy to see that a thermoelectric circuit may be used as a power-producing device. By supplying heat Q_1 at the high temperature T_1 and rejecting heat Q_2 at the low temperature T_2, a heat engine is produced. Electrical work is delivered by connecting an external load to the circuit. Conversely, the circuit may be employed as a refrigerator by delivering an electric current to the circuit, thereby producing a temperature difference between the two junctions. Thermoelectric energy converters are discussed in more detail in Chap. 15.

REVIEW QUESTIONS

1   What is a phenomenological law?

2   Give an example of a coupled flow system.

3   What is the Onsager reciprocity relation?

4   What is the Seebeck effect?

5   Describe the Peltier effect. How does it differ from the Seebeck effect?

6   Describe the Thomson effect.

7   Describe the meaning of the terms *fluxes* and *driving forces* as applied to irreversible thermodynamics.

8   What is a thermocouple?

9   What is a reversible process?

10 What is meant by "coupled flows"?

11 Define thermoelectric power.

PROBLEMS

14-1 Derive an expression similar to Eq. (14-8a) for the entropy production per unit volume resulting from a mass diffusion of gas component A through gas component B. $\left[Ans. : \left(\dfrac{dS}{d\tau} \right)_{\text{prod/vol}} : -J_{ma} \dfrac{R_A}{p_A} \dfrac{dp_A}{dx} \right]$

14-2 Derive an expression for the net *energy* flux resulting from the diffusion process in Prob. 14-1, assuming isothermal conditions.

14-3 Suppose a temperature gradient is imposed on the system in Prob. 14-1 such that there is a heat-conduction current according to Fourier's law. Derive an expression for the total entropy production per unit volume under these circumstances. Establish a set of linear phenomenological relations based on this derivation.

14-4 For the conditions of Example 14-1 establish a quadratic relation for the open-circuit voltage. Subsequently, calculate the thermocouple reading at 150°F, the Peltier coefficient at 150 and 200°F, and the difference in the Thomson coefficients at these temperatures.

14-5 Calculate the Peltier coefficient at 0°C for a thermocouple composed of aluminum and nichrome.

14-6 A thermocouple is constructed of constantan-alumel and employs a cold junction at 32°F. Calculate the output of the thermocouple at 600°F and the Seebeck coefficient at this temperature.

REFERENCES

1 Prigogine, I.: "Introduction to Thermodynamics of Irreversible Processes," Charles C Thomas, Publisher, Springfield, Ill., 1955.

2 Degroot, S. R.: "Thermodynamics of Irreversible Processes," Interscience Publishers, New York, 1952.

3 Lee, J. F., F. W. Sears, and D. L. Turcotte: "Statistical Thermodynamics," Addison-Wesley Publishing Company, Inc., Reading, Mass., 1963.

4 Callen, H. B.: "Thermodynamics," John Wiley & Sons, Inc., New York, 1960.

5 Hatsopoulos, G. N., and J. H. Keenan: "Principles of General Thermodynamics," John Wiley & Sons, Inc., New York, 1965.

6 Tribus, M.: "Thermostatics and Thermodynamics," D. Van Nostrand Company, Inc., Princeton, N.J., 1961.

7 _____: Reference Tables for Thermocouples, *Natl. Bur. Std. (U.S.) Cir.* 561, April 1955.

chapter fifteen
Direct energy conversion

15-1 INTRODUCTION

We have already discussed a number of conventional energy-conversion cycles in Chap. 12. All of these "conventional" cycles involve intermediate steps for conversion of some primary energy source to electric energy—a vapor power cycle involves the sequence of chemical energy of fuel, heat of combustion, enthalpy of steam, mechanical work output of a turbine, and, finally, electric energy output from a generator. All of the conventional power cycles are limited by the Carnot efficiency since they basically involve the conversion of heat into work.

The term *direct energy conversion* implies that we shall be concerned with devices which may be employed to convert different types of energy *directly* into electric energy. Such devices are obviously of significant practical interest. The reader should be aware, however, of the fact that direct-energy-conversion schemes do not always offer advantages over conventional power cycles. As the exposition of this chapter progresses, we shall see how these devices fit into the overall technological scheme of energy utilization.

Not all devices are limited by the Carnot principle for efficiency of energy conversion. This principle applies only to those circumstances where *heat* is converted to other forms of energy. It does not apply, for example, to the direct conversion of chemical energy into electric energy.

In this chapter we shall be concerned with a brief discussion of the principles of four direct-energy-conversion schemes and a general discussion of some of their advantages and limitations. The interested reader who wishes to pursue the subject further may consult the specialized references at the end of the chapter.

15-2 FUEL CELLS

The first direct-energy-conversion device we shall consider is the fuel cell, a device for the direct conversion of chemical energy into electric energy. Such a conversion is not limited by the Carnot efficiency. The fuel cell is similar to a battery or electrolytic cell in that two electrodes are placed in an electrolyte and a potential difference results from the reactions at the electrode surfaces. In the conventional battery the system is given an initial charge and is exhausted once this charge is gone. The fuel cell differs from the battery in that a continuous supply of reactants is fed to the cell, thereby enabling it to remain in continuous operation as long as fuel is supplied to it. In operation, fuel is supplied to one electrode and oxygen or air to the other. The electrolyte must be of such character that it permits rapid diffusion of ions through it. This diffusion process, of course, serves as the mechanism for completing an external electrical circuit.

To illustrate the principle of the fuel cell, consider the simple hydrogen-oxygen cell shown in Fig. 15-1. In this example hydrogen is the fuel. Both electrodes are constructed of a porous material to permit passage of the gases. At the cathode the hydrogen reacts with the electrolyte to produce H_2O and electrons. The charge carriers in this case are the $(OH)^-$ ions in the electrolyte, and the reaction which occurs at the anode is

$$H_2 + 2OH^- \rightarrow 2H_2O + 2e^- \tag{15-1}$$

At the cathode the reaction is

(a)

(b)

FIG. 15-1 *Simple hydrogen-oxygen fuel cell.* (a) *Schematic;* (b) *overall energy balance.*

$$2e^- + \tfrac{1}{2}O_2 + H_2O \rightarrow 2OH^- \tag{15-2}$$

The overall reaction for the fuel cell is just the sum of Eqs. (15-1) and (15-2), or

$$H_2 + \tfrac{1}{2}O_2 \rightarrow H_2O \tag{15-3}$$

Equation (15-3) represents the combustion reaction for hydrogen or the inverse reaction for electrolysis of water. On an overall basis, hydrogen and oxygen are delivered to the cell and electrical work, heat, and water are delivered by the cell, as illustrated in Fig. 15-1b. In the ideal case, the device may be operated reversibly, isothermally, and at constant pressure. For such a process we have already shown in Chap. 11 that the maximum work output is equal to the decrease in the Gibbs function [see Eq. (11-105)]:

$$W_e = -\Delta G \tag{15-4}$$

Also, in a reversible isothermal process, the heat transfer is

$$Q = T\,\Delta S \tag{15-5}$$

The overall energy balance for the cell may be obtained from the steady-flow energy equation as

$$Q = \Delta H + W_e \tag{15-6}$$

where W_e is the work *output*.

There are two possible ways to evaluate the performance of actual fuel cells: (1) by comparison with the maximum work output given by Eq. (15-4) and (2) by comparison with the total enthalpy difference ΔH across the cell. This second basis indicates that the maximum electric energy available per unit chemical energy input is

$$\eta_{\max} = \frac{\Delta G}{\Delta H} \tag{15-7}$$

We shall call $\eta_{\max}$ the *efficiency of the fuel cell*.

The work output of the cell may also be expressed in terms of the electron (or ion) transport as

$$W_e = E_r n_e \mathscr{F} = -\Delta G \tag{15-8}$$

where E_r is the voltage output of the reversible cell, n_e is the number of electrons transferred per molecule of fuel oxidized, and $\mathscr{F}$ is Faraday's constant:

$$\mathscr{F} = 96\ 500 \text{ C/g mol} = 96\ 500 \text{ J/V·g mol}$$
$$= 23\ 060 \text{ cal/V·g mol}$$

Faraday's constant is calculated as the product of Avogadro's number N_0 and the electronic charge e:

$$\mathscr{F} = N_0 e = (6.025 \times 10^{23} \text{ molecules/g mol})(1.602 \times 10^{-19} \text{ C})$$
$$= 96\ 500 \text{ C/g mol}$$

EXAMPLE 15-1   Hydrogen-oxygen fuel cell

A hydrogen-oxygen fuel cell operates at 25°C. Calculate the voltage output of the cell, the efficiency, and the electrical work output per mole of H_2 consumed and per mole of H_2O produced. Also compute the heat transfer to the surroundings.

SOLUTION   At 25°C (298 K) the standard heat of reaction and Gibbs' function change for the reaction given by Eq. (15-3) are

$$\Delta H^\circ_{298} = -68\ 317 \text{ cal/kg mol } H_2$$
$$\Delta G^\circ_{298} = -56\ 690 \text{ cal/kg mol } H_2$$

According to Eq. (15-1) there are two electrons transferred per molecule of H_2, so that $n_e = 2$. Then the cell voltage is given as

$$E_r = \frac{-\Delta G^\circ_{298}}{n_e \mathscr{F}} = \frac{-(-56\ 690)}{(2)(23\ 060)} = 1.229 \text{ V}$$

The efficiency is

$$\eta_{max} = \frac{\Delta G^\circ_{298}}{\Delta H^\circ_{298}} = \frac{-56\ 690}{-68\ 317} = 83 \text{ percent}$$

The electrical work output per mole of fuel is

$$W_e = -\Delta G = 56\ 690 \text{ cal/mol } H_2 \qquad (237.2 \text{ kJ/kg mol})$$

Since there is 1 mol of H_2O produced for each mole of fuel, there is also a work output of 56 690 cal/mol H_2O. The heat transfer is

$$Q = T\ \Delta S = \Delta H^\circ_{298} - \Delta G^\circ_{298}$$
$$= -68\ 137 + 56\ 690 = -11\ 447 \text{ cal} \qquad (-47.89 \text{ kJ/kg mol})$$

The negative sign indicates that heat is removed from the cell and transferred *to* the surroundings.

A schematic for a fuel cell employing a hydrocarbon fuel is shown in Fig. 15-2. In this example methane, CH_4, is the fuel. The reaction at the anode is

$$CH_4 + 2H_2O \rightarrow CO_2 + 8H^+ + 8e^- \qquad (15\text{-}9)$$

and the reaction at the cathode is

(a)

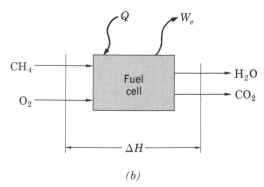

(b)

FIG. 15-2 *Simple hydrocarbon fuel cell employing methane as fuel. (a) Schematic; (b) energy balance.*

$$2O_2 + 8e^- + 8H^+ \rightarrow 2H_2O + 2H_2O \qquad (15\text{-}10)$$

The overall reaction for the cell is

$$CH_4 + 2O_2 \rightarrow CO_2 + 2H_2O \qquad (15\text{-}11)$$

It is to be noted here that the H^+ ions are the charge carriers just like the $(OH)^-$ ions were the carriers for the hydrogen-oxygen cell. For the ideal methane cell at 25°C the effectiveness is 92 percent and the cell voltage is 1.15 V.

A real fuel cell differs from the ideal cells described in the foregoing, and several important problems must be dealt with in practice. Some of these problems are described in the following:

1 The fuel must be transported to the electrode surface in contact with the electrolyte. This transport process involves the selection of a proper porous electrode material. The diffusion process at the surface of the electrode is quite complicated and subject to many losses.
2 The products of the reaction must be transported away from their respective electrodes. Again, complicated diffusion processes are involved and losses may be encountered.
3 The electrolyte must be selected so that the proper ion carriers are provided. Acid electrolytes, for example, could provide H^+ ions and alkaline electrolytes could provide $(OH)^-$ ions. An additional restriction is placed on the electrolyte in that it should not be corrosive to the electrodes and the oxygen or fuel should not dissolve in it.
4 Electrodes must be selected which do not impede the electron transport at the surface during the fuel-electrolyte-electrode interaction process.

TABLE 15-1
Performance of ideal fuel cells

Fuel	$\dfrac{-\Delta H}{n\mathscr{F}}$	$E_r = -\dfrac{\Delta G}{n\mathscr{F}}$ V			$\eta_{max} = \dfrac{\Delta G}{\Delta H}$	
		25°C	127°C	527°C	25°C	127°C
$CH_4(g)$	1.15	1.06	1.04	1.04	92.2	90.5
$C_3H_8(g)$	1.15	1.09	1.08	1.10	94.8	94.0
$C_8H_{18}(l)$	1.13	1.10	1.09	1.12	97.3	96.5
iso $C_8H_{18}(l)$	1.13	1.10	1.09	1.13	97.3	96.5
$C_{16}H_{34}(l)$	1.13	1.10	1.09	1.13	97.3	96.5
$CH_3OH(l)$	1.23	1.21	1.20	1.23	98.4	97.6
$NH_3(g)$	1.32	1.17	1.14	1.18	88.6	86.4
$H_2(g)$	1.48	1.23	1.16	1.05	83.3	78.3
$C(s)$	1.02	1.02	1.02	1.02	100.	100.

A summary of the performance of several ideal fuel cells is given in Table 15-1. It should be noted, however, that the factors associated with the transport processes first mentioned will be of primary influence in the selection of a fuel for a practical fuel cell.

15-3   THERMOELECTRIC ENERGY–CONVERSION DEVICES

The phenomenological relations describing thermoelectric effects have been discussed in Chap. 14, and this section is devoted to an analysis of the performance of an ideal thermoelectric energy converter. Consider the simple device shown in Fig. 15-3. As a result of the temperature difference imposed on the series of junctions, a potential difference is created and a current flow is generated in the external load resistance R_L. To obtain a simple indication of the performance characteristics of the device we shall perform the analysis assuming constant material properties.

There are several factors influencing the performance of the thermo-electric generator:

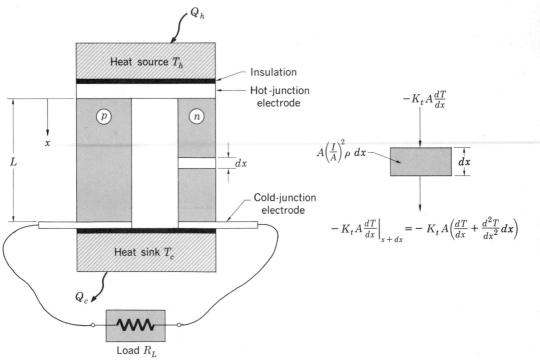

FIG. 15-3   *Thermoelectric energy converter.*

1 Heat conduction from the heat source to heat sink through the thermo-electric materials
2 Joulean heat generation resulting from the current flow in the materials
3 Voltage drops in the materials
4 Various losses resulting from contact resistance between materials, heat losses through insulation, etc.

If the Seebeck coefficient is independent of temperature [see Eq. (14-40)], the only energy generation in the p and n materials will be the result of Joulean heating. Using the energy balance indicated in Fig. 15-3, the heat conduction out of the element is equal to the Joulean heat generation plus the heat conducted into the elements. Thus

$$-K_t A \frac{dT}{dx}\bigg|_{x+dx} = -K_t A \frac{dT}{dx} + \rho A \ dx \left(\frac{I}{A}\right)^2$$

or

$$K_t \frac{d^2 T}{dx^2} + \left(\frac{I}{A}\right)^2 \rho = 0 \tag{15-12}$$

where ρ is the electric resistivity of the materials and x is measured in the direction shown. The boundary conditions are:

$$T = T_h \quad \text{at } x = 0$$
$$T = T_c \quad \text{at } x = L$$

With these boundary conditions the solution to Eq. (15-12) is

$$T - T_h = \frac{J^2 \rho}{2 K_t} \left[\frac{x}{L} - \left(\frac{x}{L}\right)^2\right] - \left(\frac{x}{L}\right)(T_h - T_c) \tag{15-13}$$

where, now, $J = I/A$ has been introduced for the current density. The heat flux at the heat source is

$$q = \frac{Q}{A} = -K_t \left(\frac{dT}{dx}\right)_{x=0} = \frac{K_t \ \Delta T}{L} - \frac{J^2 \rho L}{2} \tag{15-14}$$

where $\Delta T = T_h - T_c$.

The net conduction heat flux into the material is thus the result of a conduction flux into the material and a Joulean heat flux outward. Half the Joulean heat flows out of each end of the conductor.

For both the p and n legs the overall thermal conductance is

$$K = \left(\frac{K_t A}{L}\right)_p + \left(\frac{K_t A}{L}\right)_n \qquad (15\text{-}15)$$

Similarly, the electric resistance is

$$R = \left(\frac{\rho L}{A}\right)_p + \left(\frac{\rho L}{A}\right)_n \qquad (15\text{-}16)$$

To obtain the total heat flow into the hot junction, we must add the Peltier heat to the heat conduction given by Eq. (15-14). Thus

$$Q_h = I\alpha T_h + K\,\Delta T - \frac{I^2 R}{2} \qquad (15\text{-}17)$$

where α is the Seebeck coefficient given by

$$\alpha = \alpha_n - \alpha_p \qquad (15\text{-}18)$$

The output voltage of the device is the Seebeck voltage less the losses due to internal resistance:

$$E_{\text{out}} = \alpha\,\Delta T - IR \qquad (15\text{-}19)$$

The output power, or electrical work, is

$$P = W_e = IE_{\text{out}} = I\alpha\,\Delta T - I^2 R \qquad (15\text{-}20)$$

The thermoelectric generator is a thermal power cycle because it converts *heat* into work. As such, its efficiency is limited by the Carnot relation. The thermal efficiency is

$$\eta = \frac{\text{work output}}{\text{heat input}} = \frac{I\alpha\,\Delta T - I^2 R}{I\alpha T_h + K\,\Delta T - I^2 R/2}$$

$$\eta = \frac{\alpha\,\Delta T - IR}{\alpha T_h + K\,\Delta T/I - IR/2} \qquad (15\text{-}21)$$

The resistance of the external circuit obviously influences the efficiency since it affects the current drawn in the circuit and Joulean heat losses. The Seebeck voltage is, in effect, driving a circuit composed of an internal resistance R and an external resistance R_L. The current is, therefore,

$$I = \frac{\alpha\,\Delta T}{R + R_L} \qquad (15\text{-}22)$$

Introducing Eq. (15-22) into (15-21) and letting $m = R_L/R$,

$$\eta = \frac{\Delta T}{T_h} \frac{m}{1 + m - \Delta T/2T_h + (1 + m)^2 RK/\alpha^2 T_h} \tag{15-23}$$

To maximize the efficiency, it is necessary that the RK product take on its minimum value. Combining Eqs. (15-15) and (15-16), this product is

$$RK = \left[\left(\frac{K_t A}{L}\right)_p + \left(\frac{K_t A}{L}\right)_n\right]\left[\left(\frac{\rho L}{A}\right)_p + \left(\frac{\rho L}{A}\right)_n\right] \tag{15-24}$$

Once the thermoelectric materials are selected, the properties ρ and K_t are determined and RK must be minimized with respect to the relative sizes of the A/L ratios for the two legs. We therefore use the following condition for optimization:

$$\frac{d(RK)}{d[(L/A)_n/(L/A_p)]} = 0 \tag{15-25}$$

When this condition is applied to Eq. (15-24), it is found that

$$\left.\frac{(L/A)_n}{(L/A)_p}\right|_{opt} = \left(\frac{K_{tn}\rho_p}{K_{tp}\rho_n}\right)^{1/2} \tag{15-26}$$

and the optimum RK product becomes

$$(RK)_{opt} = [(\rho K_t)_n^{1/2} + (\rho K_t)_p^{1/2}]^2 \tag{15-27}$$

For given temperature limits the efficiency depends on the load-resistance ratio m and the grouping

$$Z = \frac{\alpha^2}{RK} \tag{15-28}$$

where Z is called the *figure of merit*. A differentiation of Eq. (15-23) with respect to m shows that the optimum-resistance ratio is given by

$$m_{opt} = \sqrt{1 + Z\overline{T}}$$
$$\overline{T} = \frac{T_h + T_c}{2} \tag{15-29}$$

After some algebraic manipulations the efficiency of the generator corresponding to the optimum value of m is given as

$$\eta = \eta_c \frac{m_{opt} - 1}{m_{opt} + T_c/T_h} \tag{15-30}$$

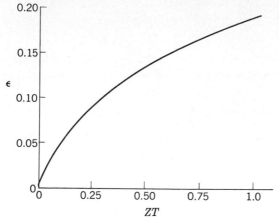

FIG. 15-4 *Material efficiency as a function of ZT.*

where $\eta_c = \Delta T/T_h$ is the Carnot efficiency. Clearly, the Carnot efficiency can only be obtained for the case of zero internal resistance (infinite Z).

For p- and n-type materials having equal electric resistivities and thermal conductivities, the figure of merit becomes

$$Z = \frac{\alpha^2}{\rho K_t} \tag{15-31}$$

The larger the value of the figure of merit, the larger is the efficiency, so this parameter is generally regarded as the most significant variable for selection of materials for thermoelectric power generation.

The modifying term in Eq. (15-30) on the Carnot efficiency is called the *material efficiency* ϵ, so that

$$\eta = \eta_c \epsilon \tag{15-32}$$

The dependence of ϵ on the ZT product is shown in Fig. 15-4. It is particularly clear from this plot that the ZT parameter controls the efficiency of the thermoelectric generator.

The foregoing theory is oversimplified in that material properties are not constant, and proper account must be taken of these variations in the analysis of a practical problem. Figures 15-5 and 15-6 illustrate the property variations of two typical thermoelectric materials.

EXAMPLE 15-2 Performance of thermoelectric generator
A thermoelectric generator uses the materials shown in Figs. 15-5 and 15-6 and operates between the temperature limits of 100 and 300°C. The cross-sectional area of the n-type element is 1 cm² and the length is

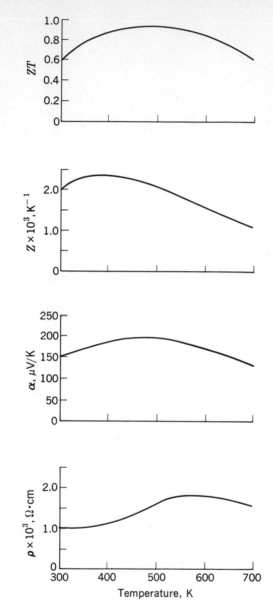

FIG. 15-5 *Temperature dependence of the thermoelectric properties of n-type 75 percent Bi_2Te_3–25 percent Bi_2Se_3.*

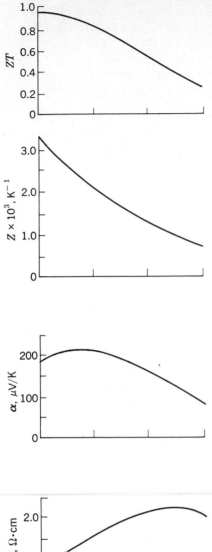

FIG. 15-6 *Temperature dependence of the thermoelectric properties of p-type 25 percent Bi_2Te_3–75 percent Sb_2Te_3 (1.75 percent excess Se).*

1 cm. Using the optimum value for the RK product, calculate the maximum generator efficiency and the efficiency for maximum power, i.e., for $R_L = R$. Also calculate the power output for both these cases.

SOLUTION We evaluate material properties at the mean temperature of 200°C and then pursue a constant-properties analysis. From Figs. 15-5 and 15-6 the properties are:

n-type	p-type
75 percent Bi_2Te_3	25 percent Bi_2Te_3
25 percent Bi_2Se_3	75 percent Sb_2Te_3
	(1.75 percent excess Se)
$\alpha = -190 \times 10^{-6}$ V/°C	$\alpha = 190 \times 10^{-6}$ V/°C
$\rho = 145 \times 10^{-3}$ Ω·cm	$\rho = 1.8 \times 10^{-3}$ Ω·cm
$Z = 2.0 \times 10^{-3}$ K^{-1}	$Z = 1.7 \times 10^{-3}$ K^{-1}

The thermal conductivities are calculated as:

$$K_{tn} = \frac{\alpha_n^2}{\rho_n Z_n} = \frac{(-190 \times 10^{-6})^2}{(1.45 \times 10^{-3})(2.0 \times 10^{-3})} = 0.0124 \text{ W/cm·°C}$$

$$K_{tp} = \frac{\alpha_p^2}{\rho_p Z_p} = \frac{(190 \times 10^{-6})^2}{(1.8 \times 10^{-3})(1.7 \times 10^{-3})} = 0.0118 \text{ W/cm·°C}$$

The figure of merit for the generator is now calculated from

$$Z = \frac{(\alpha_n - \alpha_p)^2}{(RK)_{opt}} = \frac{(\alpha_n - \alpha_p)^2}{(\sqrt{(\rho K_t)_n} + \sqrt{(\rho K_t)_p})^2}$$

$$= 1.85 \times 10^{-3} \text{ K}^{-1}$$

For the optimum value of the RK product the area-length relations of the n- and p-type materials are calculated from

$$\left. \frac{(L/A)_n}{(L/A)_p} \right|_{opt} = \left(\frac{K_{tn}\rho_p}{K_{tp}\rho_n} \right)^{1/2}$$

We have $L_n = L_p = 1$ cm and $A_n = 1$ cm², so that A_p is calculated as

$$A_p = \left[\frac{(0.0124)(1.8 \times 10^{-3})}{(0.0118)(1.45 \times 10^{-3})} \right]^{1/2} = 1.14 \text{ cm}^2$$

The total thermal and electric conductances are now calculated as

$$K = K_{tn}\frac{A_n}{L_n} + K_{tp}\frac{A_p}{L_p} = (0.0124)\frac{1}{1} + (0.0118)\frac{1.14}{1}$$

$$= 0.0258 \text{ W/°C}$$

$$R = \rho_n \frac{L_n}{A_n} + \rho_p \frac{L_p}{A_p} = (1.45 \times 10^{-3})\frac{1}{1} + (1.8 \times 10^{-3})\frac{1}{1.14}$$
$$= 3.03 \times 10^{-3} \ \Omega$$

For optimum efficiency the resistance ratio is

$$m_{\text{opt}} = (1 + Z\overline{T})^{1/2} = [1 + (1.85 \times 10^{-3})(200 + 273)]^{1/2}$$
$$= 1.37$$

Thus

$$R_L = (1.37)(3.03 \times 10^{-3}) = 4.15 \times 10^{-3} \ \Omega$$

The optimum efficiency is calculated from Eq. (15-30) as

$$\eta = \frac{\Delta T}{T_h} \frac{m_{\text{opt}} - 1}{m_{\text{opt}} + T_c/T_h}$$
$$= \frac{200}{573} \frac{1.37 - 1}{1.37 + \frac{373}{573}} = 6.39 \text{ percent}$$

This compares with the Carnot efficiency of $\eta_c = \Delta T/T_h = 34.9$ percent.
For maximum power output, $R_L = R$, $m = 1$, and Eq. (15-23) may be used to calculate the efficiency

$$\eta = \frac{200}{573} \frac{1.0}{1 + 1 - 200/(2)(573) + (1 + 1)^2/(1.85 \times 10^{-3})(573)}$$
$$= 6.23 \text{ percent}$$

In this case the optimum efficiency is not too different from the efficiency at maximum power production. The power output is calculated for both cases from

$$P = I^2 R_L = \left(\frac{\alpha \ \Delta T}{R + R_L}\right)^2 R_L = \frac{(\alpha \ \Delta T)^2}{(1 + m)^2 R_L}$$

For optimum efficiency the power output is

$$P_{\text{opt}} = \frac{(-190 - 190)^2(10^{-6})^2(200)^2}{(1 + 1.37)^2(4.15 \times 10^{-3})}$$
$$= 0.248$$

The maximum power output is

$$P_{max} = \frac{(-190 - 190)^2 (10^{-6})^2 (200)^2}{(1 + 1)^2 (3.03 \times 10^{-3})}$$

$$= 0.477 \text{ W}$$

The open-circuit voltage is $\alpha \, \Delta T$ or 0.076 V.

15-4   THERMIONIC DEVICES

A thermionic converter is basically an electronic diode which employs the principle of electronic emission from metals to convert heat directly into an electric current. A schematic of the device is shown in Fig. 15-7. The metal cathode and anode plates are maintained at temperatures T_h and T_c by the heat supplied to them. For sufficiently high temperatures electrons are liberated from the surfaces, in a manner which is dependent on the temperature and type of surface material. The higher the temperature, the larger the number of electrons liberated. Thus the converter serves as a heat engine operating between temperature limits T_h and T_c, and its efficiency is limited by the Carnot principle. It is possible† to employ the Fermi-Dirac statistical model for electrons to predict the number that will break through the surface-potential barrier which exists in all metals. The functional form of the thermionic current density is shown to be

$$J_s = AT^2 e^{-e\phi/kT} \tag{15-33}$$

† See, for example, p. 240 of Ref. [3].

FIG. 15-7   *Schematic of thermionic energy converter.*

where $A = 4\pi mek^2/h^3 = 1.20 \times 10^6$ A/m²·K² and e is the electronic charge.

Equation (15-33) is called the *Richardson-Dushman equation* and ϕ is called the *work function* of the material surface. It is possible to calculate values of ϕ based on theoretical considerations of surface potential, electron energy distribution within the solid, etc. Unfortunately these calculations are not always in agreement with experiment because of the variety of surface conditions which may be encountered in practice. Some approximate values of the work function are given in Table 15-2.

The net current flow in the thermionic diode is the difference between the emission current from the cathode and the back emission current from the anode:

$$J = J_{\text{cathode}} - J_{\text{anode}} \qquad (15\text{-}34)$$

There are several factors which prevent the thermionic converter from achieving the theoretical Carnot efficiency:

1   Impedance to electron flow resulting from the space charge between the electrodes
2   Thermal radiation transfer between the electrodes
3   Heat losses to the surroundings

The space-charge effect may be reduced by placing the electrodes very close together or by introducing positive ions into the space to neutralize the negative space charge of the electrons. Cesium vapor is commonly employed for the neutralization process because it has a low ionization potential (about 3.9 V). When a vapor is introduced, the device is called a *plasma diode*. In the *vacuum diode*, electrode spacings as low as 0.01 mm may be used to reduce the space-charge effect. Such small spacings produce very difficult fabrication problems, and for this reason plasma diodes, with

TABLE 15-2
Approximate values of work functions
for metals

Metal	Work function ϕ, V
Mo	4.2
Ta	4.2
Cu	4.5
Cs	1.8
W	4.6
Ni	5.0
Pt	5.3
Na	2.5

electrode spacings of the order of 1.0 mm, have received considerable development attention.

The voltage output of a plasma diode may be written as

$$V = \phi_c - \phi_A - \phi_p \tag{15-35}$$

where ϕ_c and ϕ_A are the cathode and anode work functions, respectively, and ϕ_p represents the plasma potential drop. The thermal efficiency is then given as

$$\eta = \frac{JV}{Q_c/A_c} \tag{15-36}$$

where Q_c/A_c is the cathode heat flux.

The total heat added to the cathode is equal to the sum of the electron-energy flux out of the cathode and the thermal radiation heat exchange:

$$\frac{Q_c}{A_c} = J\left(\phi_c + \frac{2kT_h}{e}\right) + \frac{Q_{rad}}{A_c} \tag{15-37}$$

The term $2kT_h/e$ represents the voltage equivalent to the thermal energy of the emitted electrons. The radiation heat transfer is calculated approximately from

$$\frac{Q_{rad}}{A_c} = \epsilon^* \sigma(T_h^4 - T_c^4) \tag{15-38}$$

where σ is the Stefan-Boltzmann constant and has the value 5.668×10^{-12} J/s·cm²·K⁴. ϵ^* is an appropriate constant which is a function of the radiation properties of the electrode surfaces.

It is well to caution the reader that the foregoing equations all have serious limitations, even though they may give reasonable estimates for thermionic generator performance. A precise analysis of thermionic converters is beyond the scope of our discussion, and the reader should consult the references at the end of the chapter for additional information.

EXAMPLE 15-3   Performance of thermionic generator

A thermionic generator has $\phi_c = 2.5$ V, $\phi_A = 2.0$ V, $T_h = 2000$ K, $T_c = 1000$ K, and $\phi_p = 0.1$ V at these conditions. The value of ϵ^* may be taken as 0.2 for the electrode materials used. Using these data, evaluate the efficiency of the device and compare with the theoretical Carnot value.

SOLUTION   We first calculate the voltage output with Eq. (15-35):

$$V = \phi_c - \phi_A - \phi_p = 2.5 - 2.0 - 0.1 = 0.4 \text{ V}$$

The net electron current is evaluated by applying Eq. (15-33) to the cathode and anode, and subtracting in accordance with Eq. (15-34):

$$J_C = (1.20 \times 10^6)(2000)^2 \exp\left[\frac{-(1.6 \times 10^{-19})(2.5)}{(1.38 \times 10^{-23})(2000)}\right]$$

$$= 2.438 \times 10^6 \text{ A/m}^2$$

$$J_A = (1.20 \times 10^6)(1000)^2 \exp\left[\frac{-(1.6 \times 10^{-19})(2.0)}{(1.38 \times 10^{-23})(1000)}\right]$$

$$= 1.02 \times 10^2 \text{ A/m}^2$$

The anode back current is thus negligible by comparison with the forward cathode current, and we may take

$$J = J_c = 2.438 \times 10^6 \text{ A/m}^2$$

The heat supplied to the cathode is now evaluated with Eq. (15-37):

$$\frac{Q_c}{A_c} = J\left(\phi_c + \frac{2kT_h}{e}\right) + \epsilon^* \sigma(T_h^4 - T_c^4)$$

$$= (2.438 \times 10^6)\left[2.5 + \frac{(2)(1.38 \times 10^{-23})(2000)}{(1.6 \times 10^{-19})}\right]$$

$$+ (0.2)(5.67 \times 10^{-12})(10^4)(2000^4 - 1000^4)$$

$$= 6.1 \times 10^6 + 8.41 \times 10^5 + 1.7 \times 10^5$$

$$= 7.111 \times 10^6 \text{ W/m}^2$$

We see from this calculation that about 3 percent of the heat is lost by radiation, about 12 percent to achieve the electron kinetic (thermal) energy $2kT_h/e$, and the largest portion—about 85 percent—is consumed to overcome the cathode work function.

We may now calculate the efficiency from Eq. (15-36):

$$\eta = \frac{JV}{Q_c/A_c} = \frac{(2.438 \times 10^6)(0.4)}{(7.111 \times 10^6)} = 13.7 \text{ percent}$$

The Carnot efficiency for these temperature limits is

$$\eta_c = \frac{T_h - T_c}{T_h} = \frac{1000}{2000} = 50 \text{ percent}$$

The efficiency of 13.7 percent is a reasonable value for practical plasma diodes, but we should not expect efficiencies too much greater than this value.

15-5 MAGNETOHYDRODYNAMIC GENERATORS

An electric conductor moving through a magnetic field experiences a retarding force as well as an induced electric field and current. This effect is a result of Faraday's law of electromagnetic induction. The induced emf is given by

$$E_{ind} = v \times B \tag{15-39}$$

where v is the velocity of the conductor and B is the magnetic field intensity, as shown in Fig. 15-8. The induced current density is

$$J_{ind} = \sigma E_{ind} \tag{15-40}$$

where σ is now used to designate the electric conductivity. The retarding force on the conductor is the Lorentz force given by

$$F_{ind} = J_{ind} \times B \tag{15-41}$$

From an energy point of view, the movement of the force through a displacement (mechanical work) is converted to electrical work (current flow against potential difference) by means of the electromagnetic induction principle. This is a work-work energy conversion and is not limited by the Carnot principle. Of course, electromagnetic induction forms the basis of operation for conventional electric generators.

The electromagnetic induction principle need not be limited to solid

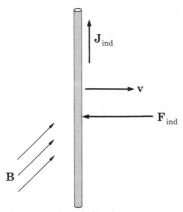

FIG. 15-8 *Faraday's law for electromagnetic induction.*

conductors; the movement of a conducting fluid through a magnetic field may also be employed for electric energy generation. When a fluid is used, the energy-conversion technique is called *magnetohydrodynamic* (MHD) *energy conversion.*

A schematic of an MHD generator is shown in Fig. 15-9. The conducting fluid is forced between the plates with a kinetic energy and pressure differential sufficient to overcome the magnetic induction force $\mathbf{F}_{ind}$. The end-view drawing illustrates the construction of the flow channel. An ionized gas is employed as the conducting fluid. Ionization is produced either by thermal means, i.e., by an elevated temperature, or by seeding with substances like cesium or potassium vapor which ionize at relatively low temperatures. In the overall power cycle the MHD converter takes the place of a turbine in a conventional vapor- or gas-turbine cycle. Still, a compressor must be used to elevate the pressure, heat is added at the high pressure, and the flow is accel-

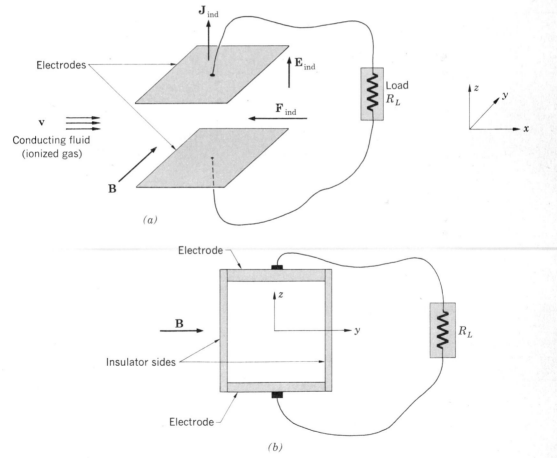

FIG. 15-9 *Schematic of magnetohydrodynamic (MHD) energy converter. (a) Flow system; (b) end view of flow channel.*

erated before entering the converter. A schematic of the MHD power cycle is illustrated in Fig. 15-10, along with a T-s diagram for the ideal processes.

In addition to the induced electric field $\mathbf{E}_{ind}$ there is also some applied field $\mathbf{E}$ at the electrodes. The total current and force are, then,

(a)

(b)

FIG. 15-10 (a) Schematic of MHD power cycle; (b) T-s diagram for cycle.

$$J = \sigma(E + v \times B) \tag{15-42}$$

$$F = J \times B = \sigma(E + v \times B) \times B \tag{15-43}$$

Clearly, the overall MHD *cycle* is a *thermal* power cycle and, as such, is limited by the Carnot efficiency. In terms of the *T-s* diagram of Fig. 15-10, the cycle thermal efficiency may be written as

$$\eta_{th} = \frac{\text{work output}}{\text{heat input}} = \frac{(h_{2_0} - h_{3_0}) - (h_{1_0} - h_{4_0})}{h_{2_0} - h_{1_0}} \tag{15-44}$$

where the indicated enthalpies are "stagnation" values which take into account the kinetic energy of the flow. The stagnation enthalpy is defined as

$$h_0 = h + \frac{V^2}{2g_c}$$

where V is the flow velocity. In practical MHD converters, high-velocity ionized gases are usually employed for the conversion process, so that the kinetic energy of the flow represents a substantial portion of the total energy.

The ideal converter process 2-3 is shown as isentropic which would assume perfect conversion of the flow pressure and kinetic energy into electric energy. Several factors can reduce the efficiency of the converter:

1 Dissipation of energy in the internal resistance of the ionized gas.
2 A space-charge barrier at the electrode surfaces.
3 Heat transfer through the electrode and insulator walls.
4 Various losses associated with fluid friction, etc.
5 Hall effect losses resulting in current induction in the direction of the flow. Such losses can be substantial when seeded combustion gases are employed as the working fluid. The Hall effect causes a voltage gradient in a direction perpendicular to the applied magnetic field and current flow.

The foregoing factors are all the subject of continuing research efforts and, although promising developments have been made, serious gaps in understanding still remain. An analysis of these factors is beyond the scope of our discussion. We may, however, indicate the parametric effects which losses may have on the nonisentropic process in the generator.

At open-circuit or no-load condition, the emf across the generator is the induced value $E_{ind} = v \times B$. For a system as shown in Fig. 15-9,

$$E_{ind} = vi \times Bj = vBk \tag{15-45}$$

and the magnitude of the voltage is simply vB. Depending on the external load, internal losses, etc., the actual emf delivered to the electrodes will be less than this value. The electrical efficiency or loading factor K is defined as

$$K = \frac{E}{vB} \tag{15-46}$$

where E is the actual voltage delivered to the electrodes. In scalar form, the current is now written as

$$J = \sigma(vB - E) = \sigma vB(1 - K) \tag{15-47}$$

The power dissipated per unit volume in the fluid is JE, or

$$JE = \sigma vB(1 - K)KvB = \sigma v^2B^2K(1 - K) \tag{15-48}$$

For a constant-velocity generator, the pressure forces must be balanced by the magnetic retarding force and, considering a channel cross-sectional area A, we have

$$pA = (\mathbf{J} \times \mathbf{B})A \, dx + \left(p + \frac{dp}{dx} \, dx\right) A \tag{15-49}$$

where p is the fluid pressure.

Using the geometry of Fig. 15-9 and Eq. (15-47) for J, Eq. (15-49) yields

$$\frac{dp}{dx} = -JB = -\sigma vB^2(1 - K) \tag{15-50}$$

In the constant-velocity generator all of the electrical work extraction from the flow must result in a decrease in the enthalpy of the gas and, consequently, as a reduction in temperature. We may form the energy balance

Enthalpy decrease of gas = electrical work generation

or

$$-\rho vA \frac{dh}{dx} \, dx = JEA \, dx \tag{15-51}$$

where ρ is now the fluid density. Assuming an ideal gas for which $dh = c_p \, dT$ and $p = \rho RT$, Eq. (15-51) may be combined with Eq. (15-48) to give

$$\frac{c_p}{T} \frac{dT}{dx} = -\frac{\sigma vB^2K(1 - K)R}{p} \tag{15-52}$$

Equation (15-52) is the energy equation for the constant-velocity generator. Recalling that $R = c_p - c_v$ and $\gamma = c_p/c_v$, Eqs. (15-50) and (15-52) may be combined to give

$$\frac{dT}{T} = \frac{K(\gamma - 1)}{\gamma} \frac{dp}{p} \qquad (15\text{-}53)$$

For constant K, Eq. (15-53) may be integrated to yield

$$\frac{T}{T_1} = \left(\frac{p}{p_1}\right)^{[K(\gamma-1)]\gamma} \qquad (15\text{-}54)$$

Equation (15-54) is identical to the conventional isentropic relation for expansion of a gas except for the factor K. The overall result of various values of K is indicated in Fig. 15-11. The smaller the value of K, the greater the losses and accompanying irreversibilities. $K = 1.0$ corresponds to isentropic flow, and $K = 0$ corresponds to a short-circuit condition. For the constant-velocity generator using an ideal gas, the adiabatic-generator efficiency may be written as

$$\eta_{ad} = \frac{\Delta h}{\Delta h_{isen}} = \frac{T_2 - T_3'}{T_2 - T_3} \qquad (15\text{-}55)$$

where the subscripts refer to the states indicated in Fig. 15-10. This equation may be rewritten with the aid of Eq. (15-54) to yield

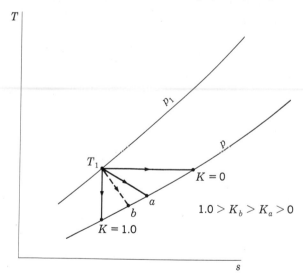

FIG. 15-11 *Effect of electrical efficiency on expansion process in constant-velocity MHD generator.*

$$\eta_{ad} = \frac{1 - (p_3/p_2)^{[K(\gamma-1)]/\gamma}}{1 - (p_3/p_2)^{(\gamma-1)/\gamma}} \tag{15-56}$$

The MHD power cycles offer promise as large-scale power-production devices at high voltages of 1000 to 10 000 V. The eventual development of such cycles is closely related to the development of high-field supercon-ducting magnets. Small-scale MHD generators do not appear to be particu-larly attractive because the losses (electrode, friction, etc.) consume a large portion of the output power. However, large-scale MHD power plants offer possibilities for significantly higher thermal efficiencies than conventional steam cycles because the MHD generator can operate at much higher tem-peratures.

REVIEW QUESTIONS

1 Why is the efficiency of a fuel cell not limited by the Carnot principle?

2 How does a fuel cell work?

3 What are some of the practical problems which determine the actual efficiency of a fuel cell?

4 Why is a thermoelectric energy-conversion device governed by the Carnot princi-ple?

5 What is the figure of merit?

6 Describe the thermionic conversion process.

7 Is the MHD generator limited by the Carnot principle? Why?

8 What possible advantage could an MHD generator have over a turbine generator?

9 Discuss some practical problems with MHD generators.

PROBLEMS

15-1 Calculate the ideal-cell emf and maximum efficiency for a fuel cell involving the reaction

$$C + O_2 \rightarrow CO_2$$

The heat of reaction and the Gibbs function change are:

$\Delta H° = -94.4$ kcal/g mol
$\Delta G° = -94.6$ kcal/g mol

15-2 Perform the necessary calculations to show that the maximum efficiency for the methane fuel cell is 92 percent and the ideal-cell voltage is 1.15 V. What flow rate, in pounds mass per hour, of methane and oxygen would be required to produce a power output of 100 kW? What heat-transfer rate would be involved under these circumstances?

15-3 What flow rates of H_2 and O_2 are required in an ideal fuel cell to produce a power output of 1 kW at 25°C?

15-4 Compare the performance of the fuel cell in Prob. 15-2 with that of the Otto cycle in Prob. 12-14. For this comparison use the same flow rate of methane as in Prob. 15-2 along with an air/fuel ratio of 15:1.

15-5 A hydrogen-oxygen fuel cell is operated at 600°C and 1 atm. Calculate the ideal-cell voltage and efficiency under these conditions.

15-6 A thermoelectric generator uses the materials shown in Figs. 15-5 and 15-6 and operates between the temperature limits of 50 and 650°C. The cross-sectional area of the n-type element is 2.3 cm² and the length is 1.5 cm. Using the optimum value of the RK product, calculate the maximum generator efficiency and the efficiency for maximum power. Calculate the power output in these cases. What is the open-circuit voltage? What is the current density in the elements?

15-7 Calculate the heat input and heat rejection for the thermoelectric generator in Prob. 15-6 under maximum power conditions and no-load conditions. Repeat the calculation for the system of Example 15-2.

15-8 A small thermoelectric generator is to be designed to produce a total power output of 15 W. The generator is constructed of the materials shown in Figs. 15-5 and 15-6 and operates between temperature limits of 20 and 100°C. How many couples are required to produce this output? Calculate the heat input and heat rejection for maximum efficiency and maximum power output conditions.

15-9 A thermionic generator has $\phi_c = 4.5$ V, $\phi_A = 2.0$ V, and $\phi_p = 0.2$ V; $T_h = 2500$ K and $T_c = 1300$ K. The value of ϵ^* is 0.7 for the electrode materials. Calculate the efficiency of the device.

15-10 What influence does the value of ϵ^* have on the efficiency of the generator in Prob. 15-9? ϵ^* takes on values between zero and unity.

15-11 Calculate the thermionic current density for tungsten at 500, 1000, and 3000 K.

15-12 Derive an expression for the entropy change in a constant-velocity MHD generator having a loading factor of K.

15-13 Derive an expression for the power output of a constant-velocity MHD generator as a function of the mass flow rate and the pressure drop across the generator.

15-14 Why does the short-circuit condition in the constant-velocity MHD generator result in a constant temperature for the flow?

15-15 A constant-velocity MHD generator operates with a pressure ratio of 3:1 using a gas for which $\gamma = 1.35$. Plot the adiabatic efficiency of this device as a function of the loading factor.

REFERENCES

1 Angrist, S. W.: "Direct Energy Conversion," Allyn and Bacon, Inc., Boston, 1965.

2 Chang, S. L.: "Energy Conversion," Prentice-Hall, Inc., Englewood Cliffs, N.J., 1963.

3 Sutton, G. W.: "Direct Energy Conversion," McGraw-Hill Book Company, New York, 1966.

4 Kaye, J., and J. A. Welsh: "Direct Conversion of Heat to Electricity," John Wiley & Sons, Inc., New York, 1960.

5 Sutton, G. W., and A. Sherman: "Engineering Magnetohydrodynamics," McGraw-Hill Book Company, New York, 1965.

6 Cambel, A. B.: "Plasma Physics and Magnetofluidmechanics," McGraw-Hill Book Company, New York, 1963.

Appendix

TABLE A-1

Important physical constants

Avogadro's number	$N_0 = 6.022\ 045 \times 10^{26}$ molecules/kg mol
Universal gas constant	$\mathscr{R} = 1545.35$ ft·lbf/lbm·mol·°R
	$= 8314.41$ J/kg mol·K
	$= 1.986$ Btu/lbm·mol·°R
	$= 1.986$ kcal/kg mol·K
Planck's constant	$h = 6.626\ 176 \times 10^{-34}$ J
Boltzmann's constant	$k = 1.380\ 662 \times 10^{-23}$ J/molecule·K
	$= 8.6173 \times 10^{-5}$ eV/molecule·K
Speed of light in vacuum	$c = 2.997\ 925 \times 10^{8}$ m/s
Standard gravitational acceleration	$g = 32.174$ ft/s²
	$= 9.806\ 65$ m/s²
Electron mass	$m_e = 9.1095 \times 10^{-31}$ kg
Charge on the electron	$e = 1.602\ 189 \times 10^{-19}$ C
Stefan-Boltzmann constant	$\sigma = 0.1714 \times 10^{-8}$ Btu/hr·ft²·R⁴
	$= 5.670\ 32 \times 10^{-8}$ W/m²·K⁴

TABLE A-2
Conversion factors

Length	1 cm	$= 0.3937$ in $= 10^4\ \mu$m $= 10^8$ A
	1 in	$= 2.540$ cm
	1 ft	$= 0.3048$ m
Mass	1 lbm	$= 0.453\ 592\ 37$ kg
	1 slug	$= 32.174$ lbm
Force	1 N	$= 10^5$ dyn
	1 lbf	$= 444\ 822$ dyn $= 4.44822$ N
	1 kgf	$= 9.806\ 65$ N
		$= 1.0$ kilopond
Pressure	1 Pa	$= 1$ N/m^2
	1 lbf/in^2	$= 2.036$ inHg at $32°$F $= 6894.76$ N/m^2
	1 inHg	$= 33\ 864$ dyn/cm$^2 = 0.0334$ atm
		$= 0.491$ lbf/in^2
	1 atm	$= 14.695\ 95$ lbf/in$^2 = 760$ mmHg at $32°$F
		$= 29.92$ inHg at $32°$F $= 2116.21$ lbf/ft^2
		$= 1.013\ 25 \times 10^5$ N/m^2
	1 kgf/cm^2	$= 9.806\ 65 \times 10^4$ N/m^2
	1 bar	$= 10^5$ N/m$^2 = 0.986\ 92$ atm
Volume	1 liter	$= 0.2642$ gal $= 0.0353$ ft$^3 = 61.03$ in^3
	1 gal	$= 231$ in^3
	1 ft^3	$= 28.3168$ liters $= 7.4805$ gal $= 0.028\ 316\ 8$ m^3
	1 in^3	$= 16.387$ cm^3
Energy	1 Btu	$= 778.16$ ft·lbf $= 252.16$ cal $= 1055.04$ J
	1 ft·lbf	$= 1.3558$ J
	1 erg	$= 1$ dyn·cm
	1 J	$= 1$ N·m $= 10^7$ ergs
	1 cal	$= 4.1854$ J (thermochemical)
Power	1 W	$= 1$ J/s $= 860.13$ cal/h $= 3.413$ Btu/h
	1 hp	$= 746$ W (electric) $= 550$ ft·lbf/s (mechanical)
		$= 2545$ Btu/h
Temperature	°F	$= 1.8°$C $+ 32$
	°R	$=$ °F $+ 459.67$
	K	$=$ °C $+ 273.15$
	°R	$= 1.8$ K
Miscellaneous	1 Btu/lbm·°F	$= 4186.8$ J/kg·°C
	1 Btu/lbm	$= 0\ 5559$ cal/g
		$= 2326$ J/kg
	1 ft^3/lbm	$= 0.062\ 427$ m^3/kg

TABLE A-3

Basic and supplemental SI units

Quantity	Unit	Symbol
Basic units		
Length	meter	m
Mass	kilogram	kg
Time	second	s
Electric current	ampere	A
Temperature	kelvin	K
Luminous intensity	candela	cd
Supplemental units		
Plane angle	radian	rad
Solid angle	steradian	sr

TABLE A-4
Derived SI units

Quantity	Name(s) of unit	Unit symbol or abbreviation, where differing from basic form	Unit expressed in terms of basic or supplementary units
Area	square meter		m^2
Volume	cubic meter		m^3
Frequency	hertz, cycle per second	Hz	s^{-1}
Density, concentration	kilogram per cubic meter		kg/m^3
Velocity	meter per second		m/s
Angular velocity	radian per second		rad/s
Acceleration	meter per second squared		m/s^2
Angular acceleration	radian per second squared		rad/s^2
Volumetric flow rate	cubic meter per second		m^3/s
Force	newton	N	$kg \cdot m/s^2$
Surface tension	newton per meter, joule per square meter	N/m, J/m^2	kg/s^2
Pressure	newton per square meter, pascal	N/m^2, Pa	$kg/m \cdot s^2$
Viscosity, dynamic	newton-second per square meter, poiseuille	$N \cdot s/m^2$, Pl	$kg/m \cdot s$
Viscosity, kinematic; diffusivity; mass conductivity	meter squared per second		m^2/s
Work, torque, energy, quantity of heat	joule, newton-meter, watt-second	J, $N \cdot m$, $W \cdot s$	$kg \cdot m^2/s^2$
Power, heat flux	watt, joule per second	W, J/s	$kg \cdot m^2/s^3$
Heat flux density	watt per square meter	W/m^2	kg/s^3
Volumetric heat release rate	watt per cubic meter	W/m^3	$kg/m \cdot s^3$
Heat transfer coefficient	watt per square meter degree	$W/m^2 \cdot deg$	$kg/s^3 \cdot deg$
Latent heat, enthalpy (specific)	joule per kilogram	J/kg	m^2/s^2
Heat capacity (specific)	joule per kilogram degree	$J/kg \cdot deg$	$m^2/s^2 \cdot deg$
Capacity rate	watt per degree	W/deg	$kg \cdot m^2/s^3 \cdot deg$
Thermal conductivity	watt per meter degree	$W/m \cdot deg$, $J \cdot m/s \cdot m^2 \cdot deg$	$kg \cdot m/s^3 \cdot deg$
Mass flux, mass flow rate	kilogram per second		kg/s
Mass flux density, mass flow rate per unit area	kilogram per square meter-second		$kg/m^2 \cdot s$
Mass-transfer coefficient	meter per second		m/s
Quantity of electricity	coulomb	C	$A \cdot s$
Electromotive force	volt	V, W/A	$kg \cdot m^2/A \cdot s^3$
Electric resistance	ohm	Ω, V/A	$kg \cdot m^2/A^2 \cdot s^3$
Electric conductivity	ampere per volt meter	$A/V \cdot m$	$A^2 \cdot s^3/kg \cdot m^3$
Electric capacitance	farad	F, $A \cdot s/V$	$A^3 \cdot s^4/kg \cdot m^2$
Magnetic flux	weber	Wb, $V \cdot s$	$kg \cdot m^2/A \cdot s^2$
Inductance	henry	H, $V \cdot s/A$	$kg \cdot m^2/A^2 \cdot s^2$
Magnetic permeability	henry per meter	H/m	$kg \cdot m/A^2 \cdot s^2$
Magnetic flux density	tesla, weber per square meter	T, Wb/m^2	$kg/A \cdot s^2$
Magnetic field strength	ampere per meter		A/m
Magnetomotive force	ampere		A
Luminous flux	lumen	lm	$cd \cdot sr$
Luminance	candela per square meter		cd/m^2
Illumination	lux, lumen per square meter	lx, lm/m^2	$cd \cdot sr/m^2$

TABLE A-5
Critical constants

Substance	Symbol	M	T_c, °K	p_c, atm	v_c, cm³/g mol (1 cm³/g mol = 10^{-3} m³/kg mol)	v_c, ft/lbm mol	z_c
Acetylene	C_2H_2	26.038	309.5	61.6	113	1.81	0.274
Air		28.967	132.41	37.25	93.25	1.49	0.320
Ammonia	NH_3	17.032	405.4	111.3	72.5	1.16	0.243
Argon	A	39 944	150.72	47.996	75	1.20	0.291
Benzene	C_6H_6	78.114	562.6	48.6	260	4.17	0.274
n-Butane	C_4H_{10}	58.124	425.17	37.47	255	4.08	0.274
Isobutane	C_4H_{10}	58.124	408.14	36.00	263	4.21	0.283
1-Butene	C_4H_8	56.108	419.6	39.7	240	3.84	0.277
Carbon dioxide	CO_2	44.011	304.20	72.90	94	1.51	0.275
Carbon monoxide	CO	28.011	132.91	34.529	93	1.49	0.294
Carbon tetrachloride	CCl_4	153 839	556.4	45.0	276	4.42	0.272
n-Deuterium	D_2	4.029	38.43	16.421			
Dodecane	$C_{12}H_{26}$	170.340	659	17.9		11.5	0.237
Ethane	C_2H_6	30.070	305.43	48.20	148	2.37	0.285
Ethyl ether	$C_4H_{10}O$	74.124	467.8	35.6	282.9	4.53	0.262
Ethylene	C_2H_4	28.054	283.06	50.50	124	1.99	0.270
Freon, F-12	CCl_2F_2	120.925	385.16	40.6	217	3.47	0.279
Helium	He	4.003	5.19	2.26	58	0.929	0.308
n-Heptane	C_7H_{16}	100.205	540.17	27.00	426	6.82	0.260
n-Hexane	C_6H_{14}	86.178	507.9	29.94	368	5.89	0.264
Hydrogen	H_2	2.016	33.24	12.797	65	1.04	0.304
Hydrogen sulfide	H_2S	34.082	373.7	88.8	98	1.57	0.284
Mercury	Hg	200.610					
Methane	CH_4	16.043	190.7	45.8	99	1.59	0.290
Methyl fluoride	CH_3F	34.035	317.71	58.0	113	1.81	0.251
Neon	Ne	20.183	44.39	26.86	41.7	0.668	0.308
Nitric oxide	NO	30.008	179.2	65.0	58	0.929	0.256
Nitrogen	N_2	28.016	126.2	33.54	90	0.144	0.291
Octane	C_8H_{18}	114.232	569.4	24.64	486	7.77	0.256
Oxygen	O_2	32.000	154.78	50.14	74	1.19	0.292
n-Pentane	C_5H_{12}	72.151	469.78	33.31	311	4.98	0.269
Isopentane	C_5H_{12}	72.151	461.0	32.92	308	4.93	0.268
Propane	C_3H_8	44.097	370.01	42.1	200	3.20	0.277
Propylene	C_3H_6	42.081	365.1	45.40	181	2.90	0.274
Sulfur dioxide	SO_2	64.066	430.7	77.8	122	1.95	0.269
Water	H_2O	18.016	647.27	218.167	56	0.897	0.230
Xenon	Xe	131.300	289.81	58.0	118.8	1.90	0.290

TABLE A-6

Enthalpy of formation, Gibbs function of formation, and absolute entropy at 25°C (77°F) and 1 atm

		Molecular	$\bar{h}_f^\circ$	
Substance	Formula	Weight	Btu/lbm·mol	kJ/kg·mol
Carbon (solid)	C(s)	12.01	0	0
Carbon (gas)	C(g)	12.01	+309 060	+718 400
Hydrogen	$H_2(g)$	2.018	0	0
Oxygen	$O_2(g)$	32.00	0	0
Carbon monoxide	CO(g)	28.01	−47 540	−110 530
Carbon dioxide	$CO_2(g)$	44.01	−169 300	−393 520
Water	$H_2O(g)$	18.02	−104 040	−241 820
Water	$H_2O(l)$	18.02	−122 970	−285 830
Hydrogen peroxide	$H_2O_2(g)$	34.02	−58 640	−136 310
Ammonia	$NH_3(g)$	17.04	−19 750	−46 190
Methane	$CH_4(g)$	16.04	−32 210	−74 850
Acetylene	$C_2H_2(g)$	26.04	+97 540	+226 730
Ethylene	$C_2H_4(g)$	28.05	+22 490	+52 280
Ethane	$C_2H_6(g)$	30.07	−36 420	−84 680
Propylene	$C_3H_6(g)$	42.08	+8 790	+20 410
Propane	$C_3H_8(g)$	44.09	−44 680	−103 850
n-Butane	$C_4H_{10}(g)$	58.12	−54 270	−126 150
n-Pentane	$C_5H_{12}(g)$	72.15	−63 000	−146 440
n-Hexane	$C_6H_{14}(g)$	86.17	−71 930	−167 200
n-Heptane	$C_7H_{16}(g)$	100.20	−80 800	−187 820
n-Octane	$C_8H_{18}(g)$	114.22	−89 680	−208 450
Benzene	$C_6H_6(g)$	78.11	+35 680	+82 930
Methyl alcohol	$CH_3OH(g)$	32.05	−86 540	−200 670
Ethyl alcohol	$C_2H_5OH(g)$	46.07	−101 230	−235 310
Sulfur dioxide	$SO_2(g)$	64.07	−127 730	−296 910

$\overline{g}_f^\circ$		$\overline{s}^\circ$	
Btu/lbm·mol	kJ/kg·mol	Btu/lbm·mol·°R	kJ/kg·mol·°K
0	0	1.36	5.74
+289 520	+672 980	37.76	157.99
0	0	31.21	130.57
0	0	49.00	205.03
−59 010	−137 150	47.21	197.56
−169 680	−394 360	51.07	213.64
−98 350	−228 590	45.11	188.72
−102 040	−237 180	16.71	69.92
−45 430	−105 600	55.60	232.63
−7 140	−16 590	45.97	192.33
−21 860	−50 790	44.49	186.16
+90 000	+209 170	48.00	200.85
+29 306	+68 120	52.54	219.83
−14 150	−32 890	54.85	229.49
+26 980	+62 720	63.80	266.94
−10 105	−23 490	64.51	269.91
−6 760	−15 710	74.11	310.12
−3 600	−8 360	83.34	348.69
−126	−293	92.83	388.40
−3 490	−8 110	102.24	427.77
+7 110	+16 530	111.55	466.73
+55 780	+129 660	64.34	269.20
−69 700	−162 000	57.29	239.70
−72 520	−168 570	67.54	282.59
−129 220	−300 370	59.40	248.53

TABLE A-7

Properties of water—saturation-temperature table (English units)*

v in ft³/lbm; h and u in Btu/lbm; s in Btu/lbm·°R

Temp., °F T	Press., psia P	Specific volume Sat. liquid v_f	Sat. vapor v_g	Internal energy Sat. liquid u_f	Sat. vapor u_g	Enthalpy Sat. liquid h_f	Evap. h_{fg}	Sat. vapor h_g	Entropy Sat. liquid s_f	Evap. s_{fg}	Sat. vapor s_g	Temp., °F T
32	0.0886	0.01602	3305	−0.01	1021.2	−0.01	1075.4	1075.4	−0.00003	2.1870	2.1870	32
35	0.0999	0.01602	2948	2.99	1022.2	3.00	1073.7	1076.7	0.00607	2.1704	2.1764	35
40	0.1217	0.01602	2445	8.02	1023.9	8.02	1070.9	1078.9	0.01617	2.1430	2.1592	40
45	0.1475	0.01602	2037	13.04	1025.5	13.04	1068.1	1081.1	0.02618	2.1162	2.1423	45
50	0.1780	0.01602	1704	18.06	1027.2	18.06	1065.2	1083.3	0.03607	2.0899	2.1259	50
55	0.2140	0.01603	1431	23.07	1028.8	23.07	1062.4	1085.5	0.04586	2.0641	2.1099	55
60	0.2563	0.01604	1207	28.08	1030.4	28.08	1059.6	1087.7	0.05555	2.0388	2.0943	60
65	0.3057	0.01604	1022	33.09	1032.1	33.09	1056.8	1089.9	0.06514	2.0140	2.1791	65
70	0.3632	0.01605	867.7	38.09	1033.7	38.09	1054.0	1092.0	0.07463	1.9896	2.0642	70
75	0.4300	0.01606	739.7	43.09	1035.4	43.09	1051.1	1094.2	0.08402	1.9657	2.0497	75
80	0.5073	0.01607	632.8	48.08	1037.0	48.09	1048.3	1096.4	0.09332	1.9423	2.0356	80
85	0.5964	0.01609	543.1	53.08	1038.6	53.08	1045.5	1098.6	0.1025	1.9193	2.0218	85
90	0.6988	0.01610	467.7	58.07	1040.2	58.07	1042.7	1100.7	0.1117	1.8966	2.0083	90
95	0.8162	0.01611	404.0	63.06	1041.9	63.06	1039.8	1102.9	0.1207	1.8744	1.9951	95
100	0.9503	0.01613	350.0	68.04	1043.5	68.05	1037.0	1105.0	0.1296	1.8526	1.9822	100
110	1.276	0.01617	265.1	78.02	1046.7	78.02	1031.3	1109.3	0.1473	1.8101	1.9574	110
120	1.695	0.01621	203.0	87.99	1049.9	88.00	1025.5	1113.5	0.1647	1.7690	1.9336	120
130	2.225	0.01625	157.2	97.97	1053.0	97.98	1019.8	1117.8	0.1817	1.7292	1.9109	130
140	2.892	0.01629	122.9	107.95	1056.2	107.96	1014.0	1121.9	0.1985	1.6907	1.8892	140
150	3.722	0.01634	97.0	117.95	1059.3	117.96	1008.1	1126.1	0.2150	1.6533	1.8684	150
160	4.745	0.01640	77.2	127.94	1062.3	127.96	1002.2	1130.1	0.2313	1.6171	1.8484	160
170	5.996	0.01645	62.0	137.95	1065.4	137.97	996.2	1134.2	0.2473	1.5819	1.8293	170
180	7.515	0.01651	50.2	147.97	1068.3	147.99	990.2	1138.2	0.2631	1.5478	1.8109	180
190	9.343	0.01657	41.0	158.00	1071.3	158.03	984.1	1142.1	0.2787	1.5146	1.7932	190
200	11.529	0.01663	33.6	168.04	1074.2	168.07	977.9	1145.9	0.2940	1.4822	1.7762	200
210	14.13	0.01670	27.82	178.1	1077.0	178.1	971.6	1149.7	0.3091	1.4508	1.7599	210
212	14.70	0.01672	26.80	180.1	1077.6	180.2	970.3	1150.5	0.3121	1.4445	1.7567	212
220	17.19	0.01677	23.15	188.2	1079.8	188.2	965.3	1153.5	0.3241	1.4201	1.7441	220
230	20.78	0.01685	19.39	198.3	1082.6	198.3	958.8	1157.1	0.3388	1.3901	1.7289	230
240	24.97	0.01692	16.33	208.4	1085.3	208.4	952.3	1160.7	0.3534	1.3609	1.7143	240

250	29.82	0.01700	13.83	218.5	1087.9	218.6	945.6	1164.2	0.3677	1.3324	1.7001	250
260	35.42	0.01708	11.77	228.6	1090.5	228.8	938.8	1167.6	0.3819	1.3044	1.6864	260
270	41.85	0.01717	10.07	238.8	1093.0	239.0	932.0	1170.9	0.3960	1.2771	1.6731	270
280	49.18	0.01726	8.65	249.0	1095.4	249.2	924.9	1174.1	0.4099	1.2504	1.6602	280
290	57.53	0.01735	7.47	259.3	1097.7	259.4	917.8	1177.2	0.4236	1.2241	1.6477	290
300	66.98	0.01745	6.472	269.5	1100.0	269.7	910.4	1180.2	0.4372	1.1984	1.6356	300
310	77.64	0.01755	5.632	279.8	1102.1	280.1	903.0	1183.0	0.4507	1.1731	1.6238	310
320	89.60	0.01765	4.919	290.1	1104.2	290.4	895.3	1185.8	0.4640	1.1483	1.6123	320
330	103.00	0.01776	4.312	300.5	1106.2	300.8	887.5	1188.4	0.4772	1.1238	1.6010	330
340	117.93	0.01787	3.792	310.9	1108.0	311.3	879.5	1190.8	0.4903	1.0997	1.5901	340
350	134.53	0.01799	3.346	321.4	1109.8	321.8	871.3	1193.1	0.5033	1.0760	1.5793	350
360	152.92	0.01811	2.961	331.8	1111.4	332.4	862.9	1195.2	0.5162	1.0526	1.5688	360
370	173.23	0.01823	2.628	342.4	1112.9	343.0	854.2	1197.2	0.5289	1.0295	1.5585	370
380	195.60	0.01836	2.339	353.0	1114.3	353.6	845.4	1199.0	0.5416	1.0067	1.5483	380
390	220.2	0.01850	2.087	363.6	1115.6	364.3	836.2	1200.6	0.5542	0.9841	1.5383	390
400	247.1	0.01864	1.866	374.3	1116.6	375.1	826.8	1202.0	0.5667	0.9617	1.5284	400
410	276.5	0.01878	1.673	385.0	1117.6	386.0	817.2	1203.1	0.5792	0.9395	1.5187	410
420	308.5	0.01894	1.502	395.8	1118.3	396.9	807.2	1204.1	0.5915	0.9175	1.5091	420
430	343.3	0.01909	1.352	406.7	1118.9	407.9	796.9	1204.8	0.6038	0.8957	1.4995	430
440	381.2	0.01926	1.219	417.6	1119.3	419.0	786.3	1205.3	0.6161	0.8740	1.4900	440
450	422.1	0.01943	1.1011	428.6	1119.5	430.2	775.4	1205.6	0.6282	0.8523	1.4806	450
460	466.3	0.01961	0.9961	439.7	1119.6	441.4	764.1	1205.5	0.6404	0.8308	1.4712	460
470	514.1	0.01980	0.9025	450.9	1119.4	452.8	752.4	1205.2	0.6525	0.8093	1.4618	470
480	565.5	0.02000	0.8187	462.2	1118.9	464.3	740.3	1204.6	0.6646	0.7878	1.4524	480
490	620.7	0.02021	0.7436	473.6	1118.3	475.9	727.8	1203.7	0.6767	0.7663	1.4430	490
500	680.0	0.02043	0.6761	485.1	1117.4	487.7	714.8	1202.5	0.6888	0.7448	1.4335	500
520	811.4	0.02091	0.5605	508.5	1114.8	511.7	687.3	1198.9	0.7130	0.7015	1.4145	520
540	961.5	0.02145	0.4658	532.6	1111.0	536.4	657.5	1193.8	0.7374	0.6576	1.3950	540
560	1131.8	0.02207	0.3877	548.4	1105.8	562.0	625.0	1187.0	0.7620	0.6129	1.3749	560
580	1324.3	0.02278	0.3225	583.1	1098.9	588.6	589.3	1178.0	0.7872	0.5668	1.3540	580
600	1541.0	0.02363	0.2677	609.9	1090.0	616.7	549.7	1166.4	0.8130	0.5187	1.3317	600
620	1784.4	0.02465	0.2209	638.3	1078.5	646.4	505.0	1151.4	0.8398	0.4677	1.3075	620
640	2057.1	0.02593	0.1805	668.7	1063.2	678.6	453.4	1131.9	0.8681	0.4122	1.2803	640
660	2362	0.02767	0.1446	702.3	1042.3	714.4	391.1	1105.5	0.8990	0.3493	1.2483	660
680	2705	0.03032	0.1113	741.7	1011.0	756.9	309.8	1066.7	0.9350	0.2718	1.2068	680
700	3090	0.03666	0.0744	801.7	947.7	822.7	167.5	990.2	0.9902	0.1444	1.1346	700
705.4	3204	0.05053	0.05053	872.6	872.6	902.5	0	902.5	1.0580	0	1.0580	705.4

* Abridged from Keenan, J. H., F. G. Keyes, P. G. Hill, and J. G. Moore, "Steam Tables," Wiley, New York, 1969.

TABLE A-7M

Properties of water—saturation-temperature table (SI units)*

v in cm³/g, 1 cm³/g = 10^{-3} m³/kg; h and u in kJ/kg; s in kJ/kg·K; p in bars, 1 bar = 10^5Pa

Temp. °C T	Press. bars P	Specific volume		Internal energy		Enthalpy			Entropy	
		Sat. liquid v_f	Sat. vapor v_g	Sat. liquid u_f	Sat. vapor u_g	Sat. liquid h_f	Evap. h_{fg}	Sat. vapor h_g	Sat. liquid s_f	Sat vapor s_g
0	0.00611	1.0002	206278	−0.03	2375.4	−0.02	2501.4	2501.3	−0.0001	9.1565
5	0.00872	1.0001	147120	20.97	2382.3	20.98	2489.6	2510.6	0.0761	9.0257
10	0.01228	1.0004	106379	42.00	2389.2	42.01	2477.7	2519.8	0.1510	8.9008
15	0.01705	1.0009	77926	62.99	2396.1	62.99	2465.9	2528.9	0.2245	8.7814
20	0.02339	1.0018	57791	83.95	2402.9	83.96	2454.1	2538.1	0.2966	8.6672
25	0.03169	1.0029	43360	104.88	2409.8	104.89	2442.3	2547.2	0.3674	8.5580
30	0.04246	1.0043	32894	125.78	2416.6	125.79	2430.5	2556.3	0.4369	8.4533
35	0.05628	1.0060	25216	146.67	2423.4	146.68	2418.6	2565.3	0.5053	8.3531
40	0.07384	1.0078	19523	167.56	2430.1	167.57	2406.7	2574.3	0.5725	8.2570
45	0.09593	1.0099	15258	188.44	2436.8	188.45	2394.8	2583.2	0.6387	8.1648
50	0.1235	1.0121	12032	209.32	2443.5	209.33	2382.7	2592.1	0.7038	8.0763
55	0.1576	1.0146	9568	230.21	2450.1	230.23	2370.7	2600.9	0.7679	7.9913
60	0.1994	1.0172	7671	251.11	2456.6	251.13	2358.5	2609.6	0.8312	7.9096
65	0.2503	1.0199	6197	272.02	2463.1	272.06	2346.2	2618.3	0.8935	7.8310
70	0.3119	1.0228	5042	292.95	2469.6	292.98	2333.8	2626.8	0.9549	7.7553
75	0.3858	1.0259	4131	313.90	2475.9	313.93	2321.4	2635.3	1.0155	7.6824
80	0.4739	1.0291	3407	334.86	2482.2	334.91	2308.8	2643.7	1.0753	7.6122
85	0.5783	1.0325	2828	355.84	2488.4	355.90	2296.0	2651.9	1.1343	7.5445
90	0.7014	1.0360	2361	376.85	2494.5	376.92	2283.2	2660.1	1.1925	7.4791
95	0.8455	1.0397	1982	397.88	2500.6	397.96	2270.2	2668.1	1.2500	7.4159
100	1.014	1.0435	1673.	418.94	2506.5	419.04	2257.0	2676.1	1.3069	7.3549
110	1.433	1.0516	1210.	461.14	2518.1	461.30	2230.2	2691.5	1.4185	7.2387
120	1.985	1.0603	891.9	503.50	2529.3	503.71	2202.6	2706.3	1.5276	7.1296
130	2.701	1.0697	668.5	546.02	2539.9	546.31	2174.2	2720.5	1.6344	7.0269
140	3.613	1.0797	508.9	588.74	2550.0	589.13	2144.7	2733.9	1.7391	6.9299
150	4.758	1.0905	392.8	631.68	2559.5	632.20	2114.3	2746.5	1.8418	6.8379
160	6.178	1.1020	307.1	674.86	2568.4	675.55	2082.6	2758.1	1.9427	6.7502
170	7.917	1.1143	242.8	718.33	2576.5	719.21	2049.5	2768.7	2.0419	6.6663
180	10.02	1.1274	194.1	762.09	2583.7	763.22	2015.0	2778.2	2.1396	6.5857
190	12.54	1.1414	156.5	806.19	2590.0	807.62	1978.8	2786.4	2.2359	6.5079
200	15.54	1.1565	127.4	850.65	2595.3	852.45	1940.7	2793.2	2.3309	6.4323
210	19.06	1.1726	104.4	895.53	2599.5	897.76	1900.7	2798.5	2.4248	6.3585
220	23.18	1.1900	86.19	940.87	2602.4	943.62	1858.5	2802.1	2.5178	6.2861
230	27.95	1.2088	71.58	986.74	2603.9	990.12	1813.8	2804.0	2.6099	6.2146
240	33.44	1.2291	59.76	1033.2	2604.0	1037.3	1766.5	2803.8	2.7015	6.1437
250	39.73	1.2512	50.13	1080.4	2602.4	1085.4	1716.2	2801.5	2.7927	6.0730
260	46.88	1.2755	42.21	1128.4	2599.0	1134.4	1662.5	2796.9	2.8838	6.0019
270	54.99	1.3023	35.64	1177.4	2593.7	1184.5	1605.2	2789.7	2.9751	5.9301
280	64.12	1.3321	30.17	1227.5	2586.1	1236.0	1543.6	2779.6	3.0668	5.8571
290	74.36	1.3656	25.57	1278.9	2576.0	1289.1	1477.1	2766.2	3.1594	5.7821
300	85.81	1.4036	21.67	1332.0	2563.0	1344.0	1404.9	2749.0	3.2534	5.7045
320	112.7	1.4988	15.49	1444.6	2525.5	1461.5	1238.6	2700.1	3.4480	5.5362
340	145.9	1.6379	10.80	1570.3	2464.6	1594.2	1027.9	2622.0	3.6594	5.3357
360	186.5	1.8925	6.945	1725.2	2351.5	1760.5	720.5	2481.0	3.9147	5.0526
374.14	220.9	3.155	3.155	2029.6	2029.6	2099.3	0	2099.3	4.4298	4.4298

* Abridged from Keenan, J. H., F. G. Keyes, P. G. Hill, and J. G. Moore, "Steam Tables," Wiley, New York, 1969.

TABLE A-8

Properties of water — saturation-pressure table (English units)*

v in ft³/lbm; h and u in Btu/lbm; s in Btu/lbm·°R

Abs. press., psi p	Temp., °F t	Specific volume		Internal energy		Enthalpy			Entropy			Abs. press., psi p
		Sat. liquid v_f	Sat. vapor v_g	Sat. liquid u_f	Sat. vapor u_g	Sat. liquid h_f	Evap. h_{fg}	Sat. vapor h_g	Sat. liquid s_f	Evap. s_{fg}	Sat. vapor s_g	
0.4	72.84	0.01606	792.0	40.94	1034.7	40.94	1052.3	1093.3	0.0800	1.9760	2.0559	0.4
0.6	85.19	0.01609	540.0	53.26	1038.7	53.27	1045.4	1098.6	0.1029	1.9184	2.0213	0.6
0.8	94.35	0.01611	411.7	62.41	1041.7	62.41	1040.2	1102.6	0.1195	1.8773	1.9968	0.8
1.0	101.70	0.01614	333.6	69.74	1044.0	69.74	1036.0	1105.8	0.1327	1.8453	1.9779	1.0
1.2	107.88	0.01616	280.9	75.90	1046.0	75.90	1032.5	1108.4	0.1436	1.8190	1.9626	1.2
1.5	115.65	0.01619	227.7	83.65	1048.5	83.65	1028.0	1111.7	0.1571	1.7867	1.9438	1.5
2.0	126.04	0.01623	173.75	94.02	1051.8	94.02	1022.1	1116.1	0.1750	1.7448	1.9198	2.0
3.0	141.43	0.01630	118.72	109.38	1056.6	109.39	1013.1	1122.5	0.2009	1.6852	1.8861	3.0
4.0	152.93	0.01636	90.64	120.88	1060.2	120.89	1006.4	1127.3	0.2198	1.6426	1.8624	4.0
5.0	162.21	0.01641	73.53	130.15	1063.0	130.17	1000.9	1131.0	0.2349	1.6093	1.8441	5.0
6.0	170.03	0.01645	61.98	137.98	1065.4	138.00	996.2	1134.2	0.2474	1.5819	1.8292	6.0
7.0	176.82	0.01649	53.65	144.78	1067.4	144.80	992.1	1136.9	0.2581	1.5585	1.8167	7.0
8.0	182.84	0.01653	47.35	150.81	1069.2	150.84	988.4	1139.3	0.2675	1.5383	1.8058	8.0
9.0	188.26	0.01656	42.41	156.25	1070.8	156.27	985.1	1141.4	0.2760	1.5203	1.7963	9.0
10	193.19	0.01659	38.42	161.20	1072.2	161.23	982.1	1143.3	0.2836	1.5041	1.7877	10
14.696	211.99	0.01672	26.80	180.10	1077.6	180.15	970.4	1150.5	0.3121	1.4446	1.7567	14.696
15	213.03	0.01672	26.29	181.14	1077.9	181.19	969.7	1150.9	0.3137	1.4414	1.7551	15
20	227.96	0.01683	20.09	196.19	1082.0	196.26	960.1	1156.4	0.3358	1.3962	1.7320	20
25	240.08	0.01692	16.31	208.44	1085.3	208.52	952.2	1160.7	0.3535	1.3607	1.7142	25
30	250.34	0.01700	13.75	218.84	1088.0	218.93	945.4	1164.3	0.3682	1.3314	1.6996	30
35	259.30	0.01708	11.90	227.93	1090.3	228.04	939.3	1167.4	0.3809	1.3064	1.6873	35
40	267.26	0.01715	10.50	236.03	1092.3	236.16	933.8	1170.0	0.3921	1.2845	1.6767	40
45	274.46	0.01721	9.40	243.37	1094.0	243.51	928.8	1172.3	0.4022	1.2651	1.6673	45
50	281.03	0.01727	8.52	250.08	1095.6	250.24	924.2	1174.4	0.4113	1.2476	1.6589	50
55	287.10	0.01733	7.79	256.28	1097.0	256.46	919.9	1176.3	0.4196	1.2317	1.6513	55

TABLE A-8
(Continued)

Abs. press psi p	Temp. °F t	Specific volume Sat. liquid v_f	Specific volume Sat. vapor v_g	Internal energy Sat. liquid u_f	Internal energy Sat. vapor u_g	Enthalpy Sat. liquid h_f	Enthalpy Evap. h_{fg}	Enthalpy Sat. vapor h_g	Entropy Sat. liquid s_f	Entropy Evap. s_{fg}	Entropy Sat. vapor s_g	Abs. press. psi p
60	292.73	0.01738	7.177	262.1	1098.3	262.2	915.8	1178.0	0.4273	1.2170	1.6443	60
65	298.00	0.01743	6.647	267.5	1099.5	267.7	911.9	1179.6	0.4345	1.2035	1.6380	65
70	302.96	0.01748	6.209	272.6	1100.6	272.8	908.3	1181.0	0.4412	1.1909	1.6321	70
75	307.63	0.01752	5.818	277.4	1101.6	277.6	904.8	1182.4	0.4475	1.1790	1.6265	75
80	312.07	0.01757	5.474	282.0	1102.6	282.2	901.4	1183.6	0.4534	1.1679	1.6213	80
85	316.29	0.01761	5.170	286.3	1103.5	286.6	898.2	1184.8	0.4591	1.1574	1.6165	85
90	320.31	0.01766	4.898	290.5	1104.3	290.8	895.1	1185.9	0.4644	1.1475	1.6119	90
95	324.16	0.01770	4.654	294.5	1105.0	294.8	892.1	1186.9	0.4695	1.1380	1.6075	95
100	327.86	0.01774	4.434	298.3	1105.8	298.6	889.2	1187.8	0.4744	1.1290	1.6034	100
110	334.82	0.01781	4.051	305.5	1107.1	305.9	883.7	1189.6	0.4836	1.1122	1.5958	110
120	341.30	0.01789	3.730	312.3	1108.3	312.7	878.5	1191.1	0.4920	1.0966	1.5886	120
130	347.37	0.01796	3.457	318.6	1109.4	319.0	873.5	1192.5	0.4999	1.0822	1.5821	130
140	353.08	0.01802	3.221	324.6	1110.3	325.1	868.7	1193.8	0.5073	1.0688	1.5761	140
150	358.48	0.01809	3.016	330.2	1111.2	330.8	864.2	1194.9	0.5142	1.0562	1.5704	150
160	363.60	0.01815	2.836	335.6	1112.0	336.2	859.8	1196.0	0.5208	1.0443	1.5651	160
170	368.47	0.01821	2.676	340.8	1112.7	341.3	855.6	1196.9	0.5270	1.0330	1.5600	170
180	373.13	0.01827	2.533	345.7	1113.4	346.3	851.5	1197.8	0.5329	1.0223	1.5552	180
190	377.59	0.01833	2.405	350.4	1114.0	351.0	847.5	1198.6	0.5386	1.0122	1.5508	190
200	381.86	0.01839	2.289	354.9	1114.6	355.6	843.7	1199.3	0.5440	1.0025	1.5465	200
250	401.04	0.01865	1.845	375.4	1116.7	376.2	825.8	1202.1	0.5680	0.9594	1.5274	250
300	417.43	0.01890	1.544	393.0	1118.2	394.1	809.8	1203.9	0.5883	0.9232	1.5115	300
350	431.82	0.01912	1.327	408.7	1119.0	409.9	795.0	1204.9	0.6060	0.8917	1.4977	350
400	444.70	0.01934	1.162	422.8	1119.5	424.2	781.2	1205.5	0.6218	0.8638	1.4856	400
450	456.39	0.01955	1.033	435.7	1119.6	437.4	768.2	1205.6	0.6360	0.8385	1.4745	450
500	467.13	0.01975	0.928	447.7	1119.4	449.5	755.8	1205.3	0.6490	0.8154	1.4644	500

550	477.07	0.01994	0.842	458.9	1119.1	460.9	743.9	1204.8	0.6611	0.7941	1.4551	550
600	486.33	0.02013	0.770	469.4	1118.6	471.7	732.4	1204.1	0.6723	0.7742	1.4464	600
700	503.23	0.02051	0.656	488.9	1117.0	491.5	710.5	1202.0	0.6927	0.7378	1.4305	700
800	518.36	0.02087	0.569	506.6	1115.0	509.7	689.6	1199.3	0.7110	0.7050	1.4160	800
900	532.12	0.02123	0.501	523.0	1112.6	526.6	669.5	1196.0	0.7277	0.6750	1.4027	900
1000	544.75	0.02159	0.446	538.4	1109.9	542.4	650.0	1192.4	0.7432	0.6471	1.3903	1000
1100	556.45	0.02195	0.401	552.9	1106.8	557.4	631.0	1188.3	0.7576	0.6209	1.3786	1100
1200	567.37	0.02232	0.362	566.7	1103.5	571.7	612.3	1183.9	0.7712	0.5961	1.3673	1200
1300	577.60	0.02269	0.330	579.9	1099.8	585.4	593.8	1179.2	0.7841	0.5724	1.3565	1300
1400	587.25	0.02307	0.302	592.7	1096.0	598.6	575.5	1174.1	0.7964	0.5497	1.3461	1400
1500	596.39	0.02346	0.277	605.0	1091.8	611.5	557.2	1168.7	0.8082	0.5276	1.3359	1500
1600	605.06	0.02386	0.255	616.9	1087.4	624.0	538.9	1162.9	0.8196	0.5062	1.3258	1600
1700	613.32	0.02428	0.236	628.6	1082.7	636.2	520.6	1156.9	0.8307	0.4852	1.3159	1700
1800	621.21	0.02472	0.218	640.0	1077.7	648.3	502.1	1150.4	0.8414	0.4645	1.3060	1800
1900	628.76	0.02517	0.203	651.3	1072.3	660.1	483.4	1143.5	0.8519	0.4441	1.2961	1900
2000	636.00	0.02565	0.188	662.4	1066.6	671.9	464.4	1136.3	0.8623	0.4238	1.2861	2000
2250	652.90	0.02698	0.157	689.9	1050.6	701.1	414.8	1115.9	0.8876	0.3728	1.2604	2250
2500	668.31	0.02860	0.131	717.7	1031.0	730.9	360.5	1091.4	0.9131	0.3196	1.2327	2500
2750	682.46	0.03077	0.107	747.3	1005.9	763.0	297.4	1060.4	0.9401	0.2604	1.2005	2750
3000	695.52	0.03431	0.084	783.4	968.8	802.5	213.0	1015.5	0.9732	0.1843	1.1575	3000
3203.6	705.44	0.05053	0.0505	872.6	872.6	902.5	0	902.5	1.0580	0	1.0580	3203.6

* Abridged from Keenan, J. H., F. G. Keyes, P. G. Hill, and J. G. Moore, "Steam Tables," Wiley, New York, 1969.

TABLE A-8M

Properties of water—saturation-pressure table (SI units)*

v in cm³/g, 1 cm³/g = 10^{-3} m³/kg; h and u in kJ/kg; s in kJ/kg·K; p in bars, 1 bar = 10^5 Pa

Press. bars P	Temp. °C T	Specific volume Sat. liquid v_f	Specific volume Sat. vapor v_g	Internal energy Sat. liquid u_f	Internal energy Sat. vapor u_g	Enthalpy Sat. liquid h_f	Enthalpy Evap. h_{fg}	Enthalpy Sat. vapor h_g	Entropy Sat. liquid s_f	Entropy Sat. vapor s_g
0.040	28.96	1.0040	34800.	121.45	2415.2	121.46	2432.9	2554.4	0.4226	8.4746
0.060	36.16	1.0064	23729.	151.53	2425.0	151.53	2415.9	2567.4	0.5210	8.3304
0.080	41.51	1.0084	18103.	173.87	2432.2	173.88	2403.1	2577.0	0.5926	8.2287
0.10	45.81	1.0102	14674.	191.82	2437.9	191.83	2392.8	2584.7	0.6493	8.1502
0.20	60.06	1.0172	7649.	251.38	2456.7	251.40	2358.3	2609.7	0.8320	7.9085
0.30	69.10	1.0223	5229.	289.20	2468.4	289.23	2336.1	2625.3	0.9439	7.7686
0.40	75.87	1.0265	3993.	317.53	2477.0	317.58	2319.2	2636.8	1.0259	7.6700
0.50	81.33	1.0300	3240.	340.44	2483.9	340.49	2305.4	2645.9	1.0910	7.5939
0.60	85.94	1.0331	2732.	359.79	2489.6	359.86	2293.6	2653.5	1.1453	7.5320
0.70	89.95	1.0360	2365.	376.63	2494.5	376.70	2283.3	2660.0	1.1919	7.4797
0.80	93.50	1.0380	2087.	391.58	2498.8	391.66	2274.1	2665.8	1.2329	7.4346
0.90	96.71	1.0410	1869.	405.06	2502.6	405.15	2265.7	2670.9	1.2695	7.3949
1.00	99.63	1.0432	1694.	417.36	2506.1	417.46	2258.0	2675.5	1.3026	7.3594
1.50	111.4	1.0528	1159.	466.94	2519.7	467.11	2226.5	2693.6	1.4336	7.2233
2.00	120.2	1.0605	885.7	504.49	2529.5	504.70	2201.9	2706.7	1.5301	7.1271
2.50	127.4	1.0672	718.7	535.10	2537.2	535.37	2181.5	2716.9	1.6072	7.0527
3.00	133.6	1.0732	605.8	561.15	2543.6	561.47	2163.8	2725.3	1.6718	6.9919
3.50	138.9	1.0786	524.3	583.95	2548.9	584.33	2148.1	2732.4	1.7275	6.9405
4.00	143.6	1.0836	462.5	604.31	2553.6	604.74	2133.8	2738.6	1.7766	6.8959
4.50	147.9	1.0882	414.0	622.77	2557.6	623.25	2120.7	2743.9	1.8207	6.8565
5.00	151.9	1.0926	374.9	639.68	2561.2	640.23	2108.5	2748.7	1.8607	6.8213
6.00	158.9	1.1006	315.7	669.90	2567.4	670.56	2086.3	2756.8	1.9312	6.7600
7.00	165.0	1.1080	272.9	696.44	2572.5	697.22	2066.3	2763.5	1.9922	6.7080
8.00	170.4	1.1148	240.4	720.22	2576.8	721.11	2048.0	2769.1	2.0462	6.6628
9.00	175.4	1.1212	215.0	741.83	2580.5	742.83	2031.1	2773.9	2.0946	6.6226
10.0	179.9	1.1273	194.4	761.68	2583.6	762.81	2015.3	2778.1	2.1387	6.5863
15.0	198.3	1.1539	131.8	843.16	2594.5	844.89	1947.3	2792.2	2.3150	6.4448
20.0	212.4	1.1767	99.63	906.44	2600.3	908.79	1890.7	2799.5	2.4474	6.3409
25.0	224.0	1.1973	79.98	959.11	2603.1	962.11	1841.0	2803.1	2.5547	6.2575
30.0	233.9	1.2165	66.68	1004.8	2604.1	1008.4	1795.7	2804.2	2.6457	6.1869
35.0	242.6	1.2347	57.07	1045.4	2603.7	1049.8	1753.7	2803.4	2.7253	6.1253
40.0	250.4	1.2522	49.78	1082.3	2602.3	1087.3	1714.1	2801.4	2.7964	6.0701
45.0	257.5	1.2692	44.06	1116.2	2600.1	1121.9	1676.4	2798.3	2.8610	6.0199
50.0	264.0	1.2859	39.44	1147.8	2597.1	1154.2	1640.1	2794.3	2.9202	5.9734
60.0	275.6	1.3187	32.44	1205.4	2589.7	1213.4	1571.0	2784.3	3.0267	5.8892
70.0	285.9	1.3513	27.37	1257.6	2580.5	1267.0	1505.1	2772.1	3.1211	5.8133
80.0	295.1	1.3842	23.52	1305.6	2569.8	1316.6	1441.3	2758.0	3.2068	5.7432
90.0	303.4	1.4178	20.48	1350.5	2557.8	1363.3	1378.9	2742.1	3.2858	5.6772
100.	311.1	1.4524	18.03	1393.0	2544.4	1407.6	1317.1	2724.7	3.3596	5.6141
110.	318.2	1.4886	15.99	1433.7	2529.8	1450.1	1255.5	2705.6	3.4295	5.5527
120.	324.8	1.5267	14.26	1473.0	2513.7	1491.3	1193.6	2684.9	3.4962	5.4924
130.	330.9	1.5671	12.78	1511.1	2496.1	1531.5	1130.7	2662.2	3.5606	5.4323
140.	336.8	1.6107	11.49	1548.6	2476.8	1571.1	1066.5	2637.6	3.6232	5.3717
150.	342.2	1.6581	10.34	1585.6	2455.5	1610.5	1000.0	2610.5	3.6848	5.3098
160.	347.4	1.7107	9.306	1622.7	2431.7	1650.1	930.6	2580.6	3.7461	5.2455
170.	352.4	1.7702	8.364	1660.2	2405.0	1690.3	856.9	2547.2	3.8079	5.1777
180.	357.1	1.8397	7.489	1698.9	2374.3	1732.0	777.1	2509.1	3.8715	5.1044
190.	361.5	1.9243	6.657	1739.9	2338.1	1776.5	688.0	2464.5	3.9388	5.0228
200.	365.8	2.036	5.834	1785.6	2293.0	1826.3	583.4	2409.7	4.0139	4.9269
220.9	374.1	3.155	3.155	2029.6	2029.6	2099.3	0	2099.3	4.4298	4.4298

* Abridged from Keenan et al., "Steam Tables," Wiley, New York, 1969.

TABLE A-9

Properties of water—superheated vapor (English units)*

v in ft³/lbm; h and u in Btu/lbm; s in Btu/lbm·°R

Temp., °F	v	u	h	s	v	u	h	s
	\multicolumn{4}{c}{1 psia (101.7°F)}				
Sat.	333.6	1044.0	1105.8	1.9779	73.53	1063.0	1131.0	1.8441
150	362.6	1060.4	1127.5	2.0151				
200	392.5	1077.5	1150.1	2.0508	78.15	1076.0	1148.6	1.8715
250	422.4	1094.7	1172.8	2.0839	84.21	1093.8	1171.7	1.9052
300	452.3	1112.0	1195.7	2.1150	90.24	1111.3	1194.8	1.9367
400	511.9	1147.0	1241.8	2.1720	102.24	1146.6	1241.2	1.9941
500	571.5	1182.8	1288.5	2.2235	114.20	1182.5	1288.2	2.0458
600	631.1	1219.3	1336.1	2.2706	126.15	1219.1	1335.8	2.0930
700	690.7	1256.7	1384.5	2.3142	138.08	1256.5	1384.3	2.1367
800	750.3	1294.4	1433.7	2.3550	150.01	1294.7	1433.5	2.1775
900	809.9	1333.9	1483.8	2.3932	161.94	1333.8	1483.7	2.2158
1000	869.5	1373.9	1534.8	2.4294	173.86	1373.9	1534.7	2.2520
	\multicolumn{4}{c}{10 psia (193.2°F)}				
Sat.	38.42	1072.2	1143.3	1.7877	26.80	1077.6	1150.5	1.7567
200	38.85	1074.7	1146.6	1.7927				
250	41.95	1092.6	1170.2	1.8272	28.42	1091.5	1168.8	1.7832
300	44.99	1110.4	1193.7	1.8592	30.52	1109.6	1192.6	1.8157
400	51.03	1146.1	1240.5	1.9171	34.67	1145.6	1239.9	1.8741
500	57.04	1182.2	1287.7	1.9690	38.77	1181.8	1287.3	1.9263
600	63.03	1218.9	1335.5	2.0164	42.86	1218.6	1335.2	1.9737
700	69.01	1256.3	1384.0	2.0601	46.93	1256.1	1383.8	2.0175
800	74.98	1294.6	1433.3	2.1009	51.00	1294.4	1433.1	2.0584
900	80.95	1333.7	1483.5	2.1393	55.07	1333.6	1483.4	2.0967
1000	86.91	1373.8	1534.6	2.1755	59.13	1373.7	1534.5	2.1330
1100	92.88	1414.7	1586.6	2.2099	63.19	1414.6	1586.4	2.1674
	\multicolumn{4}{c}{20 psia (228.0°F)}				
Sat.	20.09	1082.0	1156.4	1.7320	10.50	1093.3	1170.0	1.6767
250	20.79	1090.3	1167.2	1.7475				
300	22.36	1108.7	1191.5	1.7805	11.04	1105.1	1186.8	1.6993
350	23.90	1126.9	1215.4	1.8110	11.84	1124.2	1211.8	1.7312
400	25.43	1145.1	1239.2	1.8395	12.62	1143.0	1236.4	1.7606
500	28.46	1181.5	1286.8	1.8919	14.16	1180.1	1284.9	1.8140
600	31.47	1218.4	1334.8	1.9395	15.69	1217.3	1333.4	1.8621
700	34.47	1255.9	1383.5	1.9834	17.20	1255.1	1382.4	1.9063
800	37.46	1294.3	1432.9	2.0243	18.70	1293.7	1432.1	1.9474
900	40.45	1333.5	1483.2	2.0627	20.20	1333.0	1482.5	1.9859
1000	43.44	1373.5	1534.3	2.0989	21.70	1373.1	1533.8	2.0223
1100	46.42	1414.5	1586.3	2.1334	23.20	1414.2	1585.9	2.0568

5 psia (162.2°F) ; **14.7 psia (212.0°F)** (second set, middle block); **40 psia (267.3°F)** .

* Abridged from Keenan, J. H., F. G. Keyes, P. G. Hill, and J. G. Moore, "Steam Tables," Wiley, New York, 1969.

TABLE A-9
(Continued)

Temp., °F	v	u	h	s	v	u	h	s
	60 psia (292.7°F)				80 psia (312.1°F)			
Sat.	7.17	1098.3	1178.0	1.6444	5.47	1102.6	1183.6	1.6214
300	7.26	1101.3	1181.9	1.6496				
350	7.82	1121.4	1208.2	1.6830	5.80	1118.5	1204.3	1.6476
400	8.35	1140.8	1233.5	1.7134	6.22	1138.5	1230.6	1.6790
500	9.40	1178.6	1283.0	1.7678	7.02	1177.2	1281.1	1.7346
600	10.43	1216.3	1332.1	1.8165	7.79	1215.3	1330.7	1.7838
700	11.44	1254.4	1381.4	1.8609	8.56	1253.6	1380.3	1.8285
800	12.45	1293.0	1431.2	1.9022	9.32	1292.4	1430.4	1.8700
900	13.45	1332.5	1481.8	1.9408	10.08	1332.0	1481.2	1.9087
1000	14.45	1372.7	1533.2	1.9773	10.83	1372.3	1532.6	1.9453
1100	15.45	1413.8	1585.4	2.0119	11.58	1413.5	1584.9	1.9799
1200	16.45	1455.8	1638.5	2.0448	12.33	1455.5	1638.1	2.0130
	100 psia (327.8°F)				120 psia (341.3°F)			
Sat.	4.434	1105.8	1187.8	1.6034	3.730	1108.3	1191.1	1.5886
350	4.592	1115.4	1200.4	1.6191	3.783	1112.2	1196.2	1.5950
400	4.934	1136.2	1227.5	1.6517	4.079	1133.8	1224.4	1.6288
450	5.265	1156.2	1253.6	1.6812	4.360	1154.3	1251.2	1.6590
500	5.587	1175.7	1279.1	1.7085	4.633	1174.2	1277.1	1.6868
600	6.216	1214.2	1329.3	1.7582	5.164	1213.2	1327.8	1.7371
700	6.834	1252.8	1379.2	1.8033	5.682	1252.0	1378.2	1.7825
800	7.445	1291.8	1429.6	1.8449	6.195	1291.2	1428.7	1.8243
900	8.053	1331.5	1480.5	1.8838	6.703	1330.9	1479.8	1.8633
1000	8.657	1371.9	1532.1	1.9204	7.208	1371.5	1531.5	1.9000
1100	9.260	1413.1	1584.5	1.9551	7.711	1412.8	1584.0	1.9348
1200	9.861	1455.2	1637.7	1.9882	8.213	1454.9	1637.3	1.9679
	140 psia (353.1°F)				160 psia (363.6°F)			
Sat.	3.221	1110.3	1193.8	1.5761	2.836	1112.0	1196.0	1.5651
400	3.466	1131.4	1221.2	1.6088	3.007	1128.8	1217.8	1.5911
450	3.713	1152.4	1248.6	1.6399	3.228	1150.5	1246.1	1.6230
500	3.952	1172.7	1275.1	1.6682	3.440	1171.2	1273.0	1.6518
550	4.184	1192.5	1300.9	1.6945	3.646	1191.3	1299.2	1.6785
600	4.412	1212.1	1326.4	1.7191	3.848	1211.1	1325.0	1.7034
700	4.860	1251.2	1377.1	1.7648	4.243	1250.4	1376.0	1.7494
800	5.301	1290.5	1427.9	1.8068	4.631	1289.9	1427.0	1.7916
900	5.739	1330.4	1479.1	1.8459	5.015	1329.9	1478.4	1.8308
1000	6.173	1371.0	1531.0	1.8827	5.397	1370.6	1530.4	1.8677
1100	6.605	1412.4	1583.6	1.9176	5.776	1412.1	1583.1	1.9026
1200	7.036	1454.6	1636.9	1.9507	6.154	1454.3	1636.5	1.9358

TABLE A-9

(Continued)

Temp., °F	v	u	h	s	v	u	h	s
	\multicolumn{4}{c}{180 psia (373.1°F)}	\multicolumn{4}{c}{200 psia (381.8°F)}				
Sat.	2.533	1113.4	1197.8	1.5553	2.289	1114.6	1199.3	1.5464
400	2.648	1126.2	1214.4	1.5749	2.361	1123.5	1210.8	1.5600
450	2.850	1148.5	1243.4	1.6078	2.548	1146.4	1240.7	1.5938
500	3.042	1169.6	1270.9	1.6372	2.724	1168.0	1268.8	1.6239
550	3.228	1190.0	1297.5	1.6642	2.893	1188.7	1295.7	1.6512
600	3.409	1210.0	1323.5	1.6893	3.058	1208.9	1322.1	1.6767
700	3.763	1249.6	1374.9	1.7357	3.379	1248.8	1373.8	1.7234
800	4.110	1289.3	1426.2	1.7781	3.693	1288.6	1425.3	1.7660
900	4.453	1329.4	1477.7	1.8174	4.003	1328.9	1477.1	1.8055
1000	4.793	1370.2	1529.8	1.8545	4.310	1369.8	1529.3	1.8425
1100	5.131	1411.7	1582.6	1.8894	4.615	1411.4	1582.2	1.8776
1200	5.467	1454.0	1636.1	1.9227	4.918	1453.7	1635.7	1.9109

Temp., °F	v	u	h	s	v	u	h	s
	\multicolumn{4}{c}{250 psia (401.0°F)}	\multicolumn{4}{c}{300 psia (417.4°F)}				
Sat.	1.845	1116.7	1202.1	1.5274	1.544	1118.2	1203.9	1.5115
450	2.002	1141.1	1233.7	1.5632	1.636	1135.4	1226.2	1.5365
500	2.150	1163.8	1263.3	1.5948	1.766	1159.5	1257.5	1.5701
550	2.290	1185.3	1291.3	1.6233	1.888	1181.9	1286.7	1.5997
600	2.426	1206.1	1318.3	1.6494	2.004	1203.2	1314.5	1.6266
700	2.688	1246.7	1371.1	1.6970	2.227	1244.0	1368.3	1.6751
800	2.943	1287.0	1423.2	1.7301	2.442	1285.4	1421.0	1.7187
900	3.193	1327.6	1475.3	1.7799	2.653	1326.3	1473.6	1.7589
1000	3.440	1368.7	1527.9	1.8172	2.860	1367.7	1526.5	1.7964
1100	3.685	1410.5	1581.0	1.8524	3.066	1409.6	1579.8	1.8317
1200	3.929	1453.0	1634.8	1.8858	3.270	1452.2	1633.8	1.8653
1300	4.172	1496.3	1689.3	1.9177	3.473	1495.6	1688.4	1.8973

Temp., °F	v	u	h	s	v	u	h	s
	\multicolumn{4}{c}{350 psia (431.8°F)}	\multicolumn{4}{c}{400 psia (444.7°F)}				
Sat.	1.327	1119.0	1204.9	1.4978	1.162	1119.5	1205.5	1.4856
450	1.373	1129.2	1218.2	1.5125	1.175	1122.6	1209.6	1.4901
500	1.491	1154.9	1251.5	1.5482	1.284	1150.1	1245.2	1.5282
550	1.600	1178.3	1281.9	1.5790	1.383	1174.6	1277.0	1.5605
600	1.703	1200.3	1310.6	1.6068	1.476	1197.3	1306.6	1.5892
700	1.898	1242.5	1365.4	1.6562	1.650	1240.4	1362.5	1.6397
800	2.085	1283.8	1418.8	1.7004	1.816	1282.1	1416.6	1.6844
900	2.267	1325.0	1471.8	1.7409	1.978	1323.7	1470.1	1.7252
1000	2.446	1366.6	1525.0	1.7787	2.136	1365.5	1523.6	1.7632
1100	2.624	1408.7	1578.6	1.8142	2.292	1407.8	1577.4	1.7989
1200	2.799	1451.5	1632.8	1.8478	2.446	1450.7	1631.8	1.8327
1300	2.974	1495.0	1687.6	1.8799	2.599	1494.3	1686.8	1.8648

TABLE A-9
(Continued)

Temp., °F	v	u	h	s	v	u	h	s
	450 psia (456.4°F)				500 psia (467.1°F)			
Sat.	1.033	1119.6	1205.6	1.4746	0.928	1119.4	1205.3	1.4645
500	1.123	1145.1	1238.5	1.5097	0.992	1139.7	1231.5	1.4923
550	1.215	1170.7	1271.9	1.5436	1.079	1166.7	1266.6	1.5279
600	1.300	1194.3	1302.5	1.5732	1.158	1191.1	1298.3	1.5585
700	1.458	1238.2	1359.6	1.6248	1.304	1236.0	1356.7	1.6112
800	1.608	1280.5	1414.4	1.6701	1.441	1278.8	1412.1	1.6571
900	1.752	1322.4	1468.3	1.7113	1.572	1321.0	1466.5	1.6987
1000	1.894	1364.4	1522.2	1.7495	1.701	1363.3	1520.7	1.7471
1100	2.034	1406.9	1576.3	1.7853	1.827	1406.0	1575.1	1.7731
1200	2.172	1450.0	1630.8	1.8192	1.952	1449.2	1629.8	1.8072
1300	2.308	1493.7	1685.9	1.8515	2.075	1493.1	1685.1	1.8395
1400	2.444	1538.1	1741.7	1.8823	2.198	1537.6	1741.0	1.8704
	600 psia (486.3°F)				700 psia (503.2°F)			
Sat.	0.770	1118.6	1204.1	1.4464	0.656	1117.0	1202.0	1.4305
500	0.795	1128.0	1216.2	1.4592				
550	0.875	1158.2	1255.4	1.4990	0.728	1149.0	1243.2	1.4723
600	0.946	1184.5	1289.5	1.5320	0.793	1177.5	1280.2	1.5081
700	1.073	1231.5	1350.6	1.5872	0.907	1226.9	1344.4	1.5661
800	1.190	1275.4	1407.6	1.6343	1.011	1272.0	1402.9	1.6145
900	1.302	1318.4	1462.9	1.6766	1.109	1315.6	1459.3	1.6576
1000	1.411	1361.2	1517.8	1.7155	1.204	1358.9	1514.9	1.6970
1100	1.517	1404.2	1572.7	1.7519	1.296	1402.4	1570.2	1.7337
1200	1.622	1447.7	1627.8	1.7861	1.387	1446.2	1625.8	1.7682
1300	1.726	1491.7	1683.4	1.8186	1.476	1490.4	1681.7	1.8009
1400	1.829	1536.5	1739.5	1.8497	1.565	1535.3	1738.1	1.8321
	800 psia (518.3°F)				900 psia (532.1°F)			
Sat.	0.569	1115.0	1199.3	1.4160	0.501	1112.6	1196.0	1.4027
550	0.615	1138.8	1229.9	1.4469	0.527	1127.5	1215.2	1.4219
600	0.677	1170.1	1270.4	1.4861	0.587	1162.2	1260.0	1.4652
650	0.732	1197.2	1305.6	1.5186	0.639	1191.1	1297.5	1.4999
700	0.783	1222.1	1338.0	1.5471	0.686	1217.1	1331.4	1.5297
800	0.876	1268.5	1398.2	1.5969	0.772	1264.9	1393.4	1.5810
900	0.964	1312.9	1455.6	1.6408	0.851	1310.1	1451.9	1.6257
1000	1.048	1356.7	1511.9	1.6807	0.927	1354.5	1508.9	1.6662
1100	1.130	1400.5	1567.8	1.7178	1.001	1398.7	1565.4	1.7036
1200	1.210	1444.6	1623.8	1.7526	1.073	1443.0	1621.7	1.7386
1300	1.289	1489.1	1680.0	1.7854	1.144	1487.8	1687.3	1.7717
1400	1.367	1534.2	1736.6	1.8167	1.214	1533.0	1735.1	1.8031

TABLE A-9
(Continued)

Temp., °F	v	u	h	s	v	u	h	s
	1000 psia (544.7°F)				1200 psia (567.4°F)			
Sat.	0.446	1109.0	1192.4	1.3903	0.362	1103.5	1183.9	1.3673
600	0.514	1153.7	1248.8	1.4450	0.402	1134.4	1223.6	1.4054
650	0.564	1184.7	1289.1	1.4822	0.450	1170.9	1270.8	1.4490
700	0.608	1212.0	1324.6	1.5135	0.491	1201.3	1310.2	1.4837
800	0.688	1261.2	1388.5	1.5665	0.562	1253.7	1378.4	1.5402
900	0.761	1307.3	1448.1	1.6120	0.626	1301.5	1440.4	1.5876
1000	0.831	1352.2	1505.9	1.6530	0.685	1347.5	1499.7	1.6297
1100	0.898	1396.8	1562.9	1.6908	0.743	1393.0	1557.9	1.6682
1200	0.963	1441.5	1619.7	1.7261	0.798	1438.3	1615.5	1.7040
1300	1.027	1486.5	1676.5	1.7593	0.853	1483.8	1673.1	1.7377
1400	1.091	1531.9	1733.7	1.7909	0.906	1529.6	1730.7	1.7696
1600	1.215	1624.4	1849.3	1.8499	1.011	1622.6	1847.1	1.8290
	1400 psia (587.2°F)				1600 psia (605.1°F)			
Sat.	0.302	1096.0	1174.1	1.3461	0.255	1087.4	1162.9	1.3258
600	0.318	1110.9	1193.1	1.3641				
650	0.367	1155.5	1250.5	1.4171	0.303	1137.8	1227.4	1.3852
700	0.406	1189.6	1294.8	1.4562	0.342	1177.0	1278.1	1.4299
800	0.471	1245.8	1367.9	1.5168	0.403	1237.7	1357.0	1.4953
900	0.529	1295.6	1432.5	1.5661	0.456	1289.5	1424.4	1.5468
1000	0.582	1342.8	1493.5	1.6094	0.504	1338.0	1487.1	1.5913
1100	0.632	1389.1	1552.8	1.6487	0.549	1385.2	1547.7	1.6315
1200	0.681	1435.1	1611.4	1.6851	0.592	1431.8	1607.1	1.6684
1300	0.728	1481.1	1669.6	1.7192	0.634	1478.3	1666.1	1.7029
1400	0.774	1527.2	1727.8	1.7513	0.675	1524.9	1724.8	1.7354
1600	0.865	1620.8	1844.8	1.8111	0.755	1619.0	1842.6	1.7955
	1800 psia (621.2°F)				2000 psia (636.0°F)			
Sat.	0.218	1077.7	1150.4	1.3060	0.188	1066.6	1136.3	1.2861
650	0.251	1117.0	1200.4	1.3517	0.206	1091.1	1167.2	1.3141
700	0.291	1163.1	1259.9	1.4042	0.249	1147.7	1239.8	1.3782
750	0.322	1198.6	1305.9	1.4430	0.280	1187.3	1291.1	1.4216
800	0.350	1229.1	1345.7	1.4753	0.307	1220.1	1333.8	1.4562
900	0.399	1283.2	1416.1	1.5291	0.353	1276.8	1407.6	1.5126
1000	0.443	1333.1	1480.7	1.5749	0.395	1328.1	1474.1	1.5598
1100	0.484	1381.2	1542.5	1.6159	0.433	1377.2	1537.2	1.6017
1200	0.524	1428.5	1602.9	1.6534	0.469	1425.2	1598.6	1.6398
1300	0.561	1475.5	1662.5	1.6883	0.503	1472.7	1659.0	1.6751
1400	0.598	1522.5	1721.8	1.7211	0.537	1520.2	1718.8	1.7082
1600	0.670	1617.2	1840.4	1.7817	0.602	1615.4	1838.2	1.7692

TABLE A-9
(Continued)

Temp., °F	v	u	h	s	v	u	h	s
	2500 psia (668.3°F)				3000 psia (695.5°F)			
Sat.	0.1306	1031.0	1091.4	1.2327	0.0840	968.8	1015.5	1.1575
700	0.1684	1098.7	1176.6	1.3073	0.0977	1003.9	1058.1	1.1944
750	0.2030	1155.2	1249.1	1.3686	0.1483	1114.7	1197.1	1.3122
800	0.2291	1195.7	1301.7	1.4112	0.1757	1167.6	1265.2	1.3675
900	0.2712	1259.9	1385.4	1.4752	0.2160	1241.8	1361.7	1.4414
1000	0.3069	1315.2	1457.2	1.5262	0.2485	1301.7	1439.6	1.4967
1100	0.3393	1366.8	1523.8	1.5704	0.2772	1356.2	1510.1	1.5434
1200	0.3696	1416.7	1587.7	1.6101	0.3086	1408.0	1576.6	1.5848
1300	0.3984	1465.7	1650.0	1.6465	0.3285	1458.5	1640.9	1.6224
1400	0.4261	1514.2	1711.3	1.6804	0.3524	1508.1	1703.7	1.6571
1500	0.4531	1562.5	1772.1	1.7123	0.3754	1557.3	1765.7	1.6896
1600	0.4795	1610.8	1832.6	1.7424	0.3978	1606.3	1827.1	1.7201
	3500 psia				4000 psia			
650	0.0249	663.5	679.7	0.8630	0.0245	657.7	675.8	0.8574
700	0.0306	759.5	779.3	0.9506	0.0287	742.1	763.4	0.9345
750	0.1046	1058.4	1126.1	1.2440	0.0633	960.7	1007.5	1.1395
800	0.1363	1134.7	1223.0	1.3226	0.1052	1095.0	1172.9	1.2740
900	0.1763	1222.4	1336.5	1.4096	0.1462	1201.5	1309.7	1.3789
1000	0.2066	1287.6	1421.4	1.4699	0.1752	1272.9	1402.6	1.4449
1100	0.2328	1345.2	1496.0	1.5193	0.1995	1333.9	1481.6	1.4973
1200	0.2566	1399.2	1565.3	1.5624	0.2213	1390.1	1553.9	1.5423
1300	0.2787	1451.1	1631.7	1.6012	0.2414	1443.7	1622.4	1.5823
1400	0.2997	1501.9	1696.1	1.6368	0.2603	1495.7	1688.4	1.6188
1500	0.3199	1552.0	1759.2	1.6699	0.2784	1546.7	1752.8	1.6526
1600	0.3395	1601.7	1821.6	1.7010	0.2959	1597.1	1816.1	1.6841
	4400 psia				4800 psia			
650	0.0242	653.6	673.3	0.8535	0.0237	649.8	671.0	0.8499
700	0.0278	732.7	755.3	0.9257	0.0271	725.1	749.1	0.9187
750	0.0415	870.8	904.6	1.0513	0.0352	832.6	863.9	1.0154
800	0.0844	1056.5	1125.3	1.2306	0.0668	1011.2	1070.5	1.1827
900	0.1270	1183.7	1287.1	1.3548	0.1109	1164.8	1263.4	1.3310
1000	0.1552	1260.8	1387.2	1.4260	0.1385	1248.3	1317.4	1.4078
1100	0.1784	1324.7	1469.9	1.4809	0.1608	1315.3	1458.1	1.4653
1200	0.1989	1382.8	1544.7	1.5274	0.1802	1375.4	1535.4	1.5133
1300	0.2176	1437.7	1614.9	1.5685	0.1979	1431.7	1607.4	1.5555
1400	0.2352	1490.7	1682.3	1.6057	0.2143	1485.7	1676.1	1.5934
1500	0.2520	1542.7	1747.6	1.6399	0.2300	1538.2	1742.5	1.6282
1600	0.2681	1593.4	1811.7	1.6718	0.2450	1589.8	1807.4	1.6605

TABLE A-9M
Properties of water—superheated vapor (SI units)*

v in cm³/g, 1 cm³/g = 10^{-3} m³/kg; h and u in kJ/kg; s in kJ/kg·K

Temp. °C	v	u	h	s	v	u	h	s
	6 kPa (36.16°C)				35 kPa (72.69°C)			
Sat.	23739	2425.0	2546.4	8.3304	4526.	2473.0	2631.4	7.7158
80	27132	2487.3	2650.1	8.5804	4625.	2483.7	2645.6	7.7564
120	30219	2544.7	2726.0	8.7840	5163.	2542.4	2723.1	7.9644
160	33302	2602.7	2802.5	8.9693	5696.	2601.2	2800.6	8.1519
200	36383	2661.4	2879.7	9.1398	6228.	2660.4	2878.4	8.3237
240	39462	2721.0	2957.8	9.2982	6758.	2720.3	2956.8	8.4828
280	42540	2781.5	3036.8	9.4464	7287.	2780.9	3036.0	8.6314
320	45618	2843.0	3116.7	9.5859	7815.	2842.5	3116.1	8.7712
360	48696	2905.5	3197.7	9.7180	8344.	2905.1	3197.1	8.9034
400	51774	2969.0	3279.6	9.8435	8872.	2968.6	3279.2	9.0291
440	54851	3033.5	3362.6	9.9633	9400.	3033.2	3362.2	9.1490
500	59467	3132.3	3489.1	10.134	10192.	3132.1	3488.8	9.3194
	70 kPa (89.95°C)				100 kPa (99.63°C)			
Sat.	2365.	2494.5	2660.0	7.4797	1694.	2506.1	2675.5	7.3594
100	2434.	2509.7	2680.0	7.5341	1696.	2506.7	2676.2	7.3614
120	2571.	2539.7	2719.6	7.6375	1793.	2537.3	2716.6	7.4668
160	2841.	2599.4	2798.2	7.8279	1984.	2597.8	2796.2	7.6597
200	3108.	2659.1	2876.7	8.0012	2172.	2658.1	2875.3	7.8343
240	3374.	2719.3	2955.5	8.1611	2359.	2718.5	2954.5	7.9949
280	3640.	2780.2	3035.0	8.3162	2546.	2779.6	3034.2	8.1445
320	3905.	2842.0	3115.3	8.4504	2732.	2841.5	3114.6	8.2849
360	4170.	2904.6	3196.5	8.5828	2917.	2904.2	3195.9	8.4175
400	4434.	2968.2	3278.6	8.7086	3103.	2967.9	3278.2	8.5435
440	4698.	3032.9	3361.8	8.8286	3288.	3032.6	3361.4	8.6636
500	5095.	3131.8	3488.5	8.9991	3565.	3131.6	3488.1	8.8342
	150 kPa (111.37°C)				300 kPa (133.55°C)			
Sat.	1159.	2519.7	2693.6	7.2233	606.	2543.6	2725.3	6.9919
120	1188.	2533.3	2711.4	7.2693				
160	1317.	2595.2	2792.8	7.4665	651.	2587.1	2782.3	7.1276
200	1444.	2656.2	2872.9	7.6433	716.	2650.7	2865.5	7.3115
240	1570.	2717.2	2952.7	7.8052	781.	2713.1	2947.3	7.4774
280	1695.	2778.6	3032.8	7.9555	844.	2775.4	3028.6	7.6299
320	1819.	2840.6	3113.5	8.0964	907.	2838.1	3110.1	7.7722
360	1943.	2903.5	3195.0	8.2293	969.	2901.4	3192.2	7.9061
400	2067.	2967.3	3277.4	8.3555	1032.	2965.6	3275.0	8.0330
440	2191.	3032.1	3360.7	8.4757	1094.	3030.6	3358.7	8.1538
500	2376.	3131.2	3487.6	8.6466	1187.	3130.0	3486.0	8.3251
600	2685.	3301.7	3704.3	8.9101	1341.	3300.8	3703.2	8.5892

* Abridged from Keenan, J. H., F. G. Keyes, P. G. Hill, and J. G. Moore, "Steam Tables," Wiley, New York, 1969.

TABLE A-9M

(Continued)

Temp. °C	v	u	h	s	v	u	h	s
	500 kPa (151.86°C)				700 kPa (164.97°C)			
Sat.	374.9	2561.2	2748.7	6.8213	272.9	2572.5	2763.5	6.7080
180	404.5	2609.7	2812.0	6.9656	284.7	2599.8	2799.1	6.7880
200	424.9	2642.9	2855.4	7.0592	299.9	2634.8	2844.8	6.8865
240	464.6	2707.6	2939.9	7.2307	329.2	2701.8	2932.2	7.0641
280	503.4	2771.2	3022.9	7.3865	357.4	2766.9	3017.1	7.2233
320	541.6	2834.7	3105.6	7.5308	385.2	2831.3	3100.9	7.3697
360	579.6	2898.7	3188.4	7.6660	412.6	2895.8	3184.7	7.5063
400	617.3	2963.2	3271.9	7.7938	439.7	2960.9	3268.7	7.6350
440	654.8	3028.6	3356.0	7.9152	466.7	3026.6	3353.3	7.7571
500	710.9	3128.4	3483.9	8.0873	507.0	3126.8	3481.7	7.9299
600	804.1	3299.6	3701.7	8.3522	573.8	3298.5	3700.2	8.1956
700	896.9	3477.5	3925.9	8.5952	640.3	3476.6	3924.8	8.4391
	1.0 MPa (179.91°C)				1.5 MPa (198.32°C)			
Sat.	194.4	2583.6	2778.1	6.5865	131.8	2594.5	2792.2	6.4448
200	206.0	2621.9	2827.9	6.6940	132.5	2598.1	2796.8	6.4546
240	227.5	2692.9	2920.4	6.8817	148.3	2676.9	2899.3	6.6628
280	248.0	2760.2	3008.2	7.0465	162.7	2748.6	2992.7	6.8381
320	267.8	2826.1	3093.9	7.1962	176.5	2817.1	3081.9	6.9938
360	287.3	2891.6	3178.9	7.3349	189.9	2884.4	3169.2	7.1363
400	306.6	2957.3	3263.9	7.4651	203.0	2951.3	3255.8	7.2690
440	325.7	3023.6	3349.3	7.5883	216.0	3018.5	3342.5	7.3940
500	354.1	3124.4	3478.5	7.7622	235.2	3120.3	3473.1	7.5698
540	372.9	3192.6	3565.6	7.8720	247.8	3189.1	3560.9	7.6805
600	401.1	3296.8	3697.9	8.0290	266.8	3293.9	3694.0	7.8385
640	419.8	3367.4	3787.2	8.1290	279.3	3364.8	3783.8	7.9391
	2.0 MPa (212.42°C)				3.0 MPa (233.90°C)			
Sat.	99.6	2600.3	2799.5	6.3409	66.7	2604.1	2804.2	6.1869
240	108.5	2659.6	2876.5	6.4952	68.2	2619.7	2824.3	6.2265
280	120.0	2736.4	2976.4	6.6828	77.1	2709.9	2941.3	6.4462
320	130.8	2807.9	3069.5	6.8452	85.0	2788.4	3043.4	6.6245
360	141.1	2877.0	3159.3	6.9917	92.3	2861.7	3138.7	6.7801
400	151.2	2945.2	3247.6	7.1271	99.4	2932.8	3230.9	6.9212
440	161.1	3013.4	3335.5	7.2540	106.2	3002.9	3321.5	7.0520
500	175.7	3116.2	3467.6	7.4317	116.2	3108.0	3456.5	7.2338
540	185.3	3185.6	3556.1	7.5434	122.7	3178.4	3546.6	7.3474
600	199.6	3290.9	3690.1	7.7024	132.4	3285.0	3682.3	7.5085
640	209.1	3362.2	3780.4	7.8035	138.8	3357.0	3773.5	7.6106
700	223.2	3470.9	3917.4	7.9487	148.4	3466.5	3911.7	7.7571

TABLE A-9M
(Continued)

Temp. °C	v	u	h	s	v	u	h	s
	4.0 MPa (250.40°C)				6.0 MPa (275.64°C)			
Sat.	49.78	2602.3	2801.4	6.0701	32.44	2589.7	2784.3	5.8892
280	55.46	2680.0	2901.8	6.2568	33.17	2605.2	2804.2	5.9252
320	61.99	2767.4	3015.4	6.4553	38.76	2720.0	2952.6	6.1846
360	67.88	2845.7	3117.2	6.6215	43.31	2811.2	3071.1	6.3782
400	73.41	2919.9	3213.6	6.7690	47.39	2892.9	3177.2	6.5408
440	78.72	2992.2	3307.1	6.9041	51.22	2970.0	3277.3	6.6853
500	86.43	3099.5	3445.3	7.0901	56.65	3082.2	3422.2	6.8803
540	91.45	3171.1	3536.9	7.2056	60.15	3156.1	3517.0	6.9999
600	98.85	3279.1	3674.4	7.3688	65.25	3266.9	3658.4	7.1677
640	103.7	3351.8	3766.6	7.4720	68.59	3341.0	3752.6	7.2731
700	111.0	3462.1	3905.9	7.6198	73.52	3453.1	3894.1	7.4234
740	115.7	3536.6	3999.6	7.7141	76.77	3528.3	3989.2	7.5190
	8.0 MPa (295.06°C)				10.0 MPa (311.06°C)			
Sat.	23.52	2569.8	2758.0	5.7432	18.03	2544.4	2724.7	5.6141
320	26.82	2662.7	2877.2	5.9489	19.25	2588.8	2781.3	5.7103
360	30.89	2772.7	3019.8	6.1819	23.31	2729.1	2962.1	6.0060
400	34.32	2863.8	3138.3	6.3634	26.41	2832.4	3096.5	6.2120
440	37.42	2946.7	3246.1	6.5190	29.11	2922.1	3213.2	6.3805
480	40.34	3025.7	3348.4	6.6586	31.60	3005.4	3321.4	6.5282
520	43.13	3102.7	3447.7	6.7871	33.94	3085.6	3425.1	6.6622
560	45.82	3178.7	3545.3	6.9072	36.19	3164.1	3526.0	6.7864
600	48.45	3254.4	3642.0	7.0206	38.37	3241.7	3625.3	6.9029
640	51.02	3330.1	3738.3	7.1283	40.48	3318.9	3723.7	7.0131
700	54.81	3443.9	3882.4	7.2812	43.58	3434.7	3870.5	7.1687
740	57.29	3520.4	3978.7	7.3782	45.60	3512.1	3968.1	7.2670
	12.0 MPa (324.75°C)				14.0 MPa (336.75°C)			
Sat.	14.26	2513.7	2684.9	5.4924	11.49	2476.8	2637.6	5.3717
360	18.11	2678.4	2895.7	5.8361	14.22	2617.4	2816.5	5.6602
400	21.08	2798.3	3051.3	6.0747	17.22	2760.9	3001.9	5.9448
440	23.55	2896.1	3178.7	6.2586	19.54	2868.6	3142.2	6.1474
480	25.76	2984.4	3293.5	6.4154	21.57	2962.5	3264.5	6.3143
520	27.81	3068.0	3401.8	6.5555	23.43	3049.8	3377.8	6.4610
560	29.77	3149.0	3506.2	6.6840	25.17	3133.6	3486.0	6.5941
600	31.64	3228.7	3608.3	6.8037	26.83	3215.4	3591.1	6.7172
640	33.45	3307.5	3709.0	6.9164	28.43	3296.0	3694.1	6.8326
700	36.10	3425.2	3858.4	7.0749	30.75	3415.7	3846.2	6.9939
740	37.81	3503.7	3957.4	7.1746	32.25	3495.2	3946.7	7.0952

TABLE A-9M

(Continued)

Temp. °C	v	u	h	s	v	u	h	s
	16.0 MPa (347.44°C)				18.0 MPa (357.06°C)			
Sat.	9.31	2431.7	2580.6	5.2455	7.49	2374.3	2509.1	5.1044
360	11.05	2539.0	2715.8	5.4614	8.09	2418.9	2564.5	5.1922
400	14.26	2719.4	2947.6	5.8175	11.90	2672.8	2887.0	5.6887
440	16.52	2839.4	3103.7	6.0429	14.14	2808.2	3062.8	5.9428
480	18.42	2939.7	3234.4	6.2215	15.96	2915.9	3203.2	6.1345
520	20.13	3031.1	3353.3	6.3752	17.57	3011.8	3378.0	6.2960
560	21.72	3117.8	3465.4	6.5132	19.04	3101.7	3444.4	6.4392
600	23.23	3201.8	3573.5	6.6399	20.42	3188.0	3555.6	6.5696
640	24.67	3284.2	3678.9	6.7580	21.74	3272.3	3663.6	6.6905
700	26.74	3406.0	3833.9	6.9224	23.62	3396.3	3821.5	6.8580
740	28.08	3486.7	3935.9	7.0251	24.83	3478.0	3925.0	6.9623
	20.0 MPa (365.81°C)				24.0 MPa			
Sat.	5.83	2293.0	2409.7	4.9269				
400	9.94	2619.3	2818.1	5.5540	6.73	2477.8	2639.4	5.2393
440	12.22	2774.9	3019.4	5.8450	9.29	2700.6	2923.4	5.6506
480	13.99	2891.2	3170.8	6.0518	11.00	2838.3	3102.3	5.8950
520	15.51	2992.0	3302.2	6.2218	12.41	2950.5	3248.5	6.0842
560	16.89	3085.2	3423.0	6.3705	13.66	3051.1	3379.0	6.2448
600	18.18	3174.0	3537.6	6.5048	14.81	3145.2	3500.7	6.3875
640	19.40	3260.2	3648.1	6.6286	15.88	3235.5	3616.7	6.5174
700	21.13	3386.4	3809.0	6.7993	17.39	3366.4	3783.8	6.6947
740	22.24	3469.3	3914.1	6.9052	18.35	3451.7	3892.1	6.8038
800	23.85	3592.7	4069.7	7.0544	19.74	3578.0	4051.6	6.9567
	28.0 MPa				32.0 MPa			
400	3.83	2223.5	2330.7	4.7494	2.36	1980.4	2055.9	4.3239
440	7.12	2613.2	2812.6	5.4494	5.44	2509.0	2683.0	5.2327
480	8.85	2780.8	3028.5	5.7446	7.22	2718.1	2949.2	5.5968
520	10.20	2906.8	3192.3	5.9566	8.53	2860.7	3133.7	5.8357
560	11.36	3015.7	3333.7	6.1307	9.63	2979.0	3287.2	6.0246
600	12.41	3115.6	3463.0	6.2823	10.61	3085.3	3424.6	6.1858
640	13.38	3210.3	3584.8	6.4187	11.50	3184.5	3552.5	6.3290
700	14.73	3346.1	3758.4	6.6029	12.73	3325.4	3732.8	6.5203
740	15.58	3433.9	3870.0	6.7153	13.50	3415.9	3847.8	6.6361
800	16.80	3563.1	4033.4	6.8720	14.60	3548.0	4015.1	6.7966
900	18.73	3774.3	4298.8	7.1084	16.33	3762.7	4285.1	7.0372

TABLE A-10

Properties of water—compressed-liquid table (English units)*

v in ft³/lbm; h and u in Btu/lbm; s in Btu/lbm·°R

Temp., °F	500 psia (T_{sat} = 467.1°F)				1000 psia (T_{sat} = 544.7°F)			
	v	u	h	s	v	u	h	s
32	0.015994	0.00	1.49	0.00000	0.015967	0.03	2.99	0.00005
50	0.015998	18.02	19.50	0.03599	0.015972	17.99	20.94	0.03592
100	0.016106	67.87	69.36	0.12932	0.016082	67.70	70.68	0.12901
150	0.016318	117.66	119.17	0.21457	0.016293	117.38	120.40	0.21410
200	0.016608	167.65	169.19	0.29341	0.016580	167.26	170.32	0.29281
300	0.017416	268.92	270.53	0.43641	0.017379	268.24	271.46	0.43552
400	0.018608	373.68	375.40	0.56604	0.018550	372.55	375.98	0.56472
Sat.	0.019748	447.70	449.53	0.64904	0.021591	538.39	542.38	0.74320

Temp., °F	1500 psia (T_{sat} = 596.4°F)				2000 psia (T_{sat} = 636.0°F)			
	v	u	h	s	v	u	h	s
32	0.015939	0.05	4.47	0.00007	0.015912	0.06	5.95	0.00008
50	0.015946	17.95	22.38	0.03584	0.015920	17.91	23.81	0.03575
100	0.016058	67.53	71.99	0.12870	0.016034	67.37	73.30	0.12839
150	0.016268	117.10	121.62	0.21364	0.016244	116.83	122.84	0.21318
200	0.016554	166.87	171.46	0.29221	0.016527	166.49	172.60	0.29162
300	0.017343	267.58	272.39	0.43463	0.017308	266.93	273.33	0.43376
400	0.018493	371.45	376.59	0.56343	0.018439	370.38	377.21	0.56216
500	0.02024	481.8	487.4	0.6853	0.02014	479.8	487.3	0.6832
Sat.	0.02346	605.0	611.5	0.8082	0.02565	662.4	671.9	0.8623

Temp., °F	3000 psia (T_{sat} = 695.5°F)				4000 psia			
	v	u	h	s	v	u	h	s
32	0.015859	0.09	8.90	0.00009	0.015807	0.10	11.80	0.00005
50	0.015870	17.84	26.65	0.03555	0.015821	17.76	29.47	0.03534
100	0.015987	67.04	75.91	0.12777	0.015942	66.72	78.52	0.12714
150	0.016196	116.30	125.29	0.21226	0.016150	115.77	127.73	0.21136
200	0.016476	165.74	174.89	0.29046	0.016425	165.02	177.18	0.28931
300	0.017240	265.66	275.23	0.43205	0.017174	264.43	277.15	0.43038
400	0.018334	368.32	378.50	0.55970	0.018235	366.35	379.85	0.55734
500	0.019944	476.2	487.3	0.6794	0.019766	472.9	487.5	0.6758
Sat.	0.034310	783.5	802.5	0.9732				

* Abridged from Keenan, J. H., F. G. Keyes, P. G. Hill, and J. G. Moore, "Steam Tables," Wiley, New York, 1969.

TABLE A-10M

Properties of water—compressed-liquid table (SI units)*

v in cm³/g, 1 cm³/g = 10^{-3} m³/kg; h and u in kJ/kg; s in kJ/kg·K

Temp. C	v	u	h	s	v	u	h	s
	2.5 MPa (223.99°C)				5.0 MPa (263.99°C)			
20	1.0006	83.80	86.30	0.2961	0.9995	83.65	88.65	0 2956
40	1.0067	167.25	169.77	0.5715	1.0056	166.95	171.97	0.5705
80	1.0280	334.29	336.86	1.0737	1.0268	333.72	338.85	1.0720
120	1.0590	502.68	505.33	1.5255	1.0576	501.80	507.09	1.5233
160	1.1006	673.90	676.65	1.9404	1.0988	672.62	678.12	1.9375
200	1.1555	849.9	852.8	2.3294	1.1530	848.1	848.1	2.3255
220	1.1898	940.7	943.7	2.5174	1.1866	938.4	944.4	2.5128
Sat.	1.1973	959.1	962.1	2.5546	1.2859	1147.8	1154.2	2.9202
	7.5 MPa (290.59°C)				10.0 MPa (311.06°C)			
20	0.9984	83.50	90.99	0.2950	0.9972	83.36	93.33	0.2945
40	1.0045	166.64	174.18	0.5696	1.0034	166.35	176.38	0.5686
80	1.0256	333.15	340.84	1.0704	1.0245	332.59	342.83	1.0688
100	1.0397	416.81	424.62	1.3011	1.0385	416.12	426.50	1.2992
140	1.0752	585.72	593.78	1.7317	1.0737	584.68	595.42	1.7292
180	1.1219	758.13	766.55	2.1308	1.1199	756.65	767.84	2.1275
220	1.1835	936.2	945.1	2.5083	1.1805	934.1	945.9	2.5039
260	1.2696	1124.4	1134.0	2.8763	1.2645	1121.1	1133.7	2.8699
Sat.	1.3677	1282.0	1292.2	3.1649	1.4524	1393.0	1407.6	3.3596
	15.0 MPa (342.24°C)				20.0 MPa (365.81°C)			
20	0.9950	83.06	97.99	0.2934	0.9928	82.77	102.62	0.2923
40	1.0013	165.76	180.78	0.5666	0.9992	165.17	185.16	0.5646
100	1.0361	414.75	430.28	1.2955	1.0337	413.39	434.06	1.2917
180	1.1159	753.76	770.50	2.1210	1.1120	750.95	773.20	2.1147
220	1.1748	929.9	947.5	2.4953	1.1693	925.9	949.3	2.4870
260	1.2550	1114.6	1133.4	2.8576	1.2462	1108.6	1133.5	2.8459
300	1.3770	1316.6	1337.3	3.2260	1.3596	1306.1	1333.3	3.2071
Sat.	1.6581	1585.6	1610.5	3.6848	2.036	1785.6	1826.3	4.0139
	25.0 MPa				30.0 MPa			
20	0.9907	82.47	107.24	0.2911	0.9886	82.17	111.84	0.2899
40	0.9971	164.60	189.52	0.5626	0.9951	164.04	193.89	0.5607
100	1.0313	412.08	437.85	1.2881	1.0290	410.78	441.66	1.2844
200	1.1344	834.5	862.8	2.2961	1.1302	831.4	865.3	2.2893
300	1.3442	1296.6	1330.2	3.1900	1.3304	1287.9	1327.8	3.1741

* Abridged from Keenan, J. H., F. G. Keyes, P. G. Hill, and J. G. Moore, "Steam Tables," Wiley, New York, 1969.

TABLE A-11
Properties of water—compressed-liquid deviations from saturated liquid (English units)*

Abs. press. lbf/in² (sat. temp.)	Saturated liquid		Temperature, °F				
			32	100	200	300	400	500
		P	0.08854	0.9492	11.526	67.013	247.31	680.8
		v_f	0.016022	0.016132	0.016634	0.017449	0.018639	0.020432
		h_f	0	67.97	167.99	269.59	374.97	487.82
		s_f	0	0.12948	0.29382	0.43694	0.56638	0.68871
200 (381.79)	$(v - v_f) \times 10^5$ $h - h_f$ $(s - s_f) \times 10^3$		-1.2 +0.57 0	-1.2 +0.53 -0.08	-0.4 +0.50 -0.20	-0.9 +0.35 -0.20		
400 (444.59)	$(v - v_f) \times 10^5$ $h - h_f$ $(s - s_f) \times 10^3$		-2.2 +1.17 0	-2.2 +1.06 -0.18	-1.4 +0.96 -0.32	-1.9 +0.72 -0.34	-1.9 +0.28 -0.08	
800 (518.23)	$(v - v_f) \times 10^5$ $h - h_f$ $(s - s_f) \times 10^3$		-4.2 +2.38 +0.1	-4.2 +2.12 -0.48	-4.4 +1.87 -0.82	-4.9 +1.46 -1.14	-6.9 +0.74 -1.18	-2.2 +0.04 -0.21
1000 (544.61)	$(v - v_f) \times 10^5$ $h - h_f$ $(s - s_f) \times 10^3$		-5.2 +2.98 +0.1	-5.2 +2.64 -0.58	-5.4 +2.32 -1.02	-6.9 +1.83 -1.44	-8.9 +0.97 -1.68	-7.2 -0.10 -1.11
1500 (586.23)	$(v - v_f) \times 10^5$ $h - h_f$ $(s - s_f) \times 10^3$		-8.2 +4.48 +0.1	-7.2 +3.96 -0.88	-8.4 +3.46 -1.62	-9.9 +2.77 -2.34	-14.9 +1.57 -2.98	-18.2 -0.19 -3.21

* Differential values calculated from tabulated data for compressed water in "Thermodynamics and Transport Properties of Steam," American Society of Mechanical Engineers, New York, 1967.

TABLE A-12

Properties of saturated ammonia—temperature table (English units)*

Temp. t, °F	Press. p, psia	Specific volume ft³/lbm Sat. liquid v_f	Sat. vapor v_g	Enthalpy, Btu/lbm Sat. liquid h_f	Evap. h_{fg}	Sat. vapor h_g	Entropy, Btu/lbm·°R Sat. liquid s_f	Evap. s_{fg}	Sat. vapor s_g
−60	5.55	0.02278	44.73	−21.2	610.8	589.6	−0.0517	1.5286	1.4769
−50	7.67	0.02299	33.08	−10.6	604.3	593.7	−0.0256	1.4753	1.4497
−40	10.41	0.02322	24.86	0.0	597.6	597.6	0.0000	1.4242	1.4242
−30	13.90	0.02345	18.97	10.7	590.7	601.4	0.0250	1.3751	1.4001
−20	18.30	0.02369	14.68	21.4	583.6	605.0	0.0497	1.3277	1.3774
−10	23.74	0.02393	11.50	32.1	576.4	608.5	0.0738	1.2820	1.3558
0	30.42	0.02419	9.116	42.9	568.9	611.8	0.0975	2.2377	1.3352
5	34.27	0.02432	8.150	48.3	565.0	613.3	0.1092	1.2161	1.3253
10	38.51	0.02446	7.304	53.8	561.1	614.9	0.1208	1.1949	1.3157
20	48.21	0.02474	5.910	64.7	553.1	617.8	0.1437	1.1532	1.2969
30	59.74	0.02503	4.825	75.7	544.8	620.5	0.1663	1.1127	1.2790
40	73.32	0.02533	3.971	86.8	536.2	623.0	0.1885	1.0733	1.2618
50	89.19	0.02564	3.294	97.9	527.3	625.2	0.2105	1.0348	1.2453
60	107.6	0.02597	2.751	109.2	518.1	627.3	0.2322	0.9972	1.2294
70	128.8	0.02632	2.312	120.5	508.6	629.1	0.2537	0.9603	1.2140
80	153.0	0.02668	1.955	132.0	498.7	630.7	0.2749	0.9242	1.1991
86	169.2	0.02691	1.772	138.9	492.6	631.5	0.2875	0.9029	1.1904
90	180.6	0.02707	1.661	143.5	488.5	632.0	0.2958	0.8888	1.1846
100	211.9	0.02747	1.419	155.2	477.8	633.0	0.3166	0.8539	1.1705
110	247.0	0.02790	1.217	167.0	466.7	633.7	0.3372	0.8194	1.1566
120	286.4	0.02836	1.047	179.0	455.0	634.0	0.3576	0.7851	1.1427

* Data from Tables of Thermodynamic Properties of Ammonia, *Natl. Bur. Std. Circ.* 142, 1945.

TABLE A-15
Properties of Freon 12 (dichlorodifluoromethane)—saturation-temperature table (English units)

Temp. t, °F	Abs. press. P, psia	Specific volume, ft³/lbm Sat. liquid v_f	Evap. v_{fg}	Sat. vapor v_g	Enthalpy, Btu/lbm Sat. liquid h_f	Evap. h_{fg}	Sat. vapor h_g	Entropy, Btu/lbm·°R Sat. liquid s_f	Evap. s_{fg}	Sat. vapor s_g
-130	0.41224	0.009736	70.7203	70.730	-18.609	81.577	62.968	-0.04983	0.24743	0.19760
-120	0.64190	0.009816	46.7312	46.741	-16.565	80.617	64.052	-0.04372	0.23731	0.19359
-110	0.97034	0.009899	31.7671	31.777	-14.518	79.663	65.145	-0.03779	0.22780	0.19002
-100	1.4280	0.009985	21.1541	21.164	-12.466	78.714	66.248	-0.03200	0.21883	0.18683
-90	2.0509	0.010073	15.8109	15.821	-10.409	77.764	67.355	-0.02637	0.21034	0.18398
-80	2.8807	0.010164	11.5228	11.533	-8.3451	76.812	68.467	-0.02086	0.20229	0.18143
-70	3.9651	0.010259	8.5584	8.5687	-6.2730	75.853	69.580	-0.01548	0.19464	0.17916
-60	5.3575	0.010357	6.4670	6.4774	-4.1919	74.885	70.693	-0.01021	0.18716	0.17714
-50	7.1168	0.010459	4.9637	4.9742	-2.1011	73.906	71.805	-0.00506	0.18038	0.17533
-40	9.3076	0.010564	3.8644	3.8750	0	72.913	72.913	0	0.17373	0.17373
-30	11.999	0.010674	3.0478	3.0585	2.1120	71.903	74.015	0.00496	0.16733	0.17229
-20	15.267	0.010788	2.4321	2.4429	4.2357	70.874	75.110	0.00983	0.16119	0.17102
-10	19.189	0.010906	1.9628	1.9727	6.3716	69.824	76.196	0.01462	0.15527	0.16989
0	23.849	0.011030	1.5979	1.6089	8.5207	68.750	77.271	0.01932	0.14956	0.16888
10	29.335	0.011160	1.3129	1.3241	10.684	67.651	78.335	0.02395	0.14403	0.16798
20	35.736	0.011296	1.0875	1.0988	12.863	66.522	79.385	0.02852	0.13867	0.16719
30	43.148	0.011438	0.90736	0.91880	15.058	65.361	80.419	0.03301	0.13347	0.16648
40	51.667	0.011588	0.76198	0.77357	17.273	64.163	81.436	0.03745	0.12841	0.16586
50	61.394	0.011746	0.64362	0.65537	19.507	62.926	82.433	0.04184	0.12346	0.16530
60	72.433	0.011913	0.54648	0.55839	21.766	61.643	83.409	0.04618	0.11861	0.16479
70	84.888	0.012089	0.46609	0.47818	24.050	60.309	84.359	0.05048	0.11386	0.16434
80	98.870	0.012277	0.39907	0.41135	26.365	58.917	85.282	0.05475	0.10917	0.16392
90	114.49	0.012478	0.34281	0.35529	28.713	57.461	86.174	0.05900	0.10453	0.16353
100	131.86	0.012693	0.29525	0.30794	31.100	55.929	87.029	0.06323	0.09992	0.16315

TABLE A-13
Properties of saturated ammonia—pressure table (English units)*

Press. p, psia	Temp. t, °F	Specific volume ft³/lbm Sat. liquid v_f	Sat. vapor v_g	Enthalpy, Btu/lbm Sat. liquid h_f	Evap. h_{fg}	Sat. vapor h_g	Entropy, Btu/lbm·°R Sat. liquid s_f	Evap. s_{fg}	Sat. vapor s_g
5	-63.11	0.02271	49.31	-24.5	612.8	588.3	-0.0599	1.5456	1.4857
10	-41.34	0.02319	25.81	-1.4	598.5	597.1	-0.0034	1.4310	1.4276
15	-27.29	0.02351	17.67	13.6	588.8	602.4	0.0318	1.3620	1.3938
20	-16.64	0.02377	13.50	25.0	581.2	606.2	0.0578	1.3122	1.3700
30	0.57	0.02417	9.236	42.3	569.3	611.6	0.0962	1.2402	1.3364
40	11.66	0.02451	7.047	55.6	559.8	615.4	0.1246	1.1879	1.3125
50	21.67	0.02479	5.710	66.5	551.7	618.2	0.1475	1.1464	1.2939
60	30.21	0.02504	4.805	75.9	544.6	620.5	0.1668	1.1119	1.2787
80	44.40	0.02546	3.655	91.7	532.3	624.0	0.1982	1.0563	1.2545
100	56.05	0.02584	2.952	104.7	521.8	626.5	0.2237	1.0119	1.2356
120	66.02	0.02618	2.476	116.0	512.4	628.4	0.2452	0.9749	1.2201
140	74.79	0.02649	2.132	126.0	503.9	629.9	0.2638	0.9430	1.2068
170	86.29	0.02692	1.764	139.3	492.3	631.6	0.2881	0.9019	1.1900
200	96.34	0.02732	1.502	150.9	481.8	632.7	0.3090	0.8666	1.1756
230	105.30	0.02770	1.307	161.4	472.0	633.4	0.3275	0.8356	1.1631
260	113.42	0.02806	1.155	171.1	462.8	633.9	0.3441	0.8077	1.1518

* Data from Tables of Thermodynamic Properties of Ammonia, *Natl. Bur. Std. Circ.* 142, 1945.

TABLE A-14

Properties of superheated ammonia (English units)*

Abs. press. psia (sat. temp.)		0	20	40	60	80	100	120	140	160	180	200	220
						Temperature, °F							
10 (−41.34)	v	28.58	29.90	31.20	32.49	33.78	35.07	36.35	37.62	38.90	40.17	41.45	
	h	618.9	629.1	639.3	649.5	659.7	670.0	680.3	690.6	701.1	711.6	722.2	
	s	1.477	1.499	1.520	1.540	1.559	1.578	1.596	1.614	1.631	1.647	1.664	
15 (−27.29)	v	18.92	19.82	20.70	21.58	22.44	23.31	24.17	25.03	25.88	26.74	27.59	
	h	617.2	627.8	638.2	648.5	658.9	669.2	679.6	690.0	700.5	711.1	721.7	
	s	1.427	1.450	1.471	1.491	1.511	1.529	1.548	1.566	1.583	1.599	1.616	
20 (−16.64)	v	14.09	14.78	15.45	16.12	16.78	17.43	18.08	18.73	19.37	20.02	20.66	21.3
	h	615.5	626.4	637.0	647.5	658.0	668.5	678.9	689.4	700.0	710.6	721.2	732.0
	s	1.391	1.414	1.436	1.456	1.476	1.495	1.513	1.531	1.549	1.565	1.582	1.598
25 (−7.96)	v	11.19	11.75	12.30	12.84	13.37	13.90	14.43	14.95	15.47	15.99	16.50	17.02
	h	613.8	625.0	635.8	646.5	657.1	667.7	678.2	688.8	699.4	710.1	720.8	731.6
	s	1.362	1.386	1.408	1.429	1.449	1.468	1.486	1.504	1.522	1.539	1.555	1.571
30 (−0.57)	v	9.25	9.731	10.20	10.65	11.10	11.55	11.99	12.43	12.87	13.30	13.73	14.16
	h	611.9	623.5	634.6	645.5	656.2	666.9	677.5	688.2	698.8	709.6	720.3	731.1
	s	1.337	1.362	1.385	1.406	1.426	1.446	1.464	1.482	1.500	1.517	1.533	1.550
35 (5.89)	v		8.287	8.695	9.093	9.484	9.869	10.25	10.63	11.00	11.38	11.75	12.12
	h		622.0	633.4	644.4	655.3	666.1	676.8	687.6	698.1	709.1	719.9	730.7
	s		1.341	1.365	1.386	1.407	1.427	1.445	1.464	1.481	1.498	1.515	1.531
40 (11.66)	v		7.203	7.568	7.922	8.268	8.609	8.945	9.278	9.609	9.938	10.27	10.59
	h		620.4	632.1	643.4	654.4	665.3	676.1	686.9	697.7	708.5	719.4	730.3
	s		1.323	1.347	1.369	1.390	1.410	1.429	1.447	1.465	1.482	1.499	1.515
45 (17.87)	v		6.213	6.538	6.851	7.157	7.457	7.753	8.045	8.335	8.623	8.909	9.194
	h		618.5	630.5	642.1	653.3	664.4	675.3	686.2	697.1	707.9	718.8	729.8
	s		1.304	1.328	1.351	1.372	1.392	1.411	1.430	1.448	1.465	1.482	1.498
50 (21.67)	v			5.988	6.280	6.564	6.843	7.117	7.387	7.655	7.921	8.185	8.448
	h			629.5	641.2	652.6	663.7	674.7	685.7	696.6	707.5	718.5	729.4
	s			1.317	1.340	1.361	1.382	1.401	1.420	1.437	1.455	1.472	1.488
60 (30.21)	v			4.933	5.184	5.428	5.665	5.897	6.126	6.352	6.576	6.798	7.019
	h			626.8	639.0	650.7	662.1	673.3	684.4	695.5	706.5	717.5	738.6
	s			1.2913	1.3152	1.3373	1.3581	1.3778	1.3966	1.4148	1.4323	1.4493	1.4658

		60	80	100	120	140	160	180	200	240	280	320	360
							Temperature, °F						
70 (37.7)	v	4.401	4.615	4.822	5.025	5.224	5.420	5.615	5.807	6.187	6.563		
	h	636.6	648.7	660.4	671.8	683.1	694.3	705.5	716.6	738.9	761.4		
	s	1.294	1.317	1.338	1.358	1.377	1.395	1.413	1.430	1.463	1.494		
80 (44.4)	v	3.812	4.005	4.190	4.371	4.548	4.722	4.893	5.063	5.398	5.73		
	h	634.3	646.7	658.7	670.4	681.8	693.2	704.4	715.6	738.1	760.7		
	s	1.275	1.298	1.320	1.340	1.360	1.378	1.396	1.414	1.447	1.478		
90 (50.47)	v	3.353	3.529	3.698	3.862	4.021	4.178	4.332	4.484	4.785	5.081		
	h	631.8	644.7	657.0	668.9	680.5	692.0	703.4	714.7	737.3	760.0		
	s	1.257	1.281	1.304	1.325	1.344	1.363	1.381	1.400	1.432	1.464		
100 (56.05)	v	2.985	3.149	3.304	3.454	3.600	3.743	3.883	4.021	4.294	4.562		
	h	629.3	642.6	655.2	667.3	679.2	690.8	702.3	713.7	736.5	759.4		
	s	1.241	1.266	1.289	1.310	1.331	1.349	1.368	1.385	1.419	1.451		
140 (74.79)	v		2.166	2.288	2.404	2.515	2.622	2.727	2.830	3.030	3.227	3.420	
	h		633.8	647.8	661.1	673.7	686.0	698.0	709.9	733.3	756.7	780.0	
	s		1.214	1.240	1.263	1.284	1.305	1.324	1.342	1.376	1.409	1.440	
180 (89.78)	v			1.720	1.818	1.910	1.999	2.084	2.167	2.328	2.484	2.637	
	h			639.9	654.4	668.0	681.0	693.6	705.9	730.1	753.9	777.7	
	s			1.199	1.225	1.248	1.269	1.289	1.308	1.344	1.377	1.408	
220 (102.42)	v				1.443	1.525	1.601	1.675	1.745	1.881	2.012	2.140	2.265
	h				647.3	662.0	675.8	689.1	701.9	726.8	751.1	775.3	799.5
	s				1.192	1.217	1.239	1.260	1.280	1.317	1.351	1.383	1.413
240 (108.09)	v				1.302	1.380	1.452	1.521	1.587	1.714	1.835	1.954	2.069
	h				643.5	658.8	673.1	686.7	699.8	725.1	749.8	774.1	798.4
	s				1.176	1.203	1.226	1.248	1.268	1.305	1.339	1.371	1.402
260 (113.42)	v				1.182	1.257	1.326	1.391	1.453	1.572	1.686	1.796	1.904
	h				639.5	655.6	670.4	684.4	697.7	723.4	748.4	772.9	797.4
	s				1.162	1.189	1.213	1.235	1.256	1.294	1.329	1.361	1.391
280 (118.45)	v				1.078	1.151	1.217	1.279	1.339	1.451	1.558	1.661	1.762
	h				635.4	652.2	667.6	681.9	695.6	721.8	747.0	771.7	796.3
	s				1.147	1.176	1.201	1.224	1.245	1.283	1.318	1.351	1.382

* Data from Tables of Thermodynamic Properties of Ammonia. *Natl. Bur. Std. Circ.* 142, 1945.

110	151.11	0.012924	0.25577	0.26769	33.531	54.313	87.844	0.06745	0.09534	0.16279
120	172.35	0.013174	0.22019	0.23326	36.013	52.597	88.610	0.07168	0.09073	0.16241
130	195.71	0.013447	0.19019	0.20364	38.553	50.768	89.321	0.07583	0.08609	0.16202
140	221.32	0.013746	0.16424	0.17799	41.162	48.805	89.967	0.08021	0.08138	0.16159
150	249.31	0.014078	0.14156	0.15564	43.850	46.684	90.534	0.08453	0.07657	0.16110
160	279.82	0.014449	0.12159	0.13604	46.633	44.373	91.006	0.08893	0.07260	0.16053
170	313.00	0.014871	0.10386	0.11873	49.529	41.830	91.359	0.09342	0.06643	0.15985
180	349.00	0.015360	0.08794	0.10330	52.562	38.999	91.561	0.09804	0.06096	0.15900
190	387.98	0.015942	0.073476	0.089418	55.769	35.792	91.561	0.10284	0.05511	0.15793
200	430.09	0.016659	0.060069	0.076728	59.203	32.075	91.278	0.10789	0.04862	0.15651
210	475.52	0.017601	0.047242	0.064843	62.959	27.599	90.558	0.11332	0.03921	0.15453
220	524.43	0.018986	0.035154	0.053140	67.246	21.790	89.036	0.11943	0.03206	0.15149
230	577.03	0.021854	0.017581	0.039435	72.893	12.229	85.122	0.12739	0.01773	0.14512
233.6 (critical)	596.9	0.02870	0	0.02870	78.86	0	78.86	0.1359	0	0.1359

* Copyright 1955 and 1956, E. I. du Pont de Nemours & Company, Inc.

TABLE A-16

Properties of superheated Freon 12 (English units)*

Temp., °F	v	h	s	v	h	s	v	h	s
	5 psia			10 psia			15 psia		
0	8.0611	78.582	0.19663	3.9809	78.246	0.18471	2.6201	77.902	0.17751
20	8.4265	81.309	0.20244	4.1691	81.014	0.19061	2.7494	80.712	0.18349
40	8.7903	84.090	0.20812	4.3556	83.828	0.19635	2.8770	83.561	0.18931
60	9.1528	86.922	0.21367	4.5408	86.689	0.20197	3.0031	86.451	0.19498
80	9.5142	89.806	0.21912	4.7248	89.596	0.20746	3.1281	89.383	0.20051
100	9.8747	92.738	0.22445	4.9079	92.548	0.21283	3.2521	92.357	0.20593
120	10.234	95.717	0.22968	5.0903	95.546	0.21809	3.3754	95.373	0.21122
140	10.594	98.743	0.23481	5.2720	98.586	0.22325	3.4981	98.429	0.21640
160	10.952	101.812	0.23985	5.4533	101.669	0.22830	3.6202	101.525	0.22148
180	11.311	104.925	0.24479	5.6341	104.793	0.23326	3.7419	104.661	0.22646
200	11.668	108.079	0.24964	5.8145	107.957	0.23813	3.8632	107.835	0.23135
220	12.026	111.272	0.25441	5.9946	111.159	0.24291	3.9841	111.046	0.23614
	20 psia			25 psia			30 psia		
20	2.0391	80.403	0.17829	1.6125	80.088	0.17414	1.3278	79.765	0.17065
40	2.1373	83.289	0.18419	1.6932	83.012	0.18012	1.3969	82.730	0.17671
60	2.2340	86.210	0.18992	1.7723	85.965	0.18591	1.4644	85.716	0.18257
80	2.3295	89.168	0.19550	1.8502	88.950	0.19155	1.5306	88.729	0.18826
100	2.4241	92.164	0.20095	1.9271	91.968	0.19704	1.5957	91.770	0.19379
120	2.5179	95.198	0.20628	2.0032	95.021	0.20240	1.6600	94.843	0.19918
140	2.6110	98.270	0.21149	2.0786	98.110	0.20763	1.7237	97.948	0.20445
160	2.7036	101.380	0.21659	2.1535	101.234	0.21276	1.7868	101.086	0.20960
180	2.7957	104.528	0.22159	2.2279	104.393	0.21778	1.8494	104.258	0.21463
200	2.8874	107.712	0.22649	2.3019	107.588	0.22269	1.9116	107.464	0.21957
220	2.9789	110.932	0.23130	2.3756	110.817	0.22752	1.9735	110.702	0.22440
240	3.0700	114.186	0.23602	2.4491	114.080	0.23225	2.0351	113.973	0.22915
	35 psia			40 psia			50 psia		
40	1.1850	82.442	0.17375	1.0258	82.148	0.17112	0.80248	81.540	0.16655
60	1.2442	85.463	0.17968	1.0789	85.206	0.17712	0.84713	84.676	0.17271
80	1.3021	88.504	0.18542	1.1306	88.277	0.18292	0.89025	87.811	0.17862
100	1.3589	91.570	0.19100	1.1812	91.367	0.18854	0.93216	90.953	0.18434
120	1.4148	94.663	0.19643	1.2309	94.480	0.19401	0.97313	94.110	0.18988
140	1.4701	97.785	0.20172	1.2798	97.620	0.19933	1.0133	97.286	0.19527
160	1.5248	100.938	0.20689	1.3282	100.788	0.20453	1.0529	100.485	0.20051
180	1.5789	104.122	0.21195	1.3761	103.985	0.20961	1.0920	103.708	0.20563
200	1.6327	107.338	0.21690	1.4236	107.212	0.21457	1.1307	106.958	0.21064
220	1.6862	110.586	0.22175	1.4707	110.469	0.21944	1.1690	110.235	0.21553
240	1.7394	113.865	0.22651	1.5176	113.757	0.22420	1.2070	113.539	0.22032
260	1.7923	117.175	0.23117	1.5642	117.074	0.22888	1.2447	116.871	0.22502

* Copyright 1955 and 1956, E. I. du Pont de Nemours & Company, Inc.

TABLE A-16
(Continued)

Temp., °F	v	h	s	v	h	s	v	h	s
	60 psia			70 psia			80 psia		
60	0.69210	84.126	0.16892	0.58088	83.552	0.16556			
80	0.72964	87.330	0.17497	0.61458	86.832	0.17175	0.52795	86.316	0.16885
100	0.76588	90.528	0.18079	0.64685	90.091	0.17768	0.55734	89.640	0.17489
120	0.80110	93.731	0.18641	0.67803	93.343	0.18339	0.58556	92.945	0.18070
140	0.83551	96.945	0.19186	0.70836	96.597	0.18891	0.61286	96.242	0.18629
160	0.86928	100.776	0.19716	0.73800	99.862	0.19427	0.63943	99.542	0.19170
180	0.90252	103.427	0.20233	0.76708	103.141	0.19948	0.66543	102.851	0.19696
200	0.93531	106.700	0.20736	0.79571	106.439	0.20455	0.69095	106.174	0.20207
220	0.96775	109.997	0.21229	0.82397	109.756	0.20951	0.71609	109.513	0.20706
240	0.99988	113.319	0.21710	0.85191	113.096	0.21435	0.74090	112.872	0.21193
260	1.0318	116.666	0.22182	0.87959	116.459	0.21909	0.76544	116.251	0.21669
280	1.0634	120.039	0.22644	0.90705	119.846	0.22373	0.78975	119.652	0.22135
	90 psia			100 psia			125 psia		
100	0.48749	89.175	0.17234	0.43138	88.694	0.16996	0.32943	87.407	0.16455
120	0.51346	92.536	0.17824	0.45562	92.116	0.17597	0.35086	91.008	0.17087
140	0.53845	95.879	0.18391	0.47881	95.507	0.18172	0.37098	94.537	0.17686
160	0.56268	99.216	0.18938	0.50118	98.884	0.18726	0.39015	98.023	0.18258
180	0.58629	102.557	0.19469	0.52291	102.257	0.19262	0.40857	101.484	0.18807
200	0.60941	105.905	0.19984	0.54413	105.633	0.19782	0.42642	104.934	0.19338
220	0.63213	109.267	0.20486	0.56492	109.018	0.20287	0.44380	108.380	0.19853
240	0.65451	112.644	0.20976	0.58538	112.415	0.20780	0.46081	111.829	0.20353
260	0.67662	116.040	0.21455	0.60554	115.828	0.21261	0.47750	115.287	0.20840
280	0.69849	119.456	0.21923	0.62546	119.258	0.21731	0.49394	118.756	0.21316
300	0.72016	122.892	0.22381	0.64518	122.707	0.22191	0.51016	122.238	0.21780
320	0.74166	126.349	0.22830	0.66472	126.176	0.22641	0.52619	125.737	0.22235
	150 psia			175 psia			200 psia		
120	0.28007	89.800	0.16629						
140	0.29845	93.498	0.17256	0.24595	92.373	0.16859	0.20579	91.137	0.16480
160	0.31566	97.112	0.17849	0.26198	96.142	0.17478	0.22121	95.100	0.17130
180	0.33200	100.675	0.18415	0.27697	99.823	0.18062	0.23535	98.921	0.17737
200	0.34769	104.206	0.18958	0.29120	103.447	0.18620	0.24860	102.652	0.18311
220	0.36285	107.720	0.19483	0.30485	107.036	0.19156	0.26117	106.325	0.18860
240	0.37761	111.226	0.19992	0.31804	110.605	0.19674	0.27323	109.962	0.19387
260	0.39203	114.732	0.20485	0.33087	114.162	0.20175	0.28489	113.576	0.19896
280	0.40617	118.242	0.20967	0.34339	117.717	0.20662	0.29623	117.178	0.20390
300	0.42008	121.761	0.21436	0.35567	121.273	0.21137	0.30730	120.775	0.20870
320	0.43379	125.290	0.21894	0.36773	124.835	0.21599	0.31815	124.373	0.21337
340	0.44733	128.833	0.22343	0.37963	128.407	0.22052	0.32881	127.974	0.21793
	250 psia			300 psia			400 psia		
160	0.16249	92.717	0.16462						
180	0.17605	96.925	0.17130	0.13482	94.556	0.16537			
200	0.18824	100.930	0.17747	0.14697	98.975	0.17217	0.091005	93.718	0.16092

TABLE A-16

(Continued)

	250 psia			300 psia			400 psia		
220	0.19952	104.809	0.18326	0.15774	103.136	0.17838	0.10316	99.046	0.16888
240	0.21014	108.607	0.18877	0.16761	107.140	0.18419	0.11300	103.735	0.17568
260	0.22027	112.351	0.19404	0.17685	111.043	0.18969	0.12163	108.105	0.18183
280	0.23001	116.060	0.19913	0.18562	114.879	0.19495	0.12949	112.286	0.18756
300	0.23944	119.747	0.20405	0.19402	118.670	0.20000	0.13680	116.343	0.19298
320	0.24862	123.420	0.20882	0.20214	122.430	0.20489	0.14372	120.318	0.19814
340	0.25759	127.088	0.21346	0.21002	126.171	0.20963	0.15032	124.235	0.20310
360	0.26639	130.754	0.21799	0.21770	129.900	0.21423	0.15668	128.112	0.20789
380	0.27504	134.423	0.22241	0.22522	133.624	0.21872	0.16285	131.961	0.21253

	500 psia			600 psia		
220	0.064207	92.397	0.15683	———	———	———
240	0.077620	99.218	0.16672	0.047488	91.024	0.15335
260	0.087054	104.526	0.17421	0.061922	99.741	0.16566
280	0.094923	109.277	0.18072	0.070859	105.637	0.17374
300	0.10190	113.729	0.18666	0.078059	110.729	0.18053
320	0.10829	117.997	0.19221	0.084333	115.420	0.18663
340	0.11426	122.143	0.19746	0.090017	119.871	0.19227
360	0.11992	126.205	0.20247	0.095289	124.167	0.19757
380	0.12533	130.207	0.20730	0.10025	128.355	0.20262
400	0.13054	134.166	0.21196	0.10498	132.466	0.20746
420	0.13559	138.096	0.21648	0.10952	136.523	0.21213
440	0.14051	142.004	0.22087	0.11391	140.539	0.21664

TABLE A-17
Thermodynamic properties of air at low pressure (English units)*

Reference level: 0°R and 1 atm
Absolute entropies may be calculated from

$$s° = \phi - R \ln (p/p_0) + 1.0 \text{ Btu/lbm·°R}$$

where p_0 is the reference pressure of 1 atm

Temp., °R	h Btu/lbm	P_r	u Btu/lbm	v_r	ϕ Btu/lbm·°R
200	47.67	0.04320	33.96	1714.9	0.36303
220	52.46	0.06026	37.38	1352.5	0.38584
240	57.25	0.08165	40.80	1088.8	0.40666
260	62.03	0.10797	44.21	892.0	0.42582
280	66.82	0.13986	47.63	741.6	0.44356
300	71.61	0.17795	51.04	624.5	0.46007
320	76.40	0.22290	54.46	531.8	0.47550
340	81.18	0.27545	57.87	457.2	0.49002
360	85.97	0.3363	61.29	396.6	0.50369
380	90.75	0.4061	64.70	346.6	0.51663
400	95.53	0.4858	68.11	305.0	0.52890
420	100.32	0.5760	71.52	270.1	0.54058
440	105.11	0.6776	74.93	240.6	0.55172
460	109.90	0.7913	78.36	215.33	0.56235
480	114.69	0.9182	81.77	193.65	0.57255
500	119.48	1.0590	85.20	174.90	0.58233
520	124.27	1.2147	88.62	158.58	0.59173
540	129.06	1.3860	92.04	144.32	0.60078
560	133.86	1.5742	95.47	131.78	0.60950
580	138.66	1.7800	98.90	120.70	0.61793
600	143.47	2.005	102.34	110.88	0.62607
620	148.28	2.249	105.78	102.12	0.63395
640	153.09	2.514	109.21	94.30	0.64159
660	157.92	2.801	112.67	87.27	0.64902
680	162.73	3.111	116.12	80.96	0.65621
700	167.56	3.446	119.58	75.25	0.66321
720	172.39	3.806	123.04	70.07	0.67002
740	177.23	4.193	126.51	65.38	0.67665
760	182.08	4.607	129.99	61.10	0.68312
780	186.94	5.051	133.47	57.20	0.68942
800	191.81	5.526	136.97	53.63	0.69558
820	196.69	6.033	140.47	50.35	0.70160
840	201.56	6.573	143.98	47.34	0.70747
860	206.46	7.149	147.50	44.57	0.71323
880	211.35	7.761	151.02	42.01	0.71886
900	216.26	8.411	154.57	39.64	0.72438
920	221.18	9.102	158.12	37.44	0.72979
940	226.11	9.834	161.68	35.41	0.73509
960	231.06	10.610	165.26	33.52	0.74030
980	236.02	11.430	168.83	31.76	0.74540
1000	240.98	12.298	172.43	30.12	0.75042
1020	245.97	13.215	176.04	28.59	0.75536
1040	250.95	14.182	179.66	27.17	0.76019
1060	255.96	15.203	183.29	25.82	0.76496
1080	260.97	16.278	186.93	24.58	0.76964
1100	265.99	17.413	190.58	23.40	0.77426
1120	271.03	18.604	194.25	22.30	0.77880

* Abridged from J. H. Keenan and J. Kaye, "Gas Tables," John Wiley & Sons, Inc., New York, 1948, by permission.

TABLE A-17
(Continued)

Temp., °R	h Btu/lbm	P_r	u Btu/lbm	v_r	ϕ Btu/lbm·°R
1140	276.08	19.858	197.94	21.27	0.78326
1160	281.14	21.18	201.63	20.293	0.78767
1180	286.21	22.56	205.33	19.377	0.79201
1200	291.30	24.01	209.05	18.514	0.79628
1220	296.41	25.53	212.78	17.700	0.80050
1240	301.52	27.13	216.53	16.932	0.80466
1260	306.65	28.80	220.28	16.205	0.80876
1280	311.79	30.55	224.05	15.518	0.81280
1300	316.94	32.39	227.83	14.868	0.81680
1320	322.11	34.31	231.63	14.253	0.82075
1340	327.29	36.31	235.43	13.670	0.82464
1360	332.48	38.41	239.25	13.118	0.82848
1380	337.68	40.59	243.08	12.593	0.83229
1400	342.90	42.88	246.93	12.095	0.83604
1420	348.14	45.26	250.79	11.622	0.83975
1440	353.37	47.75	254.66	11.172	0.84341
1460	358.63	50.34	258.54	10.743	0.84704
1480	363.89	53.04	262.44	10.336	0.85062
1500	369.17	55.86	266.34	9.948	0.85416
1520	374.47	58.78	270.26	9.578	0.85767
1540	379.77	61.83	274.20	9.226	0.86113
1560	385.08	65.00	278.13	8.890	0.86456
1580	390.40	68.30	282.09	8.569	0.86794
1600	395.74	71.73	286.06	8.263	0.87130
1620	401.09	75.29	290.04	7.971	0.87462
1640	406.45	78.99	294.03	7.691	0.87791
1660	411.82	82.83	298.02	7.424	0.88116
1680	417.20	86.82	302.04	7.168	0.88439
1700	422.59	90.95	306.06	6.924	0.88758
1720	428.00	95.24	310.09	6.690	0.89074
1740	433.41	99.69	314.13	6.465	0.89387
1760	438.83	104.30	318.18	6.251	0.89697
1780	444.26	109.08	322.24	6.045	0.90003
1800	449.71	114.03	326.32	5.847	0.90308
1820	455.17	119.16	330.40	5.658	0.90609
1840	460.63	124.47	334.50	5.476	0.90908
1860	466.12	129.95	338.61	5.302	0.91203
1880	471.60	135.64	342.73	5.134	0.91497
1900	477.09	141.51	346.85	4.974	0.91788
1920	482.60	147.59	350.98	4.819	0.92076
1940	488.12	153.87	355.12	4.670	0.92362
1960	493.64	160.37	359.28	4.527	0.92645
1980	499.17	167.07	363.43	4.390	0.92926
2000	504.71	174.00	367.61	4.258	0.93205
2020	510.26	181.16	371.79	4.130	0.93481
2040	515.82	188.54	375.98	4.008	0.93756
2060	521.39	196.16	380.18	3.890	0.94026
2080	526.97	204.02	384.39	3.777	0.94296
2100	532.55	212.1	388.60	3.667	0.94564

TABLE A-17

(Continued)

Temp., °R	h Btu/lbm	P_r	u Btu/lbm	v_r	ϕ Btu/lbm·°R
2120	538.15	220.5	392.83	3.561	0.94829
2140	543.74	229.1	397.05	3.460	0.95092
2160	549.35	238.0	401.29	3.362	0.95352
2180	554.97	247.2	405.53	3.267	0.95611
2200	560.59	256.6	409.78	3.176	0.95868
2220	566.23	266.3	414.05	3.088	0.96123
2240	571.86	276.3	418.31	3.003	0.96376
2260	577.51	286.6	422.59	2.921	0.96626
2280	583.16	297.2	426.87	2.841	0.96876
2300	588.82	308.1	431.16	2.765	0.97123
2320	594.49	319.4	435.46	2.691	0.97369
2340	600.16	330.9	439.76	2.619	0.97611
2360	605.84	342.8	444.07	2.550	0.97853
2380	611.53	355.0	448.38	2.483	0.98092
2400	617.22	367.6	452.70	2.419	0.98331

TABLE A-17M

Thermodynamic properties of air at low pressure (SI units)

Reference level: 0 K and 1 atm

Absolute entropies may be calculated from

$$s° = \phi - R \ln (p/p_0) + 4.1869 \text{ kJ/kg·K}$$

where p_0 is the reference pressure of 1 atm

T, K	h, kJ/kg	P_r	u, kJ/kg	v_r	ϕ kJ/kg K
100	99.76	0.029 90	71.06	2230	1.4143
110	109.77	0.041 71	78.20	1758.4	1.5098
120	119.79	0.056 52	85.34	1415.7	1.5971
130	129.81	0.074 74	92.51	1159.8	1.6773
140	139.84	0.096 81	99.67	964.2	1.7515
150	149.86	0.123 18	106.81	812.0	1.8206
160	159.87	0.154 31	113.95	691.4	1.8853
170	169.89	0.190 68	121.11	594.5	1.9461
180	179.92	0.232 79	128.28	515.6	2.0033
190	189.94	0.281 14	135.40	450.6	2.0575
200	199.96	0.3363	142.56	396.6	2.1088
210	209.97	0.3987	149.70	351.2	2.1577
220	219.99	0.4690	156.84	312.8	2.2043
230	230.01	0.5477	163.98	280.0	2.2489
240	240.03	0.6355	171.15	251.8	2.2915
250	250.05	0.7329	178.29	227.45	2.3325
260	260.09	0.8405	185.45	206.26	2.3717
270	270.12	0.9590	192.59	187.74	2.4096
280	280.14	1.0889	199.78	171.45	2.4461
290	290.17	1.2311	206.92	157.07	2.4813
300	300.19	1.3860	214.09	144.32	2.5153
310	310.24	1.5546	221.27	132.96	2.5483
320	320.29	1.7375	228.45	122.81	2.5802
330	330.34	1.9352	235.65	113.70	2.6111
340	340.43	2.149	242.86	105.51	2.6412
350	350.48	2.379	250.05	98.11	2.6704
360	360.58	2.626	257.23	91.40	2.6987
370	370.67	2.892	264.47	85.31	2.7264
380	380.77	3.176	271.72	79.77	2.7534
390	390.88	3.481	278.96	74.71	2.7796
400	400.98	3.806	286.19	70.07	2.8052
410	411.12	4.153	293.45	65.83	2.8302
420	421.26	4.522	300.73	61.93	2.8547
430	431.43	4.915	308.03	58.34	2.8786
440	441.61	5.332	315.34	55.02	2.9020

* Adapted to SI units from *Gas Tables*, by Joseph H. Keenan and Joseph Kaye. Copyright by John Wiley & Sons, Inc., New York, 1948.

TABLE A-17M
(Continued)

T, K	h, kJ/kg	P_r	u, kJ/kg	v_r	ϕ kJ/kg K
450	451.83	5.775	322.66	51.96	2.9249
460	462.01	6.245	329.99	49.11	2.9473
470	472.25	6.742	337.34	46.48	2.9693
480	482.48	7.268	344.74	44.04	2.9909
490	492.74	7.824	352.11	41.76	3.0120
500	503.02	8.411	359.53	39.64	3.0328
510	513.32	9.031	366.97	37.65	3.0532
520	523.63	9.684	374.39	35.80	3.0733
530	533 98	10.372	381.88	34.07	3.0930
540	544.35	11.097	389.40	32.45	3.1124
550	554.75	11.858	396.89	30.92	3.1314
560	565.17	12.659	404.44	29.50	3.1502
570	575.57	13.500	411.98	28.15	3.1686
580	586.04	14.382	419.56	26.89	3.1868
590	596.53	15.309	427.17	25.70	3.2047
600	607.02	16.278	434.80	24.58	3.2223
610	617.53	17.297	442.43	23.51	3.2397
620	628.07	18.360	450.13	22.52	3.2569
630	638.65	19.475	457.83	21.57	3.2738
640	649.21	20.64	465.55	20.674	3.2905
650	659.84	21.86	473.32	19.828	3.3069
660	670.47	23.13	481.06	19.026	3.3232
670	681.15	24.46	488.88	18.266	3.3392
680	691.82	25.85	496.65	17.543	3.3551
690	702.52	27.29	504.51	16.857	3.3707
700	713.27	28.80	512.37	16.205	3.3861
710	724.01	30.38	520.26	15.585	3.4014
720	734.20	31.92	527.72	15.027	3.4156
730	745.62	33.72	536.12	14.434	3.4314
740	756.44	35.50	544.05	13.900	3.4461
750	767.30	37.35	552.05	13.391	3.4607
760	778.21	39.27	560.08	12.905	3.4751
770	789.10	41.27	568.10	12.440	3.4894
780	800.03	43.35	576.15	11.998	3.5035
790	810.98	45.51	584.22	11.575	3.5174
800	821.94	47.75	592.34	11.172	3.5312
810	832.96	50.08	600.46	10.785	3.5449
820	843.97	52.49	608.62	10.416	3.5584
830	855.01	55.00	616.79	10.062	3.5718
840	866.09	57.60	624.97	9.724	3.5850
850	877.16	60.29	633.21	9.400	3.5981
860	888.28	63.09	641.44	9.090	3.6111
870	899.42	65.98	649.70	8.792	3.6240

TABLE A-17M

(Continued)

T, K	h, kJ/kg	P_r	u, kJ/kg	v_r	ϕ kJ/kg K
880	910.56	68.98	658.00	8.507	3.6367
890	921.75	72.08	666.31	8.233	3.6493
900	932.94	75.29	674.63	7.971	3.6619
910	944.15	78.61	682.98	7.718	3.6743
920	955.38	82.05	691.33	7.476	3.6865
930	966.64	85.60	699.73	7.244	3.6987
940	977.92	89.28	708.13	7.020	3.7108
950	989.22	93.08	716.57	6.805	3.7227
960	1000.53	97.00	725.01	6.599	3.7346
970	1011.88	101.06	733.48	6.400	3.7463
980	1023.25	105.24	741.99	6.209	3.7580
990	1034.63	109.57	750.48	6.025	3.7695
1000	1046.03	114.03	759.02	5.847	3.7810
1020	1068.89	123.12	775.67	5.521	3.8030
1040	1091.85	133.34	793.35	5.201	3.8259
1060	1114.85	143.91	810.61	4.911	3.8478
1080	1137.93	155.15	827.94	4.641	3.8694
1100	1161.07	167.07	845.34	4.390	3.8906
1120	1184.28	179.71	862.85	4.156	3.9116
1140	1207.54	193.07	880.37	3.937	3.9322
1160	1230.90	207.24	897.98	3.732	3.9525
1180	1254.34	222.2	915.68	3.541	3.9725
1200	1277.79	238.0	933.40	3.362	3.9922
1220	1301.33	254.7	951.19	3.194	4.0117
1240	1324.89	272.3	969.01	3.037	4.0308
1260	1348.55	290.8	986.92	2.889	4.0497
1280	1372.25	310.4	1004.88	2.750	4.0684
1300	1395.97	330.9	1022.88	2.619	4.0868
1320	1419.77	352.5	1040.93	2.497	4.1049
1340	1443.61	375.3	1059.03	2.381	4.1229
1360	1467.50	399.1	1077.17	2.272	4.1406
1380	1491.43	424.2	1095.36	2.169	4.1580
1400	1515.41	450.5	1113.62	2.072	4.1753
1420	1539.44	478.0	1131.90	1.9808	4.1923
1440	1563.49	506.9	1150.23	1.8942	4.2092
1460	1587.61	537.1	1168.61	1.8124	4.2258
1480	1611.80	568.8	1187.03	1.7350	4.2422
1500	1635.99	601.9	1205.47	1.6617	4.2585

TABLE A-18

Thermodynamic properties of gases at low pressure (English units)*

Reference level: 0°R and 1 atm; h in Btu/lbm mol; ϕ in Btu/lbm·mol·°R

Absolute entropies at other pressures may be calculated from

$$\bar{s}° = \bar{\phi} - \mathcal{R}\ln(p/p_0) \text{ Btu/lbm·°R}$$

where p_0 is the reference pressure of 1 atm

Temp., °R	Products of combustion, 400% theoretical air		Products of combustion, 200% theoretical air		Nitrogen		Oxygen		Water vapor		Carbon dioxide		Hydrogen		Carbon monoxide	
	$\bar{h}$	$\bar{\phi}$	$\bar{h}$	$\bar{\phi}$	$\bar{h}$	$\bar{\phi}$	$\bar{h}$	$\bar{\phi}$	$\bar{h}$	$\bar{\phi}$	$\bar{h}$	$\bar{\phi}$	$\bar{h}$	$\bar{\phi}$	$\bar{h}$	$\bar{\phi}$
537	3746.8	46.318	3774.9	46.300	3729.5	45.755	3725.1	48.986	4258.3	45.079	4030.2	51.032	3640.3	31.194	3729.5	47.272
600	4191.9	47.101	4226.3	47.094	4167.9	46.514	4168.3	49.762	4764.7	45.970	4600.9	52.038	4075.6	31.959	4168.0	48.044
700	4901.7	48.195	4947.7	48.207	4864.9	47.588	4879.3	50.858	5575.4	47.219	5552.0	53.503	4770.2	33.031	4866.0	49.120
800	5617.5	49.150	5676.3	49.179	5564.4	48.522	5602.0	51.821	6396.9	48.316	6552.9	54.839	5467.1	33.961	5568.2	50.058
900	6340.3	50.002	6413.0	50.047	6268.1	49.352	6337.9	52.688	7230.9	49.298	7597.6	56.070	6165.3	34.784	6276.4	50.892
1000	7072.1	50.773	7159.8	50.833	6977.9	50.099	7087.5	53.477	8078.9	50.191	8682.1	57.212	6864.5	35.520	6992.2	51.646
1100	7812.9	51.479	7916.4	51.555	7695.0	50.783	7850.4	54.204	8942.0	51.013	9802.6	58.281	7564.6	36.188	7716.8	52.337
1200	8563.4	52.132	8683.6	52.222	8420.0	51.413	8625.8	54.879	9820.4	51.777	10955.3	59.283	8265.8	36.798	8450.8	52.976
1300	9324.1	52.741	9461.7	52.845	9153.9	52.001	9412.9	55.508	10714.5	52.494	12136.9	60.229	8968.7	37.360	9194.6	53.571
1400	10095.0	53.312	10250.7	53.430	9896.9	52.551	10210.4	56.099	11624.8	53.168	13344.7	61.124	9673.8	37.883	9948.1	54.129
1500	10875.6	53.851	11050.2	53.981	10648.9	53.071	11017.1	56.656	12551.4	53.808	14576.0	61.974	10381.5	38.372	10711.1	54.655
1600	11665.6	54.360	11859.6	54.504	11409.7	53.561	11832.5	57.182	13494.9	54.418	15829.0	62.783	11092.5	38.830	11483.4	55.154
1700	12464.3	54.844	12678.6	55.000	12178.9	54.028	12655.6	57.680	14455.4	54.999	17101.4	63.555	11807.4	39.264	12264.3	55.628
1800	13271.7	55.306	13507.0	55.473	12956.3	54.472	13485.8	58.155	15433.0	55.559	18391.5	64.292	12526.8	39.675	13053.2	56.078
1900	14087.2	55.747	14344.1	55.926	13741.6	54.896	14322.1	58.607	16427.5	56.097	19697.8	64.999	13250.9	40.067	13849.8	56.509
2000	14910.3	56.169	15189.3	56.360	14534.4	55.303	15164.0	59.039	17439.0	56.617	21018.7	65.676	13980.1	40.441	14653.2	56.922
2100	15740.5	56.574	16042.4	56.777	15334.0	55.694	16010.9	59.451	18466.9	57.119	22352.7	66.327	14714.5	40.799	15463.3	57.317
2200	16577.1	56.964	16902.5	57.177	16139.8	56.068	16862.6	59.848	19510.8	57.605	23699.0	66.953	15454.4	41.143	16279.4	57.696
2300	17419.8	57.338	17769.3	57.562	16951.2	56.429	17718.8	60.228	20570.6	58.077	25056.3	67.557	16199.8	41.475	17101.0	58.062
2400	18268.0	57.699	18642.1	57.933	17767.9	56.777	18579.2	60.594	21645.7	58.535	26424.0	68.139	16950.6	41.794	17927.4	58.414
2500	19121.4	58.048	19520.7	58.292	18589.5	57.112	19443.4	60.946	22735.4	58.980	27801.2	68.702	17707.3	42.104	18758.8	58.754
2600	19979.7	58.384	20404.6	58.633	19415.8	57.436	20311.4	61.287	23839.5	59.414	29187.1	69.245	18469.7	42.403	19594.3	59.081
2700	20842.8	58.710	21293.8	58.974	20246.4	57.750	21182.9	61.616	24957.2	59.837	30581.2	69.771	19237.8	42.692	20434.0	59.398
2800	21709.8	59.026	22187.5	59.300	21081.1	58.053	22057.8	61.934	26088.0	60.248	31982.8	70.282	20011.8	42.973	21277.2	59.705
2900	22581.4	59.331	23086.0	59.615	21919.5	58.348	22936.1	62.242	27231.2	60.650	33391.5	70.776	20791.5	43.247	22123.8	60.002
3000	23456.6	59.628	23988.5	59.921	22761.5	58.632	23817.7	62.540	28386.3	61.043	34806.6	71.255	21576.9	43.514	22973.4	60.290
3100	24335.5	59.916	24895.3	60.218	23606.8	58.910	24702.5	62.831	29552.8	61.426	36227.9	71.722	22367.7	43.773	23826.0	60.569
3200	25217.8	60.196	25805.6	60.507	24455.0	59.179	25590.5	63.113	30730.2	61.801	37654.7	72.175	23164.1	44.026	24681.2	60.841
3300	26102.9	60.469	26719.2	60.789	25306.0	59.442	26481.6	63.386	31918.2	62.167	39086.7	72.616	23965.5	44.273	25539.0	61.105

TABLE A-18
(Continued)

Temp., °R	Products of combustion, 400% theoretical air h̄	φ̄	Products of combustion, 200% theoretical air h̄	φ̄	Nitrogen h̄	φ̄	Oxygen h̄	φ̄	Water vapor h̄	φ̄	Carbon dioxide h̄	φ̄	Hydrogen h̄	φ̄	Carbon monoxide h̄	φ̄
3400	26991.4	60.734	27636.4	61.063	26159.7	59.697	27375.9	63.654	33116.0	62.526	40523.6	73.045	24771.9	44.513	26399.3	61.362
3500			28556.8	61.329	27015.9	59.944	28273.3	63.914	34323.5	62.876	41965.2	73.462	25582.9	44.748	27261.8	61.612
3600			29479.9	61.590	27874.4	60.186	29173.9	64.168	35540.1	63.221	43411.0	73.870	26398.5	44.978	28126.6	61.855
3700			30406.0	61.843	28735.1	60.422	30077.5	64.415	36765.4	63.557	44860.6	74.267	27218.5	45.203	28993.5	62.093
3800			31334.8	62.091	29597.9	60.652	30984.1	64.657	37998.9	63.887	46314.0	74.655	28042.8	45.423	29862.3	62.325
3900			32266.2	62.333	30462.8	60.877	31893.6	64.893	39240.2	64.210	47771.0	75.033	28871.1	45.638	30732.9	62.551
4000					31329.4	61.097	32806.1	65.123	40489.1	64.528	49231.4	75.404	29703.5	45.849	31605.2	62.772
4100					32198.0	61.310	33721.6	65.350	41745.4	64.839	50695.1	75.765	30539.8	46.056	32479.1	62.988
4200					33068.1	61.520	34639.9	65.571	43008.4	65.144	52162.0	76.119	31379.8	46.257	33354.4	63.198
4300					33939.9	61.726	35561.1	65.788	44278.0	65.444	53632.1	76.464	32223.5	46.456	34231.2	63.405
4400					34813.1	61.927	36485.0	66.000	45553.9	65.738	55105.1	76.803	33070.9	46.651	35109.2	63.607
4500					35687.8	62.123	37411.8	66.208	46835.9	66.028	56581.0	77.135	33921.6	46.842	35988.6	63.805
4600					36563.8	62.316	38341.4	66.413	48123.6	66.312	58059.7	77.460	34775.7	47.030	36869.3	63.998
4700					37441.1	62.504	39273.6	66.613	49416.9	66.591	59541.1	77.779	35633.0	47.215	37751.0	64.188
4800					38319.5	62.689	40208.6	66.809	50715.5	66.866	61024.9	78.091	36493.4	47.396	38633.9	64.374
4900					39199.1	62.870	41146.1	67.003	52019.0	67.135	62511.3	78.398	37356.9	47.574	39517.8	64.556
5000					40079.8	63.049	42086.3	67.193	53327.4	67.401	64000.0	78.698	38223.3	47.749	40402.7	64.735
5100					40961.6	63.223	43029.1	67.380	54640.3	67.662	65490.9	78.994	39092.8	47.921	41288.6	64.910
5200					41844.4	63.395	43974.3	67.562	55957.4	67.918	66984.0	79.284	39965.1	48.090	42175.5	65.082
5300					42728.3	63.563	44922.2	67.743	57278.7	68.172	68479.1	79.569	40840.2	48.257	43063.2	65.252

* Abridged from J. H. Keenan and J. Kaye, "Gas Tables," John Wiley & Sons, Inc., New York, 1948, by permission.

TABLE A-18M

Thermodynamic properties of gases at low pressures (SI units)

Reference levels: Enthalpy and internal energy at 298 K; entropy at 0 K and 0.1 MPa; $\bar{h}$ and $\bar{u}$ in kJ/kg mol; $\bar{s}°$ in kJ/kg mol·K

Absolute entropies at other pressures may be calculated from

$$\bar{s}° = \bar{s}°_{ref} - \mathcal{R}\, \ln(p/p_0) \text{ kJ/kg mol·K}$$

where p_0 is the reference pressure of 0.1 MPa and $\bar{s}°_{ref}$ is the tabular value

T, K	Water vapor, H_2O			Carbon dioxide, CO_2		
	$h - \bar{h}_{298}$	$\bar{u} - \bar{u}_{298}$	$\bar{s}°$	$\bar{h} - \bar{h}_{298}$	$\bar{u} - \bar{u}_{298}$	$\bar{s}°$
0	−9904	−7425	0.0	−9364	−6885	0.0
100	−6615	−4939	152.390	−6456	−4666	179.109
200	−3280	−2453	175.486	−3414	−2447	199.975
250	−1610	−1210	182.940	−1737	−1337	207.446
298	0.0	0.0	188.833	0.0	0.0	213.795
300	63	47	189.038	67	54	214.025
350	1748	1317	194.234	1987	1554	219.940
400	3452	2605	198.783	4008	3161	225.334
450	5176	3914	202.843	6119	4857	230.303
500	6920	5246	206.523	8314	6636	234.924
550	8697	6603	209.904	10581	8487	239.244
600	10498	7988	213.037	12916	10406	243.309
650	12326	9401	215.965	15310	12385	247.141
700	14184	10843	218.719	17761	14420	250.773
750	16073	12316	221.324	20265	16508	254.226
800	17991	13820	223.803	22815	18642	257.517
850	19942	15354	226.166	25409	20821	260.660
900	21924	16920	228.430	28041	23037	263.668
950	23937	18518	230.608	30706	25286	266.553
1000	25978	20143	232.706	33405	27570	269.325
1050	28057	21806	234.731	36138	29887	271.986
1100	30167	23500	236.694	38894	32227	274.555
1150	32307	25224	238.594	41679	32118	277.023
1200	34476	26978	240.443	44484	36986	279.417
1250	36676	28762	242.237	47312	39399	281.721
1300	38903	30575	243.986	50158	41828	283.956
1350	41163	32417	245.688	53024	44279	286.122
1400	43447	34286	247.350	55907	46746	288.216
1450	45759	36183	248.973	58803	49227	290.252
1500	48095	38103	250.560	61714	51721	292.224
1550	50459	40051	252.108	64640	54232	293.642
1600	52844	42020	253.622	67580	56856	296.010
1650	55255	44015	255.106	70531	59245	297.823
1700	57685	46030	256.559	73492	61836	299.592
1750	60139	48068	257.979	76462	64391	301.312
1800	62609	50122	259.371	79442	66955	302.993
1850	65102	52200	260.737	82431	69528	304.631
1900	67613	54295	262.078	85429	72111	306.232
1950	70144	55986	263.391	88435	74701	307.792
2000	72689	58540	264.681	91450	77300	309.320
2050	75252	60686	265.947	94471	79906	310.810
2100	77831	62850	267.191	97500	82519	312.269
2150	80426	65029	268.410	100534	85138	313.698
2200	83036	67224	269.609	103575	87763	315.098
2250	85658	69430	270.788	106620	90392	316.465

TABLE A-18M
(Continued)

T, K	Water vapor, H_2O			Carbon dioxide, CO_2		
	$h - \overline{h}_{298}$	$\overline{u} - \overline{u}_{298}$	$\overline{s}°$	$\overline{h} - \overline{h}_{298}$	$\overline{u} - \overline{u}_{298}$	$\overline{s}°$
2300	88295	71651	271.948	109671	93027	317.805
2350	90942	73883	273.087	112727	95667	319.120
2400	93604	76128	274.207	115788	98312	320.411
2450	96279	78386	275.310	118855	100964	321.675
2500	98964	80657	276.396	121926	103619	322.918
2550	101661	82939	277.463	125004	106281	324.135
2600	104370	85231	278.517	128085	108947	325.332
2650	107087	87533	279.550	131169	111615	326.505
2700	109813	89844	280.571	134256	114287	327.658
2750	112549	92163	281.573	137349	116964	328.793
2800	115294	94492	282.563	140444	119643	329.909
2850	118048	96831	283.538	143544	122327	331.005
2900	120813	99180	284.500	146645	125013	332.085
2950	123582	101534	285.447	149753	127704	333.146
3000	126361	103896	286.383	152862	130398	334.193
3050	129147	106267	287.303	155977	133097	335.223
3100	131942	108647	288.211	159092	135796	336.235
3150	134744	111033	289.108	162212	138500	337.233
3200	137553	113426	289.994	165331	141204	338.218
3250	140368	115825	290.865	168458	143916	339.178
3300	143197	118224	291.715	171588	146628	340.123

T, K	Hydrogen, H_2			Carbon monoxide, CO		
	$\overline{h} - \overline{h}_{298}$	$\overline{u} - \overline{u}_{298}$	$\overline{s}°$	$\overline{h} - \overline{h}_{298}$	$\overline{u} - \overline{u}_{298}$	$\overline{s}°$
0	−8468	−5989	0.0	−8669	−6190	0.0
100	−5293	−3979	102.145	−5770	−4115	165.850
200	−2770	−1969	119.437	−2858	−2040	186.025
250	−1357	−965	125.175	−1403	−1002	192.520
298	0.0	0.0	130.684	0.0	0.0	197.653
300	54	38	130.864	54	39	197.833
350	1503	1072	135.318	1512	1081	202.326
400	2958	2111	139.215	2975	2129	206.234
450	4419	3157	142.647	4447	3185	209.702
500	5883	4204	145.738	5929	4253	212.828
550	7346	5252	148.512	7428	5334	215.681
600	8812	6302	151.077	8941	6432	218.313
650	10279	7355	153.418	10472	7546	220.765
700	11749	8410	155.608	12021	8680	223.062
750	13225	9468	157.633	13589	9832	225.224
800	14703	10531	159.549	15175	11003	227.271
850	16189	11601	161.347	16777	12189	229.215
900	17682	12678	163.060	18397	13393	231.066
950	19180	13760	164.674	20034	14615	232.836
1000	20686	14850	166.223	21686	15851	234.531
1050	22200	15949	167.696	23354	17103	236.155
1100	23723	17057	169.118	25033	18367	237.719
1150	25254	18172	170.476	26725	19642	239.221
1200	26794	19295	171.792	28426	20928	240.673
1250	28346	20432	173.054	30141	22362	242.070
1300	29907	21578	174.281	31865	23535	243.426

TABLE A-18M
(Continued)

T, K	Hydrogen, H_2			Carbon monoxide, CO		
	$\bar{h} - \bar{h}_{298}$	$\bar{u} - \bar{u}_{298}$	$\bar{s}°$	$\bar{h} - \bar{h}_{298}$	$\bar{u} - \bar{u}_{298}$	$\bar{s}°$
1350	31480	22734	175.467	33597	24852	244.732
1400	33062	23900	176.620	35338	26177	245.999
1450	34661	25084	177.739	37090	27513	247.227
1500	36267	26277	178.833	38848	28856	248.421
1550	37890	27482	179.894	40613	30205	249.577
1600	39522	28698	180.929	42384	31560	250.702
1650	41165	29926	181.940	44159	32920	251.795
1700	42815	31161	182.929	45940	34284	252.861
1750	44478	32406	183.891	47729	35658	253.896
1800	46150	33663	184.833	49522	37035	254.907
1850	47831	34928	185.754	51321	38419	255.891
1900	49522	36204	186.657	53124	39807	256.852
1950	51222	37489	187.540	54931	41197	257.792
2000	52932	38782	188.406	56739	42590	258.710
2050	54651	40085	189.257	58555	43989	259.603
2100	56379	41397	190.088	60375	45394	260.480
2150	58116	42719	190.905	62195	46798	261.335
2200	59860	44048	191.707	64019	48206	262.174
2250	61612	45384	192.494	65847	49619	262.996
2300	63371	46727	193.268	67676	51032	263.802
2350	65140	48080	194.030	69509	52450	264.589
2400	66915	49440	194.778	71346	53870	265.362
2450	68700	50809	195.512	73183	55292	266.121
2500	70492	52186	196.234	75023	56716	266.865
2550	72287	53565	196.946	76868	58145	267.594
2600	74090	54952	197.649	78714	59576	268.312
2650	75900	56346	198.338	80561	61007	269.014
2700	77718	57748	199.017	82408	62438	269.705
2750	79540	59155	199.684	84261	63876	270.394
2800	81370	60569	200.343	86115	65314	271.053
2850	83203	61987	200.994	87970	66755	271.711
2900	85044	63412	201.636	89826	68193	272.358
2950	86890	64842	202.266	91683	69635	272.993
3000	88743	66279	202.887	93542	71077	273.618
3050	90597	67718	203.500	95404	72525	274.232
3100	92458	69163	204.104	97270	73974	274.839
3150	94325	70615	204.701	99133	75422	275.435
3200	96199	72072	205.293	100998	76871	276.023
3250	98077	73534	205.874	102865	78323	276.603
3300	99964	74996	206.442	104736	79775	277.166

T, K	Nitrogen, N_2			Oxygen, O_2		
	$\bar{h} - \bar{h}_{298}$	$\bar{u} - \bar{u}_{298}$	$\bar{s}°$	$\bar{h} - \bar{h}_{298}$	$\bar{u} - \bar{u}_{298}$	$\bar{s}°$
0	−8669	−6190	0.0	−8682	−6203	0.0
100	−5770	−4115	159.813	−5778	−4124	173.306
200	−2858	−2040	179.988	−2866	−2045	193.486
250	−1403	−1002	186.479	−1407	−1006	199.994
298	0.0	0.0	191.611	0.0	0.0	205.142
300	54	39	191.791	54	39	205.322
350	1511	1080	196.282	1531	1100	209.874
400	2971	2124	200.180	3029	2181	213.874
450	4436	3173	203.632	4546	3284	217.451
500	5912	4233	206.739	6088	4411	220.698

TABLE A-18M
(Continued)

	Nitrogen, N_2			Oxygen, O_2		
T, K	$\bar{h} - \bar{h}_{298}$	$\bar{u} - \bar{u}_{298}$	$\bar{s}°$	$\bar{h} - \bar{h}_{298}$	$\bar{u} - \bar{u}_{298}$	$\bar{s}°$
550	7395	5302	209.570	7656	5562	223.685
600	8891	6384	212.175	9247	6737	226.455
650	10406	7481	214.598	10862	7937	229.041
700	11937	8594	216.865	12502	9161	231.272
750	13480	9723	218.998	14162	10404	233.758
800	15046	10871	221.016	15841	11669	235.924
850	16623	12034	222.931	17536	12947	237.973
900	18221	13217	224.756	19246	14242	239.936
950	19832	14413	226.498	20970	15551	241.798
1000	21460	15625	228.166	22707	16872	243.585
1050	23103	16852	229.768	24458	18207	245.288
1100	24757	18090	231.308	26217	19550	246.928
1150	26426	19344	232.791	27987	20905	248.499
1200	28108	20609	234.264	29765	22266	250.016
1250	29799	21885	235.604	31554	23640	251.475
1300	31501	23171	236.940	33351	25021	252.886
1350	33214	24469	238.231	35155	26410	254.245
1400	34936	25774	239.484	36966	27805	255.564
1450	36666	27089	240.698	38785	29208	256.836
1500	38405	28411	241.877	40610	30618	258.078
1550	40150	29741	243.022	42441	32033	259.274
1600	41903	31078	244.137	44279	33455	260.446
1650	43664	32424	245.219	46121	34882	261.575
1700	45430	33775	246.275	47970	36314	262.685
1750	47203	35132	247.302	49826	37755	263.757
1800	48982	36495	248.304	51689	39202	264.810
1850	50764	37861	249.281	53559	40657	265.832
1900	52551	39233	250.237	55434	42116	266.835
1950	54349	40610	251.167	57315	43581	267.807
2000	56141	41991	252.078	59199	45050	268.764
2050	57943	43377	252.967	61090	46524	269.697
2100	59748	44767	253.836	62986	48005	270.613
2150	61557	46161	254.687	64891	49494	271.508
2200	63371	47559	255.522	66802	50898	272.387
2250	65187	48959	256.336	68715	52487	273.245
2300	67007	50363	257.137	70634	53990	274.090
2350	68827	51768	257.919	72561	55501	274.918
2400	70651	53176	258.689	74492	57016	275.735
2450	72480	54589	259.441	76430	58539	276.533
2500	74312	56005	260.183	78375	60068	277.316
2550	76145	57423	260.908	80322	61599	278.088
2600	77973	58843	261.622	82274	63136	278.848
2650	79819	60265	262.322	84234	64680	279.594
2700	81659	61690	263.011	86199	66230	280.329
2750	83502	63116	263.686	88170	67784	281.051
2800	85345	64544	264.350	90144	69343	281.764
2850	87190	65973	265.004	92126	70909	282.466
2900	89036	67403	265.647	94111	72479	283.157
2950	90887	68838	266.279	96103	74055	283.837
3000	92738	70274	266.902	98098	75634	284.508
3050	94591	71712	267.513	100096	77216	285.169
3100	96446	73151	268.116	102102	78806	285.822
3150	98303	74592	268.710	104113	80398	286.464
3200	100161	76034	269.295	106127	82000	287.098
3250	102021	77478	269.872	108145	83601	287.723
3300	103883	78922	270.433	110174	85202	288.333

TABLE A-19

One-dimensional isentropic compressible-flow functions for an ideal gas with constant specific heat and molecular weight and $\gamma = 1.4$*

M	M*	$\dfrac{A}{A^*}$	$\dfrac{p}{p_0}$	$\dfrac{\rho}{\rho_0}$	$\dfrac{T}{T_0}$
0	0	∞	1.00000	1.00000	1.00000
0.10	0.10943	5.8218	0.99303	0.99502	0.99800
0.20	0.21822	2.9635	0.97250	0.98027	0.99206
0.30	0.32572	2.0351	0.93947	0.95638	0.98232
0.40	0.43133	1.5901	0.89562	0.92428	0.96899
0.50	0.53452	1.3398	0.84302	0.88517	0.95238
0.60	0.63480	1.1882	0.78400	0.84045	0.93284
0.70	0.73179	1.09437	0.72092	0.79158	0.91075
0.80	0.82514	1.03823	0.65602	0.74000	0.88652
0.90	0.91460	1.00886	0.59126	0.68704	0.86058
1.00	1.00000	1.00000	0.52828	0.63394	0.83333
1.10	1.08124	1.00793	0.46835	0.58169	0.80515
1.20	1.1583	1.03044	0.41238	0.53114	0.77640
1.30	1.2311	1.06631	0.36092	0.48291	0.74738
1.40	1.2999	1.1149	0.31424	0.43742	0.71839
1.50	1.3646	1.1762	0.27240	0.39498	0.68965
1.60	1.4254	1.2502	0.23527	0.35573	0.66138
1.70	1.4825	1.3376	0.20259	0.31969	0.63372
1.80	1.5360	1.4390	0.17404	0.28682	0.60680
1.90	1.5861	1.5552	0.14924	0.25699	0.58072
2.00	1.6330	1.6875	0.12780	0.23005	0.55556
2.10	1.6769	1.8369	0.10935	0.20580	0.53135
2.20	1.7179	2.0050	0.09352	0.18405	0.50813
2.30	1.7563	2.1931	0.07997	0.16458	0.48591
2.40	1.7922	2.4031	0.06840	0.14720	0.46468
2.50	1.8258	2.6367	0.05853	0.13169	0.44444
2.60	1.8572	2.8960	0.05012	0.11787	0.42517
2.70	1.8865	3.1830	0.04295	0.10557	0.40684
2.80	1.9140	3.5001	0.03685	0.09462	0.38941
2.90	1.9398	3.8498	0.03165	0.08489	0.37286
3.00	1.9640	4.2346	0.02722	0.07623	0.35714
3.50	2.0642	6.7896	0.01311	0.04523	0.28986
4.00	2.1381	10.719	0.00658	0.02766	0.23810
4.50	2.1936	16.562	0.00346	0.01745	0.19802
5.00	2.2361	25.000	$189(10)^{-5}$	0.01134	0.16667
6.00	2.2953	53.180	$633(10)^{-6}$	0.00519	0.12195
7.00	2.3333	104.143	$242(10)^{-6}$	0.00261	0.09259
8.00	2.3591	190.109	$102(10)^{-6}$	0.00141	0.07246
9.00	2.3772	327.189	$474(10)^{-7}$	0.000815	0.05814
10.00	2.3904	535.938	$236(10)^{-7}$	0.000495	0.04762
∞	2.4495	∞	0	0	0

* Abridged from J. H. Keenan and J. Kaye, "Gas Tables," John Wiley & Sons, Inc., New York, 1948, by permission.

TABLE A-20

One-dimensional normal-shock functions for an ideal gas with constant specific heat and molecular weight and $\gamma = 1.4$*

M_x	M_y	$\dfrac{p_y}{p_x}$	$\dfrac{\rho_y}{\rho_x}$	$\dfrac{T_y}{T_x}$	$\dfrac{p_{0y}}{p_{0x}}$	$\dfrac{p_{0y}}{p_x}$
1.00	1.00000	1.00000	1.00000	1.00000	1.00000	1.8929
1.10	0.91177	1.2450	1.1691	1.06494	0.99892	2.1328
1.20	0.84217	1.5133	1.3416	1.1280	0.99280	2.4075
1.30	0.78596	1.8050	1.5157	1.1909	0.97935	2.7135
1.40	0.73971	2.1200	1.6896	1.2547	0.95819	3.0493
1.50	0.70109	2.4583	1.8621	1.3202	0.92978	3.4133
1.60	0.66844	2.8201	2.0317	1.3880	0.89520	3.8049
1.70	0.64055	3.2050	2.1977	1.4583	0.85573	4.2238
1.80	0.61650	3.6133	2.3592	1.5316	0.81268	4.6695
1.90	0.59562	4.0450	2.5157	1.6079	0.76735	5.1417
2.00	0.57735	4.5000	2.6666	1.6875	0.72088	5.6405
2.10	0.56128	4.9784	2.8119	1.7704	0.67422	6.1655
2.20	0.54706	5.4800	2.9512	1.8569	0.62812	6.7163
2.30	0.53441	6.0050	3.0846	1.9468	0.58331	7.2937
2.40	0.52312	6.5533	3.2119	2.0403	0.54015	7.8969
2.50	0.51299	7.1250	3.3333	2.1375	0.49902	8.5262
2.60	0.50387	7.7200	3.4489	2.2383	0.46012	9.1813
2.70	0.49563	8.3383	3.5590	2.3429	0.42359	9.8625
2.80	0.48817	8.9800	3.6635	2.4512	0.38946	10.569
2.90	0.48138	9.6450	3.7629	2.5632	0.35773	11.302
3.00	0.47519	10.333	3.8571	2.6790	0.32834	12.061
4.00	0.43496	18.500	4.5714	4.0469	0.13876	21.068
5.00	0.41523	29.000	5.0000	5.8000	0.06172	32.654
10.00	0.38757	116.50	5.7143	20.388	0.00304	129.217
∞	0.37796	∞	6.000	∞	0	∞

* Abridged from J. H. Keenan and J. Kaye, "Gas Tables," John Wiley & Sons, Inc., New York, 1948, by permission.

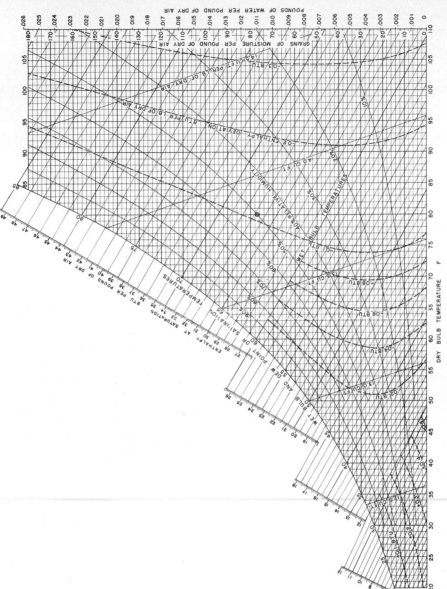

FIG. A-1 *Psychometric chart, normal temperatures, English units,* $p = 1$ *atm.* (*Carrier Corporation.*)

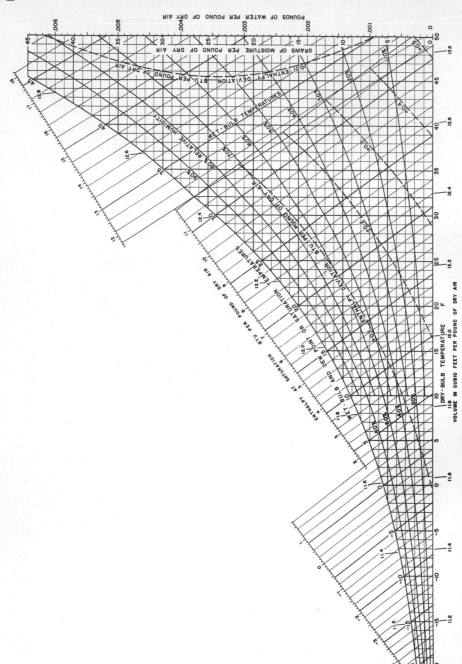

FIG. A-2 *Psychrometric chart, low temperatures, English units, p = 1 atm. (Carrier Corporation.)*

FIG. A-3 *Psychometric chart, normal temperatures, SI units,* $p = 1$ *atm.* (*Carrier Corporation.*)

FIG. A-4 *Mollier diagram for water vapor, English units.*

FIG. A-4 (cont.)

FIG. A-4M *Mollier diagram for water vapor, SI units. (By permission of American Society of Mechanical Engineers.)*

FIG. A-4M *(cont.)*

index